Atherosclerosis

Atherosclerosis

Gene Expression,
Cell Interactions, and Oxidation

edited by

Roger T. Dean and David T. Kelly

The Heart Research Institute, Sydney

OXFORD

UNIVERSITY PRESS

Great Clarendon Street, Oxford OX2 6DP

Oxford University Press is a department of the University of Oxford. It furthers the University's objective of excellence in research, scholarship, and education by publishing worldwide in

Oxford New York

Athens Auckland Bangkok Bogotá Buenos Aires Calcutta Cape Town Chennai Dar es Salaam Delhi Florence Hong Kong Istanbul Karachi Kuala Lumpur Madrid Melbourne Mexico City Mumbai Nairobi Paris São Paulo Singapore Taipei Tokyo Toronto Warsaw

with associated companies in Berlin Ibadan

Oxford is a registered trade mark of Oxford University Press in the UK and in certain other countries

Published in the United States by Oxford University Press Inc., New York

A catalogue record for this book is available from the British Library

Library of Congress Cataloguing in Publication Data
(Data available)

ISBN 0 19 850637 6

Typeset by Newgen Imaging Systems Pvt. Ltd.

Printed in Great Britain by
Biddles Ltd.,
Guildford & King's Lynn

Contents

Contributors

Mark R. Adams   The Heart Research Institute, Camperdown, Sydney, NSW 2050, Australia

Paul Baker   Department of Medicine, University of Adelaide and the Cardiovascular Investigation Unit, Royal Adelaide Hospital, Adelaide, Australia

Paul Bannon   Cardiovascular Surgery, Royal Prince Alfred Hospital, NSW 2050, Sydney, Australia

Philip Barter   Department of Medicine, University of Adelaide and the Cardiovascular Investigation Unit, Royal Adelaide Hospital, Adelaide, Australia

John A. Bingley   Centre for Research in Vascular Biology, Department of Anatomical Sciences, University of Queensland, Brisbane, Qld 4072, Australia

Andrew J. Brown   Cell Biology Group, The Heart Research Institute, Camperdown, Sydney, NSW 2050, Australia

Ulf T. Brunk   Pathology II, Linköping University, Linköping, S-58185, Sweden

Gordon R. Campbell   Centre for Research in Vascular Biology, Department of Anatomical Sciences, University of Queensland, Brisbane, Qld 4072, Australia

Julie H. Campbell   Centre for Research in Vascular Biology, Department of Anatomical Sciences, University of Queensland, Brisbane, Qld 4072, Australia

David S. Celermajer   Clinical Research Group, The Heart Research Institute, Camperdown, Sydney, NSW 2050, Australia

Moira Clay   Department of Medicine, University of Adelaide and the Cardiovascular Investigation Unit, Royal Adelaide Hospital, Adelaide, Australia

Michael J. Davies   EPR Group, The Heart Research Institute, Camperdown, Sydney, NSW 2050, Australia

Alan Daugherty   Gill Heart Institute, University of Kentucky, L 543 Kentucky Clinic, Lexington, KY 40536-0284, USA

Roger T. Dean   The Heart Research Institute, Camperdown, Sydney, NSW 2050, Australia

John W. Eaton   Department of Pediatrics, Baylor College of Medicine, Houston, Texas, USA

Jane E. Freedman Department of Medicine and Pharmacology, Georgetown University Medical Center, Washington, DC 20007, USA

Shanlin Fu Cell Biology Group, The Heart Research Institute, Camperdown, Sydney, NSW 2050, Australia

Carolyn L. Geczy Cytokine Research Unit, School of Pathology, The University of New South Wales, Sydney, NSW 2052, Australia

Göran K. Hansson Center for Molecular Medicine, Karolinska Institute, Karolinska Hospital, S-17176 Stockholm, Sweden

Ian P. Hayward Centre for Research in Vascular Biology, Department of Anatomical Sciences, University of Queensland, Brisbane, Qld 4072, Australia

Linda Hazell The Heart Research Institute, Camperdown, Sydney, NSW 2050, Australia

Jay W. Heinecke Division of Atherosclerosis, Nutrition and Lipid Research, Washington University School of Medicine, St Louis, Missouri, 63110, USA

Natalie James Micromedical Industries Ltd, Sydney, Australia

Wendy Jessup Cell Biology Group, The Heart Research Institute, Camperdown, Sydney, NSW 2050, Australia

John F. Keaney Jr. Evans Memorial Department of Medicine and Whitaker Cardiovascular Institute, Boston University School of Medicine, 715 Albany Street, W507, Boston, Massachuselts, USA

Anthony Keech NH & MRC Clinical Trials Centre, University of Sydney, NSW 1450, and Camperdown, Department of Cardiology, Royal Prince Alfred Hospital, NSW 2050, Sydney, Australia

David T. Kelly The Heart Research Institute, Camperdown, Sydney, NSW 2050, Australia

Levon M. Khachigian Centre for Thrombosis and Vascular Research, School of Pathology, The University of New South Wales, Sydney, NSW 2052, Australia

Leonard Kritharides Clinical Research Group, The Heart Research Institute, Camperdown, Sydney, NSW 2050, and Department of Cardiology, Concord Hospital, University of Sydney, NSW 2139, Australia

Malcolm A. Lyons Cell Biology Group, The Heart Research Institute, Camperdown, Sydney, NSW 2050, Australia

Anushka Patel NH & MRC Clinical Trials Centre, University of Sydney, Camperdown, NSW 1450, and Department of Cardiology, Royal Prince Alfred Hospital, NSW 2050, Sydney, Australia

Knut S. Pettersson Cardiovascular Pharmacology, AstraZeneca R & D, SE-43183 Mölndal, Sweden

Serena Pringle The Biotechnology Centre of Oslo, University of Oslo, Norway

Hans Prydz The Biotechnology Centre of Oslo, University of Oslo, Norway

John E.J. Rasko Gene Therapy Research Unit, Centenary Institute of Cancer Medicine & Cell Biology, and Sydney Cancer Centre, Central Sydney Area Health Service, Locked Bag 6, Newtown, NSW 2042, Australia

Trevor G. Redgrave Department of Physiology, The University of Western Australia, Nedlands, WA 6907, Australia

Kerry-Anne Rye Department of Medicine, University of Adelaide and the Cardiovascular Investigation Unit, Royal Adelaide Hospital, Adelaide, Australia

Roland Stocker Biochemistry Group, The Heart Research Institute, Camperdown, Sydney, NSW 2050, Australia

Geoffrey H. Tofler Cardiology Department, Royal North Shore Hospital, St Leonards, NSW 2065, Australia

Joanne M. Upston Biochemistry Group, The Heart Research Institute, Camperdown, Sydney, NSW 2050, Australia

Joseph A. Vita Whitaker Cardiovascular Institute, Boston University, School of Medicine, 715 Albany Street, W507, Massachusetts, USA

Paul K. Witting Biochemistry Group, The Heart Research Institute, Camperdown, Sydney, NSW 2050, Australia

Xi Ming Yuan Pathology II, Linköping University, Linköping, S-58185, Sweden

Abbreviations

AAV	adeno-associated virus
ABP	actin binding protein
ACAT	acetyl CoA cholesteryl acyl transferase
ACE	angiotensin-converting enzyme
AGE	advanced glycation end product
AMI	acute myocardial infarction
APC	antigen-presenting cell
bFGF	basic fibroblast growth factor
BHT	butylated hydroxytoluene
CAD	coronary artery disease
CAM	cell adhesion molecule
cDNA	complementary deoxyribonucleic acid
CE	cholesteryl ester
CETP	cholesteryl ester transfer protein
CHD	coronary heart disease
CIMT	carotid intima-media thickness
CSPG	chondroitin sulphate proteoglycan
CT	computed tomography
DHA	docosahexaenoic acid
DOPA	3,4-dihydroxyphenylalanine
DPPD	N,N'-diphenyl-phenylenediamine
DSPG	dermatan sulphate proteoglycan
DTH	delayed-type hypersensitivity
EC	endothelial cell
ECGS	endothelial cell growth supplement
EC-SOD	extracellular superoxide dismutase
EDRF	endothelial-derived relaxing factor
EGF	epidermal growth factor
Egr	early growth response factor
EL	endothelial lipase
eNOS	endothelial nitric oxide synthase
ET	endothelin
FAK	focal adhesion kinase
FGF	fibroblast growth factor
GAG	glycosaminoglycan
HDL	high-density lipoprotein
HMDM	human monocyte-derived macrophage
HMG CoA	3-hydroxy-3-methylglutaryl coenzyme A

HRP	horse radish peroxidase
HSL	hormone-sensitive lipase
HSPG	heparan sulphate proteoglycan
HUVEC	human umbilical vein endothelial cell
ICE	IL-1-converting enzyme
IEL	internal elastic lamina
KSPG	Keratin sulphate proteoglycan
LAL	lysosomal acid lipase
LCAT	lecithin–cholesterol acyltransferase
LDL	low-density lipoprotein
LPL	lipoprotein lipase
LPS	lipopolysaccharide
LTR	long terminal repeat
LXR	liver X receptor
MCP	monocyte chemoattractant protein
M-CSF	macrophage colony-stimulating factor
MHC	major histocompatibility complex
MI	myocardial infarction
MMP	matrix metalloproteinase
MRI	magnetic resonance imaging
nCE	neutral cholesteryl esterase
nCEH	neutral cholesteryl esterase hydrolase
NEFA	non-esterified fatty acid
NIDDM	non-insulin-dependent diabetes mellitus
NK	natural killer
PAI-1	plasminogen activator inhibitor
PAPC	palmitoyl-arachidonyl PC
PAR	protease-activated receptor
PC	phosphatidylcholine
PCNA	proliferating cell nuclear antigen
PDGF	platelet-derived growth factor
PDPC	palmityl-docosahexaenoyl PC
PI-PLC	phosphatidylinositol-specific phospholipase C
PKC	protein kinase C
PLPC	palmitoyl-linoleyl PC
PMA	phorbol 12-myristate 13-acetate
POPC	palmitoyl-oleyl PC
PPAR	peroxisome proliferator-activated receptor
PS	phosphatidylserine
RAGE	receptor for advanced glycation end product
RBC	red blood cell
RCT	reverse cholesterol transport
rHDL	reconstituted HDL

RXR	retinoid X receptor
SAA	serum amyloid A protein
SCAP	SREBP cleavage activating protein
SCF	stem cell factor
SOD	superoxide dismutase
SMase	sphingomyelinase
SREBP	sterol regulatory element binding protein
STATs	signal transducers and activators of transcription
TCR	T cell antigen receptor
TD	Tangier disease
TF	tissue factor
TFPI-1	tissue factor pathway inhibitor-1
TGF	transforming growth factor
TMP	tocopherol-mediated peroxidation
TPA	tissue plasminogen activator
uPAR	urokinase type plasminogen activator receptor
USP	United States Pharmacopeia
VCAM-1	vascular cell adhesion molecule-1
VEGF	vascular endothelial growth factor
VLDL	very low density lipoprotein
VSMC	vascular smooth muscle cell

Preface

Since the discovery of the low-density lipoprotein and scavenger receptors, and subsequent important observations of Dan Steinberg and colleagues implicating oxidation in the production of ligands for the latter, atherosclerosis research has had a strong focus on oxidative events. Results are now beginning to emerge from mechanistic studies and from intervention studies in both animals and humans which address the implied possibility that antioxidants might be able to prevent or restrict the development of the disease.

The book assesses these developments critically. It shows that the simple notion that antioxidant prevention of lipid oxidation might be central, is probably insufficient. While aspects of oxidation, from oxidation of proteins to oxidative events in gene signalling, remain important research targets, there is a movement towards seeing oxidation as a component of a larger response to injury mechanism. This revisits the classic ideas on atherosclerosis promoted and developed by colleagues such as the late Russell Ross. The book also discusses future trends in the field, from basic research to diagnostics, imaging, and clinical aspects.

The book originates from the Heart Research Institute, Sydney. But in order to maximize the stringency of assessment of the ideas covered, we have adopted a somewhat unusual strategy. This has been to enlist authors from different institutions, cooperating and competing, to argue the literature in their field. The authors have produced timely contributions, creating a book we hope will be both readable and valuable. We thank the contributors for their responsiveness, and Oxford University Press and our editor, Liz Owen, for their commitment.

The Heart Research Institute, Sydney

Roger Dean
David Kelly
February, 2000

1 The morphology and natural history of atherosclerosis

Xi Ming Yuan
Pathology II, Linköping University, S-58185 Linköping, Sweden

Ulf T. Brunk
Pathology II, Linköping University, S-58185 Linköping, Sweden

Linda Hazell
The Heart Research Institute, Camperdown, Sydney, NSW 2050, Australia

1.1 Risk factors and clinical significance

Atherosclerosis is a pathological process leading to several important human vascular disorders, including coronary artery disease, cerebrovascular disease, and diseases of the aorta and the peripheral arterial circulation. These disorders are responsible for more deaths than any other diseases in humans.

Atherosclerosis, alternatively termed as atheroma, is the most common type of arteriosclerosis. The latter refers to all of those lesions in which there is an increase in the thickness of the vessel walls. The lesion of atherosclerosis has been characterized as a soft, lipid-rich part ('athere', porridge) and a hard (sclerotic) fibrous or calcified component and differs from other forms of arteriosclerosis, e.g. hyaline arteriosclerosis and malignant arteriosclerosis in hypertension patients.

The fundamental pathology of an atherosclerotic lesion is the accumulation of lipid and connective tissue that eventually may obstruct blood flow through the lumen (stenosis), predispose the vessel to thrombosis, and impair the vessel's ability to respond elastically and via muscle contraction to hydrodynamic stresses. Chronic inflammation, a central component of atherosclerosis, is basically protective, but may lead to tissue damage due to excessive inflammatory responses. Basic elements of the inflammatory process in atherosclerosis include possible injury-causing agents [e.g. oxidized low-density lipoprotein (LDL) and, perhaps, infectious microorganisms], interaction between leukocytes and dysfunctional endothelium, infiltration of chronic inflammatory cells, involvement of chemokines and

cytokines, the uptake of oxidized lipids and elimination of cellular debris by phagocytes, and autoimmune response against antigens in lesions.

Epidemiological investigations have identified a number of risk factors related
to the incidence of atherosclerosis and its clinical complications. They currently
include: hypertension, increased LDL cholesterol, cigarette smoking, diabetes,
increased age and male sex, an elevated level of homocysteine, physical inactivity,
stressful life patterns, having a personal or family history of heart disease, and dietary
factors such as a lack of fibre or fruit and vegetables. Even though tremendous efforts
have been made to find the causes of atherosclerosis, the above risk factors may only
explain two-thirds of the incidence of atherosclerosis. Therefore other potential risk
factors are now being investigated, such as body iron stores (see Section 1.7) and
infectious diseases.

1.2 Distribution and development of atherosclerotic lesions

Histology of arteries   Arteries can be histologically divided into larger or elastic
arteries, medium or muscular arteries, and small arteries. The walls of arteries are
composed of intima, media and adventitia which are separated by the internal elastic
lamina and external elastic lamina.

The *intima* consists of connective tissue, smooth muscle cells and a few isolated
macrophages. It is defined from the luminal surface endothelium to the internal elastic lamina. The arterial intima can be further divided into two layers. The inner layer,
called the proteoglycan layer, is composed of abundant proteoglycans, spaced single
smooth muscle cells, and macrophages. The lower layer, called the musculoelastic
layer, is composed of abundant smooth muscle cells and elastic fibres. Under normal conditions the two layers of the intima are barely visible by light microscopy;
however, they are distinct and prominent when *adaptive intimal thickening* occurs.

The *media* is the muscular part of arterial walls, composed of smooth muscle cells,
elastin, collagen fibrils, and proteoglycans. The *adventitia* is the outer and highly
microvascular layer and contains collagen and elastic fibrils, smooth muscle cells,
and lymphatic channels. A recent study suggests that adventitial neovascularization
by vasa vasorum occurs in experimental hypercholesterolemic coronary arteries
before the development of vascular lesions.[1]

Distribution of lesions within the human vasculature   Atherosclerosis occurs mainly
in larger and medium-size arteries, i.e. arteries of sizes larger than 2 mm in diameter.
As the pathological effect of atheroma largely depends on the degree of luminal
narrowing, lesion formation in small arteries has more vital clinical significance.

In humans, lesions are most frequently observed in the abdominal aorta. Other
arteries with a high likelihood of atherosclerosis include the coronary arteries, arteries
of the legs (femoral, iliac, etc.), the arch and descending thoracic aorta, the internal
carotids, and the Circle of Willis (arteries of the brain). Lesions are often associated

with branch points, and it is thought that areas with low and/or oscillating shear stress are most susceptible.[2] The severity of lesions in different locations varies, resulting in only weak correlations between different sites. For instance, diabetics are more prone to peripheral vascular disease than normal individuals. In contrast to humans, in many animal models (e.g. rabbits and mice) the thoracic aorta is more susceptible to atherosclerosis than the abdominal aorta.

Developmental stages of lesions Atherosclerotic lesions in humans can start at a very young age. In the Pathobiological Determinants of Atherosclerosis in Youth Study (PDAY), a multi-institutional autopsy study conducted in US medical centres, a total of 2876 study subjects were examined. The authors concluded that athero-sclerosis begins in youth.[3] Fatty streak formation has even been demonstrated in human fetal aortas.[4] Fatty streaks and clinically significant raised lesions then increase rapidly in prevalence and extent during the 15–34-year age span.[3] Thus, although the clinical symptoms of atherosclerosis are not usually manifest until the later years of adulthood, these studies, and others, suggest that primary prevention of atherosclerosis should begin in childhood or adolescence.

The identification of the sequence and nature of changes in the arterial wall asso-ciated with atherosclerosis is still controversial in many regards. Even the definition of a healthy artery is not universally agreed upon. Bearing this in mind, we will first discuss some of the histopathological features that precede the formation of an advanced atherosclerotic lesion, such as intimal thickening, lipid accumulation and cellular infiltration, and then proceed to describe the developmental stages of atherosclerotic lesions.

In its gross appearance, a healthy artery appears pinkish grey with a smooth unblemished surface. In some arteries, there are areas where the intima consists only of the monolayer of endothelial cells that sit directly on the internal elastic lamina with little or no intervening space. However, in most areas, there is a layer of con-nective tissue between the endothelial cells and the internal elastic lamina, often referred to as an intimal thickening.[5] Intimal thickness increases with age of the subject, blood pressure,[6] and specific patterns of local hydrodynamic stresses.[2] It is controversial whether intimal thickening is an 'atherosclerotic' change or simply an adaptive response to normal vessel stress,[5] or even whether there is any point in making the distinction. The importance of intimal thickening in the sequence of events leading to atherosclerosis is supported by studies in humans and animals that show that atherosclerosis is more likely to develop in those areas where the intima thickens more rapidly (sometimes referred to as lesion-prone areas) than in sur-rounding areas.[5] Although the presence of an intimal thickening does not guaran-tee further atherosclerotic change, such as foam cell accumulation, further changes are usually accompanied by increased intimal thickening.[7] These studies suggest that intimal thickening, while not by itself sufficient to trigger the sequences lead-ing to clinical events, is likely to contribute to the process.

Lipid accumulation appears to occur concurrently with the formation of an intima into which it can infiltrate (see Section 1.4 for more details on the types and location of the lipid). It does not accumulate evenly throughout the artery but is concentrated in those areas with intimal thickenings (lesion-prone areas). The causes of deposition of lipid in the intimal space have not been established (discussed in Section 1.4). Both lipid deposition and intimal thickening can occur prior to any extensive cellular infiltration or formation of fatty streak lesions.[8, 9] Vessels with 'the first microscopically and chemically detectable lipid deposits', where lesions are 'not visible to the naked eye', are defined as type I lesions.[10]

The intima of grossly normal vessels contains predominantly smooth muscle cells, with rare scattered monocyte/lymphocyte cells.[11] In normal intima, smooth muscle cells of both synthetic (rough-endoplastic reticulum rich) and contractile (rough-endoplastic reticulum poor) types may be identified.[5] In some type I lesions, small isolated groups of macrophages containing lipid droplets (foam cells) can be seen.[10] The cells of an atheroma are discussed in more detail in Section 1.3.

The fatty streak (type II lesion[10]) is the first macroscopically visible stage of atherosclerosis. It consists of small raised yellowish lumps, with surrounding tissue that may be pinkish grey or yellowish. Macrophage foam cells are stratified in adjacent layers, and some smooth muscle cells may also contain lipid droplets.[10] In comparison with more advanced lesions, the proportion of cholesteryl ester 18:1 to 18:2 in fatty streaks suggests a significant level of esterification by macrophages.[12] Consistent with this, much of the lipid in the intima now becomes intracellular.[7] In the past, there has been some controversy about whether fatty streaks and other 'early lesions' are in fact precursors to advanced atherosclerotic lesions or not. A systematic study by Guyton and Klemp[7] has demonstrated the presence of features (e.g. cholesterol clefts) in some fatty streaks that are normally associated with more advanced lesions. Some other features of advanced lesions, such as calcification, can also be observed in fatty streaks by using electron microscopy to detect deposits before they are visible macroscopically or by light microscopy.[8] Thus, it is now generally accepted that fatty streaks correspond to an early stage of advanced atherosclerotic lesions. Type III lesions represent a transitional stage between fatty streaks and advanced lesions, their most characteristic feature being multiple extracellular lipid pools that do not yet form a clearly defined acellular lipid core.[10]

Advanced atherosclerotic lesions (types IV–VI[13]) frequently contain cholesterol clefts, calcification, and an acellular necrotic core with defined shoulder regions. Cholesterol clefts are boat-shaped holes in the connective tissue caused by cholesterol crystals. The cholesterol crystals form at the deepest point of the intima, usually the musculoelastic layer. Calcification can also be found at this location and is usually indicated in microscope sections by an irregular hole (cutting artefact) surrounded by fragmented tissue that stains purple with hematoxylin. Surrounding the region containing cholesterol clefts, there is often an acellular core. It is unclear whether the formation of the acellular core requires cell death or lipid deposition, or both. The content of cholesteryl esters and cholesterol suggests that the core region develops from extracellular lipid deposition in the musculoelastic layer of the intima rather than from a cluster of dying foam cells.[7, 12] However, dying cells are often seen around the core, and the core itself contains monocyte/macrophage markers detectable by immunohistochemistry.[14] Since the two proposed mechanisms (extracellular lipid accumulation versus cell death) are not mutually exclusive, they may both contribute to the formation of an acellular core. Advanced atherosclerotic lesions often have a fibrous cap (predominantly smooth muscle cells) over the core that is flanked by clusters of macrophages and lymphocytes, called 'shoulders' of the lesion. This fibrous cap distinguishes type V from type IV lesions.[13]

As the lesion gets larger, the surface integrity may fail, causing the formation of ulcers, hematomas, and/or thrombi (type VI).[13] Thrombi and hematomas are sometimes incorporated into the lesion as connective tissue covers the surface of the split. Some factors that contribute to lesion instability, and hence thrombus formation, are discussed in Section 1.8. In very large lesions neovascularization can occur through the lesion, providing an additional potential point of interaction between blood components and the lesion.

1.3 Cell biology of atherosclerotic lesions

Endothelial cells and endothelial dysfunction Endothelial cells are connected by junctional complexes (tight- and gap-junctions) and form the endothelium, which lines the entire vascular system. The cells have plasma membrane receptors for LDL, insulin, and histamine. Organelle characteristics of endothelial cells are plasmalemmal vesicles, caveolae, and Weibel-Palade bodies containing von Willebrand's factor (vWF or factor VIII), which is commonly used as a marker antigen for normal and neoplastic endothelial cells. Arterial endothelium has several important properties and functions: (i) a selectively permeable barrier, for instance, permeable to all plasma proteins; (ii) a non-thrombogenic blood–tissue interface due to the formation of prostacyclin (PGI2), heparan sulphate, and thrombo-modulin–thrombin complex; (iii) as a regulator of vascular tone by making endothelial-derived relaxing factor (EDRF), PGI2, endothelin (ET), and angiotensin II (AII); (iv) as a modulator for immune and inflammatory reactions via cytokines and growth factors, including IL-1, IL-6, IL-8, PDGF, FGF, CSF, TGF-β; (v) as a producer of extracellular matrix, including collagen, fibronectin, laminin, and proteoglycans; (vi) in modifying lipoproteins and enhancing foam cell formation; and (vii) by expressing adhesion molecules, e.g. monocyte chemoattractant protein, MCP-1.

In response to different pathological stimuli, endothelial cells may adjust their constitutive functions resulting in several types of reversible changes of their functional state, referred to as *endothelial dysfunction*. Such changes may be induced by e.g. histamine, serotonin, inflammatory cytokines, hemodynamic stress, products of lipid oxidation, glycosylation, and viruses.[15] An important transcription factor, NF-κB, has been associated with endothelial activation in atherogenesis.[16]

Smooth muscle cells and their changes in atherogenesis Smooth muscle cells, as the major cellular element of arterial media, have attracted extensive studies during the last three decades. In addition to vasoconstriction and arterial tone regulation, the cells have several other important functions in atherogenesis, including migration from media to intima, proliferation in response to inflammatory factors, and lipid and phospholipid metabolism by uptake and degradation of lipoproteins. In intermediate and advanced atherosclerotic lesions, due to increased uptake and decreased degradation of LDL, large amounts of lipid accumulate in smooth muscle cells, resulting in foam cell formation. Smooth muscle cells have two different phenotypes: contractile and synthetic. The synthetic cells not only show proliferation responses to a variety of mitogens, but also synthesize extracellular matrix molecules, such as collagen, elastin, and proteoglycans. The promoters for cell proliferation include platelet-derived growth factor (PDGF), basic fibroblast growth factor (bFGF), and interleukin-1 (IL-1).[17] The inhibitors include heparan sulphates, nitric oxide/endothelial relaxing factor (NO/EDRF), interferon-γ (IFN-γ), and transforming growth factor-β (TGF-β).[17] Interestingly, there are data suggesting that smooth muscle cells may also synthesize and secrete the mitogen PDGF in an autocrine fashion.[17]

Macrophages and early lipid deposition in atherogenesis   As the principal cells in initial lesions and fatty streaks, macrophages have an important role in early atheroma formation and in the stability of advanced atheroma as well. The cells are all originally derived from blood monocytes, and after migration into the intima, they differentiate into macrophages residing in arterial walls. Atherogenic migration of monocytes is induced by several adhesion molecules, such as vascular adhesion molecule-1 (VCAM-1), β_2 integrin, intercellular cell adhesion molecule-1 (ICAM-1), and selectins. Furthermore, a number of chemoattractants are responsible for the recruitment of monocytes/macrophages and their accumulation in atherogenesis, including MCP-1, tumour necrosis factor (TNF-α), IL-1, and M-CSF. The cells have active scavenging capacity to remove lipid and cell debris. Substantial uptake of LDL via the macrophage 'scavenger' receptors results in 'foam cell' formation within atherosclerotic lesions. Macrophages also synthesize and secrete a large number of biological substances, including proteases, hydrolases, complement, enzyme inhibitors, chemotactic agents, and growth factors.[18] Macrophages also serve as important antigen-presenting cells in atherosclerotic lesions and participate in immune modulation in atherogenesis. As under other chronic inflammatory conditions, macrophages may either die due to their scavenger function or proliferate due to autostimulation by their own production of CSF, PDGF, heparin-binding BGF-like growth factor (HB-EGF), and lymphocyte-derived IL-2.[18] There is still, however, a debate if the observed increased uptake of thymidine really represents cell proliferation or, perhaps, DNA repair following inflammatory injury.[18]

There are other inflammatory cells that are associated with the formation and development of atheroma as well. The coexistence of a large number of T lymphocytes (CD-4 and CD-8) and macrophages has been reported in atherosclerotic lesions (more details are provided in Chapter 11). Also, mast cells were recently described as components of atheroma. These cells are relatively rare or absent in normal arterial tissues.[19]

1.4 Lipid

The accumulation of lipid into the arterial intima is one of the characteristic features of atherosclerotic lesions. In developmental terms, accumulation of intimal lipid is one of the earliest events: studies using electron microscopy have identified extracellular lipid in grossly normal intima,[8, 9] LDL-like particles have been isolated from grossly normal regions of the human aorta,[20] and antibodies directed against apoB recognize material in lesion-free parts of the intima.[21, 22]

The lipid appears to accumulate in many different forms, including intracellular vesicles (monocyte/macrophages, smooth muscle cells, and endothelial cells), cholesterol crystals, and lipoprotein-like particles and extracellular droplets (including neutral lipid and vesicular lipid).[7] The ratio of cholesteryl ester to cholesterol varies as the stage of disease progresses, with fatty streaks having a higher ratio than more advanced lesions.[12] There was no significant difference in extracellular neutral lipid deposition (primarily cholesteryl esters) in the musculoelastic layer between fatty streaks and normal intima, suggesting that the massive lipid uptake by macrophages and smooth muscle cells (SMCs) in fatty streaks does not influence the deep extracellular neutral lipid deposits.[7] In contrast, vesicular-granular lipid deposits (primarily cholesterol and phospholipids) were significantly increased below fatty streaks.[7]

Some factors thought to contribute to lipid accumulation include increased plasma concentrations of lipoproteins, concentration of lipoproteins within the intima by size exclusion,[23] and retention of lipid due to interactions with connective tissue matrix.[24] In addition, alterations in physical characteristics of lipoproteins such as aggregation or increased negative charge (perhaps due to oxidation) could contribute to their entrapment. Increased permeability of the endothelium has also been suggested as an important factor;[25] however the studies cited do not appear to separate endothelial permeability from accumulation,[25] which is an important distinction, considering the known affinity of lipoproteins for some components of the extracellular matrix. In addition, there is some evidence that increased permeability at sites of lesions does not explain the extent of lipid accumulation.[26, 27]

The effects of macrophages on intimal lipid metabolism and efflux are dealt with in later chapters and are, therefore, not included here.

1.5 Calcification

Calcification has been recognized for a long time as a characteristic of advanced human atherosclerotic lesions. As such, it may contribute to lesion rupture and, thus, to clinical events caused by the formation of thrombi. For instance, calcium deposits have been shown to be significantly associated with both the location and the size of lesion fissures following angioplasty.[28] As well as its more obvious effects on plaque stability in advanced lesions, calcification, like lipid accumulation, may have subtle effects during lesion growth, as many cell signalling pathways are calcium dependent. Early stages of lesion development have been shown to contain the first evidence of calcification.[7, 8]

Calcium is a normal and essential component of human artery walls and it increases with age. Risk factors for cardiovascular disease, such as diabetes, smoking and hypertension, also increase the arterial content of calcium. The relationship between 'normal' calcium and the deposits of calcium that are so characteristic of the atherosclerotic lesion is not clear.[29]

The mechanism for the formation of focal deposits of insoluble calcium (dystrophic calcification) remains to be elucidated. The mineral deposits from human atherosclerotic aorta are primarily made of calcium apatite (71%), with small amounts of protein (15%) and carbonate (9%).[30] Initially, it was thought that such mineral deposits were formed through a physical process of nucleation of hydroxyapatite in the vesicles from dead cells, followed by crystal growth into the extracellular matrix [Ref. (31) cited in Ref. (29)]. More recently, it has become apparent that calcium deposition is at least partially controlled via cells and cellular proteins.

Matrix Gla protein (a calcium-binding protein) is constitutively expressed by vascular smooth muscle cells[32] and substantially upregulated adjacent to intimal or medial calcification. In mice with the matrix Gla protein knocked out, calcification occurs unchecked in the arterial walls and in various inappropriate cartilaginous sites.[33] Luo *et al.* suggested that the matrix Gla protein is required to actively inhibit extracellular matrix calcification in soft tissues. Consistent with the above hypothesis, calcification in human atherosclerotic lesions has been shown to be associated with the presence of macrophages and mast cells and a relative absence of smooth muscle cells.[34]

In early lesions, most of the calcified deposits are located in the musculoelastic layer and associated with elastic fibres, but very rarely with SMC.[8] In the proteoglycan-rich intimal layer, calcified deposits are observed only in regions with many foam cells.[8] In the musculoelastic layer, it has been suggested that the following sequence occurs: lipid associates with elastin, elastin calcifies and is then destroyed.[8] Treatment of rabbit and guinea pig arteries with soluble calcium salts can cause the disintegration of the elastic lamina [Ref. (35) cited in Ref. (29)], and such disruption may have pro-atherogenic effects (discussed in Section 1.6). It is likely that matrix Gla protein is only one of many proteins that help control calcification *in vivo*. For instance, collagens and fibronectin have been shown to promote or inhibit calcification of smooth muscle cells *in vitro*[36] and proteoglycans can also influence calcification.[37] In addition, other proteins associated with bone formation, such as osteopontin and osteoglycin, have been identified in atherosclerotic lesions.

1.6   Changes of the extracellular matrix in atherosclerotic lesions

The major components of connective tissue in the artery wall include collagens and other glycoproteins (such as fibronectin), elastins, and proteoglycans.

In normal thickened intima, the predominant forms of collagen are types I, III, and VI. Type V collagen is detected neither in intimal thickening nor in fatty streaks, but increases with age and lesion severity.[38] Type IV collagen is also increased in advanced lesions and is associated with the basement membrane of smooth muscle cells, as is the glycoprotein fibronectin. Basement membrane-rich smooth muscle cells are encountered when the normal extracellular matrix is disrupted.

In general, elastin appears to have an anti-atherogenic function. Expression of matrix metalloproteinase inhibitors can increase the amount of elastin in the vessel wall and reduce the extent of intimal hyperplasia after balloon injury.[39] When mice lacking elastin were generated, they died from an obstructive artery disease due to sub-endothelial cellular proliferation and reorganization of smooth muscle cells. These changes somewhat resemble atherosclerosis, although there was no inflammation, or thrombosis.[40] Elastin has often been observed to be degraded in and around atherosclerotic lesions. Consistent with elastin degradation, an elastolytic cathepsin is increased in atherosclerotic intima and is produced by monocyte/macrophages and intimal SMC[41] (proteases are also discussed in Section 1.9). In addition, elastin-derived peptides and anti-elastin antibodies, but not circulating elastase, are significantly raised in children from families with a high risk of atherosclerosis,[42] suggesting a link between elastin degradation and the immune response in those individuals that are most susceptible to atherosclerosis.

Chondroitin sulphate and dermatan sulphate proteoglycans accumulate within intimal lesions. In contrast, heparan sulphate proteoglycan shows little change or even a decrease.[37] In addition, some minor structural differences between proteoglycans from atherosclerotic tissue and normal tissue have been described.[37] With very severe disease, chondroitin sulphate, heparan sulphate, and hyaluronic acid decrease in the vessel. Increases in sulphated glycosaminoglycans enhance the LDL binding affinity of the extracellular matrix and may form complexes with LDL that cause foam cell formation. The LDL-proteoglycan affinity can be modified by LDL size, as well as by dietary and pharmacological interventions.[43]

Interestingly, extracellular matrix degradation may be enhanced by oxidation, both directly and indirectly. For instance, direct oxidation of hyaluronate proteoglycan aggregate makes the core proteins unable to associate with the central polysaccharide chain.[44] Oxidants may have an indirect action by making proteins more susceptible to proteolysis (e.g. elastin[45]) or by activating proteases (e.g. metalloproteinase[46]). Conversely, oxidants can inactivate proteinases and make highly oxidized proteins recalcitrant to proteolysis. The net effect of oxidants on matrix degradation remains to be elucidated.

1.7 Oxidants in atherosclerosis

Oxidants and antioxidants gained special significance for atherosclerosis research with the discovery that oxidized low-density lipoprotein bypassed the normal negative feedback loops in macrophages that inhibit excessive uptake of native LDL (reviewed in Ref. (47)). This discovery focused research on oxidation of lipoproteins, and other potential targets attracted much less interest. There is now increasing recognition that, in addition to lipoproteins, other oxidized targets may contribute to atherosclerosis as well.

The significance of oxidants in atherosclerotic lesions primarily depends on circumstantial evidence, since the oxidants of interest are mostly reactive with relatively short half-lives, making direct isolation or detection of them very difficult. Most studies therefore focus on products of oxidation reactions. When interpreting the results, it is essential to consider the evidence supporting specificity of the product to the hypothesized oxidant or reaction mechanism. Is the proposed oxidation reaction the only source of the product measured, or are there other alternative ways explaining why such a product could be present in a lesion? Oxysterols provide a good example of a potential alternative origin, as some oxidized products found in lesions are present in the diet. Thus, the presence of oxysterols in lesions is not sufficient evidence on its own to demonstrate oxidation in lesions (further information on oxysterols is found in Chapter 17). In a similar way, studies that use antibodies to detect oxidized products also suffer from specificity problems. Many antibodies raised against copper-oxidized LDL, or malondialdehyde-modified LDL, are cross-reactive with other forms of oxidized or modified LDL.[48] Bearing these shortcomings in mind, indirect evidence for oxidation by various different mechanisms is summarized below.

Most oxidative reactions require the presence of a reactive oxygen species, such as superoxide, hydrogen peroxide, or nitric oxide. The source of such species is assumed to be the cells present in lesions, either as a by-product of their normal role in the vessel or as a specific response to the inflammatory stimuli that accompany atherosclerosis. The reactive oxygen species listed above are relatively poor oxidants in isolation; however, when combined with the factors listed below they become much more reactive.

Oxidative enzymes in atherosclerosis Several oxidative enzymes have been proposed to have a role in atherosclerosis, including myeloperoxidase, lipoxygenase, and nitric oxide synthase. Myeloperoxidase, a heme-containing peroxidase, uses hydrogen peroxide and chloride as substrates, but may also react with superoxide and nitric oxide under some conditions. As well as being an oxidant itself, myeloperoxidase can produce other oxidants, such as hypochlorous acid/hypochlorite, radicals,[49] and reactive nitrogen species. Active myeloperoxidase[50] and some of its products, hypochlorite-oxidized proteins,[51] and chlorinated tyrosine[52] have been identified in human atherosclerotic lesions. In addition, myeloperoxidase- and hypochlorite-treated LDL have characteristics that resemble the modified LDL found in lesions.[53, 54]

Lipoxygenases are a family of non-heme, iron-containing dioxygenases that peroxidize fatty acids. 15-Lipoxygenase co-localizes with epitopes of oxidized LDL in human and rabbit atherosclerotic lesions, and stereospecific lipoxygenase products have been demonstrated in early atherosclerotic lesions of humans.[55] Interestingly, there is evidence for both pro- and anti-atherogenic activity of lipoxygenase. Lipoxygenase knockout mice on an apoE KO background showed reduced atherosclerosis;[56] however, selective over-expression of lipoxygenase in macrophages of cholesterol-fed rabbits has also been shown to reduce atherosclerosis.[57] While most evidence appears to support a pro-atherogenic effect of lipoxygenase, the above study indicates that its role(s) in atherosclerosis is likely to be more complex than first thought.

Nitric oxide synthase has both constitutive and inducible forms, and converts arginine to citrulline with the production of nitric oxide. Nitric oxide has many actions that are relevant to atherosclerosis, including regulation of vascular homeostasis by controlling vascular resistance, blood pressure, cell–cell adhesion, and proliferation.[58] The major oxidant produced from nitric oxide is thought to be peroxynitrite. There are several potential mechanisms for the formation of peroxynitrite: nitric oxide can react directly with superoxide,[59] or nitrate, a degradation product of nitric oxide, can be activated by reaction with hypochlorite or myeloperoxidase.[60, 61] Nitrated tyrosine has been detected in human atherosclerotic lesions[62] and in LDL isolated from the intima of such lesions.[63] In addition, peroxynitrite-treated LDL has pro-atherogenic characteristics.[64]

Transition metals in atherosclerosis   Transition metals have been proposed to have a role in atherosclerosis, mainly free and bound forms of iron. The presence of excess iron in lesions is supported by upregulation of the gene for ferritin (an iron storage protein) and the presence of material that stains with Prussian blue in advanced lesions.[65]

Body iron and related iron-containing proteins, including ferritin and hemoglobin, have been considered as candidate risk factors, although inconsistent data has been reported from time to time. The Bruneck study[66] demonstrated that the incidence of atherosclerosis in premenopausal women was less than half that observed in men of equal age. The sex difference disappeared within 5 years after menopause, and this was attributed to sex variations in body iron stores. In 1999, Roest *et al.* and Tuomainen *et al.* presented landmark studies on body iron and the genetics of atherosclerosis. Male carriers of the Cys282Tyr mutation of the human hemochromatosis-associated gene (HFE) are at two-fold risk for acute myocardial infarction compared with non-carriers.[67] In a cohort study of 12 239 women from Utrecht, cardiovascular death was found to be significantly associated with heterozygosity for HFE, indicating that iron metabolism is involved in cardiovascular death in postmenopausal women.[68]

Since there are several pieces of evidence indicating the role of iron in atherogenesis, iron deposition in atherosclerotic lesions may be an important type of atherosclerotic mineralization. Experimental studies have presented, and at least partially answered, the following questions:

Is tissue-iron related to development of atherosclerosis in animal models?   Araujo *et al.*[69] found that iron-overloading in rabbits augments the formation of atherosclerotic lesions that result from feeding a hypercholesterolemic diet, and that iron deposits were present in such experimental animal lesions. Using nuclear microscopy, Thong *et al.*[70] reported that there is on average a seven-fold increase in iron and a two-fold increase in phosphorus in atherosclerotic lesions from hypercholesterolemic rabbits, compared to healthy artery tissue. Using infrared spectroscopy and Raman spectroscopy techniques, Yuan *et al.* (unpublished data) have obtained similar results on advanced human atherosclerotic lesions. In 1999, again using nuclear microscopy, Ponraj *et al.*[71] demonstrated that iron accumulation occurs at the onset of lesion formation. In addition, weekly bleeding of the animals decreased iron uptake in the artery walls and delayed the onset of atherogenesis.

Clinical-pathological observations indicate the presence of iron in atherosclerotic lesions in humans. Such lesions contain redox-active iron that promotes lipid peroxidation that is inhibited by the iron-chelator desferrioxamine.[72] In 1995, Swain and Gutteridge reconfirmed a previous study[73] and for the first time detected ferroxidase I in human atherosclerotic material.[74]

Although the presence of both copper and iron, even in 'catalytically active form', has been reported in atheromatous lesions, it has not been possible to determine its origin *in vivo*. It is well known that oxidized or aged red blood cells (RBCs) are rapidly removed from the circulation, which provide a pathway for iron re-use in humans. It is interesting that ligands on oxidatively aged erythrocytes and apoptotic cells may bind to the same scavenger receptor as is used by oxidized LDL (OxLDL).[75] These findings indicate that oxidatively modified LDL not only competes for scavenger receptors of macrophages with aged RBCs, but also with other aged cells. Human monocyte-derived macrophages and J774 cells have been found to readily phagocytose oxidized erythrocytes in culture, with resulting lysosomal iron accumulation.[76] Release of low molecular weight iron and iron-containing proteins, following degradation of phagocytosed erythrocytes by cells of the reticuloendothelial system, may account for the majority of iron turnover in humans.[77] It was recently documented that lipid peroxidation of erythrocyte membranes was significantly higher in patients with CHD, whereas antioxidant activities, including SOD and GSH-Px, were lower than in controls.[78] The findings indicate that erythrophagocytosis may be an important event in the progression of atherogenesis. Lee *et al.*[79] recently provided histological evidence for the localization of hemoglobin, ferritin, heme oxygenase-1, iron, and erythrocytes in atherosclerotic lesions from humans and experimental animals. They suggested that hemoglobin/heme released from phagocytosed erythrocytes may importantly contribute to iron accumulation in atherosclerotic lesions.

What would be the pathological effects of increased lysosomal low molecular weight iron in arterial cells, particularly in macrophages? This question has been mainly studied using cultured cells. Iron-laden macrophages may partially exocytose their lysosomal contents, including low molecular weight iron and ferritin, contributing to extracellular LDL oxidation, its ensuing uptake and foam cell formation.[80] Low molecular weight iron also enhances cytotoxicity and apoptosis induced by oxidized lipids.[81] Furthermore, apoptotic cells, low molecular weight iron, and ferritin have been co-localized in human atherosclerotic lesions.[82] Iron, whether free or bound, also has the potential of degrading lipid hydroperoxides, thus amplifying free radical chain reactions and causing accelerating lipid peroxidation.[83] On the other hand, lipids may affect cellular iron homeostasis also. The modification of LDL by cultured endothelial cells or smooth muscle cells was enhanced by iron loading, and this could be counteracted by the exposure of the cells to apoferritin. Moreover, low-density lipoprotein induced ferritin synthesis in the system. Based on these observations, the authors proposed that intracellular, rather than extracellular, free iron may play a role in LDL oxidation and the development of atherosclerosis.[84]

Antioxidants in lesions   Assuming that oxidation contributes to atherosclerosis, the presence or absence of antioxidants in lesions is likely to be equally important. Despite the difficulty of preventing oxidation during sample work-up, there has been significant progress in the task of quantifying antioxidants present in atherosclerotic lesions (see Chapter 16 for more details). These studies have shown that the intimal environment does not appear to be depleted of critical antioxidants (such as ascorbate and vitamin E).[85]

1.8   Mechanisms behind arterial cell death

An acellular lipid core is an important feature of atherosclerotic lesions, which was already recognized by Virchow in 1858, cited in Ref. (86). The composition and formation of the lipid core largely remains unknown, though it has been considered a result of foam cell necrosis (and/or extracellular lipid deposition, see Section 1.2). Based on *in vitro* and *in vivo* observations, the remnants of dead macrophage/foam cells and ceroid are major components of such cores, while no smooth muscle cell antigens have been demonstrated.[14]

The mechanisms behind arterial cell turnover have recently attracted much attention.[87] Cell replication, cell death, and cell recruitment may all contribute to the turnover and cell number in atherosclerotic plaques. Lutgens *et al.* studied DNA synthesis and apoptosis during several stages (I–VI) of human atherosclerotic lesions. DNA synthesis first peaked in type II and again in type VI lesions, mainly in macrophage-derived foam cells, while apoptosis was elevated only in the advanced lesions (IV–VI), predominately in macrophages of lipid core areas. Similar results have been obtained in APOE3-Leiden transgenic mice[88] where DNA synthesis, mainly in type II lesions, and apoptosis, mainly in types IV and V, are primarily confined to macrophage-derived foam cells. Cell replication can be induced by a number of colony stimulating factors and growth factors formed by endothelium, smooth muscle cells, and macrophages. Research of the last decade indicates that, compared to cell proliferation, arterial cell death in atheroma may have important influences on the clinical consequences of atherosclerosis, such as atherosclerotic plaque stability.

There are two basic, although probably quite related, forms of cell death in atheroma, namely apoptosis and necrosis, and apoptosis seems to dominate. If apoptotic cells are not removed by professional scavengers, or other neighbouring cells, they may turn necrotic. Although several factors have been linked to arterial cell death in atherogenesis, including OxLDL and its toxic hydroxylipid compounds (e.g. 7β-hydroxycholesterol), cytokines, and cytotoxic T lymphocytes, the mechanisms behind cell death in atherosclerosis are largely unknown.

Oxidized LDL and cell death   Though it is well known that LDL is degraded in the lysosomal compartment, the effects of trapped OxLDL on the function of macrophage lysosomes and its influence on cell death have only recently been studied in some detail. In 1993 OxLDL-induced apoptosis of macrophages was first reported by Reid *et al.*[89] Following a 24 h period of exposure to OxLDL, mouse peritoneal macrophages showed ultrastructural alterations typical of apoptosis. Dose-dependent DNA fragmentation in P388D$_1$ macrophages was induced by incubation with OxLDL, being to some degree inhibited by Zn^{2+}. Fossel *et al.*[90] first demonstrated that peroxidized LDL induced cell death through lysosomal disruption in a number of tumour cell lines. Lysosomal leakage induced by peroxidized LDL was found to precede mitochondrial dysfunction and apoptosis. The damage could be inhibited, through competitive receptor-mediated LDL uptake, by the addition of 10–30-fold amounts of native LDL. Based on these observations, the authors suggested a new term – endopepsis – to describe this form of cell death. In 1997, lysosomal damage induced by OxLDL was proposed by Yuan *et al.*[91] as one of the pathways for macrophage and foam cell death. In cell culture model systems, OxLDL-induced macrophage damage was found to be associated with iron-mediated partial inactivation of lysosomal enzymes and their relocation to the cytosol. Pre-exposing cells to the iron-chelator, desferroxamine (DFO), HDL, or α-tocopherol partly prevented these events, whereas pretreating the cells with a low molecular weight iron complex that entered their lysosomal compartment augmented it. DFO was found to be endocytosed, did not permeate membranes and, thus, is selectively transported to the acidic vacuolar apparatus.[81, 92] The results suggest that the normal content of lysosomal low molecular weight iron may play an important role in OxLDL-induced cell damage. In addition to the study of lysosomal OxLDL degradation, Heeren *et al.*[93] also investigated the intracellular fate of triglyceride-rich lipoproteins using immunofluorescence methods. They found that apoE, apoC, and lipoprotein lipase follow a recycling pathway, whereas lipids and high molecular mass core proteins are degraded in late endosomal compartments and lysosomes. These findings further suggest that a better understanding of intra-lysosomal lipid metabolism is important for the understanding of atherogenesis-related cellular events.[93]

It should be noted that oxidized LDL is a poorly defined term. Various authors have used the term to describe material that has significantly different biological activities, including some forms of oxidized LDL that are not cytotoxic. In addition, the question regarding the forms of LDL oxidized *in vitro* that are most relevant to LDL oxidized *in vivo* continues to be controversial.

1.9   Stability of atherosclerotic plaques

Stable/unstable plaques and clinical relevance   Successful lipid-lowering trials have resulted in a marked reduction of clinical cardiac events, but in no major changes in lesion severity as shown by angiography. This suggests that the stability of atherosclerotic plaques, rather than the progression of them, is more closely related to the clinical consequences of atherosclerosis, such as unstable angina, myocardial infarction, and sudden death. It is generally believed that type IV and Va lesions are most often prone to rupture; consequently they are called unstable plaques. The most vulnerable areas are at the shoulders of types IV and Va, which are infiltrated by more than 60% macrophages and 9–18% T lymphocytes.[94] Morphological studies have defined characteristics of the unstable atheromatous plaque, including a thin and collagen-poor fibrous cap, an eccentric large lipid core rich in macrophages and T lymphocytes, and a few smooth muscle cells. Mechanical forces, synthesis and degradation of the extracellular matrix, and degree of inflammation may determine the stability of such plaques.[95] Among the factors that influence stability of the atherosclerotic plaque, infiltration of macrophages and lymphocytes may be the most important ones. Bauriedel *et al.*[96] compared coronary atherectomy specimens from 25 patients with unstable angina to those of another 25 patients with stable angina. They found that there was more cell death due to necrosis and apoptosis in plaques of the unstable angina group, and that these plaques contained significantly more macrophages/lymphocytes and less smooth muscle cells than the stable ones. Matrix degradation by proteolytic enzymes secreted from macrophages has been suggested as an important factor in plaque disruption. The enzymes involved in the process include serine proteases, cysteine proteases, and matrix metalloproteinases (MMPs).

The role of MMPs in plaque stability   Substantial studies on MMPs and plaque stability have indicated that interstitial collagenase (MMP-1), stromelysin (MMP-3),[97] and gelatinases (MMP-2 and MMP-9) initiate plaque disintegration.[98] Another study demonstrated that atheromatous, rather than fibrous, plaques may be prone to rupture due to increased collagenolysis by macrophages, being mediated by the MMP-1 and interstitial collagenase 3 (MMP-13).[99] There are several mechanisms for controlling the MMP activity. The activation of smooth muscle cells and macrophages by pro-inflammatory cytokines may enhance the expression of type I matrix metalloproteinases (MT1-MMP), an activator of pro-MMP-2, and influence the matrix remodelling in atherosclerosis.[100] Several specific tissue inhibitors of metalloproteinases (TIMPs) may control the activity of MMPs. The activity of MMP-9, from macrophage-derived foam cells, may be inhibited by *N*-acetyl-cysteine, also a scavenger of reactive oxygen species.[101] Finally, successful lipid-lowering trials have been associated with regulation of MMP activity in the experimental hypercholesterolemic rabbit model.[102]

The role of serine and cysteine proteases in atherogenesis has still received little attention. Treatment of macrophages with oxidized LDL, but not acetylated LDL, resulted in labilization of the lysosomal membrane and ensuing relocation of lysosomal enzymes, including *N*-acetyl-β-glucosaminidase and cathepsin L and D, while pretreatment with HDL and vitamin E diminished the lysosomal damage.[92] In contrast to findings in normal arterial tissues, abundant immunoreactive cathepsins K and S have been demonstrated in human atherosclerotic lesions, mainly associated with macrophages and smooth muscle cells.[41] These elastolytic enzymes may play an important role in vascular remodelling and plaque stability; this hypothesis is supported by recent findings from the same group, that endogenous inhibitors of the relevant enzymes are severely reduced in atherosclerotic lesions.[103]

1.10 Prospective remarks

For many years, investigations into atherosclerosis have relied on clinical endpoints due to a lack of techniques for measuring atherosclerosis directly. The current rapid improvement of imaging techniques (such as magnetic resonance and ultrasonic imaging) is allowing more reliable assessment of the extent of atherosclerosis *in vivo*. As these methods provide more sensitive detection of atherosclerosis *in vivo*, they will reduce the time required for many clinical trials and allow intervention in the earliest stages of atherosclerosis to be more easily investigated.

The availability of different fluorescent markers, specific antibodies, *in situ* hybridization, and deconvolution methods (for deciphering three-dimensional images from confocal microscopy) are now allowing the components of atherosclerotic lesions to be investigated in more detail than ever before. In addition, the recent development of laser microdissection has enabled histological features to be directly linked to biochemical and genetic characteristics, by allowing a single desired cell (or sub-class of cells) to be easily isolated from a tissue section for further analysis (e.g. using PCR).

Molecular techniques, especially gene microarrays and gene knockout methods, are also rapidly improving our understanding of cell signalling pathways and gene regulation. Thus, we will gradually determine the factors controlling arterial cell behaviour and how these cells participate in the build-up of atheromatous lesions.

As our knowledge of lesion formation, composition, and clinical-pathological consequences improves, we hope that new risk factors and novel interventions for atherosclerosis will be identified, thereby providing increasingly effective strategies against this ubiquitous, multifactorial and life-threatening disease.

References

1. Kwon, H.M., Sangiorgi, G., Ritman, E.L., McKenna, C., Holmes, D.R., Jr, Schwartz, R.S., and Lerman, A. (1998). Enhanced coronary vasa vasorum neovascularization in experimental hypercholesterolemia. *J. Clin. Invest.*, **101**, 1551.

2. Giddens, D.P., Zarins, C.K., and Glagov, S. (1993). The role of fluid mechanics in the localization and detection of atherosclerosis. *J. Biomech. Eng.*, **115**, 588.

3. Strong, J.P., Malcom, G.T., McMahan, C.A., Tracy, R.E., Newman, W.P., 3rd, Herderick, E.E., and Cornhill, J.F. (1999). Prevalence and extent of atherosclerosis in adolescents and young adults: implications for prevention from the Pathobiological Determinants of Atherosclerosis in Youth Study. *JAMA*, **281**, 727.

4. Napoli, C., D'Armiento, F.P., Mancini, F.P., Postiglione, A., Witztum, J.L., Palumbo, G., and Palinski, W. (1997). Fatty streak formation occurs in human fetal aortas and is greatly enhanced by maternal hypercholesterolemia – intimal accumulation of low density lipoprotein and its oxidation precede monocyte recruitment into early atherosclerotic lesions. *J. Clin. Invest.*, **100**, 2680.

5. Stary, H.C., Blankenhorn, D.H., Chandler, A.B., Glagov, S., Insull, W.J., Richardson, M., Rosenfeld, M.E., Schaffer, S.A., Schwartz, C.J., Wagner, W.D., and Wissler, M.D. (1992). A definition of the intima of human arteries and of its atherosclerosis-prone regions. A report from the Committee on Vascular Lesions of the Council on Arteriosclerosis, American Heart Association. *Arterioscler. Thromb.*, **12**, 120.

6. Homma, S., Ishii, T., Tsugane, S., and Hirose, N. (1997). Different effects of hypertension and hypercholesterolemia on the natural history of aortic atherosclerosis by the stage of intimal lesions. *Atherosclerosis*, **128**, 85.

7. Guyton, J.R. and Klemp, K.F. (1993). Transitional features in human atherosclerosis. Intimal thickening, cholesterol clefts, and cell loss in human aortic fatty streaks. *Am. J. Pathol.*, **143**, 1444.

8. Bobryshev, Y.V., Lord, R.S., and Warren, B.A. (1995). Calcified deposit formation in intimal thickenings of the human aorta. *Atherosclerosis*, **118**, 9.

9. Tirziu, D., Dobrian, A., Tasca, C., Simionescu, M., and Simionescu, N. (1995). Intimal thickenings of human aorta contain modified reassembled lipoproteins. *Atherosclerosis*, **112**, 101.

10. Stary, H.C., Chandler, A.B., Glagov, S., Guyton, J.R., Insull, W., Rosenfeld, M.E., Schaffer, S.A., Schwartz, C.J., Wagner, W.D., and Wissler, R.W. (1994). A definition of initial, fatty streak, and intermediate lesions of atherosclerosis. A report from the Committee on Vascular Lesions of the Council on Arteriosclerosis, American Heart Association. *Arterioscler. Thromb.*, **14**, 840.

11. Gown, A.M., Tsukada, T., and Ross, R. (1986). Human atherosclerosis. II. Immunocytochemical analysis of the cellular composition of human atherosclerotic lesions. *Am. J. Pathol.*, **125**, 191.

12. Guyton, J.R. and Klemp, K.F. (1994). Development of the atherosclerotic core region. Chemical and ultrastructural analysis of microdissected atherosclerotic lesions from human aorta. *Arterioscler. Thromb.*, **14**, 1305.

13. Stary, H.C., Chandler, A.B., Dinsmore, R.E., Fuster, V., Glagov, S., Insull, W.J., Rosenfeld, M.E., Schwartz, C.J., Wagner, W.D., and Wissler, R.W. (1995). A definition of advanced types of atherosclerotic lesions and a histological classification of atherosclerosis. A report from the Committee on Vascular Lesions of the Council on Arteriosclerosis, American Heart Association. *Arterioscler. Thromb. Vasc. Biol.*, **15**, 1512.

14. Ball, R.Y., Stowers, E.C., Burton, J.H., Cary, N.R., Skepper, J.N., and Mitchinson, M.J. (1995). Evidence that the death of macrophage foam cells contributes to the lipid core of atheroma. *Atherosclerosis*, **114**, 45.

15. Cotran, R.S., Kumar, V., and Robbins, S.L. (1994). Pathologic basis of disease. In *Blood vessels* (ed. F.J. Schoen), p. 470. W.B. Saunders Company, Pennsylvania.
16. Thurberg, B.L. and Collins, T. (1998). The nuclear factor-kappa B/inhibitor of kappa B autoregulatory system and atherosclerosis. *Curr. Opin. Lipidol.*, **9**, 387.
17. Ross, R. (1999). Atherosclerosis – an inflammatory disease. *New Engl. J. Med.*, **340**, 115.
18. Raines, E.W., Rosenfeld, M.E., and Ross, R. (1996). The role of macrophages. In *Atherosclerosis and coronary artery disease* (ed. V. Fuster and E.J. Topol), Vol. 30, p. 539. Lippincott-Raven Publishers, Philadelphia.
19. Kovanen, P.T. (1996). Mast cells in human fatty streaks and atheromas: implications for intimal lipid accumulation. *Curr. Opin. Lipidol.*, **7**, 281.
20. Hoff, H.F., Bradley, W.A., Heideman, C.L., Gaubatz, J.W., Karagas, M.D., and Gotto, A.M., Jr (1979). Characterization of low density lipoprotein-like particle in the human aorta from grossly normal and atherosclerotic regions. *Biochim. Biophys. Acta*, **573**, 361.
21. Yla-Herttuala, S., Solakivi, T., Hirvonen, J., Laaksonen, H., Mottonen, M., Pesonen, E., Raekallio, J., Åkerblom, H.K., and Nikkari, T. (1987). Glycosaminoglycans and apolipoproteins B and A-I in human aortas. Chemical and immunological analysis of lesion-free aortas from children and adults. *Arteriosclerosis*, **7**, 333.
22. Babaev, V.R., Bobryshev, Y.V., Sukhova, G.K., and Kasantseva, I.A. (1993). Monocyte/macrophage accumulation and smooth muscle cell phenotypes in early atherosclerotic lesions of human aorta. *Atherosclerosis*, **100**, 237.
23. Smith, E.B. (1990). Transport, interactions and retention of plasma proteins in the intima: the barrier function of the internal elastic lamina. *Eur. Heart J.*, **11** (Suppl. E), 72.
24. Camejo, G. (1982). The interaction of lipids and lipoproteins with the intercellular matrix of arterial tissue: its possible role in atherogenesis. *Adv. Lipid Res.*, **19**, 1.
25. Nielsen, L.B. (1996). Transfer of low density lipoprotein into the arterial wall and risk of atherosclerosis. *Atherosclerosis*, **123**, 1.
26. Schwenke, D.C. and Carew, T.E. (1989). Initiation of atherosclerotic lesions in cholesterol-fed rabbits. II. Selective retention of LDL vs selective increases in LDL permeability in susceptible sites of arteries. *Arteriosclerosis*, **9**, 908.
27. Smith, E.B. and Ashall, C. (1983). Low-density lipoprotein concentration in interstitial fluid from human atherosclerotic lesions. Relation to theories of endothelial damage and lipoprotein binding. *Biochim. Biophys. Acta*, **754**, 249.
28. Fitzgerald, P.J., Ports, T.A., and Yock, P.G. (1992). Contribution of localized calcium deposits to dissection after angioplasty. An observational study using intravascular ultrasound. *Circulation*, **86**, 64.
29. Born, G.V.R. (1993). Calcium and atherosclerosis. In *New horizons in coronary heart disease* (ed. G.V.R. Born and C.J. Schwartz), p. 22.1. Current Science, London.
30. Schmid, K., McSharry, W.O., Pameijer, C.H., and Binette, J.P. (1980). Chemical and physicochemical studies on the mineral deposits of the human atherosclerotic aorta. *Atherosclerosis*, **37**, 199.
31. Anderson, H.C., McGregor, D.H., and Tanimura, A. (1990). Mechanisms of calcification in atherosclerosis. In *Pathobiology of the human atherosclerotic plaque* (ed. S. Glagov, W.P. Newman, and S.A. Schaffer), p. 235. Springer, New York.

32. Shanahan, C.M., Proudfoot, D., Farzaneh-Far, A., and Weissberg, P.L. (1998). The role of Gla proteins in vascular calcification. *Crit. Rev. Euk. Gene Exp.*, **8**, 357.
33. Luo, G., Ducy, P., McKee, M.D., Pinero, G.J., Loyer, E., Behringer, R.R., and Karsenty, G. (1997). Spontaneous calcification of arteries and cartilage in mice lacking matrix Gla protein. *Nature*, **386**, 78.
34. Jeziorska, M., McCollum, C., and Woolley, D.E. (1998). Calcification in atherosclerotic plaque of human carotid arteries: associations with mast cells and macrophages. *J. Pathol.*, **185**, 10.
35. Katase, A. (1913). Experimentelle verkalkung am gesunden tiere. *Beitr. Pathol. Anat. Allg. Pathol.*, **57**, 516.
36. Watson, K.E., Parhami, F., Shin, V., and Demer, L.L. (1998). Fibronectin and collagen I matrixes promote calcification of vascular cells *in vitro*, whereas collagen IV matrix is inhibitory. *Arterioscler. Thromb. Vasc. Biol.*, **18**, 1964.
37. Wight, T.N. (1989). Cell biology of arterial proteoglycans. *Arteriosclerosis*, **9**, 1.
38. Katsuda, S., Okada, Y., Minamoto, T., Oda, Y., Matsui, Y., and Nakanishi, I. (1992). Collagens in human atherosclerosis. Immunohistochemical analysis using collagen type-specific antibodies. *Arterioscler. Thromb.*, **12**, 494.
39. Forough, R., Lea, H., Starcher, B., Allaire, E., Clowes, M., Hasenstab, D., and Clowes, A.W. (1998). Metalloproteinase blockade by local overexpression of TIMP-1 increases elastin accumulation in rat carotid artery intima. *Arterioscler. Thromb. Vasc. Biol.*, **18**, 803.
40. Li, D.Y., Brooke, B., Davis, E.C., Mecham, R.P., Sorensen, L.K., Boak, B.B., Eichwald, E., and Keating, M.T. (1998). Elastin is an essential determinant of arterial morphogenesis. *Nature*, **393**, 276.
41. Sukhova, G.K., Shi, G.P., Simon, D.I., Chapman, H.A., and Libby, P. (1998). Expression of the elastolytic cathepsins S and K in human atheroma and regulation of their production in smooth muscle cells. *J. Clin. Invest.*, **102**, 576.
42. Gminski, J., Drozdz, M., Ulfig-Maslanka, R., and Najda, J. (1991). Evaluation of elastin metabolism in children from families with high risk of atherosclerosis. *Atherosclerosis*, **91**, 185.
43. Camejo, G., Hurt-Camejo, E., Wiklund, O., and Bondjers, G. (1998). Association of apo B lipoproteins with arterial proteoglycans: pathological significance and molecular basis. *Atherosclerosis*, **139**, 205.
44. Katrantzis, M., Baker, M.S., Handley, C.J., and Lowther, D.A. (1991). The oxidant hypochlorite (OCl^-), a product of the myeloperoxidase system, degrades articular cartilage proteoglycan aggregate. *Free Rad. Biol. Med.*, **10**, 101.
45. Olszowska, E., Olszowski, S., Zgliczynski, J.M., and Stelmaszynska, T. (1989). Enhancement of proteinase-mediated degradation of proteins modified by chlorination. *Int. J. Biochem.*, **21**, 799.
46. Ottonello, L., Dapino, P., Scirocco, M., Dallegri, F., and Sacchetti, C. (1994). Proteolytic inactivation of alpha-1-antitrypsin by human neutrophils: involvement of multiple and interlinked cell responses to phagocytosable targets. *Eur. J. Clin. Invest.*, **24**, 42.
47. Steinbrecher, U.P., Zhang, H., and Lougheed, M. (1990). Role of oxidatively modified LDL in atherosclerosis. *Free Rad. Biol. Med.*, **9**, 155.
48. Malle, E., Hazell, L., Stocker, R., Sattler, W., Esterbauer, H., and Waeg, G. (1995). Immunologic detection and measurement of hypochlorite-modified LDL with specific monoclonal antibodies. *Arterioscler. Thromb. Vasc. Biol.*, **15**, 982.

49. Hazell, L.J., Davies, M.J., and Stocker, R. (1999). Secondary radicals derived from chloramines of apolipoprotein B-100 contribute to HOCl-induced lipid peroxidation of low-density lipoproteins. *Biochem. J.*, **339**, 489.

50. Daugherty, A., Dunn, J.L., Rateri, D.L., and Heinecke, J.W. (1994). Myeloperoxidase, a catalyst for lipoprotein oxidation, is expressed in human atherosclerotic lesions. *J. Clin. Invest.*, **94**, 437.

51. Hazell, L.J., Arnold, L., Flowers, D., Waeg, G., Malle, E., and Stocker, R. (1996). Presence of hypochlorite-modified proteins in human atherosclerotic lesions. *J. Clin. Invest.*, **97**, 1535.

52. Hazen, S.L. and Heinecke, J.W. (1997). 3-Chlorotyrosine, a specific marker of myeloperoxidase-catalyzed oxidation, is markedly elevated in low density lipoprotein isolated from human atherosclerotic intima. *J. Clin. Invest.*, **99**, 2075.

53. Hazell, L.J. and Stocker, R. (1993). Oxidation of low-density lipoprotein with hypochlorite causes transformation of the lipoprotein into a high-uptake form for macrophages. *Biochem. J.*, **290**, 165.

54. Hazell, L.J., van den Berg, J.J.M., and Stocker, R. (1994). Oxidation of low-density lipoprotein causes aggregation that is mediated by modification of lysine residues rather than lipid oxidation. *Biochem. J.*, **302**, 297.

55. Folcik, V.A., Nivararisty, R.A., Krajewski, L.P., and Cathcart, M.K. (1995). Lipoxygenase contributes to the oxidation of lipids in human atherosclerotic plaques. *J. Clin. Invest.*, **96**, 504.

56. Cyrus, T., Witztum, J.L., Rader, D.J., Tangirala, R., Fazio, S., Linton, M.F., and Funk, C.D. (1999). Disruption of the 12/15-lipoxygenase gene diminishes atherosclerosis in apo E-deficient mice. *J. Clin. Invest.*, **103**, 1597.

57. Shen, J., Herderick, E., Cornhill, J.F., Zsigmond, E., Kim, H.S., Kuhn, H., Guevara, N.V., and Chan, L. (1996). Macrophage-mediated 15-lipoxygenase expression protects against atherosclerosis development. *J. Clin. Invest.*, **98**, 2201.

58. Radomski, M.W. and Salas, E. (1995). Nitric oxide – biological mediator, modulator and factor of injury: its role in the pathogenesis of atherosclerosis. *Atherosclerosis*, **118**, S69.

59. Beckman, J.S., Beckman, T.W., Chen, J., Marshall, P.A., and Freeman, B.A. (1990). Apparent hydroxyl radical production by peroxynitrite: implications for endothelial injury from nitric oxide and superoxide. *Proc. Natl. Acad. Sci.*, **87**, 1620.

60. Van der Vliet, A., Eiserich, J.P., Halliwell, B., and Cross, C.E. (1997). Formation of reactive nitrogen species during peroxidase-catalyzed oxidation of nitrite. A potential additional mechanism of nitric oxide-dependent toxicity. *J. Biol. Chem.*, **272**, 7617.

61. Eiserich, J.P., Hristova, M., Cross, C.E., Jones, A.D., Freeman, B.A., Halliwell, B., and van der Vliet, A. (1998). Formation of nitric oxide-derived inflammatory oxidants by myeloperoxidase in neutrophils. *Nature*, **391**, 393.

62. Beckmann, J.S., Ye, Y.Z., Anderson, P.G., Chen, J., Accavitti, M.A., Tarpey, M.M., and White, C.R. (1994). Extensive nitration of protein tyrosines in human atherosclerosis detected by immunohistochemistry. *Biol. Chem. Hoppe-Seyler*, **375**, 81.

63. Leeuwenburgh, C., Hardy, M.M., Hazen, S.L., Wagner, P., Ohishi, S., Steinbrecher, U.P., and Heinecke, J.W. (1997). Reactive nitrogen intermediates promote low density lipoprotein oxidation in human atherosclerotic intima. *J. Biol. Chem.*, **272**, 1433.

64. Graham, A., Hogg, N., Kalyanaraman, B., O'Leary, V., Darley-Usmar, V., and Moncada, S. (1993). Peroxynitrite modification of low-density lipoprotein leads to recognition by the macrophage scavenger receptor. *FEBS Lett.*, **330**, 181.

65. Pang, J.H., Jiang, M.J., Chen, Y.L., Wang, F.W., Wang, D.L., Chu, S.H., and Chau, L.Y. (1996). Increased ferritin gene expression in atherosclerotic lesions. *J. Clin. Invest.*, **97**, 2204.

66. Kiechl, S. and Willeit, J. (1999). The natural course of atherosclerosis. Part I: incidence and progression. *Arterioscler. Thromb. Vasc. Biol.*, **19**, 1484.

67. Tuomainen, T.P., Kontula, K., Nyyssonen, K., Lakka, T.A., Helio, T., and Salonen, J.T. (1999). Increased risk of acute myocardial infarction in carriers of the hemochromatosis gene Cys282Tyr mutation: a prospective cohort study in men in eastern Finland. *Circulation*, **100**, 1274.

68. Roest, M., van der Schouw, Y.T., de Valk, B., Marx, J.J., Tempelman, M.J., de Groot, P.G., Sixma, J.J., and Banga, J.D. (1999). Heterozygosity for a hereditary hemochromatosis gene is associated with cardiovascular death in women. *Circulation*, **100**, 1268.

69. Araujo, J.A., Romano, E.L., Brito, B.E., Parthe, V., Romano, M., Bracho, M., Montano, R.F., and Cardier, J. (1995). Iron overload augments the development of atherosclerotic lesions in rabbits. *Arterioscler. Thromb. Vasc. Biol.*, **15**, 1172.

70. Thong, P.S., Selley, M., and Watt, F. (1996). Elemental changes in atherosclerotic lesions using nuclear microscopy. *Cell. Mol. Biol.*, **42**, 103.

71. Ponraj, D., Makjanic, J., Thong, P.S., Tan, B.K., and Watt, F. (1999). The onset of atherosclerotic lesion formation in hypercholesterolemic rabbits is delayed by iron depletion. *FEBS Lett.*, **459**, 218.

72. Yuan, X.M., Li, W., Olsson, A.G., and Brunk, U.T. (1996). Iron in human atheroma and LDL oxidation by macrophages following erythrophagocytosis. *Atherosclerosis*, **124**, 61.

73. Smith, C., Mitchinson, M.J., Aruoma, O.I., and Halliwell, B. (1992). Stimulation of lipid peroxidation and hydroxyl-radical generation by the contents of human atherosclerotic lesions. *Biochem. J.*, **286**, 901.

74. Swain, J. and Gutteridge, J.M. (1995). Prooxidant iron and copper, with ferroxidase and xanthine oxidase activities in human atherosclerotic material. *FEBS Lett.*, **368**, 513.

75. Sambrano, G.R. and Steinberg, D. (1995). Recognition of oxidatively damaged and apoptotic cells by an oxidized low density lipoprotein receptor on mouse peritoneal macrophages: role of membrane phosphatidylserine. *Proc. Natl. Acad. Sci.*, **92**, 1396.

76. Yuan, X.M., Olsson, A.G., and Brunk, U.T. (1996). Macrophage erythrophagocytosis and iron exocytosis. *Redox Report*, **2**, 9.

77. Moura, E., Noordermeer, M.A., Verhoeven, N., Verheul, A.F., and Marx, J.J. (1998). Iron release from human monocytes after erythrophagocytosis *in vitro*: an investigation in normal subjects and hereditary hemochromatosis patients. *Blood*, **92**, 2511.

78. Akkus, I., Saglam, N.I., Caglayan, O., Vural, H., Kalak, S., and Saglam, M. (1996). Investigation of erythrocyte membrane lipid peroxidation and antioxidant defense systems of patients with coronary artery disease (CAD) documented by angiography. *Clin. Chim. Acta*, **244**, 173.

79. Lee, T.S., Lee, F.Y., Pang, J.H., and Chau, L.Y. (1999). Erythrophagocytosis and iron deposition in atherosclerotic lesions. *Chin. J. Physiol.*, **42**, 17.

80. Yuan, X.M., Brunk, U.T., and Olsson, A.G. (1995). Effects of iron- and hemoglobin-loaded human monocyte-derived macrophages on oxidation and uptake of LDL. *Arterioscler. Thromb. Vasc. Biol.*, **15**, 1345.

81. Li, W., Yuan, X.M., and Brunk, U.T. (1998). OxLDL-induced macrophage cytotoxicity is mediated by lysosomal rupture and modified by intralysosomal redox-active iron. *Free Radic. Res.*, **29**, 389.

82. Yuan, X.M. (1999). Apoptotic macrophage-derived foam cells of human atheromas are rich in iron and ferritin, suggesting iron-catalysed reactions to be involved in apoptosis. *Free Radic. Res.*, **30**, 221.

83. Halliwell, B., Gutteridge, J.M., and Cross, C.E. (1992). Free radicals, antioxidants, and human disease: where are we now? *J. Lab. Clin. Med.*, **119**, 598.

84. Van Lenten, B.J., Prieve, J., Navab, M., Hama, S., Lusis, A.J., and Fogelman, A.M. (1995). Lipid-induced changes in intracellular iron homeostasis *in vitro* and *in vivo*. *J. Clin. Invest.*, **95**, 2104.

85. Suarna, C., Dean, R.T., May, J., and Stocker, R. (1995). Human atherosclerotic plaque contains both oxidized lipids and relatively large amounts of alpha-tocopherol and ascorbate. *Arterioscler. Thromb. Vasc. Biol.*, **15**, 1616.

86. Schwartz, S.M. and Bennett, M.R. (1995). Death by any other name. *Am. J. Pathol.*, **147**, 229.

87. Kockx, M.M. (1998). Apoptosis in the atherosclerotic plaque: quantitative and qualitative aspects. *Arterioscler. Thromb. Vasc. Biol.*, **18**, 1519.

88. Lutgens, E., Daemen, M., Kockx, M., Doevendans, P., Hofker, M., Havekes, L., Wellens, H., and de Muinck, E.D. (1999). Atherosclerosis in APOE*3-Leiden transgenic mice: from proliferative to atheromatous stage. *Circulation*, **99**, 276.

89. Reid, V.C., Mitchinson, M.J., and Skepper, J.N. (1993). Cytotoxicity of oxidized low-density lipoprotein to mouse peritoneal macrophages: an ultrastructural study. *J. Pathol.*, **171**, 321.

90. Fossel, E.T., Zanella, C.L., Fletcher, J.G., and Hui, K.K. (1994). Cell death induced by peroxidized low-density lipoprotein: endopepsis. *Cancer Res.*, **54**, 1240.

91. Yuan, X.M., Li, W., Olsson, A.G., and Brunk, U.T. (1997). The toxicity to macrophages of oxidized low-density lipoprotein is mediated through lysosomal damage. *Atherosclerosis*, **133**, 153.

92. Li, W., Yuan, X.M., Olsson, A.G., and Brunk, U.T. (1998). Uptake of oxidized LDL by macrophages results in partial lysosomal enzyme inactivation and relocation. *Arterioscler. Thromb. Vasc. Biol.*, **18**, 177.

93. Heeren, J., Weber, W., and Beisiegel, U. (1999). Intracellular processing of endocytosed triglyceride-rich lipoproteins comprises both recycling and degradation. *J. Cell. Sci.*, **112**, 349.

94. Hansson, G.K., Jonasson, L., Lojsthed, B., Stemme, S., Kocher, O., and Gabbiani, G. (1988). Localization of T lymphocytes and macrophages in fibrous and complicated human atherosclerotic plaques. *Atherosclerosis*, **72**, 135.

95. Rabbani, R. and Topol, E.J. (1999). Strategies to achieve coronary arterial plaque stabilization. *Cardiovasc. Res.*, **41**, 402.

96. Bauriedel, G., Hutter, R., Welsch, U., Bach, R., Sievert, H., and Luderitz, B. (1999). Role of smooth muscle cell death in advanced coronary primary lesions: implications for plaque instability. *Cardiovasc. Res.*, **41**, 480.

97. Henney, A.M., Wakeley, P.R., Davies, M.J., Foster, K., Hembry, R., Murphy, G., and Humphries, S. (1991). Localization of stromelysin gene expression in atherosclerotic plaques by *in situ* hybridization. *Proc. Natl. Acad. Sci.*, **88**, 8154.

98. Galis, Z.S., Sukhova, G.K., Lark, M.W., and Libby, P. (1994). Increased expression of matrix metalloproteinases and matrix degrading activity in vulnerable regions of human atherosclerotic plaques. *J. Clin. Invest.*, **94**, 2493.

99. Sukhova, G.K., Schonbeck, U., Rabkin, E., Schoen, F.J., Poole, A.R., Billinghurst, R.C., and Libby, P. (1999). Evidence for increased collagenolysis by interstitial collagenases-1 and -3 in vulnerable human atheromatous plaques. *Circulation*, **99**, 2503.

100. Rajavashisth, T.B., Xu, X.P., Jovinge, S., Meisel, S., Xu, X.O., Chai, N.N., Fishbein, M.C., Kaul, S., Cercek, B., Sharifi, B., and Shah, P.K. (1999). Membrane type 1 matrix metalloproteinase expression in human atherosclerotic plaques: evidence for activation by proinflammatory mediators. *Circulation*, **99**, 3103.

101. Galis, Z.S., Asanuma, K., Godin, D., and Meng, X. (1998). *N*-Acetyl-cysteine decreases the matrix-degrading capacity of macrophage-derived foam cells: new target for antioxidant therapy? *Circulation*, **97**, 2445.

102. Aikawa, M., Rabkin, E., Okada, Y., Voglic, S.J., Clinton, S.K., Brinckerhoff, C.E., Sukhova, G.K., and Libby, P. (1998). Lipid lowering by diet reduces matrix metalloproteinase activity and increases collagen content of rabbit atheroma: a potential mechanism of lesion stabilization. *Circulation*, **97**, 2433.

103. Shi, G.P., Sukhova, G.K., Grubb, A., Ducharme, A., Rhode, L.H., Lee, R.T., Ridker, P.M., Libby, P., and Chapman, H.A. (1999). Cystatin C deficiency in human atherosclerosis and aortic aneurysms. *J. Clin. Invest.*, **104**, 1191.

2 Diabetes and atherosclerosis

John W. Eaton
Department of Pediatrics, Baylor College of Medicine, Houston,
Texas, USA

Roger T. Dean
The Heart Research Institute, Camperdown, Sydney,
NSW 2050, Australia

2.1 Introduction

This brief review addresses a topic of considerable public health importance – the greatly increased incidence of vascular disease in patients with diabetes. Because the great majority of patients with diabetes have so-called type 2 diabetes (also variably known as adult onset or non-insulin-dependent diabetes mellitus (NIDDM)), unless otherwise noted we have restricted our coverage to this form of diabetes. The literature is both vast and inconclusive, and we have no doubt failed to mention a number of potentially important pathophysiologic mechanisms involved in atherogenesis in diabetes (not to mention numerous important publications by important investigators). This review represents an attempt to highlight possible contributory factors other than the obvious risk factors which typically coincide with the diabetic condition. The reader should be warned, however, that the involvement of many of these suggested factors will eventually be disproven. This is the natural course of science.

2.2 Risk for atherosclerosis in diabetes

Regardless of the population under study, it has been consistently found that diabetes increases the risk of coronary heart disease, cerebrovascular disease and peripheral vascular disease. In most western countries (but not in some eastern countries such as Japan), atherosclerotic disease is the leading cause of death in patients with type 2 diabetes.[1] In most series, the relative risk of atherosclerosis and/or cardiovascular mortality for women exceeds that for men[2] with some estimates for risk being in the range of 4–6-fold.[3] Even after multivariate analysis to correct for associated risk factors (*vide infra*), the risk ratio for cardiovascular disease has been estimated at 3 or higher for diabetic men and the risk for women once again being somewhat higher.[4] This, coupled with the high frequency of cerebrovascular and microvascular disease, takes a serious toll on patients with diabetes. Unfortunately, the pathogenic mechanisms underlying this diabetes-related predisposition to vascular diseases are not well understood.

The appearance of overt type 2 diabetes is typically preceded by a pre-diabetic period in which the patient displays insulin resistance and hyperinsulinemia but relative normoglycemia. This prodrome is followed by the appearance of the hallmarks of diabetes, hypoinsulinemia, and hyperglycemia. The pre-diabetic stage of type 2 diabetes is similar – or identical – to so-called syndrome X comprised of a tetrad of characteristics including insulin resistance, hypertension, android obesity and dyslipidemia.[5] We should note, however, that despite its popularity, the existence of syndrome X as a true syndrome has been called into question.[6]

2.3 Dyslipidemias and hemostatic abnormalities associated with diabetes

Given that people with diabetes clearly have accelerated atherogenesis, students of the disease have had no problem finding multiple factors which might help explain this. Some diabetes-associated abnormalities in lipid metabolism and hemostasis are listed in Table 2.1, taken from a recent review of this area.[4] The reader should keep

Table 2.1
Abnormalities of lipid metabolism and hemostasis associated with diabetes which may promote atherogenesis. Modified from Sowers and Lester.[4]

Dyslipidemia
- Increased plasma VLDL, LDL and lipoprotein(a)
- Decreased plasma HDL cholesterol
- Increased small dense LDL and chylomicra
- Decreased activity of lipoprotein lipase
- Increased lipoprotein glycation and oxidation

Coagulation/Fibrinolytic Abnormalities
- Elevated levels of factors VII and VIII
- Increased fibrinogen
- Increased plasminogen activator inhibitor
- Decreased plasminogen activation and fibrinolytic activity
- Increased plasma thrombin–antithrombin complexes
- Decreased antithrombin III, protein C and protein S
- Abnormalities of platelet function

in mind that some of these abnormalities can be rather subtle and some may be apparent only under conditions of stress.

The causal relationships between diabetes and atherosclerotic disease are confounded by the fact that diabetic patients are liable to have a number of risk factors which, in non-diabetic individuals, are known to predispose to atherosclerosis.[1] These include higher frequencies of hypertension, which is twice as prevalent in diabetic people,[4] visceral obesity and hypercoagulability. With regard to hemostatic abnormalities, diabetic patients (especially following myocardial infarction) tend to exhibit increased levels of von Willebrand factor, fibrinogen, plasminogen activator inhibitor-1 and factor VII, along with hyperaggregability of platelets.

Type 2 diabetes is well known to be associated with dyslipidemias, including abnormalities in circulating lipoprotein and triglyceride levels. The visceral obesity may be particularly unfortunate in this regard because fat in this location is subject to enhanced lipolytic activity which is only poorly suppressed by insulin. The resultant liberation of free fatty acids into the liver is thought to enhance the production of triglycerides and very low density lipoprotein (VLDL).[7] There is increasing evidence that elevated triglycerides and associated lipid abnormalities constitute a risk factor at least as important as cholesterol levels/distribution.[8]

The reasons for this are not yet clear. There is some evidence (from a rabbit model) that triglyceride-rich lipoproteins readily cross the endothelial barrier.[9] Furthermore, the lipolytic remnants of triglyceride-rich lipoproteins impair the endothelial barrier function.[10] This latter may be due to the actions of associated free fatty acids, known to be increased in the plasma of diabetic patients. Even modest (micromolar) concentrations of free fatty acids will (i) increase the permeability of endothelial monolayers, (ii) enhance lipid accumulation in human arterial smooth muscle,[11] (iii) promote endothelial production of inflammatory mediators (NF-κB activation, interleukin-8 production and increased intercellular adhesion molecule-1 expression),[12] (iv) profoundly alter extracellular matrix synthesis in cultured human vascular smooth muscle cells,[13] (v) cause insulin resistance,[14] and (vi) upregulate adherence molecules on human phagocytes.[15] However, there is not yet firm proof that any of these free fatty acid-mediated perturbations are contributory to the accelerated atherogenesis in diabetes.

On average, plasma cholesterol in diabetic patients is equivalent to that in non-diabetic individuals. However, the distribution of cholesterol may frequently be abnormal, with a shift toward smaller denser low-density lipoprotein (LDL) particles, elevated triglycerides and chylomicra, and modest decreases in high-density lipoprotein (HDL) cholesterol. Elevated small, dense LDL has been indicated as a particularly important risk factor. Possible interactions between abnormalities in lipid metabolism and atherosclerosis in diabetes have recently been reviewed.[16] To the extent that these variables are involved in the accelerated atherogenesis in diabetes, their importance is well described in other chapters in this volume.

It is, however, generally acknowledged that the above abnormalities alone fall far short of explaining the increased risk of atherosclerosis in diabetes. Therefore, the following represents an attempt to pull together possible alternative mechanisms which could be involved in the accelerated atherogenesis of diabetes.

2.4   Insulin and atherogenesis

Diabetes is usually thought of as a condition of insulin insufficiency. However, patients destined to develop non-insulin-dependent diabetes most frequently have, in the pre-diabetic period, elevated serum insulin levels due to insulin resistance. It has been proposed that elevated insulin *per se* may be an independent risk factor for atherosclerosis in these patients.[17] It is conceivable that hyperinsulinemia might promote the development of hypertension inasmuch as insulin increases the production of endothelin, a powerful vasoconstrictor. The possibility that these high levels of insulin might somehow promote atherogenesis is further supported by the investigations of Abe,[18] who caused chronic hyperinsulinemia in otherwise normal rats by transplantation of a second pancreas. These animals showed enhanced accumulation of cholesterol esters in the arterial wall without abnormalities in plasma lipids or increased blood pressure. These hyperinsulinemic rats did not, however, exhibit any histologic abnormalities indicative of full-blown atherosclerosis. In fact, in humans with normal insulin sensitivity, insulin itself has an endothelium-dependent vasodilatory effect but this, as well as an insulin-induced decrease in arterial stiffness, is severely decreased in obese hyperinsulinemic (probably pre-type 2 diabetic) patients.[19, 20] We should add that there is no evidence that the *exogenous* insulin sometimes used in therapy of type 2 diabetes has similar pro-atherogenic effects.

There is, in addition, some evidence from a mouse model that insulin resistance may engender increased blood pressure and elevated plasma triglycerides, both well-established risk factors for atherosclerosis.[21] The mice in question had a targeted disruption for insulin receptor substrate-1, a major intracellular substrate phosphorylated by tyrosine kinase following ligation of the insulin receptor or insulin-like growth factor-1 receptor. Because this discrete change increased both blood pressure and triglycerides, these observations do support the idea that blunted responses to insulin may secondarily engender dyslipidemia and increased blood pressure.[22]

2.5   Protein–sugar reactions: glycosylation and glycoxidation

Perhaps the majority view is that many of the unfortunate pathologic side effects of diabetes derive from hyperglycemia. Several of the following sections deal with the possible importance of glycation and glycoxidation events promoted by hyperglycemia in the pathogenesis of atherosclerosis in diabetes. However, it should be noted that large-vessel disease appears in most diabetic patients near the time of first diagnosis of diabetes. This means that large-vessel disease has been developing in these patients during the early, euglycemic phase of the evolution of type 2 diabetes.

Therefore, other aspects of the pre-diabetic state may be very importantly involved in the accelerated atherogenesis.

There is, nonetheless, substantial interest in the pathogenic importance of events mediated by reactions between proteins and sugars in diabetes-associated atherogenesis. These reactions produce a variety of early (e.g. simple Schiff base and Amadori product formation) and late products (e.g. carbonyls, dicarbonyls, pentosidine, carboxymethyl lysine, and pyrraline). The latter compounds occur in normal humans, especially in advanced age, but their formation is greatly enhanced in patients with diabetes. Because the production of many – but not all – of these glycated species involves oxidation reactions, the term glycoxidation has been proposed to describe reactions involving oxidation and glycation.[23] The later products of reactions between proteins, peptides or phospholipids with sugars have been termed 'advanced glycation end products' (AGEs) and it has been argued that their accumulation is the basis for many diabetic complications.[24] The actual structures of these glycated products and their mode of formation are incompletely known. There are, however, numerous reports indicating that some (e.g. carboxymethyl lysine) require oxidative and metal-catalysed reactions involving early glycation products.[25] In fact, we have earlier reported the generation of oxidants during reactions between glucose and proteins, especially in the presence of trace amounts of transition metals.[26–29]

Glycoxidation can be broken down into two components: the oxidation of free sugars, and that of protein-bound sugar derivatives. Baynes and colleagues have characterized many of the primary products of autoxidation of free sugars, notably the formation of glyoxal and arabinose during glucose autoxidation.[30] They and others have also demonstrated that oxidation of protein-bound sugar derivatives can be at least as efficient as that of free sugars, both types of reactions being metal-catalysed and oxygen-dependent. We have more recently shown that glycoxidation is not only accompanied by the formation of sugar adducts and their oxidation products, but also by the accumulation of a range of protein-bound oxidized amino acids (DOPA $>o$-,m-tyrosine $>$ leucine and valine hydroxides $>$ di-tyrosine) whose relative abundance is similar to that generated by oxygen-centred radical attack on proteins.[31] Furthermore, we also showed that if glycated proteins are prepared under non-oxidative conditions (in the absence of oxygen, and in the presence of chelators to restrict the redox activity of transition metals), they can subsequently autoxidize efficiently to generate the same products in the same relative abundance.[31] Thus glycoxidation reactions cannot simply be divorced from reactions of oxygen-centred radicals, and it is notable that among the products of the latter reactions are several reactive species, such as protein hydroperoxides (oxidizing) and protein-bound DOPA (reducing and metal chelating).

It can readily be envisaged, therefore, that protein glycoxidation might be a source of a range of reactive intermediates which could perturb the function of other molecules and of cells (as we have shown, for example, with DNA[32]). In accordance with this, it has also been shown that glycoxidized proteins can initiate lipid peroxidation, and oxidize NADPH. Most recently, it has also been proposed that

AGEs may enhance (intra)cellular oxidative stress, via receptor interactions at the cell surface.[33]

Thus glycoxidation generates two broad classes of reactive moieties on proteins: oxidized amino acids and carbonyls on sugar–protein adducts. It is not clear what importance can be attached to either class, but evidence has been presented[30] to suggest that diabetes does not influence the age dependence of the accumulation of *o*-tyrosine and methionine sulphoxide in human collagen. Thus while glycoxidation is accompanied by the formation of oxidized amino acids, it is not necessary to envisage that diabetes independently enhances other mechanisms of formation of oxygen-centred radicals. It is also important to note that even the best characterized glycoxidation product, carboxymethyl lysine, can be generated by reactions independent of glycoxidation, such as by lipid peroxidation;[34] while protein-bound aldehydes are also products of many other pathways of protein oxidation by oxygen-centred radicals and by hypochlorite (reviewed in Refs. (35, 36)). Since oxidized and glycated proteins accumulate as the net of production and proteolytic removal, it is difficult to make stronger deductions from these observations, even though it is surprising that enhanced glycation (in the diabetics) is not accompanied by enhanced levels of protein-bound oxidized amino acids.

We have shown that modestly oxidized proteins are degraded more rapidly than their normal counterparts, but that upon extensive oxidation they may be more difficult to handle; similarly, oxidized LDL is also poorly degradable.[37] One factor of importance is whether a protein is not only unfolded (which tends to enhance proteolytic susceptibility), but also whether it is aggregated with or without covalent cross-linking (which tends to reduce susceptibility).[38] Since even modest glycoxidation is accompanied by some cross-linking reactions, it is possible that any degree of glycoxidation of a protein causes decreased susceptibility, and perhaps retarded degradation by cells: this remains to be investigated. Furthermore, key sites of glycation on proteins, lysine and arginines, are also critical for the proteolytic action of several serine proteinases (like trypsin); thus their derivatization might in itself inhibit degradation. If glycoxidized LDL was poorly degradable, then it might facilitate foam cell formation in atherogenesis, by enhancing intracellular accumulation of both apoB and lipids. Similar effects might retard degradation of basement membrane proteins, and hence contribute to the thickening of basement membranes in diabetic nephropathy.

Even the simple products of sugar and sugar–protein reactions, such as dicarbonyls, may have pathologic effects. For example, the dicarbonyls methylglyoxal and 3-deoxyglucosone form via oxidation-independent reactions and are abnormally abundant in diabetes.[23] These have been found to induce synthesis of heparin-binding epidermal growth factor-like growth factor by rat aortic smooth muscle cells.[39] This growth factor is a potent mitogen for smooth muscle cells and its induction could well be an important event in diabetes-related atherogenesis. The proposition that AGEs might have some *in vivo* toxicity is supported by experiments in which preformed AGEs were injected into experimental animals.[40] This was accompanied by increased vascular permeability, diminished endothelium-dependent vasodilation and increased infiltration of monocytes into the sub-endothelial space. The possible importance of carbonyl/aldehyde functions in these events is suggested by the observation that aminoguanidine (an agent which can react with carbonyls and thus may prevent glycosylation-mediated protein cross-linking) blocked these events.[40]

AGEs form in LDL both *in vitro* and *in vivo*. The glycosylation end products in LDL are found on both the apoprotein and the amine functionalities of phospholipids such as phosphatidylethanolamine, with the phospholipid glycosylation products being many times more abundant than those associated with the protein. Requena *et al.*[41] have adduced some evidence that this product is actually *N*-(carboxymethyl)phosphatidylethanolamine because it reacts with an antibody prepared against that product. An interesting and potentially important consequence of the phospholipid glycosylation is concomitant oxidation of phospholipid unsaturated fatty acid residues.[42] The mechanisms involved in this glycation-enhanced oxidation are as yet unclear but may involve transition metal-catalysed reactions or other activated oxygen species. Although there is good evidence of increased lipoprotein glycation in diabetes, it is still unclear whether or not increased oxidation of plasma lipoproteins occurs.[43] Furthermore, whether these products of glycosylation of phospholipid, LDL or proteins are directly involved in the accelerated atherogenesis associated with diabetes is also unclear. There are, however, several mechanisms by which glycated materials might drive the atherogenic process. Some of these are discussed below.

2.6 AGE receptors and inflammation

The AGEs which are so abundant in people with diabetes (especially those with poor control of blood glucose levels) may interact with several cell surface receptors including p60, p90, the receptor for advanced glycation end products (RAGE), and the macrophage scavenger receptor.[44] RAGE is the best characterized of these. It is a member of the immunoglobulin superfamily and is expressed by cells such as monocyte/macrophages, endothelial cells, and smooth muscle cells.[45]

Interactions of AGEs with RAGE have a number of effects which, in sum, might be expected to promote the process of atherogenesis. When endothelial cell RAGE binds AGE, it impairs the barrier properties of endothelium and enhances the expression of vascular cell adhesion molecule-1 (VCAM-1),[46] which later can mediate phagocyte binding and accumulation in sites of inflammation. Such AGE effects may actually occur *in vivo*; at least one report indicates that diabetic patients have increased expression of VCAM-1 on aortic and internal mammary artery endothelium. In various models of cultured endothelial cells, vascular smooth muscle cells and the vasculature of experimental animals *in vivo*, AGE–RAGE interactions also cause enhanced oxidation of polyunsaturated fatty acids, activation of NF-κB, induction of mRNAs for heme oxygenase[33] and interleukin-6,[47] and induction of pro-coagulant tissue factor.[48] Furthermore, this system is subject to a positive feedback loop such that ligation of RAGE by AGE leads to the upregulation of additional RAGE, an event which is evidently dependent on NF-κB activation.[49] Both AGE and RAGE are increased in the vasculature of patients with diabetes and, therefore, this pro-inflammatory destabilization of the endothelium and upregulation of adhesion molecules (which will promote further inflammatory cell ingress) are tempting suspects in the genesis of accelerated atherogenesis in diabetes.

AGE–RAGE interactions may also have important effects on the activities of inflammatory cells within atherosclerotic lesions. Exposure of macrophages to AGEs leads to ingestion of the AGEs and activation of the macrophages such that they are more prone to produce pro-inflammatory cytokines and promote local inflammatory reactions. Indeed, not only are atheromatous lesions (whether from normal or diabetic patients) rich in AGE-like materials, but the macrophages within these lesions stain quite heavily with, e.g., anti-carboxymethyl lysine antibodies. It should, however, again be noted that the AGE-like material within these macrophages may derive more from products of lipid peroxidation reactions than from reactions with sugars.[34] Nonetheless, the net effect of these AGE–receptor interactions should be pro-inflammatory, in keeping with the view of atherosclerosis as an inflammatory disease discussed elsewhere in this volume.

A further mechanism through which AGEs might promote the progression of atherosclerotic disease involves accelerated calcification. Calcification of the vasculature is common in diabetes and a known risk factor for future serious cardiovascular problems. Yamagishi *et al.*[50] have recently reported that AGEs will drive the process of calcification mediated by cultured bovine pericytes. This is accompanied, as would be expected, by increases in mRNAs for alkaline phosphatase (indicative of early osteoblastic differentiation) and osteopontin (a marker of late osteoblastic differentiation). Whether these events occur through AGE–RAGE interactions is not yet known, but such an enhancement of vascular calcification by AGEs (if it occurs in diabetic humans) could be pathogenically quite important.

The possibility that RAGE–AGE interactions might be responsible for accelerated atherosclerosis in diabetes has recently received strong experimental support. Administration of a soluble extracellular domain of RAGE blocked the appearance of vascular hyperpermeability in diabetic rats. Furthermore, it has been found that mice deficient in apolipoprotein E spontaneously develop atherosclerotic lesions and the rate of lesion progression is markedly increased if the animals are also made diabetic. Interestingly, administration of soluble RAGE to such animals completely blocks this accelerated vessel disease without affecting either glycemia or dyslipidemia.[51] Insofar as the soluble RAGE given to these animals would not be expected to have any effects other than interacting with AGE, these observations suggest that cell membrane RAGE–AGE interactions are, in fact, important in the progression of atherosclerotic disease, particularly the accelerated disease common in diabetes.

2.7 AGE interactions with proteins and LDL

AGE modification of apolipoprotein B has been found to impair apoB dependent cellular uptake of LDL by the LDL receptor, perhaps because AGE modification hinders interactions between the receptor and the apoB LDL receptor binding site. Even when AGE-modified LDL is taken up by cells, the catabolism of the altered LDL may be affected. For example, there is evidence that the major AGE modification of LDL is actually glucosylated phosphatidylethanolamine. When macrophages are exposed to LDL containing this product, the uptake of LDL is enhanced and the cells accumulate increased amounts of cholesteryl ester and triacylglycerol.

However, AGE-mediated modifications of LDL may also occur indirectly. Thus, glycation of proteins and peptides can lead to the formation of intermediates – perhaps aldehydic/carbonyl in nature – which can further react with other targets such as LDL. Such reactions are probably the basis of glycation-induced cross-linking of structural proteins such as collagen and elastin.[52] Increased amounts of AGE-modified proteins and peptides occur in patients with end-stage renal disease and diabetes, and both conditions are associated with rapid development of atherosclerosis. Bucala *et al.*[42] have observed that reactions between LDL and AGE-peptides lead to LDL modifications which greatly slow the clearance of the modified LDL in transgenic mice expressing the human LDL receptor. It has also been argued that AGE–LDL–macrophage interactions may be an important factor in the accumulation of ceroid within atheromatous lesions. Along these same lines, in non-diabetic patients with renal failure, a glycosylated form of β_2 microglobulin accumulates (because of impaired renal clearance). Interactions between this glycosylated β_2 microglobulin and RAGE have been suggested as important in the etiology of dialysis-related amyloidosis[53] and may also play a role in the well-known acceleration of vascular disease associated with chronic renal failure.

2.8 Heightened oxidative reactions in diabetes?

For reasons not entirely clear, people with diabetes show evidence of enhanced *in vivo* oxidation processes. For example, there are suggestions that levels of fatty acid hydroperoxides are elevated by 2–3-fold with respect to the normal in patients with NIDDM[54] and those of the non-enzymatic peroxidation product of arachidonic acid, 8-epi-PGF2α, are more than three times the normal in NIDDM.[55] In addition, as mentioned above, the oxidation of LDL can be promoted by reactions with high levels of glucose.[29] Although some have questioned the relevance of LDL oxidation in the development of atherosclerosis,[56] and experimental evidence from animal models discussed elsewhere in this book is ambivalent, it remains true that such glycated and/or oxidized LDL may accumulate in atherosclerotic lesions and could well contribute to the atherogenic process.

It has also been suggested that increased amounts of autoantibodies might be produced in diabetic patients against oxidatively modified LDL. Immune complexes between modified LDL and antibody, when taken up by human macrophages, cause the accumulation of cholesterol esters and increased release of interleukin-1β and tumour necrosis factor,[57] events which would have obvious pertinence to the acceleration of atherogenesis. Furthermore, to the extent that oxidized LDL is more abundant in the diabetic vasculature, this will likely enhance the adhesion and diapedesis of monocytes into the intima.[58] The actual chemotactic activity may be due to the secondary formation, from oxidized LDL, of lysophosphatidylcholine[59] and/or increased expression of monocyte chemoattractant protein-1 (at least as indicated by a model system employing rabbit macrophages).[60]

The origins of this putative enhanced oxidation are by no means clear. The following are some of the possibilities for which there is some support in the literature:

(1) Oxidants generated consequent to reactions between proteins and sugar may be important as discussed above.[26–29]
(2) Furthermore, as described below, when certain cell types are exposed to increased glucose, cellular production of superoxide is increased dramatically.[61]
(3) There may be enhanced lipoxygenase-dependent lipid oxidation within atherosclerotic plaque[62] as well as increased production of other oxidants derived from several cell types associated with the plaque. At least one lipoxygenase product, 12-hydroxyeicosatetraenoic acid, is substantially increased in diabetic patients[63] and elevated glucose has been shown to increase both the activity and expression of 12-lipoxygenase in porcine aortic smooth muscle cells.
(4) Finally, there is substantial evidence of diabetes-associated mitochondrial dysfunction and it has been suggested that a defect involving complex III may cause increased mitochondrial generation of activated oxygen (as is observed when mitochondrial metabolism is blocked by certain inhibitors of electron transport).[64]

There is not, however, general agreement that there are heightened oxidation reactions in diabetes. For instance, there is the alternative proposition that many of these apparent footprints of oxidation may arise from processes independent of oxidation reactions.[23] Specifically, the non-oxidative modification of proteins by reactions with carbohydrates and lipids may lead to the formation of carbonyls, leading to a state of 'carbonyl stress'.[23, 53] In partial support, there is no evidence of increases in oxidized amino acids in skin collagen of patients with diabetes.[30] If, in fact, this view gains further support, it may suggest alternative ways of therapeutic moderation of some of the side effects of diabetes.

2.9 'Endothelial dysfunction' in diabetes

The possibility has been raised that diabetes predisposes to 'endothelial dysfunction'. This, according to a very popular theory, might then initiate atherogenesis.

The term endothelial dysfunction has been rather loosely applied, referring to disparate putative abnormalities such as increased vascular permeability, lack of normal endothelium-dependent relaxation of vascular smooth muscle and/or loss of anticoagulant functions. Although the mechanisms underlying the development of endothelial dysfunction in diabetes are as yet unclear, enhanced oxidation deriving from the effects, on endothelium, of abnormalities in various arms of lipid metabolism[65] or increased endothelial free radical generation[61] have been invoked. Indeed, Posch *et al.*[66] have made the very interesting observation that the addition of LDL isolated from diabetic patients (or normal LDL glycated *in vitro*) causes an ~5-fold increase in endothelial production of superoxide and diminished endothelial NO production. It has also been suggested that high glucose *per se* may not only induce 12-lipoxygenase activity but also the associated expression of vascular endothelial growth factor (which not only promotes angiogenesis but also causes increased vascular permeability and monocyte migration).[67]

In animal models there is general agreement that defective endothelium-dependent vascular relaxation occurs while endothelium-*in*dependent relaxation remains unaffected. A possible mechanism for this derives from experiments with cultured human aortic endothelial cells which, when exposed to high glucose for several days, show a > 3-fold increase in the production of superoxide (as well as a more modest increase in the production of NO).[61] A similar increase in superoxide production by both intact and endothelium-denuded uterine arteries from diabetic women has also been observed.[68] If elevated production of superoxide by vascular cells really does occur in patients with diabetes, the result could well be diminished availability of NO inasmuch as superoxide and NO react very quickly to form the non-vasoactive oxidant, peroxynitrite. Indeed, superoxide dismutase has been reported to restore the defective endothelium-dependent relaxation of diabetic rat aorta.[69] Acting on the theory that oxidants arising from unknown sources might be responsible for the lack of normal endothelium-dependent vasodilation in diabetes, Ting *et al.*[70] infused high levels of vitamin C into diabetic patients and found substantial improvements in endothelium-dependent, but not endothelium-independent, vasodilation. Interestingly, no vascular effects of vitamin C were observed in non-diabetic subjects. The amounts of vitamin C applied in these experiments were very large. Nonetheless, these results may still have pertinence to the situation *in vivo*; patients with diabetes are well known to have abnormally low levels of plasma vitamin C.[71]

Another possible mechanism whereby hyperglycemia might impair endothelial function was suggested by the observations of Kashiwagi *et al.*[72] These investigators carried out long-term cultures of human endothelial cells in elevated glucose and observed an ~50% reduction in stimulated pentose phosphate shunt activity and a similar decrease in glutathione-dependent hydrogen peroxide degradation, possibly both due to a limited supply of NADPH. Although very high concentrations of glucose were employed (up to 33 mmol l^{-1}), the effect was apparently metabolic and not osmotic because only D-glucose (and not L-) was active. Furthermore, it was subsequently found that the inclusion of pyruvate greatly reduced the magnitude of the glucose-mediated impairment of oxidative metabolism, possibly through effects on the pyridine nucleotide redox balance. Regardless of the mechanism(s) involved in this phenomenon, it would appear to be independent of non-enzymatic 'glycoxidation' reactions which should proceed at the same pace with either isomer, at least if they occur extracellularly.

Finally, Natarajan *et al.*[67] have reported that, in cultured vascular smooth muscle cells, high glucose will cause increased production of 12-hydroxyeicosatetraenoic acid (through induction of 12-lipoxygenase) and that this product as well as the high glucose environment leads to enhanced generation of vascular endothelial growth factor. This latter is implicated in the genesis of diabetic retinopathy and may also be important in the neovascularization which occurs in later-stage atherosclerotic plaque.[73]

2.10 'Glycochelates': possible involvement in accelerated atherogenesis in diabetes?

We were very impressed by recent reports[74] indicating that treatment of diabetic rats, two months following chemical induction of diabetes, with transition metal chelators would restore normal endothelium-dependent blood vessel relaxation, normalize blood flow to peripheral nerves and correct slowed nerve conduction velocity. As a result, we have hypothesized that some products of protein glycation may bind transition metals such as iron and copper. We have further suggested that such 'glycochelates' may be particularly abundant in the internal elastic laminae of arteries and could catalytically destroy endothelium-derived relaxing factor (nitric oxide (NO) or adducts of NO such as nitrosothiols).[75] This would explain why administration of chelators would normalize vascular relaxation and nerve blood flow. In partial support of this general idea, glycated proteins have been shown to bind both iron and copper in a redox active form.[76] Similar protein 'oxychelates', the result of protein oxidation, independent of glycoxidation may also explain the earlier observations of enhanced destruction of NO by certain oxidized or glycoxidized proteins; for example, it is well known that protein-bound DOPA can bind transition metals, and our evidence indicates that they can be redox transformed.[77] Indeed, since we have shown that protein glycoxidation is accompanied by direct oxidation of amino acid residues, including tyrosine,[31] such protein oxychelates may be a significant component of the envisaged glycochelates. If such glycochelates do form in the arterial tree of patients with diabetes, they could contribute to the process of atherogenesis in at least two different ways.

First, the ability of transition metals to catalyse the oxidation of LDL is well known. In fact, it appears that cell-mediated oxidation of LDL may require the conspiracy of these metals, and the addition of chelators which will block metal catalysis also blocks LDL oxidation caused by cells such as macrophages and endothelial cells.[78]

Second, metal-catalysed destruction of NO could lead to unrestrained growth of smooth muscle cells, hypertrophy of which is well known to occur in atherosclerotic lesions. Smooth muscle cell growth is normally inhibited by NO produced by endothelial cells.[61] Interruption of the normally constant flux of NO to smooth muscle cells might therefore result in smooth muscle cell proliferation. We should temper this argument, however, by pointing out that vascular smooth muscle cells from diabetics actually show enhanced proliferation *in vitro*.[79] Furthermore, although smooth muscle proliferation may be an important component of early lesion development, there is evidence of *impaired* smooth muscle cell growth in late lesions.[80]

2.11 Effects of hyperglycemia on signal transduction pathways

Elevated glucose can activate several second messenger systems in smooth muscle and other types of cells both in culture and *in vivo*. Important among messenger systems which appear to be activated by high levels of glucose are:

(1) diacylglycerol production;[81, 82]
(2) cytosolic phospholipase A2;[83]
(3) protein kinase C (PKC) especially the β_2 isoform,[84] which could be activated by either (1) or the products of (2);[85]
(4) p21(ras) and NF-κB; and
(5) p38 mitogen-activated kinase (p38 MAP kinase).[84]

The proximate cause(s) of these events is not yet known, but there is some evidence that it might involve oxidative reactions dependent on both elevated glucose and cellular metabolism which can be antagonized by α-tocopherol. It is possible that activation of these intracellular messenger systems may have powerful effects on cellular proliferation and survival. For example, it has been found that elevated glucose will cause human vascular smooth muscle cells to produce vascular permeability factor (which has been implicated in endothelial dysfunction in diabetes), probably through the intermediacy of PKC activation; production of vascular permeability factor by high glucose is prevented by inhibitors of PKC.[86] Furthermore, in diabetic rats an oral PKC inhibitor has been reported to lessen vascular dysfunction in both the kidney and retinal circulation.[87]

By the same token, elevated glucose will cause cultured smooth muscle cells to activate p38 MAP kinase and this is also observed in pancreatic β-cells and in the aortae of diabetic rats.[84] This may be an important event inasmuch as the activation of p38 MAP kinase has been related to the regulation of numerous genes, cell growth and apoptosis. This activation of p38 kinase appears also to involve PKC, probably the PKC β_2 and δ isoforms.[84] Similarly, high glucose has been reported to activate nuclear transcription factor κB in vascular smooth muscle. This also appears to be dependent on prior stimulation of PKC activity because it is blocked by calphostin C, an inhibitor of PKC.

2.12 Parallels between accelerated atherogenesis in diabetes, uremia, and cigarette smoking

On a more speculative note, there is a strange similarity between three disparate conditions, all of which present enhanced risk for the development of atherosclerosis – diabetes, chronic renal failure, and cigarette smoking. In all three instances, proteins with AGE-like modifications accumulate. It is well known that AGEs are greatly increased in the plasma and tissues of patients with end-stage renal disease.[53, 88] Less well known is the observation that tobacco smoke contains as yet unidentified compounds which react with proteins, forming products which cross-react with antibodies to AGE.[89] The plasma, ocular lenses, and coronary arteries of non-diabetic smokers contain elevated levels of anti-AGE antibody-reactive materials compared to tissues derived from non-smokers, suggesting that these AGE-like compounds could have systemic effects.[89] However, there is not yet any evidence that these AGE-like materials contribute to the accelerated atherogenesis associated with tobacco smoking or, even, that they interact with AGE receptors. Nonetheless, it is tempting to speculate that the accelerated atherogenesis in all three conditions may have at least one pathogenic mechanism in common.

2.13 Diabetes-associated atherogenesis: is it a macrovascular disease?

To the confusion over precisely how diabetes might accelerate atherogenesis, we would like to add another possibility. That is, diabetes is well known to predispose to microvascular disease and diminished microvascular function might well have secondary effects on major arteries. The microvasculature which feeds large arteries, the *vasa vasorum*, is critical to the normal functioning of the larger vessels. In an animal model, when the adventitia of the carotid artery was removed, it led to the transient formation of intimal hyperplastic lesions.[90] Interestingly, when the rabbits were also fed a hypercholesterolemic diet, the lesions persisted and, within them, macrophages were the predominant cell. This leads to the speculation that diabetes-associated abnormalities in smaller blood vessels might be important in the ultimate appearance of large-vessel disease.

The reverse is also possible – that in diabetic vasculature the proliferation of new blood vessels is enhanced. As mentioned above, there is evidence for exaggerated production of vascular endothelial growth factor (which promotes angiogenesis and also increases vascular permeability and macrophage migration).[67] More than 60 years ago, Winternitz[91] reported the presence of a 'neo *vasa vasorum*' associated specifically with atherogenic lesions in the coronary arteries. These observations were later affirmed by Barger.[92] To the extent that plaque-associated angiogenesis might be accelerated in diabetes, the proliferation of such a network of fragile new blood vessels suggests an origin for the hemorrhagic lesions which frequently co-occur with atherosclerotic plaque and which, being rich in iron, might drive oxidative events associated with atherogenesis.[93]

2.14   The chicken, the cart, the egg, and the horse

Finally, a caveat we might have neglected. We cannot rule out the possibility that the tendency to develop type 2 diabetes coincides with, but is not causal of, a tendency to develop accelerated vascular disease. That is, type 2 diabetes has a very strong genetic component (although most often not behaving as a simple monogenic trait).[94] It is possible that one or more inherited defects which predispose to the syndrome of insulin resistance and concomitant abnormalities of lipid metabolism, obesity and hypertension separately carry an associated but distinct risk for development of vascular disease. In other words, the genesis of the vascular disease might be partially independent of the diabetes *per se*. Indeed, there are multiple reports indicating that effective glycemic control does not necessarily translate into prevention or amelioration of diabetic vascular disease. Furthermore, as mentioned above, much of the development of atherosclerosis in patients with type 2 diabetes would appear to occur before the diagnosis of the diabetes, during the period characterized by simple hyperinsulinemia without concurrent hyperglycemia. In that light, many of the mechanisms discussed above may have little pertinence to diabetic atherogenesis.

2.15   Summary

Few solid conclusions can be drawn from the above review. It is probably true that, although people with diabetes also have a number of associated known risk factors for atherosclerosis, these fall far short of explaining the high frequency of vascular disease in type 2 diabetes. The additional possibilities, including effects of glycation products and elevated glucose itself on the atherogenic process, have some experimental support but there is no platinum proof. The question of causal relationships between the diabetic condition and atherogenesis is both muddled and exciting, representing as it does a fertile area for research which, we are sure, will eventually uncover interesting and important pathogenic mechanisms. At the moment, we can present only three firm conclusions. (i) Atherosclerosis is a complicated disease, the pathogenesis of which is still poorly understood. (ii) Type 2 diabetes is a complicated disease, the pathogenesis of which is still poorly understood. (iii) Together, these diseases comprise an absolute reviewers' nightmare.

References

1. Keen, H., Clark, C., and Laakso, M. (1999). Reducing the burden of diabetes: managing cardiovascular disease. *Diabetes Metab. Res. Rev.*, **15**, 186–96.
2. Meigs, J.B., Singer, D.E., Sullivan, L.M., Dukes, K.A., D'Agostino, R.B., Nathan, D.M., Wagner, E.H., Kaplan, S.H., and Greenfield, S. (1997). Metabolic control and prevalent cardiovascular disease in non-insulin-dependent diabetes mellitus (NIDDM): the NIDDM Patient Outcome Research Team. *Am. J. Med.*, **102**, 38–47.
3. Niskanen, L., Turpeinen, A., Penttila, I., and Uusitupa, M.I. (1998). Hyperglycemia and compositional lipoprotein abnormalities as predictors of cardiovascular mortality in type 2 diabetes: a 15-year follow-up from the time of diagnosis. *Diabetes Care,* **21**, 1861–69.
4. Sowers, J.R. and Lester, M.A. (1999). Diabetes and cardiovascular disease. *Diabetes Care*, **22**, (Suppl. 3), C14–20.
5. Chisholm, D.J., Campbell, L.V., and Kraegen, E.W. (1997) Pathogenesis of the insulin resistance syndrome (syndrome X). *Clin. Exp. Pharmacol. Physiol.*, **24**, 782–4.
6. Neel, J.V., Julius, S., Weder, A., Yamada, M., Kardia, S.L., and Haviland, M.B. (1998). Syndrome X: is it for real? *Genet. Epidemiol.*, **15**, 19–32.
7. Jensen, M.D., Haymond, M.W., Rizza, R.A., Cryer, P.E., and Miles, J.M. (1989). Influence of body fat distribution on free fatty acid metabolism in obesity. *J. Clin. Invest.*, **83**, 1168–73.
8. Assmann, G., Schulte, H., and von Eckardstein, A. (1996). Hypertriglyceridemia and elevated lipoprotein(a) are risk factors for major coronary events in middle-aged men. *Am. J. Cardiol.*, **77**, 1179–84.
9. Nordestgaard, B.G., Wootton, R., and Lewis, B. (1995). Selective retention of VLDL, IDL, and LDL in the arterial intima of genetically hyperlipidemic rabbits *in vivo*. Molecular size as a determinant of fractional loss from the intima-inner media. *Arterioscler. Thromb. Vasc. Biol.*, **15**, 534–42.
10. Hennig, B., Chung, B.H., Watkins, B.A., and Alvarado, A. (1992). Disruption of endothelial barrier function by lipolytic remnants of triglyceride-rich lipoproteins. *Atherosclerosis,* **95**, 235–47.
11. Laughton, C.W., Ruddle, D.L., Bedord, C.J., and Alderman, E.L. (1998). Sera containing elevated nonesterified fatty acids from patients with angiographically documented coronary atherosclerosis cause marked lipid accumulation in cultured human arterial smooth muscle-derived cells. *Atherosclerosis*, **70**, 233–46.
12. Young, V.M., Toborek, M., Yang, F., McClain, C.J., and Hennig, B. (1998). Effect of linoleic acid on endothelial cell inflammatory mediators. *Metabolism*, **47**, 566–72.
13. Olsson, U., Bondjers, G., and Camejo, G. (1999). Fatty acids modulate the composition of extracellular matrix in cultured human arterial smooth muscle cells by altering the expression of genes for proteoglycan core proteins. *Diabetes*, **48**, 616–22.
14. Roden, M., Price, T.B., Perseghin, G., Petersen, K.F., Rothman, D.L., Cline, G.W., and Shulman, G.I. (1996). Mechanism of free fatty acid-induced insulin resistance in humans. *J. Clin. Invest.*, **97**, 2859–65.

15. Mastrangelo, A.M., Jeitner, T.M., and Eaton, J.W. (1998). Oleic acid increases cell surface expression and activity of CD11b on human neutrophils. *J. Immunol.*, **161**, 4268–75.

16. Semenkovich, C.F. and Heinecke, J.W. (1997). The mystery of diabetes and atherosclerosis: time for a new plot. *Diabetes*, **46**, 327–34.

17. Despres, J.P., Lamarche, B., Mauriege, P., Cantin, B., Dagenais, G.R., Moorjani, S., and Lupien, P.J. (1996). Hyperinsulinemia as an independent risk factor for ischemic heart disease. *N. Engl. J. Med.*, **334**, 952–7.

18. Abe, H., Bandai, A., Makuuchi, M., Idezuki, Y., Nozawa, M., Oka, T., Osuga, J., Watanabe, Y., Inaba, T., and Yamada, N. (1996). Hyperinsulinaemia accelerates accumulation of cholesterol ester in aorta of rats with transplanted pancreas. *Diabetologia*, **39**, 1276–83.

19. Westerbacka, J., Vehkavaara, S., Bergholm, R., Wilkinson, I., Cockcroft, J., and Yki-Jarvinen, H. (1999). Marked resistance of the ability of insulin to decrease arterial stiffness characterizes human obesity. *Diabetes*, **48**, 821–7.

20. Steinberg, H.O., Chaker, H., Leaming, R., Johnson, A., Brechtel, G., and Baron, A.D. (1996). Obesity/insulin resistance is associated with endothelial dysfunction. Implications for the syndrome of insulin resistance. *J. Clin. Invest.*, **97**, 2601–10.

21. Abe, H., Yamada, N., Kamata, K., Kuwaki, T., Shimada, M., Osuga, J., Shionoiri, F., Yahagi, N., Kadowaki, T., Tamemoto, H. *et al.* (1998). Hypertension, hypertriglyceridemia, and impaired endothelium-dependent vascular relaxation in mice lacking insulin receptor substrate-1. *J. Clin. Invest.*, **101**, 1784–88.

22. Pinkney, J.H., Stehouwer, C.D., Coppack, S.W., and Yudkin, J.S. (1997). Endothelial dysfunction: cause of the insulin resistance syndrome. *Diabetes*, **46**, (Suppl. 2), S9–13.

23. Baynes, J.W., and Thorpe, S.R. (1999). Role of oxidative stress in diabetic complications. *Diabetes*, **48**, 1–9.

24. Brownlee, M., Cerami, A., and Vlassara, H. (1998). Advanced glycosylation endproducts in tissue and the biochemical basis of diabetic complications. *N. Engl. J. Med.*, **318**, 1315–21.

25. Ahmed, M.U., Thorpe, S.R., and Baynes, J.W. (1986). Identification of $N^{\epsilon-}$ (carboxymethyl) lysine as a degradation product of fructose lysine in glycated protein. *J. Biol. Chem.*, **261**, 4889–94.

26. Wolf, S.P. and Dean, R.T. (1987) Glucose autoxidation and protein modification. The potential role of 'autoxidative glycosylation' in diabetes. *Biochem. J.*, **245**, 243–50.

27. Wolff, S.P., Jiang, Z.Y., and Hunt, J.V. (1991). Protein glycation and oxidative stress in diabetes mellitus and ageing. *Free Radic. Biol. Med.*, **10**, 339–52.

28. Hunt, J.V., Dean, R.T., and Wolff, S.P. (1988). Hydroxyl radical production and autoxidative glycosylation. Glucose autoxidation as the cause of protein damage in the experimental glycation model of diabetes mellitus and ageing. *Biochem. J.*, **256**, 205–12.

29. Hunt, J.V., Smith, C.C., and Wolff, S.P. (1990). Autoxidative glycosylation and possible involvement of peroxides and free radicals in LDL modification by glucose. *Diabetes,* **39**, 1420–4.

30. Wells-Knecht, M.C., Lyons, T.J., McCance, D.R., Thorpe, S.R., and Baynes, J.W. (1997). Age-dependent increase in *ortho*-tyrosine and methionine sulfoxide in

human skin collagen is not accelerated in diabetes. Evidence against a generalized increase in oxidative stress in diabetes. *J. Clin. Invest*, **100**, 839–46.

31. Fu, S., Fu, M., Baynes, J.W., Thorpe, S.R., and Dean, R.T. (1998). Presence of dopa and amino acid hydroperoxides in proteins modified with advanced glycation end-products (AGEs): amino acid oxidation products as possible source of oxidative stress induced by AGE-proteins. *Biochem. J.*, **330**, 233–39.

32. Morin, B., Davies, M.J., and Dean, R.T. (1998). The protein oxidation product 3,4-dihydroxyphenylalanine (DOPA) mediates oxidative DNA damage. *Biochem. J.*, **330**, 1059–67.

33. Yan, S.D., Schmidt, A.M., Anderson, G.M., Zhang, J., Brett, J., Zou, Y.S., Pinsky, D., and Stern D. (1994). Enhanced cellular oxidant stress by the interaction of advanced glycation end products with their receptors/binding proteins. *J. Biol. Chem.*, **269**, 9889–97.

34. Fu, M.X., Requena, J.R., Jenkins, A.J., Lyons, T.J., Baynes, J.W., and Thorpe, S.R. (1996). The advanced glycation end product, $N^{\epsilon-}$ (carboxymethyl) lysine, is a product of both lipid peroxidation and glycoxidation reactions. *J. Biol. Chem.*, **271**, 9982–6.

35. Davies, M. and Dean, R.T. (1997). *Radical mediated protein oxidation: from chemistry to medicine*. Oxford University Press, London.

36. Dean, R.T., Fu, S., Stocker, R., and Davies, M.J. (1997). The biochemistry and pathology of radical-mediated protein oxidation. *Biochem. J.*, **324**, 1–18.

37. Mander, E.L., Dean, R.T., Stanley, K.K., and Jessup, W. (1994). Apolipoprotein B of oxidized LDL accumulates in the lysosomes of macrophages. *Biochim. Biophys. Acta*, **1212**, 80–92.

38. Dean, R.T., Armstrong, S., Fu, S., and Jessup, W. (1994). Oxidised proteins and their enzymatic proteolysis in eucaryotic cells: a critical appraisal. In *Free radicals in the environment, medicine and toxicology* (ed. H. Nohl, H. Esterbauer, C. Rice-Evans), pp. 47–79. Richelieu Press, London.

39. Che, W., Asahi, M., Takahashi, M., Kaneto, H., Okado, A., Higashiyama, S., and Taniguchi, N. (1997). Selective induction of heparin-binding epidermal growth factor-like growth factor by methylglyoxal and 3-deoxyglucosone in rat aortic smooth muscle cells. The involvement of reactive oxygen species formation and a possible implication for atherogenesis in diabetes. *J. Biol. Chem.*, **272**, 18453–9.

40. Vlassara, H., Fuh, H., Makita, Z., Krungkrai, S., Cerami, A., and Bucala, R. (1992). Exogenous advanced glycosylation end products induce complex vascular dysfunction in normal animals: a model for diabetic and aging complications. *Proc. Natl. Acad. Sci. USA*, **89**, 12043–7.

41. Requena, J.R., Ahmed, M.U., Fountain, C.W., Degenhardt, T.P., Reddy, S., Perez, C., Lyons, T.J., Jenkins, A.J, Baynes, J.W., and Thorpe, S.R. (1997). Carboxymethylethanolamine, a biomarker of phospholipid modification during the Maillard reaction *in vivo*. *J. Biol. Chem.*, **272**, 17473–9.

42. Bucala, R., Makita, Z., Vega, G., Grundy S., Koschinsky, T., Cerami, A., and Vlassara, H. (1994). Modification of low density lipoprotein by advanced glycation end products contributes to the dyslipidemia of diabetes and renal insufficiency. *Proc. Natl. Acad. Sci. USA*, **91**, 9441–5.

43. Lyons, T.J. and Jenkins, A.J. (1997). Lipoprotein glycation and its metabolic consequences. *Curr. Opin. Lipidol.*, **8**, 174–80.

44. Suzuki, H., Kurihara, Y., Takeya, M., Kamada, N., Kataoka, M., Jishage, K., Ueda, O., Sakaguchi, H., Higashi, T., Suzuki, T. *et al.* (1997). A role for macrophage scavenger receptors in atherosclerosis and susceptibility to infection. *Nature*, **386**, 292–6.

45. Schmidt, A.M., Hori, O., Cao, R., Yan, S.D., Brett, J., Wautier, J.L., Ogawa, S., Kuwabara, K., Matsumoto, M., and Stern, D. (1996). RAGE: a novel cellular receptor for advanced glycation end products. *Diabetes*, **45** (Suppl. 3), S77–80.

46. Schmidt, A.M., Hori, O., Chen, J.X., Li, J.F., Crandall, J., Zhang, J, Cao, R., Yan, S.D., Brett, J., and Stern D. (1995). Advanced glycation endproducts interacting with their endothelial receptor induce expression of vascular cell adhesion molecule-1 (VCAM-1) in cultured human endothelial cells and in mice. A potential mechanism for the accelerated vasculopathy of diabetes. *J. Clin. Invest.*, **96**, 1395–403.

47. Schmidt, A.M, Hasu, M., Popov, D., Zhang, J.H., Chen J., Yan, S.D, Brett, J., Cao, R., Kuwabara, K., Costache, G. *et al.* (1994). Receptor for advanced glycation end products (AGEs) has a central role in vessel wall interactions and gene activation in response to circulating AGE proteins. *Proc. Natl. Acad. Sci. USA*, **91**, 8807–11.

48. Ichikawa, K., Yoshinari, M., Iwase, M., Wakisaka, M., Doi, Y., Iino, K., Yamamoto, M., and Fujishima, M. (1998). Advanced glycosylation end products induced tissue factor expression in human monocyte-like U937 cells and increased tissue factor expression in monocytes from diabetic patients. *Atherosclerosis*, **136**, 281–7.

49. Li, J. and Schmidt, A.M. (1997). Characterization and functional analysis of the promoter of RAGE, the receptor for advanced glycation end products. *J. Biol. Chem.*, **272**, 16498–506.

50. Yamagishi, S., Fujimori, H., Yonekura, H., Tanaka, N., and Yamamoto, H. (1999). Advanced glycation endproducts accelerate calcification in microvascular pericytes. *Biochem. Biophys. Res. Commun.*, **258**, 353–7.

51. Schmidt, A.M., Yan, S.D., Wautier, J.L., and Stern, D. (1999). Activation of receptor for advanced glycation end products: a mechanism for chronic vascular dysfunction in diabetic vasculopathy and atherosclerosis. *Circ. Res.*, **84**, 489–97.

52. Sell, D.R. and Monnier, V.M. (1989). Structure elucidation of a senescence cross-link from human extracellular matrix. Implication of pentoses in the aging process. *J. Biol. Chem.*, **264**, 21597–602.

53. Miyata, T., van Ypersele de Strihou, C., Kurokawa, K., and Baynes, J.W. (1999). Alterations in nonenzymatic biochemistry in uremia: origin and significance of 'carbonyl stress' in long-term uremic complications. *Kidney Int.*, **55**, 389–99.

54. Nourooz-Zadeh, J., Tajaddini-Sarmadi, J., McCarthy, S., Betteridge, D.J., and Wolff, S.P. (1995). Elevated levels of authentic plasma hydroperoxides in NIDDM. *Diabetes*, **44**, 1054–58.

55. Gopaul, N.K., Anggard, E.E., Mallet, A.I., Betteridge, D.J., Wolff, S.P., and Nourooz-Zadeh, J. (1995). Plasma 8-epi-PGF2 alpha levels are elevated in individuals with non-insulin dependent diabetes mellitus. *FEBS Lett.* **368**, 225–9.

56. Eaton, J.W. (1996). Low-density lipoprotein oxidation and atherogenesis: we got the bandwagon, we got the band, but where's the music? *Redox Report,* **2**, 81–2.

57. Lopes-Virella, M.F. and Virella, G. (1996). Cytokines, modified lipoproteins, and arteriosclerosis in diabetes. *Diabetes*, **45** (Suppl 3), S40–4.

58. Quinn, M.T., Parthasarathy, S., Fong, L.G., and Steinberg, D. (1987). Oxidatively modified low density lipoproteins: a potential role in recruitment and retention of monocyte/macrophages during atherogenesis. *Proc. Natl. Acad. Sci. USA*, **84**, 2995–8.

59. Quinn, M.T., Parthasarathy, S., and Steinberg, D. (1988). Lysophosphatidylcholine: a chemotactic factor for human monocytes and its potential role in atherogenesis. *Proc. Natl. Acad. Sci. USA*, **85**, 2805–9.

60. Wang, G.P., Deng, Z.D., Ni, J., and Qu, Z.L. (1997). Oxidized low density lipoprotein and very low density lipoprotein enhance expression of monocyte chemoattractant protein-1 in rabbit peritoneal exudate macrophages. *Atherosclerosis*, **133**, 31–6.

61. Cosentino, F., Hishikawa, K., Katusic, Z.S., and Luscher, T.F. (1997). High glucose increases nitric oxide synthase expression and superoxide anion generation in human aortic endothelial cells. *Circulation*, **96**, 25–8.

62. Folcik, V.A., Nivar-Aristy, R.A., Krajewski, L.P., and Cathcart, M.K. (1995). Lipoxygenase contributes to the oxidation of lipids in human atherosclerotic plaques. *J. Clin. Invest.*, **96**, 504–10.

63. Antonipillai, I., Nadler, J., Vu, E.J., Bughi, S., Natarajan, R., and Horton, R. (1996). A 12-lipoxygenase product, 12-hydroxyeicosatetraenoic acid, is increased in diabetics with incipient and early renal disease. *J. Clin. Endocrinol. Metab.*, **81**, 1940–5.

64. Kristal, B.S., Koopmans, S.J., Jackson, C.T., Ikeno, Y., Park, B.J., and Yu, B.P. (1997). Oxidant-mediated repression of mitochondrial transcription in diabetic rats. *Free Radic. Biol. Med.*, **22**, 813–22.

65. Watts, G.F. and Playford, D.A. (1998). Dyslipoproteinaemia and hyperoxidative stress in the pathogenesis of endothelial dysfunction in non-insulin dependent diabetes mellitus: hypothesis. *Atherosclerosis*, **141**, 17–30.

66. Posch, K., Simecek, S., Wascher, T.C., Jurgens, G., Baumgartner-Parzer, S., Kostner, G.M., and Graier, W.F. (1999). Glycated low-density lipoprotein attenuates shear stress-induced nitric oxide synthesis by inhibition of shear stress-activated L-arginine uptake in endothelial cells. *Diabetes*, **48**, 1331–7.

67. Natarajan, R., Bai, W., Lanting, L., Gonzales, N., and Nadler, J. (1997). Effects of high glucose on vascular endothelial growth factor expression in vascular smooth muscle cells. *Am. J. Physiol.*, **273**, H2224–31.

68. Fleischhacker, E., Esenabhalu, V.E., Spitaler, M., Holzmann, S., Skrabal, F., Koidl, B., Kostner G.M., and Graier W.F. Human diabetes is associated with hyperreactivity of vascular smooth muscle cells due to altered subcellular Ca^{2+} distribution. *Diabetes*, **48**, 1323–30.

69. Hattori, Y., Kawasaki, H., Abe, K., and Kanno, M. Superoxide dismutase recovers altered endothelium-dependent relaxation in diabetic rat aorta. *Am. J. Physiol.*, **261**, H1086–94.

70. Ting, H.H., Timimi, F.K., Boles, K.S., Creager, S.J., Ganz, P., and Creager, M.A. (1998). Vitamin C improves endothelium-dependent vasodilation in patients with non-insulin-dependent diabetes mellitus. *J. Clin. Invest.*, **97**, 22–8.

71. Seghieri, G., Martinoli, L., di Felice, M., Anichini, R., Fazzini, A., Ciuti, M., Miceli, M., Gaspa, L., and Franconi, F. (1998). Plasma and platelet ascorbate pools and lipid peroxidation in insulin-dependent diabetes mellitus. *Eur. J. Clin. Invest.*, **28**, 659–63.

72. Kashiwagi, A., Asahina, T., Ikebuchi, M., Tanaka, Y., Takagi, Y., Nishio, Y., Kikkawa, R., and Shigeta, Y. (1994). Abnormal glutathione metabolism and increased cytotoxicity caused by H_2O_2 in human umbilical vein endothelial cells cultured in high glucose medium. *Diabetologia*, **37**, 264–9.

73. Kuzuya, M., Satake, S., Esaki, T., Yamada, K., Hayashi, T., Naito, M., Asai, K., and Iguchi, A. (1995). Induction of angiogenesis by smooth muscle cell-derived factor: possible role in neovascularization in atherosclerotic plaque. *J. Cell Physiol.*, **164**, 658–67.

74. Cameron, N.E. and Cotter, M.A. (1995). Neurovascular dysfunction in diabetic rats. Potential contribution of autoxidation and free radicals examined using transition metal chelating agents. *J. Clin. Invest.*, **96**, 1159–63.

75. Qian, M., and Eaton, J.W. (2000). Glycochelates: and the etiology of diabetic peripheral neuropathy. *Free Radic. Biol. Med.* **28**, 652–6.

76. Qian, M., Liu, M., and Eaton, J.W. (1998). Transition metals bind to glycated proteins forming redox active 'glycochelates': implications for the pathogenesis of certain diabetic complications. *Biochem. Biophys. Res. Commun.*, **250**, 385–9.

77. Gieseg, S.P., Simpson, J.A., Charlton, T.S., Duncan, M.W., and Dean, R.T. (1993). Protein bound 3,4-dihydroxyphenylalanine is a major reductant formed during hydroxyl radical damage to proteins. *Biochemistry*, **32**, 4780–6.

78. Kritharides, L., Jessup, W., and Dean, R.T. (1995). EDTA differentially and incompletely inhibits components of prolonged cell-mediated oxidation of low-density lipoproteins. *Free Radic. Res.*, **22**, 399–417.

79. Oikawa, S., Hayasaka, K., Hashizume, E., Kotake, H., Midorikawa, H., Sekikawa, A., Kikuchi, A., and Toyota, T. (1996). Human arterial smooth muscle cell proliferation in diabetes. *Diabetes*, **45** (Suppl 3), S114–16.

80. Libby, P. (1995). Molecular bases of the acute coronary syndromes. *Circulation*, **91**, 2844–50.

81. Kunisaki, M., Bursell, S.E., Umeda, F., Nawata, H., and King, G.L. (1994). Normalization of diacylglycerol-protein kinase C activation by vitamin E in aorta of diabetic rats and cultured rat smooth muscle cells exposed to elevated glucose levels. *Diabetes*, **43**, 1372–7.

82. Inoguchi, T., Xia, P., Kunisaki, M., Higashi, S., Feener, E.P., and King, G.L. (1994). Insulin's effect on protein kinase C and diacylglycerol induced by diabetes and glucose in vascular tissues. *Am. J. Physiol.*, **267**, E369–79.

83. Xia, P., Kramer, R.M., and King, G.L. (1995). Identification of the mechanism for the inhibition of Na+,K + -adenosine triphosphatase by hyperglycemia involving activation of protein kinase C and cytosolic phospholipase A2. *J. Clin. Invest.*, **96**, 733–40.

84. Igarashi, M., Wakasaki, H., Takahara, N., Ishii, H., Jiang, Z.Y., Yamauchi, T., Kuboki, K., Meier, M., Rhodes, C.J., and King, G.L. (1999). Glucose or diabetes activates p38 mitogen-activated protein kinase via different pathways. *J. Clin. Invest.*, **103**, 185–95.

85. Ohara, Y., Peterson, T.E., Zheng, B., Kuo, J.F., and Harrison, D.G. (1994). Lysophosphatidylcholine increases vascular superoxide anion production via protein kinase C activation. *Arterioscler. Thromb.*, **14**, 1007–13.

86. Williams, B., Gallacher, B., Patel, H., and Orme, C. (1997). Glucose-induced protein kinase C activation regulates vascular permeability factor mRNA expression and peptide production by human vascular smooth muscle cells *in vitro*. *Diabetes*, **46**, 1497–1503.

87. Ishii, H., Jirousek, M.R., Koya, D., Takagi, C., Xia, P., Clermont, A., Bursell, S.E., Kern, T.S., Ballas, L.M., Heath, W.F. *et al.* (1996). Amelioration of vascular dysfunctions in diabetic rats by an oral PKC beta inhibitor. *Science*, **272**, 728–31.

88. Sell, D.R. and Monnier, V.M. (1990). End-stage renal disease and diabetes catalyze the formation of a pentose-derived crosslink from aging human collagen. *J. Clin. Invest.*, **85**, 380–4.

89. Cerami, C., Founds, H., Nicholl, I., Mitsuhashi, T., Giordano, D., Vanpatten, S., Lee, A., Al-Abed, Y., Vlassara, H., Bucala, R. *et al.* (1997). Tobacco smoke is a source of toxic reactive glycation products. *Proc. Natl. Acad. Sci. USA*, **94**, 13915–20.

90. Barker, S.G., Tilling, L.C., Miller, G.C., Beesley, J.E., Fleetwood, G., Stavri, G.T., Baskerville, P.A., and Martin, J.F. (1994). The adventitia and atherogenesis: removal initiates intimal proliferation in the rabbit which regresses on generation of a 'neoadventitia'. *Atherosclerosis*, **105**, 131–44.

91. Winternitz, M.C., Thomas, R.M., and LeCompte, P.M. (1938). *The biology of arteriosclerosis*. Charles C. Thomas, Springfield, IL.

92. Barger, A.C., Beeuwkes, R.D., Lainey, L.L., and Silverman, K.J. (1984). Hypothesis: vasa vasorum and neovascularization of human coronary arteries. A possible role in the pathophysiology of atherosclerosis. *N. Engl. J. Med.*, **310**, 175–7.

93. Balla, J., Jacob, H.S., Balla, G., Nath, K., Eaton, J.W., and Vercellotti, G.M. (1993). Endothelial-cell heme uptake from heme proteins: induction of sensitization and desensitization to oxidant damage. *Proc. Natl. Acad. Sci. USA*, **90**, 9285–9.

94. Medici, F., Hawa, M., Ianari, A., Pyke, D.A., and Leslie, R.D. (1999). Concordance rate for type II diabetes mellitus in monozygotic twins: actuarial analysis. *Diabetologia*, **42**, 146–50.

3 Antioxidant intervention studies in animals

Knut S. Pettersson
Cardiovascular Pharmacology, AstraZeneca R&D, SE-43183 Mölndal, Sweden

Paul K. Witting
Biochemistry Group, The Heart Research Institute, Camperdown, Sydney, NSW 2050, Australia

Roland Stocker
Biochemistry Group, The Heart Research Institute, Camperdown, Sydney, NSW 2050, Australia

3.1 Introduction

The pathology of atherosclerosis is multi-factorial, with several risk factors affecting the likelihood and severity of the disease. Recently, much attention has been placed on oxidation of low-density lipoprotein (LDL) as an early event in the development of the disease. Oxidative modification of LDL is known to change the LDL particle into a more atherogenic type than native LDL.[1] Shortly after it was reported that the lipid-lowering agent probucol could protect LDL from becoming oxidized *in vitro*,[2] it was also shown that this agent can protect LDL receptor-deficient rabbits against atherosclerotic lesion development[3, 4] independent of its lipid-lowering effect.[3] This finding was the most prominent experimental support for the 'oxidation theory' of atherogenesis.[1]

Many lines of evidence support the oxidation theory (for reviews see Refs. (1, 5–7)). Briefly, oxidized LDL (OxLDL) can initiate the formation of foam cells, is cytotoxic, and assists in many additional potential pro-atherogenic events. The latter include the attraction to and retention of inflammatory cells within the vessel wall, the attenuation of endothelium-dependent vasodilation, and the proliferation of vascular smooth muscle cells. The most important observations are, however, still that a number of structurally different antioxidants, including probucol[3, 4] butylated hydroxytoluene[8] and N,N'-diphenyl-phenylenediamine,[9, 10] inhibit atherosclerosis in animal models of hypercholesterolemia.

However, not all results support the oxidation hypothesis, see e.g. Fig. 2b in Witting *et al.*[11] The appeal of the oxidation theory may be questioned on the basis that a large proportion of the available evidence is indirect and does not distinguish LDL oxidation as a consequence rather than a cause of atherosclerosis. The theory also continues to suffer from the lack of suitable surrogate markers to measure LDL oxidation in the vessel wall. In addition, and despite its common use, OxLDL prepared *in vitro* lacks chemical characterization and therefore most likely varies from one laboratory to the other, and its relevance to OxLDL formed *in vivo* remains questionable. The mechanism(s) by which lipoprotein-derived lipids in the vessel wall become oxidized, and the effects of antioxidants on atherosclerosis-related cardiovascular disease in man are reviewed in other chapters of this book. Here we summarize the effects of antioxidants on and the relation of lipid (per)oxidation to atherosclerosis in animal models. As the number of *in vivo* studies in animals is very large, we only cover a selected number of the studies published. Some comparisons to trials in humans are made, especially in the case of vitamin E, where it is possible to directly validate the experimental approaches used.

3.2   Animal models of atherosclerosis

Ultrasound techniques and magnetic resonance imaging allow the non-invasive investigation of the gross morphology of superficial arteries such as carotid and femoral arteries.[12, 13] Studies of the vessel wall in coronary arteries can be performed only by the use of intra-vascular ultrasound.[14] Although technological development is rapid, these techniques cannot as yet be employed to discriminate between the media and intima layers of arteries, nor can they give detailed information on the composition of the vessel wall. Furthermore, as the rate of lesion progression is slow and the number of antioxidants available for administration to humans is limited, the complexity and feasibility of studying the effect of antioxidants on atherogenesis in human subjects is limited.

There are many different types of animal models proposed to study atherosclerosis. Each may vary significantly from each other, yet there is no single model that completely replicates the various stages of atherosclerosis in humans. However, cholesterol feeding and mechanical injury to the endothelium are two common features shared by the majority of the models currently in use. Animal models may also differ with respect to the degree of dietary cholesterol and/or antioxidant supplementation as well as the duration of intervention and duration and degree of mechanical endothelial injury. Notwithstanding these limitations, the use of animal models has allowed the study of biochemical, morphological and morphometric changes in developing lesions, affording detailed information on the extent, structure and composition of atherosclerosis in these models.

Rabbits and mice are the two species most commonly used at present for experimental atherosclerosis research. In rabbits, atherosclerosis is normally induced by feeding the animals a cholesterol-enriched diet, commonly leading to a total plasma cholesterol concentration of 40–50 mM, or even higher. Such diets lead to a rapid development of foam cell-rich lesions. An alternative is the Watanabe rabbit (WHHL), which has a mutation in its LDL receptor, leading to spontaneous hyperlipidemia and atherosclerosis.[15] This model has the advantage that, similar to humans, it has most of its cholesterol in LDL particles rather than in β-migrating, VLDL-like particles, as is the case for cholesterol-fed rabbits. Also, WHHL rabbits, due to the lack of functional LDL receptors, do not develop severe liver disease that is often associated with diet-induced hyperlipidemia and represents a potential confounding factor in the assessment of drug efficacy. WHHL rabbits can therefore reach an advanced age and will develop severe atherosclerotic disease throughout the arterial tree. This renders WHHL rabbits suitable for studies of treatment effects of drugs on existing and advanced stage lesions.

Importantly, the rupture of atherosclerotic plaques rather than the development of the disease *per se* is the major cause of coronary thrombosis and myocardial infarct in humans. Unfortunately, at present there is no well-accepted animal model of plaque rupture that truly mimics this acute situation in humans. Recently, a rabbit model of atherosclerotic plaque rupture was reported[16] in which an inflatable balloon was imbedded into the denuded arterial wall. Upon cholesterol feeding, atherosclerotic lesions form and cover the balloon. The intra-plaque balloon is then inflated to produce a mechanical rupture of the lesion. Such a model reportedly permits the analyses of both plaque mechanical strength and the effects of rupture on thrombosis. Further evaluation of this model is necessary, however, before it can be accepted as a model for the acute rupture of plaque in humans.

The mouse model is now the most common mammalian system used for the study of atherosclerosis. The advantages of this model include the ease of breeding useful quantities of animals, the ease of housing, relatively short generation times, and importantly the availability of inbred strains. Mice, however, have virtually no LDL and are extremely resistant to developing atherosclerosis. However, certain strains of mice will develop atherosclerotic lesions in the *sinus aorticus* when fed a diet containing high concentrations of cholesterol, fat and bile salt, whereas other strains are resistant to lesion formation.[17]

Transgenic mouse models of atherosclerosis have supplanted the cholesterol-induced models of atherosclerosis. Today there are abundant genetically modified mouse strains that will develop lesions generally in the vasculature, either spontaneously or when placed on a high-fat diet. Mice deficient in apolipoprotein E (apoE$^{(-/-)}$ mice) [18, 19] or the LDL receptor (LDLr$^{(-/-)}$ mice)[20] are most commonly used.

The experimental condition normally employed in animal interventions differs from that where drug treatment of patients is studied. Therefore, the applicability of data from animal experiments to the human situation is not obvious and limited. For example, in animals drug interventions are started generally using animals without or with only minor lesions, which is different to the situation in humans. Lipid metabolism in these animals is also different from that in humans and together these differences can be a confounding factor when evaluating data from animal models. Another limitation in the use of mice (but not rabbits) is the relatively small amount of aortic material that can be harvested from individual animals for use in biochemical analyses. Importantly however, animal studies offer the possibility to directly study the relationship between lesion formation and biochemical processes in the vessel wall, which is not possible in humans. Surprisingly, this opportunity is not always used.

3.3 Antioxidant protection of LDL

The 'oxidation theory' of atherogenesis claims that it is the oxidation of LDL-derived lipids that is important in atherogenesis. Therefore, lipophilic antioxidants have most commonly been tested for a potential anti-atherogenic effect. Among the naturally occurring lipophilic antioxidants in humans, vitamin E (α-tocopherol) is the most common antioxidant in lipoproteins, with ubiquinol-10 and β-carotene being less abundant.[21] Thus, one important line of research is to investigate and compare the role of these naturally occurring antioxidants in lipid peroxidation and atherogenesis. There are also a number of different synthetic lipophilic antioxidants of different chemical structures that have been investigated for their role in atherogenesis.

3.3.1 Natural antioxidants and atherosclerosis in animals

It is commonly assumed that animal intervention studies with natural antioxidants have overall yielded positive results. However, closer inspection of all published studies shows that this is not clearly the case, at least for vitamin E supplementation alone. In a recent review of the literature in this field, only 6 of the 34 studies examined showed that pharmacological doses of vitamin E supplements decrease atherosclerosis through a mechanism that can be distinguished clearly from a lipid-lowering action.[22] This was 'countered' by 4 and 20 studies reporting a disease-promoting effect and no effect, respectively. The reasons for these findings are not clear at present, although we have suggested[23, 24] that this may reflect the complex role α-tocopherol (α-TOH) plays in the control of LDL lipid peroxidation. Thus, although α-TOH is LDL's most abundant antioxidant and a powerful radical scavenger,[25] the vitamin alone does not necessarily prevent the oxidation of lipids in emulsions like lipoproteins.[24, 26]

Experiments of Parker and associates[27] indicate that the plasma cholesterol burden may affect the ability of vitamin E to inhibit atherosclerosis in hamsters. Thus, disease-promoting factors other than those directly related to LDL oxidation may override the anti-atherogenic activity of antioxidants. This is a confounding factor that should be controlled for when evaluating antioxidants. Also, results with vitamin E are overall more promising if one considers studies on the effect of α-TOH supplements in otherwise vitamin E-deficient animals.[24] Four studies[27–30] observed a disease-attenuating effect whereas only one study reported no effect.[31]

Clearly, these are areas that deserve more detailed investigations. Unfortunately, a serious confounding factor is that 'standard chow' already contains large amounts of vitamin E.[32] Emphasis should therefore be given to studies that adjust both cholesterol and antioxidants in 'standard' and supplemented chows to concentrations more comparable to those present in human diets.

Several investigators have employed vitamin E in combination with other antioxidants, and about two in three studies show a protective effect.[22] Remarkably however, four of the five studies employing vitamins E and C together did not find significant inhibition of atherosclerosis.[22] This is surprising given that the combination of the two antioxidants has been shown to offer outstanding protection against LDL oxidation.[33] Also, animal studies overall indicate that a deficiency in vitamin C results in structural disorders of the vessel wall, and that vitamin C supplement may attenuate atherosclerosis (reviewed in Ref. (34)).

The extent to which antioxidant supplements protected arterial lipids against oxidation has not generally been investigated in these animal studies. However, vitamin E has been shown to protect isolated plasma lipoproteins against Cu^{2+}-induced oxidation. Using this *ex vivo* measure as a surrogate for antioxidant efficacy *in vivo*, Fruebis *et al.*[35] showed that supplementation of WHHL rabbits with vitamin E resulted in both high plasma concentration of the vitamin (475 versus 96 μM in controls) and an effective prolongation of the lag time. In spite of this, vitamin E failed to affect the rate of development of atherosclerotic lesions, and vitamin E was clearly inferior to probucol as an anti-atherosclerotic agent.[35] There are several possible explanations for these findings. For example, control animals may have already had high tissue levels of vitamin E, and the vitamin may have a complex role in lipid peroxidation in the vasculature.[23, 24] Thus, the study by Fruebis *et al.*[35] may simply show that the measure of lag time in response to Cu^{2+}-induced oxidation does not reflect *in vivo* aortic lipid peroxidation rather than disqualifying the 'oxidation theory'. These examples show that the interpretation of results from animal studies is difficult where indirect measures of antioxidant efficacy are related to lesion progression.

Another example of a difficult-to-interpret study is that of Kleinveld *et al.*[36] These authors studied the effects of a relatively low dose of vitamin E on lesion progression in relatively mature WHHL rabbits (treatment from approximately 6 to 12 months of age), i.e. animals that already had advanced lesions prior to the intervention. There is limited quantitative data on lesion severity in WHHL rabbits of this age. Daugherty *et al.*[37] showed that the proportion of aortic surface covered with lesion increased from 70 to 91% whereas the cholesterol content in the aorta doubled as the age of the WHHL rabbits increased from 9 to 15 months. This suggests that disease progression was reflected by thickening of the intima rather than an expansion of the affected surface area. Kleinveld *et al.*[36] used surface area as the measure of lesion severity, and vitamin E reduced this measure non-significantly from 59 to 49%. Unfortunately, the per cent lesion area at the onset of the study was not reported although it likely was of similar magnitude. To obtain a reduction in atherosclerosis, the anti-atherogenic effect of the low dose of vitamin E used would have had to be dramatic. Thus, the use of an insensitive measure of lesion progression/regression, and the absence of a measure of lesion severity at the onset of treatment[36] makes it difficult to draw conclusions on the efficacy of vitamin E as an anti-atherogenic agent in this study.

There also seems to be a high degree of variation in aortic lesion areas reported by different laboratories for WHHL rabbits of 9 to 15 months of age. For example, we observed that in 6-month-old rabbits approximately 77% of the aortic surface is covered with lesions,[11] while e.g. Fruebis *et al.*[35] reported only a 25% coverage in 7-month-old rabbits.

In a recent study, Pratico *et al.*[38] used an interesting approach to study *in vivo* antioxidant effects of vitamin E and its relation to atherosclerosis. They supplemented diets given to apoE $^{(-/-)}$ mice for 5 months with vitamin E and measured the concentration of isoprostanes in plasma, urine, and the vessel wall of the animals. Isoprostanes are products of non-enzymatic oxidation of arachidonic acid and are used as a 'marker' of non-enzymatic lipid oxidation *in vivo*. The authors reported that vitamin E decreased both atherogenesis and the accumulation of isoprostanes in the vascular tissue. This is a potentially important finding as vitamin E did not reduce plasma cholesterol and atherogenesis was inhibited in parallel with an antioxidant effect in the aorta. Interestingly, though presently not understood, the results of Pratico *et al.*[38] are not in agreement with e.g. those obtained in rabbits.[35] Thus, differences in efficacy of vitamin E in different animals may need to be considered.

Polyphenols, present in foods and beverages such as tea and wine, have antioxidant properties *in vitro*. Recent animal intervention studies employing various mixtures of natural polyphenols as supplements indicate a protective effect,[39–42] although negative outcomes have also been reported.[43] While these results are encouraging and warrant further experimentation, their interpretation is complicated and will require follow-up mechanistic investigations, such as direct studies of oxidation of lipids in the vessel wall.

3.3.2 Natural antioxidants and atherosclerosis in humans

In the case of vitamin E, it is possible to review animal experimental data with reference to information on the effects of vitamin E on cardiovascular mortality and morbidity in humans, and to some extent on the progression of atherosclerotic lesions. A large number of descriptive case-control and cohort studies have provided an early and influential indication that OxLDL may cause atherosclerosis. These studies document that the regular intake of foods rich in fat-soluble antioxidants such as vitamin E is associated with a lower frequency of clinical events and mortality due to coronary heart disease (CHD) (reviewed in Ref. (44)). Importantly, yet often overlooked, substantial risk reduction in mortality generally correlates with only a moderate increase in plasma levels and dietary intake of vitamin E. For example, Gey *et al.*[45] reported that compared to subjects with a median plasma level of α-TOH of about $20\,\mu\text{mol}\,1^{-1}$, those with about $30\,\mu\text{mol}\,1^{-1}$ had an approximately 85% lower mortality rate. A large study that followed 19 687 women without symptoms of cardiovascular disease who did not consume vitamin supplements for seven years, showed that the multivariate adjusted, relative risks for CHD decreased significantly from the lowest ($\leqslant 4.9$ IU/day) to highest quintile of vitamin E intake ($\geqslant 9.6$ IU/day). Under these conditions, dietary vitamin C was without effect.[46] In contrast to these findings, two other large studies[47, 48] failed to observe a beneficial effect of dietary vitamin E (or vitamin C) on CHD. Instead, they reported a beneficial effect of supplemented vitamin E for men[47] and women,[48] whereas the Iowa Women's Health Study failed to observe a beneficial effect of vitamin E supplements.[46]

Intervention studies with natural antioxidants in humans and animals are largely limited to supplementation with pharmacological doses of vitamin E and β-carotene alone or in combination with other natural compounds. There are several recent reviews on the effect of antioxidants and CHD in humans.[49] Overall, the results of the major human randomized trials are disappointing. They do not prove the value of antioxidant vitamins and suggest that their contribution to the benefits indicated by observational epidemiology may have been overestimated.

The recent GISSI-Prevenzione study employed 300 mg of vitamin E per day and followed 11 324 patients with a recent myocardial infarction (MI) for 3.5 years.[50] Vitamin E failed to significantly lower the risk of death, non-fatal MI and stroke (whereas a supplement with n-3 polyunsaturated fatty acids offered protection). Preliminary results from another large, prospective trial, the HOPE trial, were recently presented at the AHA meeting in Atlanta, November 1999. It was also concluded that vitamin E supplements did not reduce cardiovascular mortality or morbidity. The lack of a protective effect in the GISSI and HOPE trials questions the efficacy of vitamin E in the secondary prevention of established atherosclerosis implied earlier by the results of the comparatively much smaller CHAOS study (2002 patients followed for 1.4 years).[51] Even when considering the trend towards a beneficial outcome in this trial, the overall impact of vitamin E is less impressive than that of proven secondary prevention drugs, such as statins, aspirin, β-blockers, and ACE inhibitors. The fact that four of five large-scale trials carried out to date have failed to yield a positive result also raises doubts as to whether future trials with vitamin E will show that supplements with antioxidant vitamins truly offer protection from CHD.

The negative result of the ATBC Cancer Prevention Study[52] has been considered irrelevant as the dose of vitamin E was low,[49] although this reason could itself be questioned. The median plasma α-TOH concentration after vitamin E supplementation ($40.1\,\mu\text{M}$) in that study was substantially higher than that in the control group ($28.8\,\mu\text{M}$).[52] Such an increase is comparable to that found in observational studies that associate α-TOH intake with a reduced risk of CHD (see above). Furthermore, the 49.6% increase in plasma α-TOH (from 26.8 to $40.1\,\mu\text{M}$)[52] is not different to that reported in the CHAOS study for subjects receiving 400 IU of vitamin E per day. In that study, plasma α-TOH increased from 34.2 to 51.1 μM[51] and antioxidant supplement was associated with a decreased risk of non-fatal MI. Irrespective of the issue of vitamin E dose, neither the observed decrease in non-fatal MI[51] nor the increase in hemorrhagic stroke[52] can be explained readily by the oxidation theory, i.e. by vitamin E inhibiting LDL oxidation.

In the midst of ongoing debate over its potential benefit, the safety of vitamin E supplementation must be established, as is highlighted by the β-carotene story (reviewed in Ref. (49)). A recent report concluded that up to 800 IU of vitamin E daily was without adverse effects in healthy older adults,[53] but the number of subjects studied was small and the follow-up duration short. In contrast to this report, the intervention studies mentioned earlier were not designed to evaluate the safety of vitamin E. Therefore, appropriate studies are needed to assess the safety of long-term supplementation with large doses of vitamin E.

In human studies mortality is a common endpoint, whereas animal experiments normally employ a measure of the size of the atherosclerotic lesion(s) as endpoint. In a sub-population of the HOPE study, the intima to media thickness was measured in the carotid bifurcation to examine the effect of vitamin E supplementation on the rate of growth of lesions. Vitamin E also failed to show a beneficial effect on this measure,[54] thereby supporting the results obtained in a majority of the animal intervention studies (see above). A combination of both vitamins C and E has been shown to be more effective than vitamin E alone in protecting LDL from becoming oxidized *in vitro* (see above). In apparent support of this, Salonen *et al.*[55] recently reported that plaque growth in the carotid artery was significantly reduced in men who were given the combination, while vitamin E alone was ineffective. This is however not immediately consistent with the results from animal studies (see above). Recently, it was also reported that there is a significant negative association between lycopene intake and the thickness of the carotid artery wall measured by ultrasound in men but not in women.[56] Thus the effects of dietary antioxidants on cardiovascular disease in humans overall remain unclear.

To summarize the effects of natural antioxidants on atherosclerosis, there seems to be little support for the hypothesis that vitamin E supplementation should lead to any major protection against atherosclerosis development or its complications. Combined treatment with vitamin E and another (co)antioxidant has provided some promising results that warrant further investigation. However, the relation of the combination treatment to vessel wall lipid peroxidation remains unknown, and this issue needs to be addressed in the future.

3.3.3 Synthetic antioxidants

3.3.3.1 Probucol

Among the synthetic antioxidants, probucol (which originally was developed as a lipid-lowering drug) has been studied most extensively. The results obtained overall provide perhaps the strongest support for a beneficial effect of an antioxidant on atherosclerosis, although some contentious issues remain.

Probucol commonly inhibits atherosclerosis in hypercholesterolemic rabbits[3, 4] and non-human primates.[57] In fact, the anti-atherogenic effect of probucol, which is independent of its hypolipidemic activity,[3] is one of the cornerstones of the oxidation theory of atherosclerosis.[1] Probucol also reverses established plaques in rabbits[58] and xanthomas in humans,[59] and this adds to its potential as a drug against CHD.

Some contentious issues remain however with probucol, in addition to the reported lowering of high-density lipoproteins (HDL) in humans.[60] Thus, probucol has been reported to fail to attenuate atherosclerosis in cholesterol-fed New Zealand White rabbits maintained at comparable plasma cholesterol levels,[61] a finding that appears to contradict the results obtained with WHHL rabbits.[3] Probucol was also reported to be ineffective in reducing established lesions in mature LDL receptor-deficient rabbits[37] and non-human primates.[57] In most rabbit studies, probucol has been given at a dose of 1 g or more per day. Lower doses may not be effective in inhibiting atherosclerosis in cholesterol-fed rabbits,[36] although this study may have also suffered from limited sensitivity in detecting an anti-atherogenic effect, for the reasons discussed above. Perhaps the most puzzling finding is that probucol appears to consistently promote atherosclerosis in the aortic root of mice,[62–66] in spite of a significant cholesterol-lowering effect. The reason(s) for this paradoxical pro-atherogenic activity of probucol in mice is presently unknown.

3.3.3.2 Probucol analogues

Jackson and colleagues[67, 68] tested probucol analogues with substitutions at the disulphide-linked carbon and an additional substitution at a *tert*-butyl of each phenolic ring for their ability to prevent aortic atherosclerosis in WHHL rabbits. The ability of the analogues to prevent Cu^{2+}-induced oxidation of LDL *in vitro* and to inhibit aortic atherosclerosis *in vivo* directly related to the concentrations of the analogues in LDL.[68] This suggested that the antioxidant activity of probucol and analogues directly related to their ability to reduce atherosclerosis (see, however, below).

The probucol analogue, bis(3,5-di-*tert*-butyl-4-hydroxy-phenylether)propane (BM15.0639), failed to inhibit atherosclerosis in LDL receptor-deficient rabbits, even when supplemented at a dose that increased the resistance of LDL to *ex vivo* oxidation induced by Cu^{2+} to a greater extent than did probucol (used as positive control).[69] More recently, Witting *et al.*[11] reported that the natural probucol metabolite 3,3′-5,5′-tetra-*tert*-butyl-4,4′-bisphenol also failed to inhibit atherosclerosis in WHHL rabbits, although the bisphenol prevented lipid peroxidation in the vessel wall just as effectively as probucol (positive control).[11] These studies question a causative link between atherosclerosis and LDL oxidation in the vessel wall (see below).

3.3.3.3 Other synthetic antioxidants

Several studies have been performed, largely in rabbits and mice, with lipophilic antioxidants chemically unrelated to probucol. *N,N′*-Diphenyl-phenylenediamine

(DPPD) has been shown to reduce lesion development in both rabbits and mice, and to prolong the lag time in *ex vivo* oxidation of LDL induced by Cu^{2+}.[9, 10] Butylated hydroxytoluene (BHT) has also been reported to prevent atherosclerosis in rabbits.[8]

Noguchi *et al.*[70] developed a novel phenolic radical-scavenging antioxidant considering various factors that determine antioxidant potency. The novel compound, 2,3-dihydro-5-hydroxy-2,2-dipentyl-4,6-di-*tert*-butylbenzofuran (BO-653), rapidly reduces α-tocopheroxyl radical, effectively inhibits *in vitro* LDL oxidation, and its phenoxyl radical is reduced by ascorbate.[70] BO-653 added to the diet reaches aortic vessels and suppresses atherosclerosis in normal mice, LDL receptor-deficient mice and rabbits.[66] This study thus reports both the effect of the antioxidant on atherosclerosis in several species, and several measures of *ex vivo* antioxidant efficacy. If the relation between any of these measures to the extent to which artery wall lipids are protected against oxidation could be established, this compound could strengthen the oxidation theory, although it would not negate the findings of Witting *et al.*[11]

Synthetic antioxidants have not been investigated to any larger extent in clinical trials, and studies with reported CV mortality are lacking. The PQRST study investigated whether probucol, in addition to lipid lowering with cholestyramine, had any impact on the lumen volume of the femoral artery in hypercholesterolemic patients.[71] The results were negative, yet several factors limit the conclusions that can be drawn from this study with regard to the validity of the oxidation theory. First, probucol markedly reduced high-density lipoproteins, an effect that correlated quantitatively to the femoral artery lumen.[60] Second, the lumen volume measure used in this study is not a direct measure of the artery wall disease. As remodelling occurs in response to lesion development,[72] the lumen size may not simply reflect the processes in the vessel wall. It is thus difficult to compare the experiences made in the limited numbers of human clinical trials with probucol on atherosclerosis with the outcome of animal studies.

3.3.3.4   Summary of antioxidants in atherosclerosis

Table 3.1 attempts to summarize the effects of the lipophilic antioxidants on atherosclerosis in animals reviewed here and, where possible, its relation to the outcome of clinical trials in humans. The early studies with probucol showing antioxidant and anti-atherogenic effects had a marked impact, so that the oxidation theory has become widely accepted. The pattern that emerges from Table 3.1 is that it is not unequivocal to conclude that antioxidants inhibit atherosclerosis in animals (or humans). Not only are there differences in efficacy between different antioxidants, there also appears to be species differences in the response to antioxidant treatment, and the results of measures of LDL oxidizability do not match disease outcome. Thus, there are clearly limitations to the support to the oxidation theory resulting from experimental studies, and the overall assessment of antioxidants as putative anti-atherogenic drugs is complex.

Table 3.1 Lipophilic antioxidants and atherosclerosis

Test drug	LDL antioxidant measure		Mouse				Rabbit		Human
	Cu^{2+} oxidation (lag time)	anti-TMP	C57Bl/6	apoE $(-/-)$	LDLr $(-/-)$	apoE $(-/-)$; LDLr $(-/-)$	WHHL	NZW	
Probucol	↑	↔	↑	↑	↑	—	↓	↓(↔)	↔
BHT	↑	↑	—	—	—	—	—	↓	—
BM15.0639	↑	↑	—	—	—	—	↔	—	—
DPPD	↑	↑	—	↓	—	—	—	↓	—
BO-653	↑	↑	↓	—	↓	—	↓	—	—
Bisphenol	↔	↑	—	—	—	↓	↔	—	—
Vitamin E	↑	↓	—	↓	—	—	↔	—	↔

The table, modified from Witting et al.,[11] summarizes the effects of lipophilic antioxidants on measures of antioxidant efficacy and atherosclerosis. Arrows indicate the treatment effects: ↑, ↓ and ↔ indicate increase, decrease and no effect, respectively; — indicates that data is not available, or that the studies have not been analysed in the present review. An increase in lag time and anti-TMP reflects increased *in vitro* antioxidant activity. Clearly, different antioxidants not only have different effects on atherosclerosis in one species, but different species may also respond differently to treatment with the same antioxidants.

What are the limitations to studies performed to date to definitely prove or disprove the oxidation theory? As mentioned, the picture emerging from human antioxidant interventions indicates that some of the benefits from earlier observational epidemiological studies may have been overestimated. Several factors may contribute to this overall disappointing outlook. In general, the atheroma found in the arterial wall, and a major underlying cause of coronary artery disease, is the result of the activity of multiple atherogenic factors that likely occur to varying degrees at different times. To believe that all factors at any time can be influenced by antioxidants may be to underestimate the complexity of the disease, just as it is to assume that all populations respond to a particular antioxidant treatment in the same way. Rather than attempting to 'treat' atherosclerosis generally, it may be more rewarding to direct efforts on subsets of coronary vascular disease and/or populations. Such an approach could advance our understanding of the mechanisms underlying disease progression.

A definition of the patient populations that could benefit from antioxidant intervention is, however, still needed. One example might be restenosis after percutaneous transluminal coronary angioplasty (PTCA), where probucol repeatedly has shown efficacy, although a 'vitamin cocktail' was ineffective and, in fact, eliminated the beneficial effect observed with probucol treatment alone.[73] One interesting outcome of this study was that probucol may have maintained an adequate lumen of the dilated coronary arteries by 'positive remodelling' rather than by inhibition of neointima growth,[14] although the interpretation of the results of this study is itself complex. Remodelling of atherosclerotic arteries is a known phenomenon,[72] but it is rarely addressed in experimental studies. In fact, the commonly used measures of atherosclerosis in animals are not suitable to estimate this. The potential effect of antioxidants on remodelling during atherogenesis has not been examined in detail in animal experiments, and thus remains an interesting subject for future studies.

3.3.4 Antioxidant efficacy

3.3.4.1 Plasma versus vessel wall parameters

As indicated above, several early studies[57, 74] provided support for the implied underlying assumption that the ability of plasma LDL to resist *ex vivo* lipid oxidation reflects the *in vivo* effectiveness of an antioxidant supplement and hence the inhibition of atherosclerosis. However, it is now increasingly clear that the resistance of LDL to *ex vivo* oxidation is not a reliable surrogate for the ability of an antioxidant therapy to inhibit atherosclerosis.[11, 41, 69, 75] Furthermore, there is also recent evidence suggesting that the resistance of plasma LDL to *ex vivo* oxidation by Cu^{2+} does not reflect the extent of antioxidant protection within the vessel wall. Thus, treatment of rabbits with bisphenol effectively prevented aortic lipid oxidation, although it only marginally increased the resistance of plasma LDL to oxidation induced by Cu^{2+}.[11] Until a suitable surrogate is established, it is prudent to directly examine the efficacy of a supplemented antioxidant to inhibit oxidative damage in the tissue affected, i.e. the vessel wall rather than plasma lipoproteins.

3.3.4.2 Oxidation parameter(s) to assess lipoprotein oxidation in the vessel wall

Oxidative modification of lipoproteins is a chemically complex process that involves different types of alterations to lipid and protein moieties. These vary depending on many factors, such as the type of oxidant, the ratio of oxidant to target and the duration of oxidant exposure involved, as well as presently ill-characterized repair processes that may follow oxidative damage to lipoproteins within the vessel wall.[6, 24, 76] Unfortunately, we do not know which lipoprotein oxidation 'marker(s)' are pathologically relevant, and whether this depends on the animal species and/or disease stage examined.

One of the unanswered key issues here is to establish to what extent lesion lipoproteins are oxidized. For example, as far as we have examined,[11, 77–80] lipoproteins in animal and human lesions are not generally oxidized to an extent where α-TOH is depleted. In fact, even at the most advanced stages of human atherosclerosis, lipid-adjusted levels of α-TOH are normal[78, 79] and most of the detected lipid hydroperoxides and alcohols (LO(O)H) appear to have accumulated in the presence of vitamin E.[24] The underlying chemical mechanism of this can be explained by tocopherol-mediated peroxidation[24, 26] and the ability of methionine residues in apolipoproteins to reduce lipid hydroperoxides into alcohols.[81, 82] This may have important consequences on indices of lipoprotein lipid peroxidation. In the presence of the vitamin, accumulation of LO(O)H predominates while secondary products including fragments of lipid hydroperoxides (e.g. hydroxynonenal and malondialdehyde) are minor products.[21] Consistent with this, LO(O)H represent the major single class of oxidized lipids and accumulate time-dependently in both lipoproteins undergoing *in vitro* oxidation in the presence of tocopherol.[26] LO(O)H also represent a major class of oxidized lipids in animal[77] and human lesions[78, 79] (unpublished observation). Therefore, it seems reasonable to measure LO(O)H as an index of the overall extent of lipoprotein oxidation in the vessel wall.[11]

What we do not know at present, however, is which lipid oxidation products are atherogenic. This represents a serious limitation. It means that analysis of any lipid or, for that matter, protein oxidation product does not necessarily reflect a relevant measure of disease promoting oxidative damage. As a first step towards acquiring relevant knowledge, it may be helpful to compile an 'inventory' of known lipid (LO(O)H, F_2-isoprostanes, oxysterols, phospholipids with fragmented side chains, tocopherol oxidation products) and protein oxidation products (nitrated, chlorinated and hydroxylated amino acids, lipid–protein adducts, etc.) present in human lesions, and examine how the various parameters change depending on the severity of disease. At present, this type of information is available only in fragmented form that does not allow direct comparison between different oxidation products in a single tissue. Independent of the parameter measured, the oxidized moiety should be determined relative to the parent molecule to evaluate the extent of lipoprotein oxidation. This is important because the concentration of the parent molecule may vary, particularly if it is a lipid. If not done, interpretation of the results is unnecessarily complicated. Demonstration of the same oxidation products in the vessel wall of animal models of atherosclerosis could then be a way to verify the validity of the experimental model.

3.3.4.3 Mechanism of anti-atherogenic action of antioxidants

Animal intervention studies that directly address the fundamental question of whether lipoprotein oxidation is a cause rather than a consequence of atherosclerosis are essentially lacking. Gene transfer and/or knockout studies employing genes that encode products that promote or inhibit lipoprotein oxidation in the vessel wall have the potential to provide such direct information.

There is now good evidence that oxidants derived from lipoxygenase(s) and myeloperoxidase as well as reactive nitrogen species are involved in the oxidative modification of vessel wall components, including lipoproteins (reviewed in Ref. (6)). This and existing knowledge on the enzymatic antioxidant defences against lipid peroxidation provides clear targets to modulate, and to test the effect of this on atherogenesis. For example, initial results indicating the appearance of oxidation-specific lipid–protein adducts of OxLDL after the transfer of 15-lipoxygenase into rabbit iliac arteries[83] are encouraging. This is particularly so in light of the recent finding that disruption of the 12/15-lipoxygenase gene diminishes the propensity of apoE ($-/-$) mice to develop atherosclerosis.[84] Intriguingly however, over-expression of 15-LO in rabbit macrophages has been reported to protect against lesion formation,[85] again demonstrating the complexity of the pathology of atherosclerosis. It would be important now to establish the relationship between the extent of atherosclerosis and tissue activity of the lipoxygenase. With the increasing availability of different gene knockout mice, a similar approach can now be taken with e.g. myeloperoxidase and nitric oxide synthase, as well as antioxidant enzymes such as extracellular superoxide dismutase and glutathione peroxidase.

3.3.4.4 Dissociation of atherosclerosis from aortic lipoprotein oxidation

We have measured LO(O)H in the vessel wall of animals supplemented with antioxidants in an attempt to relate intima lipoprotein lipid oxidation to atherogenesis. Using the probucol metabolite bisphenol, we observed that in mice deficient in both apolipoprotein E and LDL receptor, inhibition of accumulation of LO(O)H was associated with inhibition of atherosclerosis.[80] Interestingly however, the same antioxidant failed to affect lesion formation in WHHL rabbits, although it effectively prevented LO(O)H accumulation in the aorta.[11] Notwithstanding the above-mentioned limitations regarding *in vivo* lipoprotein oxidation parameters, the dissociation of atherosclerosis from aortic accumulation of LO(O)H[11] suggests, for the first time, that intimal lipoprotein oxidation is not required for atherogenesis.

As suggested,[11, 86] it will be important to examine whether the above dissociation can be confirmed in and extended to other animal models. Using apoE ($-/-$) mice fed a high-fat diet for 6 months, we observed that while probucol increased lesion formation in the aortic root (see above) it increasingly and substantially inhibited disease the further distal the site examined from the aortic root (P.K. Witting, K. Pettersson, and R. Stocker, unpublished). Despite the observed inhibition of atherosclerosis however, probucol did not affect the extent to which aortic lipids were oxidized (P.K. Witting, K. Pettersson, and R. Stocker, unpublished), suggesting that inhibition of lipoprotein oxidation in the vessel wall is not required for inhibition of atherosclerosis in mice by probucol.

Intriguingly, these recent results suggest that probucol inhibits atherosclerosis by means other than the inhibition of lipoprotein oxidation. The identification of such additional effects may offer exciting new perspectives for antioxidant intervention and suggests an extension of the original oxidation theory to processes in addition to and/or separate from lipoprotein oxidation. A number of potential targets exist for antioxidant actions, including endothelial cell function, platelet aggregation, endothelial cell–leukocyte interaction and other pro-inflammatory processes, and SMC cell proliferation (for a review see Ref. (5)). These will need to be considered in future antioxidant intervention studies as well as in the development of new antioxidants for the treatment of cardiovascular disease.

3.4   Conclusions

The cardiovascular benefit of antioxidant supplements remains unproven, and therefore a recommendation to the public is not justified. Yet certain antioxidants do inhibit atherosclerosis in some situations. The challenge is to establish the underlying cause for this. Several key scientific issues need to be addressed. In particular, research efforts need to focus more directly on the relevance and contribution of lipoprotein oxidation in the vessel wall as a cause of atherosclerosis. This will require establishing both parameters for lipoprotein oxidation that are relevant to disease progression and suitable surrogates to monitor the efficacy with which antioxidants inhibit lipoprotein oxidation in the vessel wall. In addition to lipoprotein oxidation, other potential targets for antioxidant intervention likely exist, such as endothelial cells. A systematic investigation of the *in vivo* effects of antioxidants on these functions in addition to lipoprotein oxidation may eventually provide a more rational base for the design of novel antioxidant drugs.

References

1. Steinberg, D., Parthasarathy, S., Carew, T.E., Khoo, J.C., and Witztum, J.L. (1989). Beyond cholesterol: modifications of low-density lipoprotein that increase its atherogenicity. *N. Engl. J. Med.*, **320**, 915.
2. Parthasarathy, S., Young, S.G., Witztum, J.L., Pittman, R.C., and Steinberg, D. (1986). Probucol inhibits oxidative modification of low density lipoprotein. *J. Clin. Invest.*, **77**, 641.
3. Carew, T.E., Schwenke, D.C., and Steinberg, D. (1987). Antiatherogenic effect of probucol unrelated to its hypocholesterolemic effect: evidence that antioxidants *in vivo* can selectively inhibit low density lipoprotein degradation in macrophage-rich fatty streaks and slow the progression of atherosclerosis in the Watanabe heritable hyperlipidemic rabbit. *Proc. Natl. Acad. Sci. USA*, **84**, 7725.
4. Kita, T., Nagano, Y., Yokode, M., Ishii, K., Kume, N., Ooshima, A., Yoshida, H., and Kawai, C. (1987). Probucol prevents the progression of atherosclerosis in Watanabe heritable hyperlipidemic rabbit, an animal model for familial hypercholesterolemia. *Proc. Natl. Acad. Sci. USA*, **84**, 5928.
5. Diaz, M.N., Frei, B., Vita, J.A., and Keaney, J.F., Jr (1997). Antioxidants and atherosclerotic heart disease. *N. Engl. J. Med.*, **337**, 408.
6. Heinecke, J.W. (1998). Oxidants and antioxidants in the pathogenesis of atherosclerosis: implications for the oxidized low density lipoprotein hypothesis. *Atherosclerosis*, **141**, 1.
7. Navab, M., Berliner, J.A., Watson, A.D., Hama, S.Y., Territo, M.C., Lusis, A.J., Shih, D.M., VanLenten, B.J., Frank, J.S., Demer, L.L., Edwards, P.A., and Fogelman, A.M. (1996). The Yin and Yang of oxidation in the development of the fatty streak. A review based on the 1994 George Lyman Duff Memorial Lecture. *Arterioscler. Thromb. Vasc. Biol.*, **16**, 831.
8. Björkhem, I., Henriksson-Freyschuss, A., Breuer, O., Diczfalusy, U., Berglund, L., and Henriksson, P. (1991). The antioxidant butylated hydroxytoluene protects against atherosclerosis. *Arterioscler. Thromb.*, **11**, 15.
9. Sparrow, C.P., Doebber, T.W., Olszewski, J., Wu, M.S., Ventre, J., Stevens, K.A., and Chao, Y.S. (1992). Low density lipoprotein is protected from oxidation and the progression of atherosclerosis is slowed in cholesterol-fed rabbits by the antioxidant *N,N'*-diphenyl-phenylenediamine. *J. Clin. Invest.*, **89**, 1885.
10. Tangirala, R.K., Casanada, F., Miller, E., Witztum, J.L., Steinberg, D., and Palinski, W. (1995). Effect of the antioxidant *N,N'*-diphenyl 1,4-phenylenediamine (DPPD) on atherosclerosis in apoE-deficient mice. *J. Lipid Res.*, **15**, 1625.
11. Witting, P.K., Pettersson, K., Östlund-Lindqvist, A.-M., Westerlund, C., Wågberg, M., and Stocker, R. (1999). Dissociation of atherogenesis from aortic accumulation of lipid hydro(pero)xides in Watanabe heritable hyperlipidemic rabbits. *J. Clin. Invest.*, **104**, 213.
12. Hulthe, J., Wikstrand, J., Emanuelsson, H., Wiklund, O., de Feyter, P.J., and Wendelhag, I. (1997). Atherosclerotic changes in the carotid artery bulb as measured by B-mode ultrasound are associated with the extent of coronary atherosclerosis. *Stroke*, **28**, 1189.

13. Yuan, C., Beach, K.W., Smith, L.H., Jr, and Hatsukami, T.S. (1998). Measurement of atherosclerotic carotid plaque size *in vivo* using high resolution magnetic resonance imaging. *Circulation*, **98**, 2666.

14. Côté, G., Tardif, J.C., Lespérance, J., Lambert, J., Bourassa, M., Bonan, R., Gosselin, G., Joyal, M., Tanguay, J.F., Nattel, S., Gallo, R., and Crepeau, J. (1999). Effects of probucol on vascular remodelling after coronary angioplasty. Multivitamins and Protocol Study Group. *Circulation*, **99**, 30.

15. Havel, R.J., Kita, T., Kotite, L., Kane, J.P., Hamilton, R.L., Goldstein, J.L., and Brown, M.S. (1982). Concentration and composition of lipoproteins in blood plasma of the WHHL rabbit. An animal model of human familial hypercholesterolemia. *Arteriosclerosis*, **2**, 467.

16. Rekhter, M.D., Hicks, G.W., Brammer, D.W., Work, C.W., Kim, J.S., Gordon, D., Keiser, J.A., and Ryan, M.J. (1998). Animal model that mimics atherosclerotic plaque rupture. *Circ. Res.*, **83**, 705.

17. Paigen, B., Morrow, A., Holmes, P.A., Mitchell, D., and Williams, R.A. (1987). Quantitative assessment of atherosclerotic lesions in mice. *Atherosclerosis*, **68**, 231.

18. Nakashima, Y., Plump, A.S., Raines, E.W., Breslow, J.L., and Ross, R. (1994). Apo-E deficient mice develop lesions of all phases of atherosclerosis throughout the arterial tree. *Arterioscler. Thromb.*, **14**, 133.

19. Zhang, S.H., Reddick, R.L., Piedrahita, J.A., and Maeda, N. (1992). Spontaneous hypercholesterolemia and arterial lesions in mice lacking apolipoprotein E. *Science*, **258**, 468.

20. Ishibashi, S., Goldstein, J.L., Brown, M.S., Herz, J., and Burns, D.K. (1994). Massive xanthomatosis and atherosclerosis in cholesterol-fed low density lipoprotein receptor-negative mice. *J. Clin. Invest.*, **93**, 1885.

21. Esterbauer, H., Gebicki, J., Puhl, H., and Jürgens, G. (1992). The role of lipid peroxidation and antioxidants in oxidative modification of LDL. *Free Radic. Biol. Med.*, **13**, 341.

22. Stocker, R. (1999). Dietary and pharmacological antioxidants and atherosclerosis. *Curr. Opin. Lipidol.*, in press.

23. Stocker, R. (1999). The ambivalence of vitamin E in atherogenesis. *TiBS*, **24**, 219.

24. Upston, J.M., Terentis, A.C., and Stocker, R. (1999). Tocopherol-mediated peroxidation (TMP) of lipoproteins: implications for vitamin E as a potential antiatherogenic supplement. *FASEB J.*, **13**, 977.

25. Burton, G.W. and Ingold, K.U. (1986). Vitamin E: application of the principles of physical organic chemistry to the exploration of its structure and function. *Acc. Chem. Res.*, **19**, 194.

26. Bowry, V.W, and Stocker, R. (1993). Tocopherol-mediated peroxidation. The pro-oxidant effect of vitamin E on the radical-initiated oxidation of human low-density lipoprotein. *J. Am. Chem. Soc.*, **115**, 6029.

27. Parker, R.A., Sabrah, T., Cap, M., and Gill, B.T. (1995). Relation of vascular oxidative stress, α-tocopherol, and hypercholesterolemia to early atherosclerosis in hamsters. *Arterioscler. Thromb. Vasc. Biol.*, **15**, 349.

28. Özer, N.K., Sirikci, Ö., Taha, S., San, T., Moser, U., and Azzi, A. (1998). Effect of vitamin E and probucol on dietary cholesterol-induced atherosclerosis in rabbits. *Free Radic. Biol. Med.*, **24**, 226.

29. Sulkin, N.M. and Sulkin, D.F. (1960). Intimal lesions in arteries of vitamin E deficient rats. *Proc. Soc. Exp. Biol. Med.*, **103**, 111.

30. Xu, R., Yokoyama, W.H., Irving, D., Rein, D., Walzem, R.L., and German, J.B. (1998). Effect of dietary catechin and vitamin E on aortic fatty streak accumulation in hypercholesterolemic hamsters. *Atherosclerosis*, **137**, 29.

31. Stein, O., Dabach, Y., Hollander, G., Halperin, G., Thiery, J., and Stein, Y. (1996). Relative resistance of the hamster to aortic atherosclerosis in spite of prolonged vitamin E deficiency and dietary hypercholesterolemia. Putative effect of increased HDL? *Biochim. Biophys. Acta*, **1299**, 216.

32. Lehr, H.A., Vajkoczy, P., Menger, M.D., and Arfors, K.E. (1999). Do vitamin E supplements in diets for laboratory animals jeopardize findings in animal models of disease? *Free Radic. Biol. Med.*, **26**, 472.

33. Frei, B., England, L., and Ames, B.N. (1989). Ascorbate is an outstanding antioxidant in human blood plasma. *Proc. Natl. Acad. Sci. USA*, **86**, 6377.

34. Lynch, S.M., Gaziano, J.M., and Frei, B. (1996). Ascorbic acid and atherosclerotic cardiovascular disease. *Subcell. Biochem.*, **25**, 331.

35. Fruebis, J., Carew, T.E., and Palinski, W. (1995). Effect of vitamin E on atherogenesis in LDL receptor-deficient rabbits. *Atherosclerosis*, **117**, 217.

36. Kleinveld, H.A., Demacker, P.N., and Stalenhoef, A.F. (1994). Comparative study on the effect of low-dose vitamin E and probucol on the susceptibility of LDL to oxidation and the progression of atherosclerosis in Watanabe heritable hyperlipidemic rabbits. *Arterioscler. Thromb.*, **14**, 1386.

37. Daugherty, A., Zweifel, B.S., and Schonfeld, G. (1991). The effects of probucol on the progression of atherosclerosis in mature Watanabe heritable hyperlipidaemic rabbits. *Br. J. Pharmacol.*, **103**, 1013.

38. Pratico, D., Tangirala, R.K., Radar, D., Rokach, J., and FitzGerald, G.A. (1998). Vitamin E suppresses isoprostane generation *in vivo* and reduces atherosclerosis in apoE-deficient mice. *Nature Med.*, **4**, 1189.

39. Hayek, T., Fuhrman, B., Vaya, J., Rosenblat, M., Belinky, P., Coleman, R., Elis, A., and Aviram, M. (1997). Reduced progression of atherosclerosis in apolipoprotein E-deficient mice following consumption of red wine, or its polyphenols quercetin or catechin, is associated with reduced susceptibility of LDL to oxidation and aggregation. *Arterioscler. Thromb. Vasc. Biol.*, **17**, 2744.

40. Wu, Y.J., Hong, C.Y., Lin, S.J., Wu, P., and Shiao, M.S. (1998). Increase of vitamin E content in LDL and reduction of atherosclerosis in cholesterol-fed rabbits by a water-soluble antioxidant-rich fraction of *Salvia miltiorrhiza*. *Arterioscler. Thromb. Vasc. Biol.*, **18**, 481.

41. Aviram, M. and Fuhrman, B. (1998). Polyphenolic flavonoids inhibit macrophage-mediated oxidation of LDL and attenuate atherogenesis. *Atherosclerosis*, **137** (Suppl.), S45.

42. Yamakoshi, J., Kataoka, S., Koga, T., and Ariga, T. (1999). Proanthocyanidin-rich extract from grape seeds attenuates the development of aortic atherosclerosis in cholesterol-fed rabbits. *Atherosclerosis*, **142**, 139.

43. Tijburg, L.B., Wiseman, S.A., Meijer, G.W., and Weststrate, J.A. (1997). Effects of green tea, black tea and dietary lipophilic antioxidants on LDL oxidizability and atherosclerosis in hypercholesterolaemic rabbits. *Atherosclerosis*, **135**, 37.

44. Gaziano, J.M. (1996). Antioxidants in cardiovascular disease: randomized trials. *Nutrition*, **12**, 583.

45. Gey, K.F., Puska, P., Jordan, P., and Moser, U.K. (1991). Inverse correlation between plasma vitamin E and mortality from ischemic heart disease in cross-cultural epidemiology. *Am. J. Clin. Nutr.*, **53**, 326S.

46. Kushi, L.H., Folsom, A.R., Prineas, R.J., Mink, P.J., Wu, Y., and Bostick, R.M. (1996). Dietary antioxidant vitamins and death from coronary heart disease in postmenopausal women. *N. Engl. J. Med.*, **334**, 1156.

47. Rimm, E.B., Stampfer, M.J., Ascherio, A., Giovannucci, E., Colditz, G.A., and Willett, W.C. (1993). Vitamin E consumption and the risk of coronary heart disease in men. *N. Engl. J. Med.*, **328**, 1450.

48. Stampfer, M.J., Hennekens, C.H., Manson, J.E., Colditz, G.A., Rosner, B., and Willett, W.C. (1993). Vitamin E consumption and the risk of coronary disease in women. *N. Engl. J. Med.*, **328**, 1444.

49. Faggiotto, A., Poli, A., and Catapano, A.L. (1998). Antioxidants and coronary artery disease. *Curr. Opin. Lipidol.*, **9**, 541.

50. GISSI-Prevenzione Investigators (1999). Dietary supplementation with n-3 polyunsaturated fatty acids and vitamin E after myocardial infarction: results of the GISSI-Prevenzione trial. *Lancet*, **354**, 447.

51. Stephens, N.G., Parsons, A., Schofield, P.M., Kelly, F., Cheeseman, K., Mitchinson, M.J., and Brown, M.J. (1996). Randomised controlled trial of vitamin E in patients with coronary disease: Cambridge Heart Antioxidant Study (CHAOS). *Lancet*, **347**, 781.

52. The Alpha-Tocopherol Beta Carotene Cancer Prevention Study Group (1994). The effect of vitamin E and beta carotene on the incidence of lung cancer and other cancers in male smokers. *N. Engl. J. Med.*, **330**, 1029.

53. Meydani, S.N., Meydani, M., Blumberg, J.B., Leka, L.S., Pedrosa, M., Diamond, R., and Schaefer, E.J. (1998). Assessment of the safety of supplementation with different amounts of vitamin E in healthy older adults. *Am. J. Clin. Nutr.*, **68**, 311.

54. Lonn, E.M., Yusuf, S., Smith, S., Moore-Cox, A., Derksen, C., Magi, A., Musseau, A., Bosch, J., Pogue, J., and Doris, I. (1999). Results of the study to evaluate carotid ultrasound changes in patients treated with ramipril and vitamin E (SECURE). In *American Heart Association, 72nd Scientific Sessions*, Atlanta, GA. *Circulation*, **100**, I-185.

55. Salonen, J.T., Nyyssonen, K., Salonen, R., Lakka, H.-M., Kaikkonen, J., Porkkala-Sarataho, E., Voutilainen, S., Lakka, T.A., Rissanen, T.H., Leskinen, L., Tuomainen, T.-P., and Poulsen, H.E. (1999). The effect of vitamin E and vitamin C on carotid atherosclerosic progression: the Antioxidant Supplementation in Atherosclerosis Prevention (ASAP) study. In *American Heart Association, 72nd Scientific Sessions*, Atlanta, GA. *Circulation*, **100**, I-238.

56. Rissanen, T.H., Voutilainen, S., Nyyssonen, K., Salonen, R., and Salonen, J.T. (1999). Association between low plasma lycopene concentration and increased common carotid artery wall thickness. In *American Heart Association, 72nd Scientific Sessions*, Atlanta, GA. *Circulation*, **100**, I-710.

57. Sasahara, M., Raines, E.W., Chait, A., Carew, T.E., Steinberg, D., Wahl, P.W., and Ross, R. (1994). Inhibition of hypercholesterolemia-induced atherosclerosis in the nonhuman primate by probucol. I. Is the extent of atherosclerosis related to resistance of LDL to oxidation? *J. Clin. Invest.*, **94**, 155.

58. Nagano, Y., Nakamura, T., Matsuzawa, Y., Cho, M., Ueda, Y., and Kita, T. (1992). Probucol and atherosclerosis in the Watanabe heritable hyperlipedemic

rabbit – long-term antiatherogenic effect and effects on established plaques. *Atheroslcerosis*, **92**, 131.

59. Kajinami, K., Nishitsuji, M., Takeda, Y., Shimizu, M., Koizumi, J., and Mabuchi, H. (1996). Long-term probucol treatment results in regression of xanthomas, but in progression of coronary atherosclerosis in a heterozygous patient with familial hypercholesterolemia. *Atherosclerosis*, **120**, 181.

60. Johansson, J., Olsson, A.G., Bergstrand, L., Elinder, L.S., Nilsson, S., Erikson, U., Molgaard, J., Holme, I., and Walldius, G. (1995). Lowering of HDL2b by probucol partly explains the failure of the drug to affect femoral atherosclerosis in subjects with hypercholesterolemia. A Probucol Quantitative Regression Swedish Trial (PQRST) Report. *Arterioscler. Thromb. Vasc. Biol.*, **15**, 1049.

61. Stein, Y., Stein, O., Delplanque, B., Fesmire, J.D., Lee, D.M., and Alaupovic, P. (1989). Lack of effect of probucol on atheroma formation in cholesterol-fed rabbits kept at comparable plasma cholesterol levels. *Atherosclerosis*, **75**, 145.

62. Zhang, S.H., Reddick, R.L., Avdievich, E., Surles, L.K., Jones, R.G., Reynolds, J.B., Quarfordt, S.H., and Maeda, N. (1997). Paradoxical enhancement of atherosclerosis by probucol treatment in apolipoprotein E-deficient mice. *J. Clin. Invest.*, **99**, 2858.

63. Moghadasian, M.H., McManus, B.M., Godin, D.V., Rodrigues, B., and Frohlich, J.J. (1999). Proatherogenic and antiatherogenic effects of probucol and phytosterols in apolipoprotein E-deficient mice: possible mechanisms of action. *Circulation*, **99**, 1733.

64. Benson, G.M., Schiffelers, R., Nicols, C., Latchman, J., Vidgeon-Hart, M., Toseland, C.D.N., Suckling, K.E., and Groot, P.H.E. (1998). Effect of probucol on serum lipids, atherosclerosis and toxicology in fat-fed LDL receptor deficient mice. *Atherosclerosis*, **141**, 237.

65. Bird, D.A., Tangirala, R.K., Fruebis, J., Steinberg, D., Witztum, J.L., and Palinski, W. (1998). Effect of probucol on LDL oxidation and atherosclerosis in LDL receptor deficient mice. *J. Lipid Res.*, **39**, 1079.

66. Cynshi, O., Kawabe, Y., Suzuki, T., Takashima, Y., Kaise, H., Nakamura, M., Ohba, Y., Kato, Y., Tamura, K., Hayasaka, A., Higashida, A., Sakaguchi, H., Takeya, M., Takahashi, K., Inoue, K., Noguchi, N., Niki, E., and Kodama, T. (1998). Antiatherogenic effects of the antioxidant BO-653 in three different animal models. *Proc. Natl. Acad. Sci. USA*, **95**, 10123.

67. Mao, S.J., Yates, M.T., Parker, R.A., Chi, E.M., and Jackson, R.L. (1991). Attenuation of atherosclerosis in a modified strain of hypercholesterolemic Watanabe rabbits with use of a probucol analogue (MDL 29 311) that does not lower serum cholesterol. *Arterioscler. Thromb.*, **11**, 1266.

68. Mao, S.J., Yates, M.T., Rechtin, A.E., Jackson, R.L., and Van-Sickle W.A. (1991). Antioxidant activity of probucol and its analogues in hypercholesterolemic Watanabe rabbits. *J. Med. Chem.*, **34**, 298.

69. Fruebis, J., Steinberg, D., Dresel, H.A., and Carew, T.A. (1994). A comparison of the antiatherogenic effects of probucol and a structural analogue of probucol in low density lipoprotein receptor-deficient rabbits. *J. Clin. Invest.*, **94**, 392.

70. Noguchi, N., Iwaki, Y., Takahashi, M., Komuro, E., Kato, Y., Tamura, K., Cynshi, O., Kodama, T., and Niki, E. (1997). 2,3-Dihydro-5-hydroxy-2,

2-dipentyl-4,6-di-*tert*-butylbenzofuran: design and evaluation as a novel radical-scavenging antioxidant against lipid peroxidation. *Arch. Biochem. Biophys.*, **342**, 236.

71. Walldius, G., Erikson,U., Olsson, A.G., Bergstrand, L., Hadell, K., Johansson, J., Kaijser, L., Lassvik, C., Molgaard, J., Nilsson, S. *et al.* (1994). The effect of probucol on femoral atherosclerosis: the Probucol Quantitative Regression Swedish Trial (PQRST). *Am. J. Cardiol.*, **74**, 875.

72. Glagov, S., Weisenberg, E., Zarins, C.K., Stankunavicius, R., and Kolettis, G.J. (1987). Compensatory enlargement of human atherosclerotic coronary arteries. *N. Engl. J. Med.*, **316**, 1371.

73. Tardif, J.-C., Côté, G., Lespérance, J., Bourassa, M., Lambert, J., Doucet, S., Bilodeau, L., Nattel, S., and de Guise, P. (1997). Probucol and multivitamins in the prevention of restenosis after coronary angioplasty. *N. Engl. J. Med.*, **337**, 365.

74. Mao, S.J., Yates, M.T., and Jackson, R.L. (1994). Antioxidant activity and serum levels of probucol and probucal metabolites. *Methods Enzymol.*, **234**, 505.

75. Fruebis, J., Bird, D.A., Pattison, J., and Palinski, W. (1997). Extent of antioxidant protection of plasma LDL is not a predictor of the antiatherogenic effect of antioxidants. *J. Lipid. Res.*, **38**, 2455.

76. Stocker, R. (1999). *Antioxidant defenses in the vascular wall*. In *Oxidative stress and vascular disease*. (ed. J.F. Jr Keaney), p. 27. Kluwer Academic Publishers, Boston.

77. Letters, J.M., Witting, P.K., Christison, J.K., Westin Eriksson, A., Pettersson, K., and Stocker, R. (1999). Changes to lipids and antioxidants in plasma and aortae of apoE-deficient mice. *J. Lipid Res.*, **40**, 1104.

78. Niu, X., Zammit, V., Upston, J.M., Dean, R.T., and Stocker, R. (1999). Co-existence of oxidized lipids and α-tocopherol in all lipoprotein fractions isolated from advanced human atherosclerotic plaques. *Arterioscl. Thromb. Vasc. Biol.*, **19**, 1708.

79. Suarna, C., Dean, R.T., May, J., and Stocker, R. (1995). Human atherosclerotic plaque contains both oxidized lipids and relatively large amounts of α-tocopherol and ascorbate. *Arterioscler. Thromb. Vasc. Biol.*, **15**, 1616.

80. Witting, P.K., Pettersson, K., Östlund-Lindqvist, A.-M., Westerlund, C., Westin Eriksson, A., and Stocker, R. (1999). Inhibition by a co-antioxidant of aortic lipoprotein lipid peroxidation and atherosclerosis in apolipoprotein E and low density lipoprotein receptor gene double knockout mice. *FASEB J.*, **13**, 667.

81. Sattler, W., Christison, J.K., and Stocker, R. (1995). Cholesterylester hydroperoxide reducing activity associated with isolated high- and low-density lipoproteins. *Free Radic. Biol. Med.*, **18**, 421.

82. Garner, B., Waldeck, A.R., Witting, P.K., Rye, K.-A., and Stocker, R. (1998). Oxidation of high density lipoproteins. II. Evidence for direct reduction of HDL lipid hydroperoxides by methionine residues of apolipoproteins AI and AII. *J. Biol. Chem.*, **273**, 6088.

83. Ylä-Herttuala, S., Luoma, J., Viita, H., Hiltunen, T., Sisto, T., and Nikkari, T. (1995). Transfer of 15-lipoxygenase gene into rabbit iliac arteries results in the appearance of oxidation-specific lipid–protein adducts characteristic of oxidized low density lipoprotein. *J. Clin. Invest.*, **95**, 2692.

84. Cyrus, T., Witztum, J.L., Rader, D.J., Tangirala, R., Fazio, S., Linton, M.F., and Funk, C.D. (1999). Disruption of the 12/15-lipoxygenase gene diminishes atherosclerosis in apo E-deficient mice. *J. Clin. Invest.*, **103**, 1597.

85. Shen, J., Herderick, E., Cornhill, J.F., Zsigmond, E., Kim, H.-S., Kühn, H., Guevara, N.V., and Chan, L. (1996). Macrophage-mediated 15-lipoxygenase expression protects against atherosclerosis development. *J. Clin. Invest.*, **98**, 2201.

86. Heinecke, J.W. (1999). Is lipid peroxidation relevant to atherogenesis? *J. Clin. Invest.*, **104**, 135.

4 Antioxidant intervention studies in humans

Anushka Patel
NH&MRC Clinical Trials Centre, University of Sydney, Camperdown, NSW 1450,
Australia
Department of Cardiology, Royal Prince Alfred Hospital, Missenden Road, Camperdown,
Sydney, NSW 2050, Australia

Anthony Keech
NH&MRC Clinical Trials Centre, University of Sydney, Camperdown, NSW 1450,
Australia
Department of Cardiology, Royal Prince Alfred Hospital, Sydney, Australia.
Faculty of Medicine, University of Sydney, Camperdown, NSW 2050, Australia

4.1 Introduction

A great deal of interest has arisen as to the role of antioxidant vitamins in the development of atherosclerosis over the last two to three decades. From initial cross-sectional and case–control studies suggesting a protective effect of certain vitamins against vascular disease, considerable biological data has been produced which supports the concept in principle. Subsequently, much larger observational cohort studies have mostly supported similar findings that one or more of the vitamins said to function as antioxidants in man appear to be associated with less development of clinical atherosclerosis and, particularly, the development of coronary heart disease and stroke. Nevertheless, to date the large-scale randomized controlled trials designed to directly evaluate the benefits of chronic oral vitamin supplementation have been almost universally disappointing. While one smaller trial has reported benefits that were conferred so rapidly as to cast uncertainty on its design quality, the reasons as to why the larger trials of longer duration have failed to demonstrate an advantage of vitamin supplementation remain unclear. The possibilities range from those of too little supplementation to achieve a sufficient change in chronic vitamin level in plasma, tissue or LDL, or of too short a duration of intervention to affect the progression of vascular disease, to that of consideration that perhaps these vitamins may fail to act as antioxidants in the atherosclerotic process. Further possibilities include that supplementation with racemic mixtures, rather than the natural form, of each vitamin is associated with some disadvantage or that combinations of vitamin supplementation are needed for material benefit. Threshold effects, whereby dietary amounts above a certain commonly attained quantity provide maximum benefit, could also apply. As far as is possible, the former scenarios continue to be addressed by randomized controlled trials which remain ongoing. The results of these trials, together with the continuing information arising from careful laboratory experimentation, will help to provide the answers which have become critically important to our current understanding of the mechanisms of vascular disease.

4.2 Epidemiological studies

While there have been some inconsistencies, the weight of the epidemiological evidence favours an inverse association between antioxidant vitamin intake and the incidence of cardiovascular disease. The considerable interest in this topic generated over the last decade has principally arisen from reports of marked benefits with antioxidant supplementation in these studies.

4.2.1 Cross-sectional and correlational studies

An early indication of the possible benefits of antioxidant vitamins arose from correlational and cross-sectional studies demonstrating an association between regional and temporal variations in fruit and vegetable consumption, and the incidence of coronary heart disease.[1, 2] Others have identified a similar inverse relationship between serum antioxidant vitamin levels and coronary disease between regions.[3, 4] Although by design such studies are potentially seriously confounded, the work of Gey,[3] for example, suggested that naturally occurring differences in population-average levels of plasma vitamin E might be associated with more than a two-fold difference in rates of coronary events between countries.

4.2.2 Case–control studies

Case–control studies have also provided evidence of an association between antioxidant vitamin intake and cardiovascular disease. The hospital-based European Study on Antioxidants, Myocardial Infarction and Cancer of the Breast (EURAMIC)[5] found no association between vitamin E levels in adipose tissue and myocardial infarction, but did report a significant inverse relationship with adipose tissue β-carotene levels. Another population-based study[6] compared plasma levels of antioxidant vitamins in angina patients with controls, and found a significant inverse association for vitamin E, vitamin C and β-carotene levels, although the effect of vitamin C was substantially reduced after adjustment for cigarette smoking.

A difficulty with case–control studies that rely on recall of diet or of other confounding factors such as smoking, is potential bias that arises from differential recall between cases and controls. Similarly, in those studies where concurrent biologic measurements are used to assess exposure, it is not possible to determine to what extent exposure (i.e. plasma or tissue vitamin level) was modified by the outcome of interest. A design that overcomes some of these problems is the nested case–control study, where plasma levels are taken at baseline, with subsequent follow-up. A nested case–control study within the Multiple Risk Factor Intervention Trial (MRFIT)[7] failed to find any evidence of an association between serum vitamin E levels taken at baseline and the risk of death due to coronary disease or non-fatal myocardial infarction over the next 20 years. Other similar nested case–control studies using stored serum samples have also not found an association between vitamin A and E levels with subsequent coronary events,[8, 9] although one study with 7–15 years follow-up did find such an association with β-carotene levels.[10] However, an important problem with nested case–control studies that depend upon analysis of plasma that has been stored for several years, is artefactual lowering of measured antioxidant vitamin levels due to degradation over time and possible differential degradation between samples leading to difficulties in interpretation.

4.2.3 Prospective cohort studies

Several large prospective cohort studies evaluating the relationship between antioxidant vitamins and cardiovascular outcomes have been conducted and are outlined below (see also Table 4.1). The Nurses Health Study[11] is a large cohort of 87 245 US female nurses which has been followed since 1980. Based on dietary questionnaire at baseline, the consumption of vitamin E was estimated and compared with cardiovascular outcomes after 8 years of follow-up. Compared to the lowest quintile of vitamin E consumption, the highest quintile had a 34% lower (95% CI, 13–50%) risk of major coronary events following adjustment for a number of potential confounding factors. This high intake quintile was largely comprised of patients taking nutritional supplements of at least 100 IU vitamin E daily, with a significantly lower risk observed among those having had at least 2 years of continuous consumption. No effect was observed within quintiles comprised of those with a high natural dietary intake of vitamin E or users of multivitamin preparations containing only smaller doses of vitamin E.

Similar results were observed in a follow-up study of 39 910 US male health professionals[12] free of documented cardiovascular disease at baseline. Men who consumed at least 100 IU of vitamin E for at least 2 years had an adjusted relative risk for a major coronary event of 0.63 (95% CI, 0.47–0.84), compared with non-users. Once again, support for an inverse association between non-supplemental dietary vitamin E intake and coronary artery disease was weak at best. This study reported an inverse relationship between β-carotene and coronary disease for current and former smokers, but not for those who had never smoked. After adjustment for vitamin E intake, there was no association between vitamin C and coronary disease. In the same cohort of patients, no association was observed between intake of any of the antioxidant vitamins and risk of ischaemic or total stroke.[13]

Table 4.1 Major prospective cohort studies of antioxidant vitamins and cardiovascular disease

Study	Size of cohort (n)	Vitamins evaluated*	Measure of vitamin status	Duration of follow-up[†] (years)	Major outcomes assessed	Associations with major outcomes; RR (highest versus lowest group[‡]; 95% CI)
Nurses Health Study[11]	87 245	E	Dietary questionnaire	Up to 8	Major coronary events	0.66 (0.50–0.87)
Health Professionals Follow-up Study[12]	39 910	E β-c[§] C	Dietary questionnaire	4	Coronary events	Vit E 0.60 (0.44–0.81) β-c 0.71 (0.53–0.86) Vit C 1.25 (0.91–1.71)
Iowa Women's Health Study[15]	34 486	E β-c C	Dietary questionnaire	7	Coronary death	Vit E 0.96(0.62–1.51) β-c 1.03 (0.63–1.70) Vit C 1.49 (0.96–2.30)
NHANES-I[14]	11 348	C	Dietary questionnaire	10	Cardiovascular death	Men – 0.58 (0.41–0.78) Women – 0.75 (0.55–0.99)
Basel Prospective Study[18]	2974	β-c C	Plasma levels	12	Coronary death	β-c 0.65 (0.45–0.93) Vit C 0.80 (0.50–1.30) Both 0.51 (0.29–0.91)
LRC-CPPT[19]	1899	β-c	Plasma levels	13	Coronary events	0.64 (0.44–0.92)

RR: relative risk; CI: confidence interval.

* See text for study design.

† Duration of follow-up as scheduled.

‡ Cardiovascular disease, comparison between the highest and lowest intake or levels groups (e.g. quintiles or quartiles).

§ β-carotene.

A more generalized population-based prospective cohort study using baseline dietary questionnaire was the First National Health and Nutrition Examination Survey (NHANES-I) Epidemiological Follow-up Study cohort,[14] where vitamin C consumption in 11 348 non-institutionalized adults aged 25–74 years was compared with cardiovascular and cancer outcomes after a median of 10 years follow-up. A strong inverse relationship for men and a weak inverse correlation for women was found between vitamin C intake and all-cause standardized mortality ratio. The adjusted relative risk of cardiovascular disease for men with the highest vitamin C intake was 0.58 (95% CI, 0.41–0.78). Other prospective observational studies examining different populations, such as postmenopausal women in the Iowa Women's Health Study,[15] a population with a high incidence of coronary heart disease in Finland,[16] and an elderly population in the Rotterdam Study,[17] have also found inverse relationships between various levels of reported antioxidant vitamin intake and cardiovascular disease.

As with some nested case–control studies, prospective cohort studies using stored plasma samples to measure antioxidant vitamin levels have been conducted. The Basel Prospective Study,[18] comprising a population with high baseline vitamin E levels, reported an inverse association between vitamin C and β-carotene levels and cardiovascular mortality after 12 years follow-up, although this was not significant for vitamin C. Those with simultaneously low baseline levels of both these vitamins had a significantly increased risk of cardiovascular death. The Lipid Research Clinics Coronary Primary Prevention Trial and Follow-up Study[19] (LRC-CPPT) reported a significantly lower risk of non-fatal myocardial infarction or coronary death within the highest quartile of baseline β-carotene level, among hyperlipidaemic men with no history of coronary heart disease.

A number of mechanistic prospective cohort studies have been conducted, examining the relationship between dietary antioxidant vitamin intake and the progression of coronary atherosclerosis and carotid intima–media thickness (CIMT). In the Cholesterol Lowering Atherosclerosis Study (CLAS),[20] a randomized trial of cholesterol lowering with colestipol-niacin versus placebo, subjects reporting spontaneous vitamin E supplementation of greater than 100 IU per day over 2 years had significantly less progression of atherosclerosis based on quantitative coronary angiography. Significantly less CIMT progression was also observed among the group of high vitamin E supplementation, but in this case only among the group randomized to placebo cholesterol-lowering treatment.[21] No such effect was observed in the active drug group, nor for vitamin C intake overall. A similar study conducted on the Atherosclerosis Risk in Communities (ARIC) Study cohort[22] reported a significant inverse relationship between vitamin C intake and CIMT progression only for those aged > 55 years. Again, only in this age group, vitamin E was inversely associated with CIMT, although this was weaker and significant only in women.

In summary, the epidemiological data do provide some evidence for an inverse association between intake of antioxidant vitamins and cardiovascular disease. While consistent data from large, well designed and conducted observational studies are highly encouraging, they can only be considered hypothesis generating as the potential for bias with these designs is significant. The most important limitation of these studies is the likelihood of significant lifestyle differences between those with high antioxidant vitamin intake (especially those who regularly take nutritional supplements) and those with low intake. While attempts are always made to adjust for such potentially confounding factors, it is never possible to fully adjust for these or to adjust at all for the many important but unmeasured and unknown factors. Furthermore, it is not possible to determine from these observational studies whether any noted benefit, if real, is due to the vitamin in question, or whether these are simply markers of some other nutritional factor. The fact that differential effects are seen between the different vitamins does seem to argue against this, but the inability to isolate an exposure of interest always leaves an element of uncertainty. For all these reasons, specific randomized interventional trials remain the only rigorous method for testing the antioxidant vitamin hypothesis.

4.3 Randomized interventional studies of antioxidant vitamin supplementation

A number of small-scale interventional studies of antioxidant vitamin supplementation have been performed, mostly examining intermediate clinical endpoints. In a randomized double-blind placebo-controlled trial of 125 patients presenting within 24 h of a suspected acute myocardial infarction,[23] patients received intravenous vitamin C (1000 mg) and vitamin A (50 000 IU), followed by oral vitamin E (400 mg, equivalent to approximately 440 IU), vitamin C (1000 mg), β-carotene (25 mg) and vitamin A (50 000 IU) daily or matching placebos for 28 days. Based on creatine kinase release (area under the curve) and the 12-lead electrocardiogram (QRS score), antioxidant vitamin use was associated with a significant reduction in infarct size. The Multivitamins and Probucol (MVP) trial[24] was a double-blind placebo-controlled randomized trial of restenosis following elective percutaneous transluminal coronary angioplasty (PTCA). With a 2 × 2 factorial design, 317 patients were randomized to a combination of β-carotene (30 000 IU), vitamin C (500 mg) and vitamin E (700 IU), versus probucol 500 mg daily, versus all four together, versus none. Treatment was received for 4 weeks prior to PTCA, and 6 months subsequently. While probucol had a substantial and significant effect on luminal diameter reduction and restenosis rate at 6 months, the effects of the vitamin combination did not differ from placebo. It was also observed that probucol alone had a greater positive effect on outcome compared with probucol in combination with vitamins, raising the possibility of deleterious pro-oxidant effects of these vitamins at high doses. However, another small study of 100 patients[25] that randomized patients to a much higher dose of vitamin E (1200 IU per day) for 4 months after successful PTCA, did find less restenosis defined by coronary angiogram or thallium scan with vitamin E (34.6% versus 50%), although this was not statistically significant. More recently, a small study of vitamin C supplementation for 4 months after elective PTCA also provided some evidence of reduction in minimal lumen diameter and restenosis rate.[26] The antioxidant effects of these vitamins have also been examined in the context of coronary artery bypass surgery.[27] In a double-blind placebo-controlled randomized study in 76 patients, administration of vitamin E (750 IU) for 7–10 days, and vitamin C (1000 mg) 12 h prior to surgery had no effect on perioperative myocardial injury as assessed by enzyme release and thallium scan.

Observations from epidemiological studies have also raised the possibility that the antioxidant vitamins may have an effect on modifiable 'traditional' risk factors for coronary heart disease. Clearly the observational studies have potential for bias in terms of healthy participants being likely to exhibit a whole range of healthy behaviours. In a double-blind placebo-controlled randomized trial[28] of combined daily β-carotene (6 mg), vitamin C (500 mg) and vitamin E (400 IU), 297 retired men and women were followed up for 2–4 months. At the end of this short period, there was no significant effect of supplementation on blood pressure, fasting lipid levels, and fasting glucose.

While small randomized trials often do provide important mechanistic data, and information on other specific outcomes of interest, only large-scale interventional studies involving several thousand participants can provide reliable data on clinical outcomes of primary interest such as myocardial infarction and death, and on safety.

4.4 Large-scale randomized trials of antioxidant vitamin supplementation

Several large-scale randomized clinical trials of the antioxidant vitamins have been completed, some of which have generated controversial results. The earlier trials were specifically designed to test the hypothesis that the antioxidant vitamins reduce the incidence of certain cancers; however good quality data on cardiovascular outcomes was also collected. More recently, studies specifically designed to examine cardiovascular outcomes have been conducted. The major completed and published randomized antioxidant vitamin trials are described below (see also Table 4.2).

1. The Linxian Study[29] was an open-label randomized trial of 29 584 men and women aged 40–69 years in Linxian county of China comparing the effects of various nutritional supplements using a multiple arm combination treatment design, primarily on the incidence of cancer with follow-up of 5.25 years. Four combinations of supplements were examined: retinol + zinc, riboflavin + niacin, vitamin C (120 mg) + molybdenum, and β-carotene (15 mg) + vitamin E (30 mg) + selenium. Each combination was given with at least one other, and compared with placebo in a complex partial factorial design. The leading cause of death in this cohort was

Table 4.2 Large-scale randomized controlled trials of vitamin supplementation and cardiovascular disease

Study title	Size of study sample (n)	Vitamins evaluated*	Duration of follow-up (years)[†]	Major outcomes assessed	Effects of active treatment on the major outcomes; RR (95% CI)
Linxian[29]	29 584	E β-c[‡] C	5.25	Overall mortality Cerebrovascular death	Vit E + β-c + selenium 0.91 (0.84–0.99) Vit C + molybdenum 1.01 (0.93–1.10) Vit E + β-c + selenium 0.90 (0.76–1.07) Vit C + molybdenum 1.04 (0.88–1.24)
ATBC[30]	29 133	E β-c	6.1 (median)	Overall mortality	Vit E 1.02 (0.95–1.09); β-c 1.08 (1.01–1.16)
CARET[34]	18 314	β-c	4	Overall mortality Cardiovascular death	1.17 (1.03–1.33) 1.26 (0.99–1.61)
Physicians Health Study[35]	22 071	β-c	12	Overall mortality Cardiovascular death	1.02 (0.93–1.11) 1.09 (0.93–1.27)
CHAOS[36]	2002	E	1.4 (median)	Overall mortality Major cardiovascular event	1.18 (0.62–2.27) 0.53 (0.34–0.83)
GISSI-P[37]	11 324	E	3.5	Overall mortality Cardiovascular death	0.92 (0.82–1.04) 0.94 (0.81–1.10)
HOPE[39]	9541	E	4.5	Overall mortality Cardiovascular death	1.0 (0.89–1.13) 1.05 (0.90–1.22)

RR: relative risk; CI: confidence interval.

* see text for study design.

† Average follow-up unless otherwise specified.

‡ β-carotene.

cancer-related, the majority of these being due to gastro-oesophageal cancers. Cerebrovascular disease accounted for 25% of all deaths, while other causes comprised a small proportion individually. It is interesting to note that this group of subjects is known to have blood levels of vitamin E and β-carotene consistently lower than that of Western populations, but with a low incidence of coronary artery disease, highlighting the inadequacies of correlational data. The combination of β-carotene, vitamin E and selenium was associated with a small but significant reduction in overall mortality (RR 0.91, 95% CI, 0.84–0.99), which was essentially a result of a reduction in cancer death. There was no effect on cerebrovascular death, and none of the other supplement combinations was associated with an effect on overall mortality, cancer or cerebrovascular death. Other specific vascular outcomes were not examined, although the very low incidence of coronary heart disease in this population suggests probable inadequate power to investigate such outcomes even with such a large sample size. The doses employed in this study were small in relation to those associated with beneficial effect in the observational studies.

2. The Alpha-Tocopherol, Beta-Carotene (ATBC) Cancer Prevention Study[30] was a double-blind, placebo-controlled randomized trial of 29 133 male smokers aged between 50 and 69 years from Finland. It was designed as a study of the primary prevention of lung cancer, and included patients both with and without a documented history of coronary artery disease. A 2×2 factorial design was utilized: vitamin E alone, vitamin E + β-carotene, β-carotene alone, and placebo. Active treatment consisted of vitamin E 50 mg (approximately equivalent to 55 IU), or β-carotene 20 mg. Follow-up was for 5–8 years with a median 6.1 years. Overall, the study found no effect of vitamin E on the incidence of lung cancer or on overall mortality (RR 1.02, 95% CI, 0.95–1.09, $p = 0.6$). There was a slightly increased incidence of death resulting from haemorrhagic stroke with vitamin E use, but fewer deaths related to ischaemic stroke or ischaemic heart disease. β-Carotene, however, was associated with an 18% increase in lung cancer incidence which achieved statistical significance, but no increase in death due to lung cancer. β-Carotene was also associated with an 8% increase in overall mortality (RR = 1.08, 95% CI, 1.01–1.16, $p = 0.02$).

Subsequent analyses on the ATBC cohort examined cardiovascular endpoints in more detail. Neither vitamin E nor β-carotene had any effect on non-fatal myocardial infarction, cardiovascular death, or the combined endpoint of major coronary events in the 27 272 participants with no prior history of myocardial infarction.[31] Vitamin E was associated with a significant reduction in non-fatal myocardial infarction when administered alone (adjusted RR = 0.62, 95% CI, 0.41–0.96) in 1862 participants with a prior history of myocardial infarction,[32] but not when combined with β-carotene. β-Carotene was associated with an excess of fatal cardiovascular deaths in this subgroup (adjusted RR = 1.75, 95% CI, 1.16–2.64). Those patients with a history of infarction who were assigned both β-carotene and vitamin E also had an excess of cardiovascular deaths (adjusted RR = 1.58, 95% CI, 1.05–2.40). Thus the excess in overall mortality with β-carotene initially reported appears to be a result of an excess in cardiovascular mortality. Finally, 1795 patients with angina at baseline (as determined by the WHO Rose Questionnaire) were examined.[33] Vitamin E and β-carotene had no effect on angina recurrence, progression of severity of angina, and major coronary events among this group of patients.

The ATBC Study was the first major interventional study of antioxidant vitamins, with most unexpected results in view of the weight of the epidemiological evidence. The possible harmful effects of β-carotene were particularly surprising. A major limitation of this study, however, relates to the vitamin E dose employed. A dose of 50 mg, approximately equivalent to 55 IU, is much lower (and resulted in lower plasma vitamin E levels) than that associated with the greatest benefit in the observational cohort studies, and may have been inadequate to test the hypothesis. An additional problem with the study related to the fact that about one-third of the patients receiving β-carotene reported yellowing of the skin, compared to approximately 7% of the placebo group. This potentially may have resulted in some bias arising from unblinding.

3. The β-Carotene and Retinol Efficacy Trial (CARET)[34] was another large-scale study primarily designed to examine the incidence of lung cancer. In this randomized double-blind placebo-controlled trial, 18 314 men and women (who were heavy smokers or recent former smokers, or workers exposed to asbestos) were enrolled in a comparison of β-carotene 30 mg + vitamin A (retinol) 25 000 IU with placebo. The mean length of follow-up was only 4 years, as this study was terminated prematurely following an interim analysis and concurrent publication of the findings from the ATBC Cancer Prevention Study.[30] At study termination, active treatment was associated with an increased incidence of lung cancer (RR 1.28, 95% CI, 1.04–1.57, $p = 0.02$), lung cancer mortality, and all-cause mortality (RR 1.17, 95% CI, 1.03–1.33, $p = 0.02$). There was also an excess of cardiovascular death associated with active treatment, although this just failed to achieve statistical significance (RR 1.26, 95% CI, 0.99–1.61).

The results of CARET, following soon after the publication of the ATBC Trial findings, largely extinguished enthusiasm for supplemental β-carotene as an intervention to reduce the incidence of cancer and cardiovascular disease. Although the mechanism by which β-carotene may cause excess cancer and cancer deaths remains obscure, and this may be a chance finding in a high-risk population, it became apparent that β-carotene supplementation was unlikely to produce any significant benefits.

4. The Physicians Health Study[35] was a double-blind placebo-controlled randomized trial of 22 071 US physicians aged 40–84 years with no history of documented cardiovascular disease. The design was a 2×2 factorial of aspirin and β-carotene. The dose of β-carotene employed was 50 mg on alternate days. The

aspirin arm of the study was terminated prematurely due to a statistically significant 44% reduction in the risk of a first myocardial infarction; however the randomized β-carotene component continued to 12 years follow-up. At study completion, no association was found between β-carotene and incidence of lung cancer or all-malignant neoplasm, cancer death, myocardial infarction, stroke, cardiovascular death, or all-cause mortality. The heterogeneity of effect observed between current and former smokers in the earlier epidemiological data was not apparent.

While providing no evidence of any of the harm suggested by the ATBC and CARET studies, the Physicians Health Study, which had by far the longest follow-up, again failed to produce evidence of any beneficial effects of β-carotene supplementation.

5. The Cambridge Heart Antioxidant Study (CHAOS)[36] was a single-centre randomized double-blind placebo-controlled secondary prevention trial specifically designed to test the effects of vitamin E on cardiovascular outcomes. Two thousand and two patients with angiographically confirmed coronary artery disease participated, the overwhelming majority with angina or demonstrable reversible ischaemia. Approximately half the patients randomized to active treatment received 800 IU throughout the trial; this dose was changed to 400 IU for subsequent recruits based on serum vitamin E levels. Median follow-up was for 510 days (range 3–981), and the primary endpoints were non-fatal myocardial infarction, and the combined endpoint of non-fatal myocardial infarction or death, defined as a major cardiovascular event. Of note, the groups were unbalanced with respect to certain baseline characteristics, including sex ratio, serum total cholesterol, systolic blood pressure, diabetes status, and use of β-blockers. Overall, with adjustment, this study reported a significant reduction in major cardiovascular events (RR 0.53, 95% CI, 0.34–0.83, $p = 0.005$) and in non-fatal myocardial infarction (RR 0.23, 95% CI, 0.11–0.47, $p < 0.001$) with vitamin E use. There was a slight excess of cardiovascular death with vitamin E use; however, this was not significant (RR 1.18, 95% CI, 0.62–2.27, $p = 0.61$).

While this study was the first major interventional study to support the substantial benefits of vitamin E therapy observed in the epidemiological data, there are some important caveats when interpreting these results. Of particular concern is the lack of balance in some of the baseline characteristics reported. While the investigators suggest that these imbalances actually favour placebo, and therefore may lead to an underestimate of treatment effect, it does raise concerns about the randomization process and other potential imbalances in unreported and unmeasured factors. Secondly, the magnitude of effect on non-fatal myocardial infarction and the combined endpoint is highly surprising, given the follow-up period of this study was quite short. If these findings are real, the excess of cardiovascular deaths in the active group, although not statistically significant, is not easily consistent with such a marked reduction in non-fatal myocardial infarction. Inadequate power is unlikely to be the entire explanation for this finding. Of some concern, the methods used to obtain information about further coronary events allowed the possibility to miss events occurring outside the area of the Regional Health Authority, which could introduce under-reporting of events and some potential bias. Finally, the use of two different doses of vitamin E, which are then analysed as a homogeneous group, again raises some issues relating to interpretation and generalization of the findings. Nevertheless, despite these potential limitations, CHAOS was the first large-scale interventional trial to provide any evidence that supplementation of vitamin E in large doses may be of benefit in the secondary prevention of coronary heart disease.

6. The Gruppo Italiano per lo Studio della Sopravvivenza nell'Infarto miocardico – Prevenzione (GISSI-P) Study[37] was a multicentre open-label randomized trial of 11 324 patients with a recent ($\leqslant 3$ months) history of myocardial infarction. The study employed a 2×2 factorial design to examine the effects of vitamin E (300 mg daily, approximately equivalent to 330 IU) and n-3 polyunsaturated fatty acids with 3.5 years follow-up. While fatty acid supplementation was associated with a modest but significant reduction in the combined primary endpoint of death, non-fatal myocardial infarction and stroke, vitamin E use had no such effect on any cardiovascular endpoint or on all-cause mortality.

It is important to note that the GISSI-P Trial was open-label (i.e. not blinded) and this can lead to significant potential bias introduced by either the investigators or the participants. With that proviso, the findings from this secondary prevention trial are inconsistent with that of CHAOS, which also employed a high vitamin E dosage regime. These two studies were conducted on quite different populations – CHAOS in East Anglia in a population with a lower dietary intake of fruit and vegetables, and GISSI-P in Italy where a Mediterranean-style diet predominates. Baseline serum vitamin E levels were not reported in the GISSI-P Study. Another potentially important population difference relates to the recently reported increased prevalence of a polymorphism in the gene for endothelial nitric oxide synthase (eNOS) in East Anglia,[38] leading to speculation that vitamin E may exhibit greater protective effects in such a population.

7. The Heart Outcomes Prevention Evaluation Study (HOPE)[39] was recently published. This randomized double-blind placebo-controlled 2×2 factorial designed study examined the effects of the angiotensin-converting enzyme inhibitor, ramipril, and vitamin E (400 IU per day) on cardiovascular outcomes in 9541 high-risk men and women. The mean follow-up was for 4.5 years and the primary endpoint was a composite of myocardial infarction, stroke and cardiovascular death. Overall, vitamin E supplementation had no effect on this combined endpoint (RR 1.05, 95% CI, 0.95–1.16, $p = 0.33$), nor on overall mortality, incidence of myocardial infarction or stroke.

A number of other large-scale interventional studies of the antioxidant vitamins are ongoing. These are outlined below (see also Table 4.3).

Table 4.3 Some ongoing large-scale randomized controlled trials of vitamin supplementation and cardiovascular disease

Study title	Size of cohort (n)	Vitamins evaluated*	Scheduled duration of follow-up (years)	Major outcomes assessed
Heart Protection Study[40]	20 536	E, C, β-c[†] combined	5	Non-fatal MI plus fatal CHD
SU.VI.MAX[41]	12 735	E, C, β-c combined	8	Non-fatal MI plus fatal CHD
WHS[42]	40 000	E, β-c	⩾10	Non-fatal MI plus fatal CHD
WACS[43]	8000	E, C, β-c	⩾5	Non-fatal MI plus fatal CHD
PHS II[44]	15 000	E, C, β-c	⩾5	Cancer, caridovascular and eye disease

MI: myocardial infarction; CHD: coronary heart disease.

*See text for study design.

†β-Carotene.

4.5 Ongoing clinical interventional trials of antioxidant vitamins

1. The Medical Research Council / British Heart Foundation (MRC/BHF) Heart Protection Study[40] is a large-scale randomized double-blind placebo-controlled trial using a 2×2 factorial design to assess the effects of simvastatin (40 mg) and a combined nutritional supplement of vitamin E (600 mg – approximately equivalent to 660 IU), vitamin C (250 mg) and β-carotene (20 mg) in subjects at high risk of a coronary event. Recruitment of 20 536 men and women into this study was completed between 1994 and 1997, and follow-up is planned for at least 5 years. Primary outcomes are total as well as fatal coronary disease, while secondary endpoints include cause-specific mortality and stroke. The sample size of this study is sufficiently large to examine the effects of intervention within various prespecified subgroups, including any variation in the effects of vitamin supplementation according to baseline lipid levels.

2. The Supplementation en Vitamines et Minéraux AntioXydants (SU.VI.MAX) Study[41] is a randomized double-blind, placebo-controlled, primary prevention trial designed to test the efficacy of a combined supplement of vitamin C (120 mg), vitamin E (30 mg, equivalent to no more than 45 IU), β-carotene (6 mg), selenium (100 µg) and zinc (20 mg) on the

incidence of cancer and cardiovascular events. Recruitment of 12 735 healthy men and women aged 35–60 years was completed in 1994, and follow-up is planned for 8 years.

3. Despite some epidemiological evidence of possible heterogeneity in effect of vitamin supplementation by gender, most of the large-scale randomized trials published to date have included relatively small numbers of women. Both the ATBC Study and the Physicians Health Study did not include women at all. This important question is being currently addressed in two large ongoing studies. The Women's Health Study[42] is a primary intervention study of over 40 000 women designed to assess the risks and benefits of aspirin, β-carotene and vitamin E in the prevention of cardiovascular disease and cancer in healthy women. The Women's Antioxidant and Cardiovascular Study (WACS)[43] is a randomized secondary prevention companion trial to the Women's Health Study. In this study, approximately 8000 female health professionals with a prior history of cardiovascular disease will be assigned vitamin E, vitamin C, β-carotene, and/or placebo in a $2 \times 2 \times 2$ factorial design. The main cardiovascular endpoints will be non-fatal myocardial infarction, non-fatal stroke, coronary revascularization procedures, and total cardiovascular mortality.

4. The design of a Physicians' Health Study II, further evaluating antioxidants in men, has just been announced.[44] This trial, in 15 000 men, aged at least 55 years, is planning to test vitamins, E, C and β-carotene individually and in combination against placebo in a factorial design.

4.6 Summary of the evidence for the effects of vitamin E on cardiovascular disease

Epidemiological data from large prospective cohort studies strongly suggest an association between increased vitamin E intake (sufficient to induce substantial increase in plasma and LDL vitamin levels)[45] and reduced incidence of cardiovascular events. The Linxian Study[29] did find a small, statistically significant reduction in overall mortality with a vitamin E, β-carotene and selenium combination, but this was largely due to a reduction in cancer death. In addition, it is not possible to determine which one or what combination of these factors contributed to this beneficial effect. The dose of vitamin E in this study was small, as it was in the ATBC Study,[30] which failed to demonstrate any convincing evidence of benefit with vitamin E supplementation. The hope that failure of these two large intervention studies to confirm the epidemiological data reflected inadequate dosing has not been supported by the most recently completed large-scale trials – GISSI-P[37] and HOPE.[39] CHAOS[36] is the only interventional trial that has reported quite a marked benefit with vitamin E supplementation. However, the discordance between clinical outcomes (marked significant reduction in non-fatal myocardial infarction, yet small and non-significant increase in cardiovascular death) with effects observed after only a short period of observation, issues related to design and methodology, and the fact that these results have not been reproduced, increases the likelihood that these findings were due to imbalanced randomization or chance.

Overall, therefore, despite a sound theoretical foundation and strong epidemiological evidence, extensive randomized interventional testing over a period of up to 6 years has been unable to provide sufficient evidence to support the use of supplemental vitamin E, in the context of either primary or secondary prevention of coronary heart disease.

4.7 Summary of the evidence for the effects of β-carotene on cardiovascular disease

The epidemiological evidence supporting the benefits of β-carotene in preventing cardiovascular disease is inconsistent. However, the presence of stronger evidence indicating a protective role in cancer prevention led to the first large-scale interventional studies of β-carotene supplementation. As previously mentioned, the Linxian Study[29] did suggest benefit in overall mortality with a combined supplement of β-carotene, vitamin E and selenium, but this trial has proven to be the exception. This may in part reflect marked differences in the populations under study. Both the ATBC Study[30] and the CARET Study[34] found no evidence of any benefit on cancer or cardiovascular outcomes with β-carotene supplementation. The studies in fact suggested possible harm with an increased incidence of lung cancer, and in the case of CARET, cancer deaths and overall mortality as well. While not reproducing any similar evidence of harm, the Physicians Health Study[35] also failed to report any benefit associated with β-carotene supplementation. It seems extremely unlikely that the appropriate interpretation of these results should be that β-carotene (a vitamin) is carcinogenic and over a shorter time course than most known carcinogens. However, while the study of β-carotene intervention is continuing in trials using combined supplementation, current extensive data from interventional studies suggest it is unlikely that long-term β-carotene use has any significant beneficial effects on either the prevention of cancer or cardiovascular disease.

4.8 Summary of the evidence for the effects of vitamin C on cardiovascular disease

Although less extensively studied than vitamin E and β-carotene, prospective data from cohort studies such as NHANES-I[14] and the Iowa Women's Health Study[15] provide some support for an association between vitamin C intake and a reduction in cardiovascular events. Interventional studies, however, have been quite limited. The Linxian Study failed to show benefit with the vitamin C/molybdenum combination on cancer, cerebrovascular or overall mortality. Ongoing trials such as the Heart Protection Study,[40] SU.VI.MAX,[41] WHS[42] and WACS[43] have all included vitamin C supplementation; as this is in combination with vitamin E and β-carotene in two of these trials, in those, it will be difficult to specifically identify benefit attributable to any one factor.

4.9 Non-vitamin antioxidant studies

A number of non-vitamin antioxidants have also been investigated, such as probucol (The Probucol Quantitative Regression Swedish Trial [PQRST][46] and The Multivitamins and Probucol [MVP] Trial,[24] coenzyme Q,[47] and selenium.[29] Of these, only selenium, where the epidemiological evidence is less consistent,[48] has been evaluated in a single large scale clinical trial to date.[29] Various other antioxidants, including plant lycopenes, *N*-acetylcysteine and other synthetic agents are being actively pursued in a variety of cardiovascular disease settings.

4.10 Conclusions

Despite promising observational data, the benefits of antioxidant vitamin supplementation over a period of several years on atherosclerotic vascular disease in humans are doubtful, based on clinical trials that have been completed. Some of the discrepancy between the observational and interventional data may be related to bias associated with epidemiological study design. Other explanations include the presence of an unidentified co-nutrient that may be responsible for the benefit observed in observational studies. Alternately, it may be that antioxidant supplementation in combination is necessary for beneficial effects, a theory which is being investigated further in ongoing trials.

It has been suggested that the antioxidant vitamins may be effective not only in reducing the progression of atherosclerosis, but also on plaque stabilization and thus reduction in the rate of acute events in patients with established coronary disease. While it appears that the latter is not supported by the bulk of the trial data, the former has probably not been adequately tested. All the clinical trials to date enrolled middle-aged or elderly men and women who, whether or not they exhibited signs and symptoms of overt cardiovascular disease, were likely to have significant established atherosclerotic lesions. Follow-up was generally only for 3–8 years, the longest being 12 years in the Physicians Health Study. In most of these trials, participants were at high risk of a cardiovascular event, thus providing a sufficient number of events to detect a difference in acute events. Targeting subjects at a younger age with early atherosclerotic disease may be theoretically more attractive, but would require many years of treatment and follow-up at considerable expense to accumulate sufficient endpoints to enable a valid conclusion. Understanding why chronic oral supplementation with adequate doses of the antioxidant vitamins would so far appear not to protect against cardiovascular disease will eventually help us to better understand the underlying processes of atherosclerosis and find other treatments of value in its prevention.

References

1. Verlangieri, A.J., Kapeghian, J.C., el Dean, S., and Bush, M. (1985). Fruit and vegetable consumption and cardiovascular mortality. *Med. Hypotheses*, **16**, 7.
2. Ginter, E. (1979). Decline of coronary mortality in United States and vitamin C. *Am. J. Clin. Nutr.*, **32**, 511.
3. Gey, K.F., Puska, P., Jordan, P., and Moser, U.K. (1991). Inverse correlation between plasma vitamin E and mortality from ischemic heart disease in cross-cultural epidemiology. *Am. J. Clin. Nutr.*, **53**, 326S.
4. Riemersma, R.A., Oliver, M., Elton, R.A., Alfthan, G., Vartiainen, Salo, M., Rubba, P., Mancini, M., Georgi, H., Vuilleumier, J.P. *et al.* (1990). Plasma antioxidants and coronary heart disease: vitamins C and E, and selenium. *Eur. J. Clin. Nutr.*, **44**, 143.
5. Kardinaal, A.F., Aro, A., Kark, J.D., Riemersma, R.A., van't Veer, P., Gomez-Aracena, J., Kohlmeier, L., Ringstad, J., Martin, B.C., Mazaev, V.P. *et al.* (1995). Association between beta-carotene and acute myocardial infarction depends on polyunsaturated fatty acid status. The EURAMIC Study. European Study on Antioxidants, Myocardial Infarction, and Cancer of the Breast. *Arterioscler. Thromb. Vasc. Biol.*, **15**, 726.
6. Riemersma, R.A., Wood D.A., Macintyre, C.C., Elton, R.A., Gey, K.F., and Oliver, M.F. (1991). Risk of angina pectoris and plasma concentrations of vitamins A, C, and E and carotene. *Lancet*, **337**, 1.
7. Evans, R.W., Shaten, B.J., Day, B.W., and Kuller, L.H. (1998). Prospective association between lipid soluble antioxidants and coronary heart disease in men. The Multiple Risk Factor Intervention Trial. *Am. J. Epidemiology*, **147**, 180.
8. Kok, F.J., de Bruijn, A.M., Vermeeren, R., Hofman, A., van Laar, A., de Bruin, M., Hermus, R.J., and Valkenburg, H.A. (1987). Serum selenium, vitamin antioxidants, and cardiovascular mortality: a 9-year follow-up study in the Netherlands. *Am. J. Clin. Nutr.*, **45**, 462.
9. Salonen, J.T., Salonen, R., Penttila, I., Herranen, J., Jauhiainen, M., Kantola, M., Lappetelainen, R., Maenpaa, P.H., Alfthan, G., and Puska, P. (1985). Serum fatty acids, apolipoproteins, selenium and vitamin antioxidants and the risk of death from coronary artery disease. *Am. J. Cardiol.*, **56**, 226.
10. Street, D.A., Comstock, G.W., Salkeld, R.M., Schuep, W., and Klag, M.J. (1994). Serum antioxidants and myocardial infarction. Are low levels of carotenoids and alpha-tocopherol risk factors for myocardial infarction? *Circulation*, **90**, 1154.
11. Stampfer, M.J., Hennekens, C.H., Manson, J.E., Colditz, G.A., Rosner, B., and Willett, W.C. (1993). Vitamin E consumption and the risk of coronary disease in women. *N. Engl. J. Med.*, **328**, 1444.
12. Rimm, E.B., Stampfer, M.J., Ascherio, A., Giovannucci, E., Colditz, G.A., and Willett, W.C. (1993). Vitamin E consumption and the risk of coronary heart disease in men. *N. Engl. J. Med.*, **328**, 1450.
13. Ascherio, A., Rimm, E.B., Hernan, M.A., Giovannucci, E., Kawachi, I., Stampfer, M.J., and Willett, W.C. (1999). Relation of consumption of vitamin E, vitamin C, and carotenoids to risk for stroke among men in the United States. *Ann. Intern. Med.*, **130**, 963.

14. Enstrom, J.E., Kanim, L.E., and Klein, M.A. (1992). Vitamin C intake and mortality among a sample of the United States population. *Epidemiology*, **3**, 194.

15. Kushi, L.H., Folsom, A.R., Prineas, R.J., Mink, P.J., Wu, Y., and Bostick, R.M. (1996). Dietary antioxidant vitamins and death from coronary heart disease in postmenopausal women. *N. Engl. J. Med.*, **334**, 1156.

16. Knekt, P., Reunanen, A., Jarvinen, R., Seppanen, R., Heliovaara, M., and Aromaa, A. (1994). Antioxidant vitamin intake and coronary mortality in a longitudinal population study. *Am. J. Epidemiology*, **139**, 1180.

17. Klipstein-Grobusch, K., Geleijnse, J.M., den Breeijen, J.H., Boeing, H., Hofman, A., Grobbee, D.E., and Witteman, J.C. (1999). Dietary antioxidants and risk of myocardial infarction in the elderly: the Rotterdam Study. *Am. J. Clin. Nutr.*, **69**, 261.

18. Gey, K.F., Stahelin, H.B., and Eichholzer, M. (1993). Poor plasma status of carotene and vitamin C is associated with higher mortality from ischemic heart disease and stroke: Basel Prospective Study. *Clin. Invest.*, **71**, 3.

19. Morris, D.L., Kritchevsky, S.B., and Davis, C.E. (1994). Serum carotenoids and coronary heart disease. The Lipid Research Clinics Coronary Primary Prevention Trial and Follow-up Study. *JAMA*, **272**, 1439.

20. Hodis, H.N., Mack, W.J., LaBree, L., Cashin-Hemphill, L., Sevanian, A., Johnson, R., and Azen, S.P. (1995). Serial coronary angiographic evidence that antioxidant vitamin intake reduces progression of coronary artery atherosclerosis. *JAMA*, **273**, 1849.

21. Azen, S.P., Qian, D., Mack, W.J., Sevanian, A., Selzer, R.H., Liu, C.R., Liu, C.H., and Hodis, H.N. (1996). Effect of supplementary antioxidant vitamin intake on carotid arterial wall intima–media thickness in a controlled clinical trial of cholesterol lowering. *Circulation*, **94**, 2369.

22. Kritchevsky, S.B., Shimakawa, T., Tell, G.S., Dennis, B., Carpenter, M., Eckfeldt, J.H., Peacher-Ryan, H., and Heiss, G. (1995). Dietary antioxidants and carotid artery wall thickness. The ARIC Study. Atherosclerosis Risk in Communities Study. *Circulation*, **92**, 2142.

23. Singh, R.B., Niaz, M.A., Rastogi, S.S., and Rastogi, S. (1996). Usefulness of antioxidant vitamins in suspected acute myocardial infarction (the Indian experiment of infarct survival-3). *Am. J. Cardiol.*, **77**, 232.

24. Tardif, J.C., Cote, G., Lesperance, J., Bourassa, M., Lambert, J., Doucet, S., Bilodeau, L., Nattel, S., and de Guise, P. (1997). Probucol and multivitamins in the prevention of restenosis after coronary angioplasty. Multivitamins and Probucol Study Group. *N. Engl. J. Med.*, **337**, 365.

25. DeMaio, S.J., King, S.B., 3rd, Lembo, N.J., Roubin, G.S., Hearn, J.A., Bhagavan, H.N., and Sgoutas, D.S. (1992). Vitamin E supplementation, plasma lipids and incidence of restenosis after percutaneous transluminal coronary angioplasty (PTCA). *J. Am. Coll. Nutr.*, **11**, 68.

26. Tomoda, H., Yoshitake, M., Morimoto, K., and Aoki, N. (1996). Possible prevention of postangioplasty restenosis by ascorbic acid. *Am. J. Cardiol.*, **78**, 1284.

27. Westhuyzen, J., Cochrane, A.D., Tesar, P.J., Mau, T., Cross, D.B., Frenneaux, M.P., Khafagi, F.A., and Fleming, S.J. (1997). Effect of preoperative supplementation with alpha-tocopherol and ascorbic acid on myocardial injury in patients undergoing cardiac operations. *J. Thorac. Cardiovasc. Surg.*, **113**, 942.

28. Miller, E.R, III, Appel, L.J., Levander, O.A., and Levine, D.M. (1997). The effect of antioxidant vitamin supplementation on traditional cardiovascular risk factors. *J. Cardiovasc. Risk*, **4**, 19.

29. Blot, W.J., Li, J.Y., Taylor, P.R., Guo, W., Dawsey, S., Wang, G.Q., Yang, C.S., Zheng, S.F., Gail, M., Li, G.Y. *et al.* (1993). Nutrition intervention trials in Linxian, China: supplementation with specific vitamin/mineral combinations, cancer incidence, and disease-specific mortality in the general population. *J. Natl. Cancer Inst.*, **85**, 1483.

30. The Alpha-Tocopherol Beta-Carotene Cancer Prevention Study Group. (1994). The effect of vitamin E and beta carotene on the incidence of lung cancer and other cancers in male smokers. *N. Engl. J. Med.*, **330**, 1029.

31. Virtamo, J., Rapola, J.M., Ripatti, S., Heinonen, O.P., Taylor, P.R., Albanes, D., and Huttunen, J.K. (1998). Effect of vitamin E and beta carotene on the incidence of primary nonfatal myocardial infarction and fatal coronary heart disease. *Arch. Int. Med.*, **158**, 668.

32. Rapola, J.M., Virtamo, J., Ripatti, S., Huttunen, J.K., Albanes, D., Taylor, P.R., and Heinonen, O.P. (1997). Randomised trial of alpha-tocopherol and beta-carotene supplements on incidence of major coronary events in men with previous myocardial infarction. *Lancet*, **349**, 1715.

33. Rapola, J.M., Virtamo, J., Ripatti, S., Haukka, J.K., Huttunen, J.K., Albanes, D., and Heinonen, O.P. (1998). Effects of alpha tocopherol and beta carotene supplements on symptoms, progression, and prognosis of angina pectoris. *Heart*, **79**, 454.

34. Omenn, G.S., Goodman, G.E., Thornquist, M.D., Balmes, J., Cullen, M.R., Glass, A., Keogh, J.P., Meyskens, F.L., Valanis, B., Williams, J.H. *et al.* (1996). Effects of a combination of beta carotene and vitamin A on lung cancer and cardiovascular disease. *N. Engl. J. Med.*, **334**, 1150.

35. Hennekens, C.H., Buring, J.E., Manson, J.E., Stampfer, M., Rosner, B., Cook, N.R., Belanger, C., LaMotte, F., Gaziano, J.M. *et al.* (1996). Lack of effect of long-term supplementation with beta carotene on the incidence of malignant neoplasms and cardiovascular disease. *N. Engl. J. Med.*, **334**, 1145.

36. Stephens, N.G., Parsons, A., Schofield, P.M., Kelly, F., Cheeseman, K., Mitchinson, M.J., and Brown, M.J. (1996). Randomised controlled trial of vitamin E in patients with coronary disease: Cambridge Heart Antioxidant Study (CHAOS). *Lancet*, **347**, 781.

37. GISSI-Prevenzione Investigators. (1999). Dietary supplementation with n-3 polyunsaturated fatty acids and vitamin E after myocardial infarction: results of the GISSI-Prevenzione trial. Gruppo Italiano per lo Studio della Sopravvivenza nell'Infarto miocardico. *Lancet*, **354**, 447.

38. Brown, M. (1999). Do vitamin E and fish oil protect against ischaemic heart disease? *Lancet*, **354**, 441.

39. The Heart Outcomes Prevention Evaluation Study Investigators. (2000). Vitamin E supplementation and cardiovascular events in high-risk patients. *N. Engl. J. Med.*, **342**, 154.

40. MRC/BHF Heart Protection Study Collaborative Group. (1999). MRC/BHF Heart Protection Study of cholesterol-lowering therapy and of antioxidant vitamin supplementation in a wide range of patients at increased risk of coronary heart disease death: early safety and efficacy experience. *Eur. Heart J.*, **20**, 725.

41. Hercberg, S., Preziosi, P., Briancon, S., Galan, P., Triol, I., Malvy, D., Roussel, A.M., and Favier, A. (1998). A primary prevention trial using nutritional doses of antioxidant vitamins and minerals in cardiovascular diseases and cancers in a

general population: the SU.VI.MAX study – design, methods, and participant characteristics. Supplementation en VItamines et Mineraux AntioXydants. *Controlled. Clin. Trials.*, **19**, 336.

42. Buring, J.E. and Hennekens, C.H. (1994). Randomized trials of primary prevention of cardiovascular disease in women. An investigator's view. *Ann. Epidemiol.*, **4**, 111.

43. Manson, J.E., Gaziano, J.M., Spelsberg, A., Ridker, P.M., Cook, N.R., Buring, J.E., Willett, W.C., and Hennekens, C.H. (1995). A secondary prevention trial of antioxidant vitamins and cardiovascular disease in women. Rationale, design, and methods. The WACS Research Group. *Ann. Epidemiol.*, **5**, 261.

44. Christen, W.G., Gaziano, J.M., and Hennekens, C.H. (2000). Design of Physicians' Health Study II – a randomized trial of beta-carotene, vitamins E and C, and multivitamins, in prevention of cancer, cardiovascular disease, and eye disease, and review of results of complete trials. *Ann. Epidemiol.*, **10**, 125.

45. Keech, A., Collins, R., and Peto, R. (1992). Vitamin E supplementation inhibits LDL oxidation in patients at risk of coronary disease taking simvastatin or placebo. *XI Int. Symp Drugs Lip. Met.*, 59.

46. Walldius, G., Erikson, U., Olsson, A.G., Bergstrand, L., Hadell, K., Johansson, J., Kaijser, L., Lassvik, C., Molgaard, J., Nilsson, S., *et al.* (1994). The effect of probucol on femoral atherosclerosis: the Probucol Quantitative Regression Swedish Trial (PQRST). *Am. J. Cardiol.*, **74**, 875.

47. Singh, R.B., Wander, G.S., Rastogi, A., Shuckla, P.K., Mittal, A., Sharma, J.B., Mehrotra, S.K., Kapoor, R., and Chopra, R.K. (1998). Randomized, double-blind placebo-controlled trial of coenzyme Q10 in patients with acute myocardial infarction. *Cardiovasc. Drug Ther.* **12**, 347.

48. Salvini, S., Hennekens, C.H., Morris, S.J., Willett, W.C., and Stampfer, M.J. (1995). *Am. J. Cardiol.*, **76**, 1218.

5 Vascular dysfunction and its contribution to atherogenesis

David S. Celermajer
Clinical Research Group, The Heart Research Institute,
Camperdown, Sydney, NSW 2050, Australia

Joseph A. Vita
Whitaker Cardiovascular Institute, Boston University School of Medicine,
715 Albany street, W507, Boston, Massachusetts, USA

5.1 Introduction

Traditionally, attention in atherosclerosis research and clinical practice has focused on vascular structure – plaque size and composition – with particular regard to the degree of arterial obstruction. In recent years, however, there has been a realization that vessels are dynamic, especially at the sites of plaque. Further narrowing in response to exercise and mental stress, for example, may underlie many episodes of ischemia and even acute coronary syndromes.

Normal vascular function – particularly a preserved dilator response in arteries rather than a maladaptive constriction – has therefore commanded much attention when *in vivo* studies of atherosclerosis have been undertaken. Although regulation of vessel tone is complex, the homeostatic control exerted by the endothelium seems to be of primary importance. Endothelial elaboration of dilators and constrictors may also play a key role in early atherogenic events, such as monocyte adhesion and smooth muscle proliferation.[1]

This chapter will therefore deal with new insights into vascular function, as this pertains to early and late events in the atherogenic process.

5.2 The endothelium as a functional organ

The endothelium, the cell layer that lines the blood vessels, has an enormous range of important homeostatic roles. Normal endothelial functions include control over thrombosis and thrombolysis, platelet and leukocyte interactions with the vessel wall, and regulation of vascular tone and growth. Since healthy endothelial function plays a central role in cardiovascular control, it follows that endothelial dysfunction may contribute to disease states characterized by vasospasm, vasoconstriction, inflammation, leukocyte adhesion, excessive thrombosis, and/or abnormal vascular proliferation, including atherosclerosis and hypertension.[2] Indeed, it has been shown that endothelial dysfunction precedes the early morphological atherosclerotic changes in the arterial walls, and may therefore play a role in the development and growth of atherosclerotic lesions, as well as in the development of ischemia and thrombosis in the late stages of the disease.

After the realization that atherosclerosis is a process that develops over decades and therefore has a long pre-clinical (silent) phase prior to the onset of symptoms, there has been considerable research interest into developing non-invasive diagnostic tools for detecting and monitoring early vascular disease in asymptomatic subjects. As endothelial dysfunction is a key underlying factor in the atherosclerotic process, '*in vivo*' markers of endothelial abnormalities have been sought, particularly those involving disturbed endothelium-dependent vasomotion or released cellular/molecular products.

5.2.1 Normal endothelial function

The vascular endothelium is the one-cell-layer-thick lining of the vessel wall that senses changes in hemodynamic forces, and blood-borne signals. The endothelium responds to physical and chemical stimuli by synthesis and/or release of a variety of vasoactive, thromboregulatory and signal molecules, as well as growth factors. Substances released by the endothelium include nitric oxide (NO, the endothelium-derived relaxing factor (EDRF)), prostacyclin, endothelins, endothelial cell growth factor(s), interleukins, adhesion molecules, and fibrinolytic factors.[2]

The endothelium has anti-coagulant, anti-platelet, and fibrinolytic properties. Endothelial cells are the major sites for anti-coagulant reactions involving thrombin. Platelet adhesion to endothelial cells is markedly inhibited by the endothelium-derived arachidonic acid metabolite, prostacyclin.[3] The same stimuli that activate platelets, such as thrombin, also act to release prostacyclin and tissue-type plasminogen activator from the endothelium, which allows normal endothelium to limit the extent of platelet plug formation.[1]

The endothelium also controls underlying smooth muscle tone in response to certain pharmacological and physiological stimuli. This control process involves a number of membrane receptors, complex intracellular pathways, and synthesis and release of a variety of relaxing and constricting substances. Furchgott and Zawadzki[4] first suggested the existence of an endothelium-derived relaxing factor in 1980, when they observed that rabbit aortic rings relaxed to acetylcholine only in the presence of an intact endothelium. The biological effects of EDRF are mediated by NO (or by metabolite(s) closely related to NO). NO is synthesized from L-arginine by an enzyme, NO synthase.[5] The reaction is stereospecific, and L-arginine is converted to NO and L-citrulline. The generation of NO from L-arginine can also be specifically blocked by arginine analogues, such as N^G-monomethyl-L-arginine (L-NMMA).

NO maintains low arterial tone at rest, in both the systemic and pulmonary circulations.[6, 7] In addition, NO release is stimulated by increased flow (leading to increased shear stress on the endothelium) and by bradykinin, thrombin, acetylcholine, serotonin and a variety of other circulating agents, which increase NO release via activation of specific endothelial cell membrane receptors. The vasodilatory activity of NO is due to interaction with the iron atom of heme in guanylate cyclase, causing its activation and thereby an increase in intracellular c-GMP levels. In smooth muscle cells, the increase in c-GMP activates cGMP-dependent protein kinase, leading to a reduction of intracellular calcium and vasorelaxation. Nitric oxide may also cause smooth muscle relaxation by direct effects on calcium-depend-ent potassium channels.[8] The same pathways are involved in the mechanism of action of exogenous nitrovasodilators, such as sodium nitroprusside and nitroglycerin. In addition, normal endothelial function also plays an important role in vascular growth, leukocyte adhesion, immunological regulation, metabolism of circulating amines, lipoprotein metabolism, and integration and transduction of blood-borne signals.[1]

5.2.2 Endothelial injury

Endothelial damage can result in an imbalance between relaxing and contracting factors, between anti- and pro-coagulant mediators, or growth-inhibiting and -promoting factors. Such dysfunction can result from mechanical or biochemical injury to the endothelium, or even from stimulation of endothelial cells leading to inappropriate or maladaptive physiological functions. Clinically, endothelial dysfunction can manifest (for example) as vasospasm, thrombus formation and possibly atherosclerosis.

In normocholesterolemic animals, it has been shown that physical damage to the endothelium can lead to atherosclerotic-lesion formation.[9] Furthermore, experimentally induced acute hypertension has been shown to disturb endothelial integrity. Hyperhomocysteinemia, on the other hand, is believed to cause endothelial dysfunction by causing chemical injury.[10] Several cardiovascular risk factors (such as high LDL cholesterol, low HDL cholesterol, active and passive cigarette smoking, and diabetes) have been consistently associated with disturbances in normal endothelial physiology, even among young subjects without signs or symptoms of overt atherosclerosis (Fig. 5.1).[1, 11] The various mechanisms whereby these risk factors cause endothelial damage are largely unknown; however a common denominator for all these conditions is increased oxidative stress, which has therefore been suggested as an important cause of endothelial dysfunction.[12] Once the atherosclerotic process has been initiated, there are several cellular processes operating within developing atherosclerotic lesions, such as inflammation and lipoprotein oxidation, which maintain injurious stimuli to the endothelium.

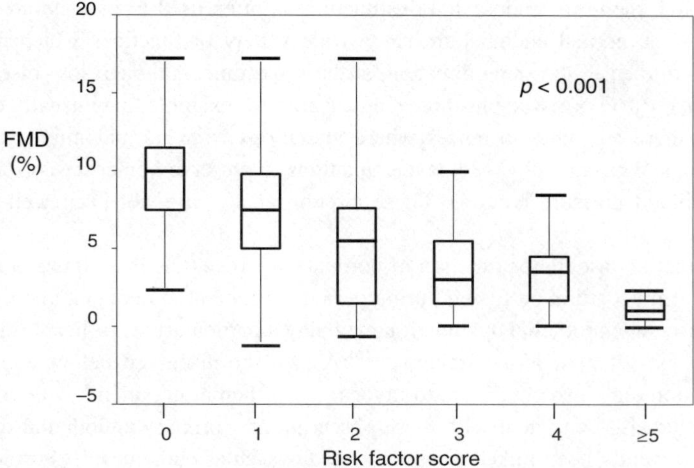

Fig. 5.1 Relationship between risk factor score and flow-mediated endothelium-dependent brachial artery vasodilation (FMD) in 500 healthy asymptomatic subjects (error bars represent the interquartile range). The composite risk factor score was calculated for each subject, with 1 point for each of the following: cholesterol ≥ 240 mg dl^{-1} (≥ 6.2 mmol l^{-1}); current or former smoker; current smoker with a consumption ≥ 10 pack-years; family history; male gender; age ≥ 50 years (after Celermajer et al.[11]).

5.2.3 Consequences of endothelial dysfunction

Early in atherogenesis, there are consequences of endothelial dysfunction that might promote the formation of atherosclerotic lesions, such as fatty streaks and fibrous plaques. These include increased adherence of monocytes (via increased expression of cell adhesion molecules),[13] enhanced permeability to monocyte/macrophages and lipoproteins, which then accumulate in the vessel wall, increased platelet adherence, and increased smooth muscle cell migration and proliferation.

Endothelial dysfunction may also be accompanied by decreased production and/or local bioavailability of NO. In the presence of increased oxidative stress, this decrease may be due to excessive production of superoxide anions, with consequent degradation of NO before it can reach its target tissues. For example, superoxide production is enhanced in the presence of hypercholesterolemia, with consequently decreased bioavailability of NO.[14] As NO acts as a vasodilator, and inhibits platelet adherence and aggregation, smooth muscle proliferation and endothelial cell–leukocyte interactions, decreased NO activity may contribute importantly to the initiation and progression of atherosclerotic lesions.[2] In animal models where endothelial dysfunction (rather than physical damage) has been produced by hypercholesterolemia[15] or hyperhomocysteinemia, for example, impaired NO-dependent vasodilation precedes plaque formation. These observations are consistent with (but do not prove) a role for endothelial dysfunction in atherogenesis.

Although the term 'endothelial dysfunction' is often used to indicate an adverse situation, the normal endothelium has a wide variety of functions which may not all be perturbed in the same direction, at the same time. Although loss of endothelium-derived NO or tissue plasminogen activator, for example, may usually be deleterious, there are other instances where alterations from normal might represent physiological or pathophysiological adaptations to preserve homeostasis (for example, in blood pressure control). These circumstances have not been well studied to date.

Endothelial vasodilator function of coronary microvessels is an important determinant of myocardial perfusion during periods of increased demand, and therefore microvascular endothelial dysfunction may play a particularly significant role in the pathogenesis of myocardial ischemia.[16] As will be discussed below, endothelial dysfunction may also contribute to myocardial ischemia in patients with advanced atherosclerosis. Consistent with these observations, coronary endothelial dysfunction has recently been linked to adverse cardiovascular outcomes.[17] Furthermore, improved endothelial function may explain in large part the reduction in myocardial ischemia and prolonged survival associated with interventions such as cholesterol-lowering therapy.[18]

5.2.4 Qualities of an ideal test for endothelial dysfunction

Ideally, testing for arterial endothelial dysfunction should involve methods that are safe, non-invasive, readily available, economical, reliable and reproducible, which correlate with the extent of sub-clinical atherosclerosis, predict the subsequent risk and respond to therapy. Moreover, since endothelial dysfunction is not a uniform event, it follows that a panel of tests would be required to measure several different aspects of normal anti-atherogenic endothelial functions. Although no such tests are currently available in clinical practice, areas of potential interest for detection of endothelial dysfunction and related early atherosclerotic events include tests of endothelium-dependent vasomotion, as well as circulating markers of endothelial function and/or oxidative stress. The latter is covered in Chapter 20.

5.2.5 Assessment of endothelial function *in vivo*

Over the last decade, a large number of studies have assessed arterial endothelial function in human health and disease. Most of these have tested the ability of normal endothelium to produce NO-dependent vasodilation in response to pharmacological and/or physiological stimuli. Although it is only one of many endothelial functions, NO release may be particularly important because of its actions on platelets, monocytes and smooth muscle cells.[5]

5.2.6 Diagnostic assays for NO

NO is formed in the endothelium from the amino acid precursor L-arginine by the activity of the constitutive endothelial NOS isoenzyme.[5] Decreased availability of biologically active NO in the vascular wall may be one of the earliest detectable findings during atherogenesis. The possible causes for impaired NO responses by the vessel wall include decreased NOS production (decreased NOS gene expression activity or impaired substrate availability) and/or increased catabolism of NO (e.g. by superoxide anions).

5.2.6.1 Assays for measuring NO in plasma and urine samples

NO production can be estimated by measuring the amount of NO directly from plasma samples, or its metabolites nitrite and nitrate, using a chemiluminescence method. Sample storage and technical difficulties with such assays, however, as well as multiple non-vascular sources of nitrates, have made such measurement unsuitable for routine clinical use. Furthermore, these measures are extremely dependent on dietary composition and may change with day-to-day alterations in nutrient intake.[19]

5.2.6.2 Functional methods – NO-dependent vasomotion

Most of the functional methods for *in vivo* endothelial testing examine the ability of endothelium to cause NO-dependent vasodilation in response to the pharmacological and physiological stimuli. Initially, these studies were invasive and involved intra-arterial infusion of vasoactive agents. More recently, non-invasive methods to examine shear stress-mediated NO release have been developed and applied to a variety of patient populations and disease states.

Invasive coronary testing   In vivo assessment of coronary endothelial function in humans was first reported by Ludmer *et al.* in 1986.[20] The coronary artery diameter was measured by quantitative angiography, before and during intracoronary infusion of acetylcholine. In arteries with preserved endothelial function, acetylcholine stimulated the release of NO from endothelial cells, resulting in vasodilation, whereas in subjects with endothelial dysfunction, vasoconstriction was observed (due to a direct smooth muscle constrictor effect of acetylcholine). Soon after, invasive testing of coronary microvascular endothelium was described, by measuring the response of coronary flow to administration of endothelium-dependent (e.g. acetylcholine) and -independent (e.g. nitroprusside) small-vessel dilator substances, using Doppler wires or catheters.[21] These techniques have provided valuable insights into the risk factors for endothelial dysfunction in coronary arteries, and into the role of endothelial dysfunction preceding overt atherosclerosis and in predisposing to plaque activation and constriction.[22] The potential for reversibility of endothelial dysfunction in the coronary arteries by novel therapeutic interventions has also been assessed using this methodology (see below).

Invasive forearm test – plethysmography method   Endothelial function can also be measured in the forearm microcirculation by intra-arterial infusion of endothelium-dependent and -independent vasodilator substances, followed by measurement of forearm flow using venous occlusion plethysmography techniques.[23] Various drugs such as acetylcholine, methacholine, carbachol, bradykinin, serotonin, and substance P have been used to assess endothelium-dependent vasodilation. Direct intra-arterial infusion in the forearm permits the use of drug doses that are 100–1000-fold lower than those required when administered systemically.[24] To take into account any variations in blood flow due to e.g. changes in alertness or any systemic effects of infused drugs, the blood flow in the contralateral non-infused forearm is often measured as a control. Results may then be expressed as the ratio between the experimental and the control forearm blood flow.[24] Commonly utilized protocols involve co-infusion of acetylcholine with an arginine analogue such as L-NMMA, which competitively inhibits NO synthesis. This procedure allows for the determination of the NO-dependent component of acetylcholine stimulated flow. Infusion of L-NMMA alone allows estimation of the fraction of basal blood flow that is NO-dependent. The plethysmography technique has been widely applied in studies investigating the effects of risk factors for peripheral endothelial function[23] and also more recently in studies assessing the reversibility of endothelial dysfunction.

Non-invasive coronary testing – positron emission tomography   The major disadvantage of intra-arterial tests is their invasive nature, which makes them generally unsuitable for routine use in subjects without overt disease and for studies requiring multiple determinations of vascular function. Therefore, non-invasive or minimally invasive clinical tests of endothelial function have been developed.

Positron emission tomography is a unique method that enables non-invasive quantification of myocardial blood flow. By measuring blood flow at rest and after pharmacological stimulation with dipyridamole or adenosine, it is possible to quantitate coronary blood flow reserve (defined as the ratio of maximal flow to basal flow). Recent studies have shown that flow reserve measured using tracers such as ^{15}O-labelled water or [^{13}N]ammonia is reduced in asymptomatic subjects with hypercholesterolemia,[25] suggesting that this impairment coronary reactivity may present an early marker of sub-clinical coronary atherosclerosis and endothelial dysfunction.

Subsequent studies have supported these findings, as coronary flow reserve becomes impaired even in apparently healthy young individuals with risk factors for atherosclerosis.[26] Furthermore, improvement in coronary flow reserve has been demonstrated after low-fat diet and cardiovascular conditioning in patients with hyperlipidemia and after short-term statin therapy in patients with established coronary artery disease.[27]

The exact defect in myocardial vasodilatory function responsible for the blunted response after dipyridamole is unknown, but may occur due to dysfunction of either endothelial cells and/or smooth muscle cells.[28] Dipyridamole increases flow via a direct effect on vascular smooth muscle cells. In addition, secondary release of NO and other vasodilators from the endothelium in response to shear stress produced by the increased flow may elicit more prominent vasodilation in the vessels with preserved endothelial function. The myocardial flow response to dipyridamole may thus be regarded as an integrated measure of coronary reactivity including vascular smooth muscle relaxation and endothelial function, both of which may become disturbed in the early stages of atherosclerosis.

The major disadvantages of the positron emission tomography technique are its high cost and radiation exposure. With the development of less costly and more readily available non-invasive imaging methods, however, the assessment of coronary flow reserve could become a useful tool in primary prevention setting, for identifying high-risk individuals and monitoring the effects of risk factor modification. Potential semi-quantitative methods for the assessment of flow reserve that are being developed include echocardiographic techniques, such as transthoracic imaging of coronary flow velocity reserve[29] and perfusion imaging with myocardial contrast agents.[30]

Non-invasive ultrasound method Endothelial function may also be tested non-invasively in the peripheral conduit arteries, using high-resolution external vascular ultrasound. In this non-invasive method,[11] arterial diameter is measured in response to an increase in shear stress, which causes endothelium-dependent dilation, and in response to sub-lingual nitroglycerin, an endothelium-independent dilator. The brachial arterial dilator response to shear stress has been shown to be due mainly to endothelial release of NO,[31] and to correlate significantly with invasive testing of coronary endothelial function[32] as well as with the extent and severity of coronary atherosclerosis.

Being non-invasive, the ultrasound method has been applied widely to asymptomatic subject groups, including children and young adults. These studies have provided important information of the influence of various risk factors on arterial physiology in childhood and young adult life. As it is accurate and relatively reproducible,[33] serial studies may be performed, allowing trials of reversibility of endothelial dysfunction in asymptomatic subjects at high risk of arterial disease. By using this test, endothelial dysfunction has been demonstrated in asymptomatic children and young adults with risk factors for atherosclerosis, such as hypercholesterolemia and cigarette smoking.[34] Active and passive cigarette smoking has been shown to be associated with impaired endothelium-dependent dilation and this impairment is partially reversible after smoking cessation.[1] Other risk factors associated with endothelial dysfunction using this technique include aging, diabetes mellitus[35] and hyperhomocysteinemia.[36] It has also been shown that the traditional vascular risk factors identified from epidemiologic studies (i.e. high cholesterol, smoking, older age) may *interact* to determine the risk of endothelial dysfunction in asymptomatic subjects, in the same way as they are known to interact in determining the risk of clinical cardiovascular endpoints (Fig. 5.1).[11]

Although factors known to impair endothelial function in the coronary circulation also are associated with impaired flow-mediated dilation in the brachial artery, the direct relevance of this technique to cardiovascular risk remains an open question. Furthermore, it is unclear whether endothelial dysfunction is a generalized process that affects all vascular beds to an equal extent in a given disease state. In experimental models of atherosclerosis, endothelium-dependent responses and mechanisms of endothelial dysfunction may differ according to the vessel studied.[37] However, human studies examining both brachial artery and coronary artery endothelial function in the same individual do suggest parallel responses.[32]

Another disadvantage of this technique is that it is relatively difficult to perform, requiring a skilled sonographer and an appropriate training period. Furthermore, the 'error' of the method probably precludes serial study of endothelial function for individuals. Therefore, it is currently not recommended for use in routine clinical practice.[1] This recommendation may change, however, after future developments in image quality during peripheral scanning and automated analytical methods for measuring arterial diameter.

5.2.7 Circulating markers of endothelial function

Endothelial injury may result in the release of various factors that can be detected in the circulation and potentially used as markers of endothelial dysfunction.

Endothelin-1   Endothelin-1 is an endothelium-derived peptide that has powerful vasoconstrictor and mitogenic properties.[38] Increased levels of endothelin-1 have been demonstrated in conditions associated with endothelial dysfunction, such as atherosclerosis, hypercholesterolemia, and cigarette smoking.[39] Furthermore, oxidized LDL has been shown to stimulate endothelin-1 production and secretion.[40] A recent study by Lerman *et al.*[41] demonstrated increased endothelin-1 immunoreactivity in subjects with coronary artery endothelial dysfunction (assessed by invasive coronary testing with intra-arterial acetylcholine infusions). Most likely, injury to vascular endothelial cells triggers the release of endothelin-1, and therefore it has been suggested that the levels of circulating endothelin in plasma may represent a marker of endothelial dysfunction. As endothelin-1 also has mitogenic properties, it may play a role in the development of premature atherosclerosis.[38] Despite these mechanistic links, endothelin-1 levels in subjects with and without atherosclerosis appear to overlap too much for this measure to be a sensitive or a specific marker of vascular disease, when used in isolation.

Von Willebrand factor   Von Willebrand factor (vWF) is a glycoprotein synthesized mainly by vascular endothelial cells. It has important functions in hemostasis, participating in the coagulation cascade and in the formation of platelet plugs at the sites of endothelial damage.[42] Recent studies have shown that elevated levels of vWF may be predictive of recurrent cardiovascular events in patients with known cardiovascular disease.[43] vWF is also elevated in patients with hypercholesterolemia, and reduction of total cholesterol has been shown to be associated with a reduction in vWF levels. Furthermore, its levels are elevated in situations characterized by vascular damage with denudation of the endothelium, and therefore it has been suggested that 'injured' endothelial cells leak vWF leading to increased plasma levels. More likely, however, increased levels of vWF indicate altered function of the endothelium associated with increased secretion of the protein. Therefore, it has been proposed that the terms 'endothelial activation' or 'stimulation' may be more appropriate in this setting than endothelial damage.[42]

TPA and PAI-1 Tissue plasminogen activator (TPA) is a protein released by endothelial cells. It activates the reaction where plasminogen is converted to plasmin (plasmin is a potent fibrinolytic enzyme that degrades all proteins). TPA regulates the fibrinolytic activity of blood in physiological balance with plasminogen activator inhibitor-1 (PAI-1), another protein product of endothelial cells, which plays a role in the regulation of fibrinolysis, serving as the primary inhibitor of TPA. These circulating endothelial products are elevated in hypercholesterolemia,[44] suggesting the presence of a thrombogenic state in hypercholesterolemia related to inhibition of normal fibrinolysis. Furthermore, prospective studies have shown that high concentrations of TPA antigen in apparently healthy men may be associated with subsequent risk of myocardial infarction and stroke.[45] Data from the ARIC Study[46] have shown that increased levels of TPA and PAI-1 are associated with sub-clinical carotid atherosclerosis, measured as increased intima–media thickness. Although one might expect that elevation of TPA would reflect increased fibrinolytic capacity and decreased cardiovascular risk, studies that distinguish TPA activity from TPA antigen level indicate that cardiovascular disease is associated with a reduction in TPA activity and an increase in PAI-1 activity. These studies have thus demonstrated that abnormal fibrinolytic balance may play a role in the early progression of atherosclerotic lesions and that the activation of the endogenous fibrinolytic system may occur several years before cardiovascular complications arise. It has also been shown that both TPA and PAI-1 production are higher in atheromatous arteries than in normal blood vessels. Therefore, it has been suggested that increased TPA and PAI-1 levels may reflect excessive endothelial cell stimulation by thrombin and/or endothelial damage during atherosclerosis.[46]

Adhesion molecules Binding of circulating leukocytes to the vascular endothelium and further leukocyte migration into the sub-endothelial space are major processes in the development of atherosclerosis. These events are mediated through a diverse family of cellular adhesion molecules that are expressed on the surface of vascular endothelial cells. Among the identified adhesion molecules, the expression and biological properties of vascular cell adhesion molecule-1 (VCAM-1), endothelial–leukocyte adhesion molecule-1 (E-selectin), intercellular adhesion molecule-1 (ICAM-1), and P-selectin are well characterized. Circulating soluble forms of these adhesion molecules have been detected in plasma and are elevated during inflammatory conditions, and can be measured using commercial immunoassays. The role of these molecules in atherogenesis is discussed in detail in Chapter 9.

Several studies have suggested that circulating adhesion molecules may serve as markers of endothelial damage and/or atherosclerosis. In an analysis by Hwang et al.[47] from the ARIC dataset, E-selectin and ICAM-1 were associated with the carotid artery intima–media thickness and were significant independent predictors of incident coronary artery disease. This study also showed higher levels of ICAM-1 in smokers and a correlation between ICAM-1 and pack-years of cigarettes. Squadrito et al.[48] observed higher levels of circulating ICAM-1 and E-selectin in patients with acute myocardial infarction compared to patients with stable angina and healthy controls. Increased levels of plasma P-selectin have been observed in unstable angina, myocardial infarction, coronary spasm, and hypercholesterolemia.[49] Blann and McCollum[50] studied patients with ischemic heart disease and peripheral vascular disease, and also found elevated levels of ICAM-1 in patients compared to controls. Furthermore, Ridker et al.[51] have also demonstrated that serumsoluble ICAM-1 levels may predict cardiovascular event rates in apparently healthy men (Fig. 5.2). The diagnostic utility of such plasma markers, therefore, is an area of continuing interest.

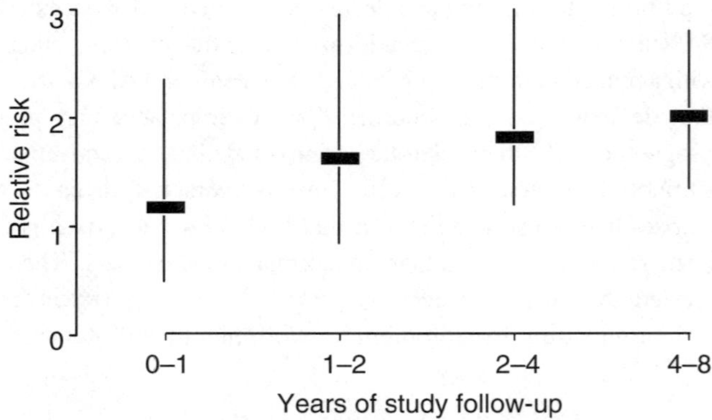

Fig. 5.2 Relative risks (95% confidence intervals) for first myocardial infarction associated with baseline concentrations of soluble ICAM-1 of more than 260 ng dl^{-1}, according to years of study follow-up. The risk of myocardial infarction increases with the length of follow-up, being significant after 2 years ($p < 0.001$), suggesting a role for ICAM-1 in early atherogenesis (after Ridker et al.[92]).

5.2.7.1 Endothelial studies in late disease

Atherosclerosis is a chronic disease that progresses over decades from fatty streaks to complex atherosclerotic plaques. As lesions develop and obstruct the arterial lumen, patients may develop exertional angina pectoris and/or have asymptomatic myocardial ischemia that is detectable by ambulatory ECG monitoring. Alternatively, the first manifestation of coronary disease may be the development of unstable angina or acute myocardial infarction. These latter conditions typically occur following erosion or rupture and thrombosis of atherosclerotic plaques, which may or may not have been previously symptomatic.

As discussed above, studies in experimental models and human subjects have made it clear that endothelial dysfunction is associated with coronary risk factors, precedes lesion development, and contributes to atherogenesis. In addition to this well-defined role in early disease, human studies suggest that endothelial dysfunction also contributes to the later symptomatic phases of coronary artery disease.

Stable coronary syndromes   In regard to stable angina, several lines of evidence suggest that vascular dysfunction plays a pathophysiological role. It is a common clinical observation that the severity of exertional angina may vary day to day for individual patients and that the level of physical activity required to produce symptoms may change over relatively short periods of time. This behaviour suggests that the degree of coronary obstruction may be dynamic rather than fixed in many patients.

Several studies provided direct evidence to support this concept. Bicycle exercise induces constriction of atherosclerotic coronary arteries, while angiographically normal coronary arteries display a vasodilator response to this common stimulus for angina.[22] Coronary segments with vasoconstrictor responses to exercise also constrict during intracoronary acetylcholine infusion, but maintain a normal dilator response to nitroglycerin, suggesting endothelial vasomotor dysfunction.[20] A similar relationship between endothelial dysfunction and pathological constriction of atherosclerotic coronary arteries has been observed for other clinically relevant stimuli, including the cold pressor test, tachycardia, and emotional stress.[52]

In addition to increasing myocardial oxygen demand, exercise and other stimuli for angina have complex effects that may influence vasomotor tone. These effects include increased coronary blood flow, activation of the sympathetic nervous system, and increased circulating catecholamines. In atherosclerotic coronary arteries, endothelial dysfunction may lead to a loss of the normal vasodilator response to increased flow[53] and worsening of the vasoconstrictor response to catecholamines due to a loss of normal nitric oxide release in response to α_2-adrenergic receptor stimulation.[54] These responses may account for the observed coronary constriction and contribute to myocardial ischemia in the setting of a partial coronary obstruction and increased demand.

Acute coronary syndromes Endothelial dysfunction may also contribute to acute coronary syndromes where plaque rupture, intraluminal thrombosis, and vaso-constriction have been implicated. Accumulation of leukocytes in plaque is associated with increased vulnerability to rupture[55] and endothelial activation with expres-sion of adhesion molecules and chemotactic factors and loss of nitric oxide may contribute to this inflammatory process. Loss of endothelium-dependent vasodilation in response to flow may increase local shear stress at the arterial wall,[56] further increasing the risk for plaque rupture. Increased expression of PAI-1, which inhibits endogenous fibrinolysis, and loss of the anti-platelet effects of nitric oxide and prostacyclin may exacerbate the thrombotic response to rupture. Substances produced by aggregating platelets and the clotting cascade, including serotonin and thrombin, have potent vasoconstrictor and pro-thrombotic effects. These products also stimulate release of nitric oxide from the endothelium, and loss of this homeostatic mechanism may exacerbate the adverse effects of these compounds.

The relevance of endothelial dysfunction to acute coronary syndromes is supported by several human studies. Serotonin has been shown to dilate angiographically normal coronary arteries, but produces constriction in patients with coronary athero-sclerosis.[57] Patients with unstable angina display more severe impairments of vascular function compared to patients with stable angina.[58] Furthermore, vaso-constrictor responses to acetylcholine are more severe in infarct-related vessels compared to non-infarct-related vessels with matched stenosis severity.[59] It remains possible that this impairment of endothelial function in culprit arteries is a consequence, rather than a cause, of acute coronary syndromes. However, a preliminary report suggests that the presence of endothelial dysfunction in the coronary circulation predicts increased risk for coronary artery disease events over the next seven years.[17] It remains unknown, however, whether the presence of vascular dysfunction in peripheral arteries provides similar prognostic information for coronary events.

5.2.7.2 Studies of reversibility of endothelial dysfunction

Although endothelial vasomotor function is impaired in atherosclerotic coronary arteries compared to normal subjects, endothelial dysfunction in this setting is not inevitable. A substantial fraction of patients with coronary disease display normal vasodilator responses to endothelium-dependent agonists in some or all coronary segments.[60] This finding raises the possibility that endothelial dysfunction in the coronary circulation may be reversible, despite the continued presence of athero-sclerotic lesions. Improvement in endothelial function has the potential to improve symptoms of angina and lessen the risk for acute coronary events, even if there is little or no change in the severity of atherosclerotic lesions. In fact, a number of acute and chronic interventions have been shown to improve endothelial function in patients with coronary atherosclerosis. These studies provide insight into the causes of endothelial dysfunction and new approaches to the management of coronary artery disease.

Cholesterol-lowering therapy   There is reasonably convincing evidence that cholesterol-lowering therapy improves endothelial dysfunction in both the coronary circulation and peripheral arteries of patients with hypercholesterolemia.[61] Although the initial coronary studies were non-randomized or poorly controlled, Treasure *et al.*[61] demonstrated a beneficial effect with 6 months of treatment in a randomized multicentre study. Furthermore, additional studies are relatively uniform in demonstrating beneficial effects in the conduit artery and resistance vessels in the forearm. In contrast, randomized studies[62] suggest that the beneficial effects of shorter term cholesterol-lowering therapy (6–12 months) are less marked or absent in the coronary circulation of patients with relatively normal cholesterol levels at baseline (see Fig. 5.3). Further studies are needed to determine which patients benefit and what treatment regimens are most effective in this regard.

The mechanism of benefit of cholesterol-lowering therapy on endothelial function is somewhat controversial. Experimental studies suggest that simply reducing serum cholesterol is sufficient to produce a benefit.[15] In humans, marked cholesterol lowering with LDL apheresis has an immediate beneficial effect.[63] However, investigators have also postulated direct effects of HMG-CoA reductase inhibitors to increase endothelial levels of nitric oxide synthase.

Converting enzyme inhibitors   Another intervention shown to improve coronary artery endothelial function in patients with coronary artery disease is angiotensin-converting enzyme (ACE) inhibition. The TREND study was a randomized, placebo-controlled double-blind study that examined the effect of the ACE inhibitor

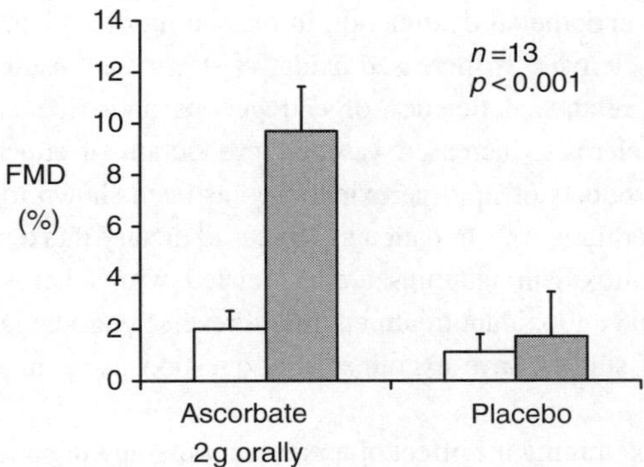

Fig. 5.3   Acute administration of ascorbic acid improves brachial artery flow-mediated dilation in patients with coronary artery disease (after Levine *et al.*, circulation 1996; 93:1107–13).

quinapril on coronary endothelial function in 129 normotensive patients with coronary disease.[64] Quinapril treatment for 6 months was associated with conversion of acetylcholine-induced constriction to a vasodilator response, suggesting improved endothelial function. Active treatment also reduced blood pressure, leaving open the question of whether another vasodilator would have had a similar effect. Supporting a specific effect, 8 weeks of treatment with quinapril improved brachial artery flow-mediated dilation in patients with coronary artery disease, while there was no improvement in patients randomized to treatment with the calcium channel antagonist amlodipine.[65] Enalapril, another ACE inhibitor, had no effect in that study, and the authors suggest that increased tissue specificity of quinapril accounts for this difference. Beneficial effects of acute treatment with ACE inhibitors on endothelium-dependent vasodilation have also been demonstrated in patients with hypertension and congestive heart failure,[66] although longer term studies with ACE inhibition have not uniformly shown a beneficial effect on endothelial function.[67]

Several possible mechanisms could explain the beneficial effects of ACE inhibition on endothelium-dependent vasodilation. Studies in a rat model of hypertension suggest that angiotensin II increases production of superoxide anion and induces endothelial dysfunction by stimulating expression of NADH/NADPH oxidase.[68] Inhibition of this process may prevent superoxide-mediated 'inactivation' of nitric oxide. In addition to converting angiotensin I to angiotensin II, ACE also plays a role in the metabolism of bradykinin. Since bradykinin stimulates endothelial production of nitric oxide, ACE inhibition may improve endothelial function by preventing this effect. In support of this mechanism, a selective bradykinin receptor antagonist has been shown to block the beneficial effect of ACE inhibition on brachial artery flow-mediated dilation.[69] A comparative study demonstrated that the angiotensin II receptor antagonist losartan does not improve brachial artery flow-mediated dilation, further implicating a non-angiotensin II dependent mechanism.[65]

Antioxidants   There is strong evidence that increased oxidative stress in the vasculature contributes to endothelial dysfunction in the setting of coronary risk factors and established atherosclerosis.[70] Increased oxidative stress may result from an excess of oxidants and/or a relative deficiency of endogenous antioxidants. In experimental models of atherosclerosis, increased vascular production of superoxide anion (and accumulation of products of lipid peroxidation) has been shown to impair endothelium-dependent vasodilation.[14] In patients, decreased dietary intake and low plasma or tissue levels of antioxidant vitamins are associated with atherosclerosis.[70] These findings suggest that antioxidant treatment might reverse endothelial dysfunction, and a large number of studies have examined this question using natural and synthetic antioxidants.

The first study examining the effect of antioxidant therapy in patients with coronary disease used the synthetic antioxidant probucol.[62] In that study, the combination of lovastatin and probucol significantly improved the coronary artery response to acetylcholine compared to lipid-lowering therapy alone or diet. This beneficial effect correlated with antioxidant protection of low-density lipoprotein. Other studies with lipid-soluble antioxidants, including α-tocopherol, have yielded mixed results. Most studies examining the effects of endothelial function in forearm microvessels of patients with risk factors or atherosclerosis were negative,[71] although one small study in hypercholesterolemic patients demonstrated a positive effect.[72] More positive results have been reported in studies that examined brachial artery flow-mediated dilation, particularly in patients with multiple risk factors, including cigarette smoking.[65] Patients with fewer risk factors seem to derive no benefit from α-tocopherol.

A more consistent beneficial effect on endothelial vasodilator function has been observed with ascorbic acid. Acute administration of physiological doses of ascorbic acid improves brachial artery flow-mediated dilation in patients with coronary artery disease, cigarette smoking, and congestive heart failure (fig. 5.3).[66] Acute ascorbic acid treatment also improves endothelium-dependent dilation in the coronary circulation of patients with hypertension and vasospastic angina. Direct intra-arterial infusion of ascorbic acid improves forearm microvascular endothelial function in patients with diabetes mellitus, hypercholesterolemia, cigarette smoking, and hypertension.[73] Chronic studies have yielded more mixed results. Chronic treatment has no effect on forearm microvascular responses of hypercholesterolemic patients or brachial flow-mediated dilation of smokers.[74] However, chronic treatment has been shown to improve flow-mediated dilation in patients with coronary artery disease and congestive heart failure.[75]

Regarding the mechanism of improvement with ascorbic acid treatment, studies involving arterial infusion produce super-physiological levels of ascorbic acid, that may preserve nitric oxide by scavenging superoxide anions. This mechanism is unlikely to be important in those studies involving oral administration and lower plasma ascorbic acid levels.[76] Recent work suggests that physiological concentrations of ascorbic acid may alter the endothelial redox state and directly increase eNOS activity.[77] Other water-soluble antioxidants that alter the cellular redox state have been shown to improve endothelial function, including glutathione and drugs that augment glutathione synthesis.[78] Ascorbic acid may also act by preventing oxidation of essential cofactors for nitric oxide synthase activity, including tetrahydrobiopterin, which has been shown to improve endothelium-dependent dilation in patients with hypercholesterolemia.[79]

Therefore, the effects of antioxidants on endothelial function have been somewhat variable to date. It is possible, furthermore, that beneficial effects (where observed) may depend not on antioxidation of lipids or other radical-dependent processes, but possibly on other factors, such as regulation of transcription.

5.2.7.3 Other interventions

Several other interventions have been shown to improve endothelial function. For example, acute and chronic estrogen treatment improves NO-dependent dilation in the coronary and peripheral arteries.[80] Investigators have postulated that a protective effect of estrogen on the endothelium may explain the markedly reduced incidence of coronary disease in premenopausal women, although the beneficial effects of hormone replacement therapy are less clear.[81]

A number of studies have shown that L-arginine, the substrate for nitric oxide synthesis, improves endothelial function in humans. Acute administration of L-arginine improves endothelium-dependent dilation in the coronary circulation of patients with atherosclerosis and in forearm microvessels of patients with hypercholesterolemia.[82] Chronic treatment also improves brachial artery flow-mediated dilation.[83] L-Arginine also has beneficial effects on other endothelial functions, including the ability to inhibit monocyte adhesion.[84] These beneficial effects of L-arginine have been termed paradoxical, because physiological L-arginine levels far exceed the known K_m of nitric oxide synthase. Recent studies have suggested an endogenous inhibitor of nitric oxide synthase that is elevated in disease states associated with endothelial dysfunction (asymmetric dimethyl arginine) may produce a relative deficiency of L-arginine that may be overcome by supplementation.[85] However, the precise mechanisms for the beneficial effects of L-arginine remain unknown.

5.2.7.4 Clinical implications

It is interesting to note that many of the therapies shown to improve endothelial vasodilator function have been proven to reduce cardiovascular risk. For example, there is unequivocal evidence that cholesterol-lowering therapy[86] and ACE inhibitor therapy[87] reduce cardiovascular disease risk in patients with vascular disease. Although the results for antioxidant supplementation are more controversial, there is also strong evidence that dietary antioxidants have a protective effect on cardiovascular disease risk.[88] Improved endothelial function following the reduction of other risk factors has also been observed, including cigarette smoking, sedentary lifestyle, and hyperhomocysteinemia.[89]

Many of these therapies exert their beneficial effect over a relatively short period of time and do not necessarily induce significant reduction of atherosclerosis. Thus, it seems likely that an improvement in the vasodilator, anti-thrombotic, and anti-inflammatory properties of the endothelium may explain, in part, the beneficial effects. A further implication of these observations is the possibility that examination of the effect of a new therapy on endothelial dysfunction may guide the development of new approaches to the treatment of cardiovascular disease.

5.2.7.5 Smooth muscle cells

Much of the foregoing discussion of the endothelium and its relevance to vascular function/atherogenesis reflects the enormous recent interest in this area of vascular biology research. Nevertheless, other cells (smooth muscle cells and fibroblasts) and vascular components (e.g. extracellular matrix) clearly also play a role in the determination of vascular function.

Dynamic smooth muscle physiology is only now becoming a focus of '*in vivo*' research. Preliminary data from our group[90] and others suggest that impaired smooth muscle dilator responses, independent of endothelial function, may also be an important event in atherogenesis. Experimental studies suggest that fibroblasts in the adventitia are a major source of superoxide anion[91] and thus may also be important determinants of vascular function. These areas will be subject to detailed study in the coming years.

5.3 Future directions

The understanding that vascular function (as well as structure) plays a determining role in the risk of atherosclerotic events has forced a paradigm shift in the research and clinical approaches to diagnosis and treatment of arterial disease. The next challenges in the areas of vascular function and control include (i) the establishment of prognostic/outcomes information in healthy subjects with arterial dysfunction, (ii) the development of user-friendly clinical tests of arterial function, and (iii) the clearer definition of interventions that improve vessel function and their links to clinical outcomes. Such research endeavours are already underway, and are likely to lead to new insights into atherosclerosis and its complications.

References

1. Celermajer, D.S. (1997). Endothelial dysfunction: does it matter? Is it reversible? *J. Am. Coll. Cardiol.*, **30**, 325–33.
2. Ross, R. (1993). The pathogenesis of atherosclerosis: a perspective for the 1990s. *Nature*, **362**, 801–9.
3. Moncada, S. (1982). Biological importance of prostacyclin. *Br. J. Pharmacol.*, **76**, 3–31.
4. Furchgott, R.F. and Zawadzki, J.V. (1980). The obligatory role of endothelial cells in the relaxation of arterial smooth muscle by acetylcholine. *Nature*, **288**, 373–6.
5. Moncada, S. and Higgs, E.A. (1993). The L-arginine-nitric oxide pathway. *N. Engl. J. Med.*, **329**, 2002–12.
6. Vallance, P., Collier, J., and Moncada, S. (1989). Effects of endothelium-derived nitric oxide on peripheral arteriolar tone in man. *Lancet*, **2**, 997–1000.
7. Celermajer, D.S., Dollery, C., Burch, M., and Deanfield, J.E. (1994). Role of endothelium in the maintenance of low pulmonary vascular tone in normal children. *Circulation*, **89**, 2041–4.
8. Bolotina, V., Najibi, S., Palacino, J.J., Pagano, P.J., and Cohen, R.A. (1994). Nitric oxide directly activates calcium-dependent potassium channels in vascular smooth muscle. *Nature*, **368**, 850–3.
9. Moore, S. (1973). Thromboatherosclerosis in normolipemic rabbits; a result of continued endothelial damage. *Lab. Invest.*, **29**, 478–87.

10. Olofssen, B.O., Dahlen, G., and Nilsson, T.K. (1989). Evidence for increased levels of plasminogen activator inhibitor and tissue plasminogen activator in plasma of patients with angigraphically verified coronary artery disease. *Eur. Heart. J.*, **10**, 77–82.

11. Celermajer, D.S., Sorensen, K.E., Bull, C., Robinson, J., and Deanfield, J.E. (1994). Endothelium-dependent dilation in the systemic arteries of asymptomatic subjects relates to coronary risk factors and their interaction. *J. Am. Coll. Cardiol.*, **24**, 1468–74.

12. Keaney, J.F. Jr and Vita, J.A. (1995). Atherosclerosis, oxidative stress, and antioxidant protection in endothelium-derived relaxing factor action. *Prog. Cardiovasc. Dis.*, **38**, 129–54.

13. Steinberg, D.S. (1987). Lipoproteins and the pathogenesis of atherosclerosis. *Circulation*, **76**, 508–14.

14. Ohara, Y., Peterson, T.E., and Harrison, D.G. (1993). Hypercholesterolemia increases endothelial superoxide anion production. *J. Clin. Invest.*, **91**, 2546–51.

15. Harrison, D.G., Armstrong, M.L., Frieman, P.C., and Heistad, D.D. (1987). Restoration of endothelium-dependent relaxation by dietary treatment of atherosclerosis. *J. Clin. Invest.*, **80**, 1808–11.

16. Zeiher, A.M., Krause, T., Schachinger, V., Minners, J., and Moser, E. (1995). Impaired endothelium-dependent vasodilation of coronary resistance vessels is associated with exercise-induced myocardial ischemia. *Circulation*, **91**, 2345–52.

17. Volker, S., Britten, M.B., Zeiher, A.M., and Goethe, J.W. (1999). Impaired epicardial coronary vasoreactivity predicts for adverse cardiovascular events during long-term follow-up. *Circulation*, **100**, I-54 (Abstract).

18. Andrews, T.C., Raby, K., Barry, J., Naimi, C.L., Alfred, E., Ganz, P., and Selwyn, A.P. (1997). Effect of cholesterol reduction on myocardial ischemia in patients with coronary disease. *Circulation*, **95**, 324–8.

19. Wang, J., Brown, M.A., Tam, S.H., Chan, M.C., and Whitworth, J.A. (1997). Effects of diet on measurements of nitric oxide metabolites. *Clin. Exp. Pharmacol. Physiol.*, **24**, 418–20.

20. Ludmer, P.L., Selwyn, A.P., Shook, T.L., Wayne, R.R., Mudge, G.H., Alexander, R.W., and Ganz, P. (1986). Paradoxical vasoconstriction induced by acetylcholine in atherosclerotic coronary arteries. *N. Engl. J. Med.*, **315**, 1046–51.

21. Drexler, H. and Zeiher, A.M. (1991). Endothelial function in human coronary arteries *in vivo*. *Hypertension*, **18**, 90–9.

22. Gordon, J.B., Ganz, P., Nabel, E.G., Fish, R.D., Zebede, J., Mudge, G.H., Alexander, R.W., and Selwyn, A.P. (1989). Atherosclerosis influences the vasomotor response of epicardial coronary arteries to exercise. *J. Clin. Invest.*, **83**, 1946–52.

23. Creager, M.A., Cooke, J.P., Mendelsohn, M.E., Gallagher, S.J., Coleman, S.M., Loscalzo, J., and Dzau, V.J. (1990). Impaired vasodilation of forearm resistance vessels in hypercholesterolemic humans. *J. Clin. Invest.*, **86**, 228–34.

24. Webb, D.J. (1995). The pharmacology of human blood vessels *in vivo*. *J. Vasc. Res.*, **32**, 2.

25. Yokoyama, I., Ohtake, T., Momomura, S., Nishikawa, J., Sasaki, Y., and Omata, M. (1996). Reduced coronary flow reserve in hypercholesterolemic patients without overt coronary stenosis. *Circulation*, **94**, 3232–8.

26. Pitkänen, O.-P., Nuutila, P., Raitakari, O. T., Porkka, K.V.K., Iida, H., Nuotio, I., Rönnemaa, T., Viikari, J.S.A., Taskinen, M.-R., Enholm, C., and Knuuti, J. (1998).

Coronary flow reserve in young men with familial combined hypercholesterolemia. *Circulation*, **99**, 1678–84.

27. Huggins, G.S., Pasternak, R.C., Alpert, N.M., Fischman, A.J., and Gewirtz, H. (1998). Effect of short term treatment of hyperlipidemia on coronary vasodilator function and myocardial perfusion in regions having substantial impairment of baseline dilator reserve. *Circulation*, **98**, 1291–6.

28. Leipert, B., Becker, B.F., and Gerlach, E. (1992). Different endothelial mechanisms involved in coronary responses to known vasodilators. *Am. J. Physiol.*, **262**, H1676–83.

29. Hozumi, T., Yoshida, K., Ogata, Y., Akasaka, T., Asami, Y., Takagi, T., and Morioka, S. (1998). Noninvasive assessment of significant left anterior descending coronary artery stenosis by coronary flow velocity reserve with transthoracic color Doppler. *Circulation*, **97**, 1557–62.

30. Broillet, A., Puginier, J., Ventrone, R., and Schneider, M. (1998). Assessment of myocardial perfusion by intermittent harmonic power Doppler using SonoVue, a new ultrasound contrast agent. *Invest. Radiol.*, **33**, 209–15.

31. Joannides, R., Haefeli, W.E., Linder, L., Richard, V., Bakkali, E.H., Thuillez, C., and Luscher, T.F. (1995). Nitric oxide is responsible for flow-dependent dilatation of human peripheral conduit arteries *in vivo*. *Circulation*, **91**, 1314–19.

32. Anderson, T.J., Uehata, A., Gerhard, M.D., Meredith, I.T., Knab, S., Delagrange, D., Lieberman, E.H., Ganz, P., Creager, M.A., Yeung, A.C., and Selwyn, A.P. (1995). Close relationship of endothelial function in the human coronary and peripheral circulations. *J. Am. Coll. Cardiol.*, **26**, 1235–41.

33. Sorensen, K.E., Celermajer, D.S., Spiegelhalter, D.J., Georgakopoulos, D., Robinson, J., Thomas, O., and Deanfield, J.E. (1995). Non-invasive measurement of endothelium-dependent arterial responses in man: accuracy and reproducibility. *Br. Heart. J.*, **74**, 247–53.

34. Celermajer, D.S., Sorensen, K.E., Gooch, V.M., Spiegelhalter, D.J., Miller, O.I., Sullivan, I.D., Lloyd, J.K., and Deanfield, J.E. (1992). Non-invasive detection of endothelial dysfunction in children and adults at risk of atherosclerosis. *Lancet*, **340**, 1111–15.

35. Clarkson, P., Celermajer, D.S., Donald, A.E., Sampson, M., Sorensen, K., Adams, M.R., Yue, D.K., Betteridge, D.J., and Deanfield, J.E. (1996). Impaired vascular reactivity in insulin-dependent diabetes mellitus is related to disease duration and LDL-cholesterol levels. *J. Am. Coll. Cardiol.*, **28**, 573–9.

36. Woo, K.S., Chook, P., Lolin, Y.I., Cheung, A.S.P., Chan, L.T., Sun, Y.Y., Sanderson, J.E., Metreweli, C., and Celermajer, D.S. (1997). Hyperhomocysteinemia is a risk factor for arterial dysfunction in humans. *Circulation*, **96**, 2642–4.

37. Cowan, C.L., Palacino, J.J., Najibi, S., and Cohen, R.A. (1993). Potassium channel-mediated relaxation to acetylcholin in rabbit arteries. *J. Phamacol. Exp. Ther.*, **266**, 1482–9.

38. Mathew, V., Hasdai, D., and Lerman, A. (1996). The role of endothelin in coronary atherosclerosis. *Mayo. Clin. Proc.*, **71**, 769–77.

39. Lerman, A., Edwards, B.S., Hallett, J.W., Heublein, D.M., Sandberg, S.M., and Burnett, J.C.J. (1991). Circulating and tissue endothelin immunoreactivity in advanced atherosclerosis. *N. Engl. J. Med.*, **325.**, 997–1001.

40. Boulanger, C.M., Tanner, F.C., Bea, M.-L., Hahn, A.W.A., Werner, A., and Luscher, T.F. (1992). Oxidized low density lipoproteins induce mRNA

expression and release of endothelin from human and porcine endothelium. *Circ. Res.*, **70**, 1191–7.

41. Lerman, A., Holmes, D.R.J., Bell, M.R., Garratt, K.N., Nishimura, R.A., and Burnett, J.C.J. (1995). Endothelin in coronary endothelial dysfunction and early atherosclerosis in humans. *Circulation*, **92**, 2426–31.

42. Mannucci, P.M. (1998). Von Willebrand factor. A marker of endothelial damage? *Arterioscler. Thromb. Vasc. Biol.*, **18**, 1359–62.

43. Jansson, J.H., Nilsson, T.K., and Johnson, O. (1991). Von Willebrand factor in plasma: a novel risk factor for recurrent myocardial infarction and death. *Br. Heart. J.*, **1991**, 351–5.

44. Geppert, A., Graf, S., Beckmann, R., Hornykewycz, S., Schuster, E., Binder, B.R., and Huber, K. (1998). Concentration of endogenous tPA antigen in coronary artery disease. Relation to thrombotic events, aspirin treatment, hyperlipidemia, and multivessel disease. *Arterioscler. Thromb. Vasc. Biol.*, **18**, 1634–42.

45. Ridker, P.M., Hennekens, C.H., Stampfer, M.J., Manson, A.E., and Vaughan, D.E. (1994). Prospective study of endogenous tissue plasminogen activator and risk of stroke. *Lancet*, **343**, 940–3.

46. Salomaa, V., Stinson, V., Kark, J.D., Folsom, A.R., Davis, C.E., and Wu, K.K. (1995). Association of fibrinolytic parameters with early atherosclerosis. The ARIC Study. *Circulation*, **91**, 284–90.

47. Hwang, S.J., Ballantyne, C.M., Sharrett, A.R., Smith, L.C., Davis, C.E., Gotto, A.M.J., and Boerwinkle, E. (1997). Circulating adhesion molecules VCAM-1, ICAM-1, and E-selectin in carotid atherosclerosis and incident coronary heart disease cases: the Atherosclerosis Risk in Communities (ARIC) Study. *Circulation*, **96**, 4219–25.

48. Squadrito, F., Saitta, A., Altavilla, D., Ioculano, M., Canale, P., Campo, G.M., Squadrito, G., Di Tano, G., Mazzu, A., and Caputi, A.P. (1996). Thrombolytic therapy with urokinase reduces increased circulating endothelial adhesion mol-ecules in acute myocardial infarction. *Inflamm. Res.*, **45**, 14–19.

49. Davi, G., Romano, M., Mezzetti, A., Procopio, A., Iacobelli, S., Antidormi, T., Bucciarelli, T., Alessandrini, P., Cuccurullo, F., and Bittolo-Bon, G. (1998). Increased levels of soluble P-selectin in hypercholesterolemic patients. *Circulation*, **97**, 953–7.

50. Blann, A.D. and McCollum, C.N. (1994). Circulating endothelial cell/leukocyte adhesion molecules in atherosclerosis. *Thromb. Haemost.*, **72**, 151–4.

51. Ridker, P.M., Hennekens, C.H., Roitman-Johnson, B., Stampfer, M.J., and Allen, J. (1998). Plasma concentrations of soluble intercellular adhesion molecule 1 and risk or future myocardial infarction in apparently healthy men. *Lancet*, **351**, 88–92.

52. Yeung, A.C., Vekshtein, V.I., Krantz, D.S., Vita, J.A., Ryan, T.J., Jr, Ganz, P., and Selwyn, A.P. (1991). The effect of atherosclerosis on the vasomotor response of coronary arteries to mental stress. *N. Engl. J. Med.*, **325**, 1551–6.

53. Cox, D.A., Vita, J.A., Treasure, C.B., Fish, R.D., Alexander, R.W., Ganz, P., and Selwyn, A.P. (1989). Atherosclerosis impairs flow-mediated dilation of coronary arteries in humans. *Circulation*, **80**, 458–65.

54. Vita, J.A., Treasure, C.B., Yeung, A.C., Vekshtein, V.I., Fantasia, G.M., Fish, R.D., Ganz, P., and Selwyn, A.P. (1992). Patients with evidence of coronary endothelial dysfunction as assessed by acetylcholine infusion demonstrate marked increase in sensitivity to constrictor effects of catecholamines. *Circulation*, **85**, 1390–7.

55. Libby, P. (1995). Molecular basis of the acute coronary syndromes. *Circulation*, **91**, 2844–50.
56. Vita, J.A., Treasure, C.B., Ganz, P., Cox, D.A., Fish, R.D., and Selwyn, A.P. (1989). Control of shear stress in the epicardial coronary arteries of humans: impairment by atherosclerosis. *J. Am. Coll. Cardiol.*, **14**, 1193–9.
57. Golino, P., Piscione, F., Willerson, J.T., Capelli-Bigazzi, M., Focaccio, A., Villari, B., Indolfi, C., Russolillo, E., Condorelli, M., and Chiarello, M. (1991). Divergent effects of serotonin on coronary artery dimensions and blood flow in patients with coronary atherosclerosis and control patients. *N. Engl. J. Med.*, **324**, 641–8.
58. Bogaty, P., Hackett, D., Davies, G., and Maseri, A. (1994). Vasoreactivity of the culprit lesion in unstable angina. *Circulation*, **90**, 5–11.
59. Okumura, K., Yasue, H., Matsuyama, K., Ogawa, H., Morikami, Y., Obata, K., and Sakaino, N. (1992). Effect of acetylcholine on the highly stenotic coronary artery: difference between the constrictor response of the infarct-related coronary artery and that of the noninfarct-related artery. *J. Am. Coll. Cardiol.*, **19**, 752–8.
60. El-Tamimi, H., Mansour, M., Wargovich, T.J., Hill, J.A., Kerensky, R.A., Conti, C.R., and Pepine, C.J. (1994). Constrictor and dilator responses to intracoronary acetylcholine in adjacent segments of the same coronary artery in patients with coronary artery disease: endothelial function revisited. *Circulation*, **89**, 45–51.
61. Treasure, C.B., Klein, J.L., Weintraub, W.S., Talley, J.D., Stillabower, M.E., Kosinski, A.S., Zhang, J., Boccuzzi, S.J., Cedarholm, J.C., and Alexander, R.W. (1995). Beneficial effects of cholesterol-lowering therapy on the coronary endothelium in patients with coronary artery disease. *N. Engl. J. Med.*, **332**, 481–7.
62. Anderson, T.J., Meredith, I.T., Yeung, A.C., Frei, B., Selwyn, A., and Ganz, P. (1995). The effect of cholesterol lowering and antioxidant therapy on endothelium-dependent coronary vasomotion. *N. Engl. J. Med.*, **332**, 488–93.
63. Tamai, O., Matsuoka, H., Itabe, H., Wada, Y., Kohno, K., and Iamaizumi, T. (1997). Single LDL apheresis improves endothelium-dependent vasodilation in hypercholesterolemic humans. *Circulation*, **95**, 76–82.
64. Mancini, G.B., Henry, G.C., Macaya, C., O'Neill, B.J., Pucillo, A.L., Carere, R.G., Wargovich, T.J., Mudra, H., Luscher, T.F., Klibaner, M.I., Haber, H.E., Uprichard, A.C., Pepine, C.J., and Pitt, B. (1996). Angiotensin-converting enzyme inhibition with quinapril improves endothelial vasomotor dysfunction in patients with coronary artery disease. The TREND (Trial on Reversing ENdothelial Dysfunction) Study. *Circulation*, **94**, 258–65.
65. Anderson, T.J., Overhiser, R.W., Haber, H.E., and Charbonneau, F. (2000). A comparative study of four anti-hypertensive agents on endothelial function in patients with coronary disease. *J. Am. Coll. Cardiol.*, **35**, 60–6.
66. Hornig, B., Arakawa, N., Haussmann, D., and Drexler, H. (1998). Differential effects of quinaprilat and enalaprilat on endothelial function of conduit arteries in patients with chronic heart failure. *Circulation*, **98**, 2842–8.
67. Mullen, M.J., Clarkson, P., Donald, A.E., Thomson, H., Thorne, S.A., Powe, A.J., Furuno, T., Bull, T., and Deanfield, J.E. (1998). Effect of enalaprilon endothelial function in young insulin-dependent diabetic patients: a randomized, double-blind study. *J. Am. Coll. Cardiol.*, **31**, 1330–5.
68. Rajagopalan, S., Kurz, S., Munzel, T., Tarpey, M., Freeman, B.A., Griendling, K.K., and Harrison, D.G. (1996). Angiotensin II-mediated hypertension in the rat increases vascular superoxide production via membrane NADH/NADPH oxidase activation. *J. Clin. Invest.*, **97**, 1916–23.

69. Horning, B., Kohler, C., and Drexler, H. (1997). Role of bradykinin in mediating vascular effects of angiotensin-converting enzyme inhibitors in humans. *Circulation*, **95**, 1115–18.

70. Diaz, M.N., Frei, B., Vita, J.A., and Keaney, J.F., Jr. (1997). Antioxidants and atherosclerotic heart disease. *N. Engl. J. Med.*, **337**, 408–17.

71. Gilligan, D.M., Sack, M.N., Guetta, V., Casino, P.R., Quyyumi, A.A., Rader, D.J., Panza, J.A., and Cannon, R.O. (1994). Effect of antioxidant vitamins on low density lipoprotein oxidation and impaired endothelium-dependent vasodilation in patients with hypercholesterolemia. *J. Am. Coll. Cardiol.*, **24**, 1611–17.

72. Green, D., O'Driscoll, G., Rankin, J.M., Maiorana, A.J., and Taylor, R.R. (1998). Beneficial effect of vitamin E administration on nitric oxide function in subjects with hypercholesterolaemia. *Clin. Sci.*, **95**, 361–7.

73. Ting, H.H., Timimi, F.K., Boles, K.S., Creager, S.J., Ganz, P., and Creager, M.A. (1996). Vitamin C improves endothelium-dependent vasodilation in patients with non-insulin-dependent diabetes mellitus. *J. Clin. Invest.*, **97**, 22–8.

74. Heitzer, T., Yla, H.S., Wild, E., Luoma, J., and Drexler, H. (1999). Effect of vitamin E on endothelial vasodilator function in patients with hypercholesterolemia, chronic smoking or both. *J. Am. Coll. Cardiol.*, **33**, 499–505.

75. Gokce, N., Keaney, J.F., Jr, Frei, B., Holbrook, M., Olesiak, M., Leeuwenburgh, C., Heinecke, J.W., and Vita, J.A. (1999). Long-term ascorbic acid administration reverses endothelial vasomotor dysfunction in patients with coronary artery disease. *Circulation*, **99**, 3234–40.

76. Jackson, T.S., Xu, A., Vita, J.A., and Keaney, J.F., Jr. (1998). Ascorbate prevents the interaction of superoxide and nitric oxide only at very high physiological concentrations. *Circ. Res.*, **83**, 916–22.

77. Heller, R., Munscher-Paulig, F., Grabner, R., and Till, U. (1999). L-Ascorbic acid potentiates nitric oxide synthesis in endothelial cells. *J. Biol. Chem.*, **274**, 8254–60.

78. Vita, J.A., Frei, B., Holbrook, M., Gokce, N., Leaf, C., and Keaney, J.F., Jr. (1998). L-2-Oxothiazolidine-4-carboxylic acid reverses endothelial dysfunction in patients with coronary artery disease. *J. Clin. Invest.*, **101**, 1408–14.

79. Stroes, E., Kastelein, J., Cosentino, F., Erkelens, W., Wever, R., Koomans, H., Luscher, T., and Rabelink, T. (1997). Tetrahydrobiopterin restores endothelial function in hypercholesterolemia. *J. Clin. Invest.*, **99**, 41–6.

80. Gilligan, D.M., Quyyumi, A.A., and Cannon, R.O. (1994). Effects of physiological levels of estrogen on coronary vasomotor function in postmenopausal women. *Circulation*, **89**, 2545–51.

81. Hulley, S., Grady, D., Bush, T., Furberg, C., Herrington, D., Riggs, B., and Vittinghoff, E. (1998). Randomized trial of estrogen plus progestin for secondary prevention of coronary heart disease in postmenopausal women. Heart and Estrogen/progestin Replacement Study (HERS) Research Group. *JAMA*, **280**, 605–13.

82. Creager, M.A., Gallagher, S.J., Girerd, X.J., Coleman, S.M., Dzau, V.J., and Cooke, J.P. (1992). L-Arginine improves endothelium-dependent vasodilation in hyper-cholesterolemic humans. *J. Clin. Invest.*, **90**, 1248–53.

83. Clarkson, P., Adams, M.R., Powe, A.J., Donald, A.E., McCredie, R., Robinson, J., McCarthy, S.N., Keech, A., Celermajer, D.S., and Deanfield, J.E. (1994). Oral

L-arginine improves endothelium-dependent dilation in hypercholesterolemic young adults. *J. Clin. Invest.*, **67**, 1989–94.

84. Adams, M.R., Jessup, W., and Celermajer, D.S. (1997). Cigarette smoking is associated with increased human monocyte adhesion to endothelial cells: reversibility with oral L-arginine but not vitamin C. *J. Am. Coll. Cardiol.*, **29**, 491–7.

85. Boger, R.H., Bode-Boger, S.M., Szuba, A., Tsao, P.S., Chan, J.R., Tangphao, O., Blaschke, T.F., and Cooke, J.P. (1998). Asymmetric dimethylarginine (ADMA): a novel risk factor for endothelium dysfunction: its role in hypercholesterolemia. *Circulation*, **98**, 1842–7.

86. Levine, G.N., Keaney, J.F., Jr, Vita, J.A. (1995). Cholesterol reduction in cardiovascular disease: clinical benefits and possible mechanisms. *N. Engl. J. Med.*, **332**, 512–21.

87. Heart Outcomes Prevention Evaluation Study Investigators. (2000). Effects of an angiotensin-converting-enzyme inhibitor, ramipril, on cardiovascular events in high-risk patients. *N. Engl. J. Med.*, **342**, 145–53.

88. Rimm, E.B., Stampfer, M.J., Ascherio, A., Giovannucci, E., Colditz, G.A., and Willett, W.C. (1993). Vitamin E consumption and the risk of coronary heart disease in men. *N. Engl. J. Med.*, **328**, 1450–6.

89. Raitakari, O.T., Adams, M.R., McCredie, R.J., Griffiths, K.A., and Celermajer, D.S. (1999). Arterial endothelial dysfunction related to passive smoking is potentially reversible in healthy young adults. *Ann. Intern. Med.*, **130**, 578–81.

90. Adams, M.R., Robinson, J., McCredie, R., Seale, P.J., Sorensen, K.E., Deanfield, J.E., and Celermajer, D.S. (1998). Smooth muscle dysfunction occurs independently of impaired endothelium-dependent dilatation in adults at risk of atherosclerosis. *J. Am. Coll. Cardiol.*, **32**, 123–7.

91. Pagano, P.J., Clark, J.K., Cifuentes-Pagano, M.E., Clark, S.M., Callis, G.M., and Quinn, M.T. (1997). Localization of a constitutively active, phagocyte-like NADPH oxidase in rabbit aortic adventitia: enhancement by angiotensin II. *Proc. Natl. Acad. Sci. USA*, **94**, 14483–8.

92. Ridker, P.M., Hennekens, C.H., Roitman-Johnson, B., Stampfer, M.J., and Allen, J. (1998). Plasma concentrations of soluble intercellular adhesion molecule 1 and risk of future myocardial infarction in apparently healthy men. *Lancet*, **351**, 88–92.

6 Gene therapy for vascular diseases: closer to delivering the goods?

John E.J. Rasko
Gene Therapy Research Unit, Centenary Institute of Cancer Medicine and Cell Biology,
and Sydney Cancer Centre, Central Sydney Area Health Service, Locked Bag 6,
Newtown, NSW 2042, Australia

David S. Celermajer
Clinical Research Group, The Heart Research Institute,
Camperdown, Sydney, NSW 2050, Australia

6.1 Introduction

The excitement surrounding the potential application of gene therapy to cardiovascular disease has been sustained since the first *in vivo* animal trials were published in 1990.[1, 2] The ideas and techniques of gene therapy build on some of the most important scientific advances of the twentieth century – molecular biology and the human genome project. However, despite steady progress, the promises of gene therapy have not yet been realized.

Atherosclerosis and its consequences (distal ischemia and infarction) remain the commonest causes of morbidity and mortality in westernized nations, with (as yet) imperfect preventative and therapeutic strategies. In this context, this disease provides an attractive target for gene therapy.

In this chapter, following a brief summary of terms and concepts, we provide a short history of gene therapy to offer perspective and highlight how young the field remains. Each of the different vector systems used in cardiovascular gene therapy are discussed in terms of their advantages and disadvantages. Having outlined the means by which new genetic information can be transferred, we next focus on the appropriate cellular targets and routes of delivery. The clinical problems of atherosclerosis, thrombosis, restenosis, and myocardial angiogenesis are presented in the context of therapeutic opportunities offered by this new technology. We conclude with a brief summary of certain strategies which could be used to overcome some of the difficulties facing this nascent but burgeoning field. We have excluded discussion of the promising gene therapies designed to target dyslipidemias and refer the interested reader to a recent thorough review.[3]

6.1.1 Gene therapy: terminology and rationale

Throughout this review, 'gene therapy' refers to the introduction of new genetic information in somatic cells to achieve a therapeutic benefit. Germ line gene therapy, which seeks to modify the subjects' reproductive cells so that vertical transmission will occur, is not discussed in this chapter. A conceptual distinction should be maintained between gene therapy versus drug or cell therapy although they are frequently combined in practice. Gene therapy is a form of pro-drug therapy inasmuch as the actual therapeutic is a protein, ribozyme or mRNA *coded for* by the introduced gene. As such, the idea of using genes instead of drugs to treat local diseases like thrombosis offers the advantages of local production with paracrine benefits and levels that may be more 'physiological' than those achieved by systemic delivery. It is worth noting that the use of oligonucleotides as therapeutics is not widely regarded as gene therapy since their efficacy typically depends on direct effects, but we have included particular examples for completeness.

There are several reasons why delivery of a gene (gene therapeutic) might offer advantages beyond delivery of the protein encoded by the same gene (recombinant biotherapeutic). Once stable expression is achieved, gene therapy results in sustained endogenous protein production by the transduced cells. For example, the short biological half-life of VEGF (less than 10 min) suggests that greater clinical benefits may be expected by sustained protein production.[4] Furthermore, in contrast to recombinant bacterially synthesized biotherapeutics,[5] endogenous production of proteins may actually offer improved efficacy due to more 'physiological' post-transcriptional modifications, for example glycosylation. In the future, therapeutic gene expression may be regulated by drug-sensitive promoters or enhancers. Indeed, a safety feature which can be readily engineered into gene therapeutics offers the possibility of killing all transduced cells if required for any reason by using 'suicide' genes like the herpes simplex virus type 1 thymidine kinase gene.[6] For non-integrating vector systems such as adenovirus, the typically self-limited therapeutic gene expression may be of benefit in acute situations where sustained but short duration therapy for less than a fortnight may be preferred.[7]

Another reason to prefer gene therapy above recombinant biotherapy is the compelling advantage offered by local or regional protein production. Using gene therapy, large amounts of protein can be produced locally, thereby reducing the chance of systemic side effects. An uncontrolled systemic angiogenic stimulus, for example, could lead to catastrophic consequences such as tumour or hemangioma promotion, proliferative retinopathy, or other pharmacological side effects.[8] As such, the therapeutic window of gene therapy may be superior to systemic drug therapies due to higher local concentrations of protein at the disease target site. Finally, an active area of basic research concerns the use of *regulatable* genes by means of drug-responsive promoters or proteins, or excisable genes. Once stable therapeutic gene delivery is achieved, regulation of therapeutic genes will be an attractive and logical development in the treatment of some specific diseases.

6.2 Vascular gene therapy – historical perspective

The following points indicate why blood vessels present attractive targets for gene therapies:

(1) The endothelium of vessels are readily accessible at almost any location in the body either following systemic injection of vector at a peripheral or surgically revealed vessel, or by intra-luminal catheter delivery.
(2) Local and/or systemic benefits could be achieved following the successful introduction of new genetic information into vascular cells.
(3) Once endothelial cells stably express the transgene, a durable benefit can be expected due to their long-term survival *in situ*.
(4) *Ex vivo* gene therapy of homologous endothelial cells followed by reimplantation is facilitated by their multipotentiality and engraftment capacities.[9]

Consequently, although the focus of this review is to describe gene therapy's promise for the treatment of vascular diseases, successful vascular gene therapy would presage new therapeutic opportunities for many other human diseases.

Although the history of gene therapy may be traced back to the earliest studies identifying DNA as the heritable material, real advances have only occurred subsequent to the revolution in molecular biology commencing in the mid-twentieth century.[10] In 1983 the first demonstration that new genetic material could be successfully introduced into live animals occurred when intravenous injection of liposomal preproinsulin gene led to a measurable production of insulin.[11] The earliest pre-clinical attempts at cardiac and vascular gene therapy were published around 1990. Transduction of endothelial cells *ex vivo* followed by arterial reimplantation was demonstrated in minipigs and dogs.[12] Following injection of plasmid DNA directly into the left ventricular wall of adult rats, expression of a reporter gene for at least 4 weeks demonstrated the needed proof-of-principle for *in vivo* gene therapy.[1] Specific, sustained expression ($\leqslant 147$ days) of a reporter gene was demonstrated in arterial endothelial and vascular smooth muscle cells of pigs *in vivo* following retroviral transduction by catheterization.[2]

The first approved human trial of gene therapy began on 14 September 1990 at the National Institutes of Health in Bethesda, USA.[13] Two children suffering from profound immunodeficiency due to adenosine deaminase deficiency received autologous

T cells which had been transduced with a retroviral vector engineered to express the enzyme they lacked. Although many useful insights and reassurances regarding safety resulted from that study, no meaningful clinical benefit was achieved. Almost a decade elapsed before a group from Hôpital Necker in France reported clinical efficacy using retroviral gene transduction of autologous bone marrow cells in the treatment of a related severe immunodeficiency due to lack of the gamma-common cytokine receptor.[14] This recent success story serves to highlight the fact that gene therapy as a serious treatment modality is in its infancy. Just as several years of pre-clinical and clinical disappointments occurred before bone marrow transplantation or thrombolytic therapies became relevant to clinical practice, so too gene therapy is currently undergoing its phase of rapid development where specific future applications remain uncertain.

6.3 Vascular and cardiac delivery systems

6.3.1 Vector systems

The discussion of specific vector systems in the following section is primarily oriented toward vascular gene therapy. Several recent overviews and general introductions to the burgeoning field of cardiovascular gene therapy are recommended.[15] As might be expected in a rapidly evolving field such as this, many competing vector systems offer diverse advantages, disadvantages, and opportunities for further research (see Table 6.1).

6.3.1.1 DNA

Perhaps the most simple conceptual form of gene delivery systems involves the use of covalently closed circular DNA called plasmids purified from bacteria. The DNA can be introduced unmodified into cells ('naked') or following complex formation with reagents designed to increase the efficiency of uptake. These chemical modifications include: many different formulations of cationic liposomes;[16] dendritic polymers ('dendrimers') which are highly branched, tree-like molecules;[17] 'standard' transfection methods related to co-precipitation of DNA with calcium phosphate DEAE-dextran; or combinations of these approaches with viral vectors.[18] There are also physical methods which enhance the efficiency of plasmid DNA delivery including electroporation and the use of microbeads coated with DNA for ejection by pressurized devices such as the 'gene gun'.[19]

When plasmid DNA is used for gene transfer, the therapeutic gene is typically represented by its coding sequence derived from messenger RNA (cDNA). Eukaryotic regulatory elements are also required, including an upstream promoter to drive transcription in mammalian cells. A careful choice of the promoter can direct tissue-specific expression of the therapeutic gene or even permit regulatable expression. In order to propagate plasmids in bacteria, an origin of replication and

Table 6.1 Gene transfer technologies

Vector	Advantages	Problems	Insert size (kb)	Approx. titre
Non-viral				
'Naked' DNA	Very large 'payload', good safety profile, no viral genes, production is easy, episomal	Very inefficient	$\geqslant 100$	
Liposomal DNA	Very large 'payload', good safety profile, broad target cell range, no viral genes, episomal	Inefficient, unstable	$\geqslant 100$	
Viral				
Retrovirus	Stable and efficient genomic integration, largest clinical experience, good safety profile, non-immunogenic	Insertional mutagen, quiescent cells resist transduction, transcriptional silencing	< 10	$10^6 – 10^8$ ffu ml^{-1}

Table 6.1 (*Cont*)

Lentivirus	Transduction of non-dividing cells, stable genomic integration, non-immunogenic	Unproven safety, currently requires high-level containment, transcriptional silencing	<10	10^6–10^8 ffu ml^{-1}
Adenovirus	Very efficient transduction of dividing and quiescent cells, episomal maintenance avoids mutagenesis	Immunogenicity prevents repeat therapy, pro-inflammatory, specific cells not readily targeted, vector manipulation was laborious	<10 <35 for 'gutless'	10^9–10^{12} ffu ml^{-1}
AAV	Genomic integration, efficient	Production is difficult,	<5	10^8–10^{11} ffu ml^{-1}

a selectable prokaryotic gene such as one that codes for ampicillin resistance is required, but may be excised from the plasmid prior to delivery *in vivo*. Plasmid DNA is taken up by cells via endocytosis and pinocytosis and transferred to lysosomal degradation pathways, which explains why only a very small fraction actually leads to therapeutic gene expression.[20] Most short-term expression of the therapeutic gene occurs from non-integrated (episomal) plasmid.

If the greatest impediment confronting gene therapy is a lack of efficiency in delivery, then non-viral DNA gene transfer systems have the most daunting technical impasse to overcome (with the exception of some vaccination scenarios). Chromosomal integration is required for sustained expression of therapeutic genes which in turn requires cell division, thereby excluding neurones and many other quiescent or mature 'end-cells' from this approach. Cardiomyocytes in adults probably divide very rarely if ever. Nevertheless, the conceptual and technical simplicity of this approach makes it attractive. Additionally, by avoiding infectious agents, the lack of viral or ancillary genes suggests that immunogenicity should be minimized, thereby facilitating repeated doses of DNA gene therapeutics. In the future, very large stretches of DNA in artificial chromosomes may circumvent the need for non-physiological control elements and the use of cDNA by transferring complete genomic regions.[21]

6.3.1.2 Retrovirus and lentivirus

An obvious first choice for gene delivery vectors is to steal the best solution nature has to offer in the form of viruses – masters of forcing a cell to do their bidding. Viruses achieve this goal by efficiently combining essential elements for gene delivery in relatively stable particles. These essential elements include the abilities to: target specific receptors, bind with high affinity and penetrate the plasma membrane; integrate stably in chromosomal DNA so that the provirus becomes part of the cell's heritable material; and enforce expression of introduced genes. The genetic material of retroviruses is stored as double-stranded, linear RNA of about 6 kb which must be reverse transcribed to DNA before genomic integration can occur. Retroviruses have been the most commonly used gene delivery system in the hundreds of clinical trials performed and they have achieved a convincing safety profile in humans.

Typically the retroviral vectors used are derived from murine leukemia-causing viruses which have had all structural genes removed and replaced by a therapeutic gene with or without a heterologous promoter. The essential viral elements retained in the vector are the flanking long terminal repeats (LTR) which mediate genomic integration and expression, and the packaging sequence 'psi' which directs the incorporation of the vector into virions.[22] All the other structural elements are supplied *in trans* in a packaging cell line. That is, the genes which encode the virion coat, internal structure, integrase and reverse transcriptase are stably expressed by plasmids which do not have a psi sequence and are therefore not incorporated into virions (a viral 'split-genome').

In this way, viruses which are produced by packaging cells cannot propagate beyond a single transduction event because the genetic material required for 'wild-type' virus propagation remains separated in the packaging cell and not in the virion. Only the therapeutic gene is transferred. Nevertheless, there is a very small risk that (i) recombination of the separated plasmids could occur to recreate the wild-type virus, thereby allowing unrestrained virus propagation, (ii) a 'helper' virus could infect the vector-transduced cell and supply the missing genes or (iii) the random insertion of the retrovirus could cause cell transformation either by enhancing expression of a nearby oncogene or by inactivating a tumour suppressor gene.[23]

The specificity of the virus-cell interaction is determined by the envelope protein expressed on the virion and its cognate receptor expressed at the target cell surface.[24] Virions cannot bind cells with high affinity if the cognate receptor is not expressed. On human cells there are about a dozen known retrovirus receptors.[25] Surprisingly, different virus vectors can be incorporated into virions with differing envelope proteins so that the same vector can be targeted to different cell types. This phenomenon of 'pseudotyping' has facilitated the use of envelope proteins such as vesicular stomatitis virus glycoprotein which offers increased stability and therefore the opportunity to concentrate virions to improve transduction rates.[26]

One of the biggest obstacles for efficient retroviral transduction of vascular endothelial cells and other important tissues is their slow replication rate. Integration of a retroviral vector requires that it gains access to chromosomal DNA in the nucleus either by penetrating the nuclear membrane or by persisting in the cytoplasm until the nuclear membrane is removed during cell division. The consequence of this physiological constraint is that retroviruses efficiently transduce only actively dividing cells.[27] This constraint is not suffered by complex viruses such as the Human Immunodeficiency Virus and other lentiviruses which express accessory proteins.[28] As such, although lentiviral vectors have some distance to go before achieving a convincing clinical safety record, they offer great promise in overcoming one of the significant problems associated with retroviral vectors.[29]

6.3.1.3 Adenovirus

Adenovirus is a non-enveloped, double-stranded DNA virus which is pathogenic in humans. There are around 50 different human serotypes classified into six sub-groups based on DNA similarities. Adenovirus 2 and 5 are most frequently used as the starting point for human gene transfer vectors. The first generation of adenovirus vectors were deleted in the early-expressed E1 genes which, at high viral concentrations, are dispensable for replication. Additional mutations of the E2 and E4 regions needed for adenoviral replication were introduced to reduce the risk of recombination which could lead to replication-competent virus.[30] Third-generation and 'gutless' adenovirus vectors coupled with greatly simplified methods of introducing therapeutic DNA into vectors have made this system increasingly useful.[31]

The development of adenoviral vectors was a major advance in gene transfer technologies for several reasons. Adenovirus can be produced at very high titres to facilitate efficient gene transfer. Since most adenovirus is non-integrating, the risk of insertional mutagenesis is negligible. Adenovirus is capable of transducing a wide variety of human cell types owing to the broad expression of its cognate cellular receptors, although some cell types probably have insignificant receptor expression and therefore would be resistant to transduction.[32] Indeed adenovirus vectors were originally intended for the treatment of cystic fibrosis due to their tropism for the respiratory epithelium. Finally, the ability of adenovirus to infect quiescent or non-dividing cells such as cardiomyocytes, neurones, and endothelial cells is one of its most attractive features.[33]

The main problem concerning adenovirus gene therapy has been its immunogenicity, although the 'gutless' vectors are devoid of immune epitopes.[34] Most of us have been exposed to, and mounted an immune response against, adenovirus. Therefore the fact that adenovirus is rapidly cleared by the immune system and does not typically integrate means that its use is currently limited to single exposures for short-term therapeutic effects. This scenario may actually be well-suited for some cardiovascular conditions where, for example, an anti-coagulant could be expressed for several weeks from a vascular graft during the period of high risk for thrombosis and then expression would rapidly taper. However, since the end of 1999 a pall hangs over adenovirus clinical trials due to the poorly explained sudden death of an otherwise-well young patient with a genetic defect who received very high doses of third-generation adenovirus.[35]

6.3.1.4 Adeno-associated virus

Adeno-associated virus (AAV) is a single-stranded defective DNA virus which is non-pathogenic in humans.[36] AAV is a *dependovirus* as its propagation requires that the host cell contains helper adenovirus. The genome of AAV is a 4.7 kb linear molecule bounded by inverted terminal repeats, and contains the required genes *rep* and *cap* which can be supplied *in trans* in transfected packaging cells. However, even in split packaging systems the propensity for recombination is remarkably high. The extent to which AAV can integrate in target cell genomes following infection *in vivo* is not entirely resolved as most expression is short-term from episomal vector.[37] Several groups have reported long-term expression of genes for many months following *in vivo* transduction.[38] An attractive feature of AAV is its ability to transduce dividing and non-dividing cells with relatively high efficiency (see above). This feature has propelled the application of AAV in vascular and systemic gene therapy by an increasing number of groups.[39]

There are a number of problems for AAV vectors. Production of clinical-grade AAV is very cumbersome and troubled by contamination with replication-competent adenovirus which has often been present in stocks, but below the limits of detection examined by some investigators.[40] Most adults have been exposed to AAV and mounted an immune response so that repeat administration of AAV vectors is currently problematical.[41] One fundamental limitation of AAV is its genome size, which limits the specific therapeutic genes which can be incorporated.

6.3.1.5 Other systems including viruses, ribozymes, and antisense

In addition to the above-mentioned viruses, there are a number of other viral and non-viral systems being developed for gene therapy. The herpes simplex virus has been used in a murine model of tissue ischemia.[42] Other viral vector systems currently under development include those based on vaccinia, Epstein-Barr virus, baculovirus, simian virus 40, alphaviruses and more. It may be that certain biological features of these viruses will make them perfect vectors for use in highly specific therapeutic scenarios in the coming years.

Ribozymes are autocatalytic single-stranded RNA molecules designed to cleave any specific target RNA and thereby reduce or abolish its expression.[43] For example, use of a ribozyme against c-myb to reduce proliferation in smooth muscle cells has shown promise in animal models of restenosis.[44] Similar goals of inhibiting gene expression are shared by 'antisense' approaches in which vectors code for RNA which is complementary to, and inhibits the stability and translation of, target genes. Many investigators have used antisense techniques combined with different delivery systems in the context of cardiovascular diseases.[45]

6.3.2 Delivery sites for gene therapeutics

The accurate delivery of genes to the appropriate target cells for cardiovascular therapy remains an important barrier to the clinical application of gene transfer strategies. The optimization of gene delivery to target cells depends on the location of these cells as well as characteristics of their cell biology, such as replicative capacity and cell turnover times.

The major target cells in the cardiovascular system for gene therapy are vascular cells (endothelial, smooth muscle, fibroblast and potentially inflammatory cells) and myocytes. The challenges for gene transfer to these two different types of target cells are quite different, as outlined below.

6.3.2.1 Vascular cells

Cells in the vessel wall are theoretically accessible through the bloodstream, and have reasonably high replication rates (especially after balloon injury). This enhances the appeal of using viral vectors for gene transfer, and also allows consideration of percutaneous site-specific gene delivery systems, to a targeted local arterial segment.

Infusion of certain viral vectors into normal arteries with an intact endothelium results in transfection of intimal cells (primarily endothelial cells); however, cells further from the lumen than one or two cell layers are rarely successfully transfected.[46] Where the vessel wall has been injured and/or pressure applied to the infused vector, DNA may then be delivered transmurally through the vessel intima, and gene expression in the media and even adventitia has been observed.[47, 48]

Non-pressurized infiltration of nitric oxide synthase gene has been assessed in a rabbit model of atherosclerosis, where a dwell time of 15 min successfully inhibited cell adhesion molecule expression and inflammatory cell infiltration in the carotid arteries of cholesterol-fed rabbits.[49] Due to the limited tissue penetration of this technique, however, and the transient nature of gene expression, several innovative catheter designs have been introduced to enhance gene transfer to the vessel wall. These include: double-balloon catheters, which isolate the segment of interest between the balloons while the vector is applied to the arterial segment between the balloons; porous balloon catheters where vectors are pressurized through the pores in the balloon and even hydrogel catheters, to enhance gene transfer efficiency.[50]

Other endovascular techniques for gene delivery have been used experimentally. Vascular grafts may be 'seeded' with genetically modified endothelial cells, which then express the gene of interest after the graft is implanted *in vivo*. Dunn *et al.*,[51] for example, have used such a strategy where venous endothelial cells were transduced with a retroviral vector encoding human tissue plasminogen activator. The arterial interposition graft was treated *ex vivo*, and the 'seeded' endothelial cells were noted to express tissue plasminogen activator (TPA) and to be less prone to thrombosis.

The treatment of *ex vivo* grafts before reimplantation has also been assessed in humans. Pressure-mediated transfection of explanted saphenous veins with a decoy oligonucleotide for E2F (a key cell-cycle stimulator) has been shown to produce effective sequence-specific inhibition of cell-cycle gene expression and DNA replication, with possibly lower failure rates of human primary bypass vein grafting.[52] *Ex vivo* pressure-mediated transfection techniques are also being applied in trials of coronary artery bypass surgery, where nitric oxide synthase gene is being transfected in an attempt to inhibit vein graft stenosis. This technique has already proven successful in a dog model.[53]

Finally, a novel strategy for vascular cell gene therapy is introduction of the gene of interest via an adventitial route. Kullo *et al.*[54] have reported gene transfer of endothelial nitric oxide synthase to the adventitia via injection of vector into the periarterial sheath of a rabbit carotid, and found successful expression of the gene in the neointima of the treated arterial segment. This technique has not yet been applied in human studies.

6.3.2.2 Muscle cells

As atherosclerosis is the commonest of the clinical cardiovascular diseases, and its consequences relate to ischemia of the muscle supplied by stenosed or occluded arteries, an alternative strategy for treatment of atherosclerotic disease is treatment of the ischemic muscle, distal to the obstructive atherosclerotic plaque. In the case of aorto-iliac disease, skeletal muscle in the lower limbs may be treated, whereas in coronary artery disease, cardiac myocytes are the potential targets for gene transfer. Biological considerations in these two muscle beds may be quite different, as cardiac but not skeletal myocytes are terminally differentiated cells, and therefore require a vector that is not dependent on cell replication for delivery and thence expression. The major challenges to date in treatment of ischemic muscle have been poor penetration of the transferred genes, and only transient gene expression after administration.

To date, the most popular technique for gene therapy of muscle has involved injection of genes directly into the affected muscle groups. Although this is a technically simple procedure, the method is limited by transfection of a small number of cells within a few millimetres of the injection site (therefore many injection sites are often used). Furthermore, expression of the transfected genes is temporally limited to a few weeks only.[55]

Isner's group have pioneered *in vivo* gene therapy of peripheral ischemic skeletal muscle in humans. They have reported direct intramuscular gene transfer of naked DNA for vascular endothelial growth factor (VEGF), which is most effective in hypoxic muscle where the protein's receptor is over-expressed (see below). Anatomic and functional efficiency has been demonstrated by increased local levels of VEGF protein, improved hemodynamic measurements, greater collateral formation as assessed by angiography and magnetic resonance, reduced leg pain, and increased healing of ischemic ulcers. Therefore direct intramuscular gene transfer of plasmid DNA does appear to stimulate collateral vessel growth effectively, despite a predicted low transfection efficiency. A similar strategy of direct intramuscular injection has been applied in the heart by Schumacher *et al.*,[56] where injection of naked DNA for fibroblast growth factor (FGF) has enhanced collateral formation in the ischemic anterior wall of the heart, when performed in association with coronary artery bypass grafting.

A theoretically more attractive approach is the use of viral vectors, to aid transfection efficiency. This technique has been applied *in vivo* by Rosengart *et al.*,[57] who have reported a phase 1 assessment of direct intra-myocardial administration of an adenovirus vector expressing VEGF cDNA to subjects with severe coronary atherosclerosis. Further clinical studies will elucidate whether naked DNA, viral-vector-assisted local gene transfer or other techniques will result in the most improved clinical outcomes.

Finally, intra-pericardial gene transfer has recently been described as a technique that might achieve much greater penetration of gene transfection into the myocardium. Fromes *et al.*[58] have described a novel strategy where replication-deficient adenovirus is injected into the pericardial sac, along with a mixture of collagenase and hyaluronidase, to enhance diffusion of the transgene. They describe transfection rates of up to 40% of the myocardium in a rat model. Zhang *et al.*[59] have also shown greater transfection success with a pericardial injection strategy compared to direct myocardial injection, in a murine model of gene therapy.

One important caveat about gene therapy and the cardiovascular system relates to the need for tissue and location specificity of any administered genes. As much recent interest has focused on growth factor gene therapy, it must be noted that growth factors may have deleterious effects if over-expressed outside the target tissue of interest. Atheromatous plaques themselves rely on a myriad of small vessels for their growth, and there is a theoretical consideration that growth factor gene therapy may be deleterious if it enhances small-vessel formation and potentially intraplaque hemorrhage, in the region of existing atherosclerotic lesions. If such therapy has systemic spill-over, there is also the theoretical risk of enhancing new vessel formation in malignant tissues, with potentially deleterious consequences. Despite this reservation, exciting developments in gene transfer technology and devices for accurate tissue delivery are enhancing the prospects of successful cardiovascular gene therapy for the treatment of atherosclerotic and other cardiovascular diseases.

6.4 Disease targets

6.4.1 Primary atherosclerosis and thrombosis

As outlined in detail elsewhere in this book, the process of atherosclerosis is extraordinarily complex and takes many decades to develop. A large number of genes and proteins are involved, both in mediating the effects of risk factors (for example, the low-density lipoprotein receptor) and in participating directly in the initiation, growth, and degradation of atherosclerotic plaques themselves (for example, cell adhesion molecules and matrix metalloproteinases). Although it is theoretically attractive to over-express or suppress certain genes that might influence the atherogenic process, such approaches have not been successful to date. The difficulties with tissue targeting, specificity of effects, the long duration of atherogenesis and the short expression time of most transfected genes, and the clinical acceptability of such treatments in pre-symptomatic subjects have all been obstacles to the successful gene therapy of atherosclerotic plaques. Nevertheless, gene therapy may impact on the predisposing factors to atherosclerosis, such as hypercholesterolemia or hypertension, even if gene therapy of primary atherosclerotic plaques themselves proves technically difficult. Endothelial cells may also be a potential target; for example, control of transmigration or shear-regulated gene expression. At this stage, however, the most likely phase of atherosclerosis to be amenable to gene therapy is that of discrete vulnerable plaques, potentially treatable by inhibition of inflammatory genes such as collagenase or stromelysin. Such approaches have not yet been reported in the literature.

Thrombosis is a more appealing target for gene therapy, as many of the targets are blood borne or reside on the luminal surface of vessels. In the last decade there has been a great increase in our understanding of the regulation of thrombosis and thrombolysis, with the description and cloning of thrombolytic agents such as TPA as well as their inhibitors, such as plasminogen activator inhibitor-1. Transfection of endothelial cells over-expressing TPA can be used to line artificial grafts before implantation. Such an approach has successfully inhibited thrombosis in an experimental animal model.[51] Viral-vector-directed secretion of hirudin, a powerful anti-thrombotic agent, from endothelial cells lining implanted grafts has also been shown to significantly reduce the formation of thrombus.[60]

Nevertheless, a major challenge to the clinical utility of anti-thrombosis gene therapy is the unpredictability of the thrombotic event. Although thrombosis is the key pathogenetic event in arterial occlusion complicating atherosclerotic plaque in the coronary and the carotid circulations, its erratic occurrence makes it difficult to prevent by gene therapy techniques which may be limited by transient expression problems. For these reasons, work to date for gene therapy of atherothrombosis has concentrated particularly on situations where cell growth or thrombosis are more rapid and/or predictable events, for example in the settings of balloon injury or surgical bypass grafting, which are considered below. By contrast, 'naturally occurring' atherothrombotic disease has proven more difficult to treat by gene therapy, for the reasons outlined above.

6.4.2 Restenosis

Restenosis is an important clinical problem in two situations: following balloon angioplasty of obstructive atherosclerotic lesions, and following bypass grafting of lesions using either endogenous vein or artificial arterial grafts. Clinically significant restenosis affects approximately 40% of balloon angioplasties by 6 months after the procedure, and has proven resistant to almost all medical therapies designed to inhibit this process to date. Even the use of stents to physically keep an artery open has only resulted in small decreases in the amount of restenosis observed. With bypass grafting, approximately 60–70% of venous grafts will be stenosed or occluded 10 years after surgery. Although anti-platelet and cholesterol lowering therapies may decrease these rates slightly, this is still an important therapeutic problem, and therefore a potentially exciting area for gene therapy.

Two major strategies of gene therapy have been applied to the problem of vascular restenosis. The first is to examine gene products that inhibit cell proliferation. The second has been to enhance endothelial regrowth, and rely on the new endothelial cells to secrete endogenous factors (such as nitric oxide) which might limit the neointimal proliferative response.

Inhibition of cell proliferation relies on the local delivery of an anti-proliferative agent during the period of neointimal proliferation and extracellular matrix synthesis after the time of injury. Successful transfection of this kind might limit expansion of the intima and reduce clinical restenosis. Approaches that have been explored in this setting include recombinant chimeric toxins,[61] gene transfer of viral thymidine kinase,[62] or antisense oligonucleotide strategies to inhibit cell proliferation.[52]

One promising approach that has been found effective in animal models is the selective inhibition of dividing cells by transfection of a herpes virus thymidine kinase gene into smooth muscle cells after balloon injury. This enzyme, when expressed in the vessel wall, converts gancyclovir (a nucleoside analogue and antiviral pharmaceutical) into an active toxic form, which subsequently causes cell death. Transduction of cells with adenoviral vectors encoding HSV thymidine kinase gene followed by a course of gancyclovir, three weeks after balloon injury, have resulted in significant reduction in neointimal proliferation. This utilizes the innovative strategy of gene transfer of an enzyme that can prompt the formation of a cytotoxic drug locally within an artery, under the appropriate conditions, thereby limiting the possibility of restenosis.

The antisense oligonucleotide strategy has recently been described in a phase 1 study in humans.[52] These investigators studied a specific 'decoy' oligodeoxynucleotide, which binds to and inactivates the pivotal cell-cycle transcription factor E2F. Effective transfection was associated with fewer graft occlusions and revisions in the treated compared with the placebo group, in this small study.

The other major strategy for inhibiting restenosis is local delivery of vascular endothelial growth factors to accelerate re-endothelialization and attenuate intimal hyperplasia in balloon-injured arteries.[63] Vascular permeability factor has also been shown to accelerate endothelial regrowth following balloon angioplasty.[64] These data, along with several studies showing the beneficial effects of vessel wall transfection after balloon injury with nitric oxide synthase gene, point the way to a potentially promising strategy for inhibiting restenosis. A similar approach is also being trialed for vein grafts in clinical coronary artery bypass surgery.

6.4.3 Myocardial angiogenesis

The straightforward idea that blocked vessels can be replaced or physically bypassed has given rise to very significant improvements in the morbidity and mortality associated with coronary artery and peripheral vascular diseases. Gene therapy strives to replace the physical interventions associated with bypass surgery with biological agents to induce local proliferation of new collateral vessels around the blockage.[65] The idea is both compelling and elegant – if only it can be made to work! The following description of two specific angiogens should be regarded as examples of a flourishing list of other gene therapeutics being investigated for applications in myocardial angiogenesis (Table 6.2). Other angiogens of potential therapeutic importance include angiopoietin-1, nitric oxide, tumour necrosis factor, platelet-derived growth factor, transforming growth factor, and hepatocyte growth factor. In order that the individual rationale underpinning the testing of each reagent is appreciated, a brief description of the normal biology of each molecule has been included.

6.4.3.1 Vascular endothelial growth factor

VEGF plays important roles in both the physiology and especially the pathology of angiogenesis.[73] Although encoded for by one gene, there are five splice variants of

Table 6.2 Results of some gene therapy studies *in vivo* for atherosclerosis or thrombosis

Gene	Vector	Target and route	Outcome	Reference
FGF-5	Adenovirus	Heart/intravascular	Improved myocardial blood flow and function	(85)
$VEGF_{121}$	Adenovirus	Myocardial injection	Induced collaterals, improved myocardial perfusion and function	(66)
$VEGF_{121,\ 165,\ 189}$	Plasmid	Leg/intra-arterial balloon catheter	Induced collaterals	(67, 79)
$VEGF_{165}$	Adenovirus	Leg muscle injection	Protected against acute vascular occlusion	(68)
Nitric oxide synthase	Sendai virus/liposome	Carotid artery	Inhibited neointimal hyperplasia	(49, 69)
Growth arrest homeobox (Gax)	Adenovirus	Carotid and iliac arteries/intravascular	Inhibited neointimal hyperplasia	(70)
Fas-ligand	Adenovirus	Carotid artery/intravascular	Inhibited neointimal hyperplasia	(70)
Herpes simplex virus thymidine kinase	Adenovirus	Iliac artery/intravascular	Inhibited neointimal hyperplasia	(71)
Kallikrein	Adenovirus	Carotid artery/intravascular	Inhibited neointimal hyperplasia	(72)

VEGF which result in proteins of 206, 189, 165, 145 and 121 amino acids. The glyco-sylated polypeptide forms a homodimer of 34–42 kDa which has a biological half-life of less than 10 min.[4] The biological response of target cells to each of the VEGF splice variants varies, but increases in cell survival, proliferation and permeability are typical.[74] VEGF is secreted both by tumourigenic cells and by normal smooth muscle, corpus luteal and adrenal cortex cells.[75] Expression of VEGF and at least one of its receptors are increased following hypoxia, attesting to their importance in ischemic conditions.[76]

The VEGF cell surface receptors (Flt-1, Flt-4, and KDR/Flk-1) are members of the large receptor tyrosine kinase family.[77] The expression of these high-affinity receptors on relatively few cell types other than endothelial cells augurs well for the specificity of VEGF therapies. $VEGF_{165}$ also binds neuropilin-1 and neuropilin-2, which are also expressed on endothelial cells.[75] In addition, the various VEGF isoforms have widely differing binding affinities for heparin and heparan sulphate proteoglycans on cell surfaces and in the extracellular matrix.[78]

The above details make a strong case for testing VEGF as a pro-angiogenic gene therapeutic. A number of clinical phase I safety studies have been reported. In most instances, the studies have involved few patients and lacked control groups, so that the results must be viewed with caution. Considerable excitement was generated by an early case report of possible efficacy following percutaneous intra-luminal arterial application of naked plasmid DNA encoding $VEGF_{165}$.[79] Objective evidence of increased local angiogenesis in the ischemic limb was demonstrated one month following the procedure. The same group from Boston subsequently reported several different approaches using VEGF. When $VEGF_{165}$ plasmid was injected in the muscles of nine patients with ischemic arterial insufficiency, a clinical benefit was documented in about half of them (see Fig. 6.1).[80]

In each of these small studies, new genetic material expressing VEGF was delivered to human somatic cells. At the very least, such studies demonstrated safety and feasibility.

6.4.3.2 Fibroblast growth factors

The family of fibroblast growth factors (FGFs) comprise at least 17 members including basic (bFGF) and acidic FGFs.[81] FGFs are the second-most extensively studied angiogens, after VEGFs, and were the first molecules used to demonstrate the feasibility of collateral vessel induction around areas of ischemia.[82] The FGFs are relatively small heparin-binding proteins which are widely expressed. They are responsible for a remarkably diverse set of functions including cell proliferation and migration, organogenesis and bone morphogenesis. bFGF is bound to, and stored in, the extracellular matrix and can be released following exposure to VEGF, thereby leading to a synergistic angiogenic drive.[83] There are at least four widely expressed high-affinity receptors for FGFs which contain tyrosine kinases (FGFR1–4).[84] Perhaps the most exciting report using an FGF was the intracoronary adenoviral

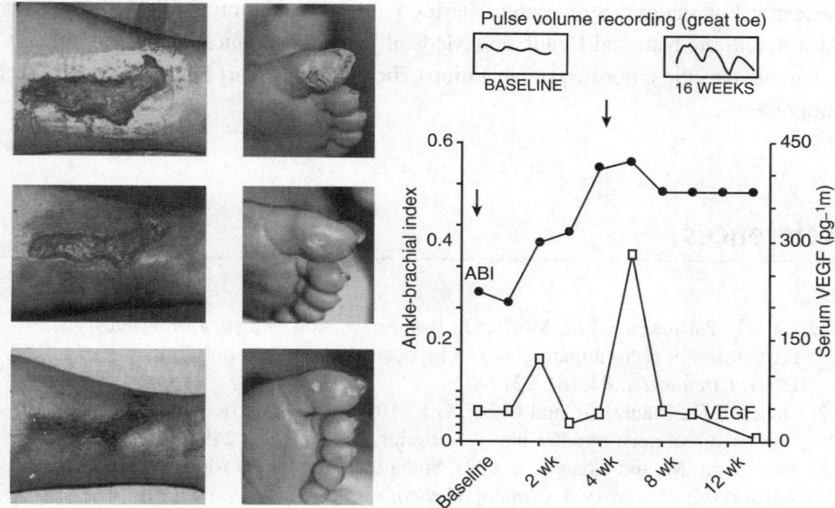

Fig. 6.1 Effects of naked VEGF DNA injections (at baseline and four weeks, *vertical arrows*) in a 33-year-old woman with severe peripheral vascular disease. (*Left*) Arterial ulcer healing over 3 months. (*Right*) Improving ankle brachial index (ABI) and also shows serum VEGF levels. Reproduced from Baumgartner *et al.*[80]

introduction of FGF-5 in a pig model of stress-induced myocardial ischemia. In that study, myocardial blood flow and function were improved for the three-month observation period.[85]

6.5 Summary

Gene therapy for cardiovascular diseases, particularly atherosclerosis and its consequences, is an appealing area which is growing rapidly, with some promising initial results. Problems remain, however, before the techniques could become widespread in clinical practice. These include choice of safe and effective vectors, appropriate delivery systems, and enhancing prolonged expression of genes in the specific desired target cells. The choice of gene is also not straightforward, although much initial attention has focused on nitric oxide and growth factors. As phase I studies address these questions with some success in terms of delivery, safety and even clinical efficacy, there is real hope for continued advances in this novel therapeutic field.

Acknowledgments

J.E.J.R. gratefully acknowledges grants from: The Royal Australasian College of Physicians' CSL Fellowship in Medical Research; The Rebecca L. Cooper Medical

Research Foundation; Sylvia and Charles Viertel Foundation Clinical Investigator Award; and the National Health and Medical Research Council of Australia. D.S.C acknowledges the latter two organizations, the National Heart Foundation and other supporters.

References

1. Lin, H., Parmacek, M.S., Morle, G., Bolling, S., and Leiden, J.M. (1990). Expression of recombinant genes in myocardium *in vivo* after direct injection of DNA. *Circulation*, **82**, (6), 2217–21.
2. Nabel, E.G., Plautz, G., and Nabel, G.J. (1990). Site-specific gene expression *in vivo* by direct gene transfer into the arterial wall. *Science*, **249**, (4974), 1285–8.
3. Belalcazar, M. and Chan, L. (1999). Somatic gene therapy for dyslipidemias. *Journal of Laboratory & Clinical Medicine*, **134**, (3), 194–214.
4. Takeshita, S., Zheng, L.P., Brogi, E., Kearney, M., Pu, L.Q., Bunting, S., Ferrara, N., Symes, J.F., and Isner, J.M. (1994). Therapeutic angiogenesis. A single intraarterial bolus of vascular endothelial growth factor augments revascularization in a rabbit ischemic hind limb model. *Journal of Clinical Investigation*, **93**, (2), 662–70.
5. Russell, C.S. and Clarke, L.A. (1999). Recombinant proteins for genetic disease. *Clinical Genetics*, **55**, (6), 389–94.
6. Bonini, C., Ferrari, G., Verzeletti, S., Servida, P., Zappone, E., Ruggieri, L., Ponzoni, M., Rossini, S., Mavilio, F., Traversari, C., and Bordignon, C. (1997). HSV-TK gene transfer into donor lymphocytes for control of allogeneic graft-versus-leukemia. *Science*, **276**, (5319), 1719–24
7. Magovern, C.J., Mack, C.A., Zhang, J., Rosengart, T.K., Isom, O.W., and Crystal, R.G. (1997). Regional angiogenesis induced in nonischemic tissue by an adenoviral vector expressing vascular endothelial growth factor. *Human Gene Therapy*, **8**, (2), 215–27.
8. Folkman, J. and Shing, Y. (1992). Angiogenesis. *Journal of Biological Chemistry*, **267**, (16), 10931–4.
9. Asahara, T., Murohara, T., Sullivan, A., Silver, M., van der Zee, R., Li, T., Witzenbichler, B., Schatteman, G., and Isner, J.M. (1997). Isolation of putative progenitor endothelial cells for angiogenesis. *Science*, **275**, (5302), 964–7.
10. Wolff, J.A. and Lederberg, J. (1994). An early history of gene transfer and therapy. *Human Gene Therapy*, **5**, (4), 469–80.
11. Nicolau, C., Le Pape, A., Soriano, P., Fargette, F., and Juhel, M.F. (1983). *In vivo* expression of rat insulin after intravenous administration of the liposome-entrapped gene for rat insulin I. *Proceedings of the National Academy of Sciences of the United States of America*, **80**, (4), 1068–72.
12. Nabel, E.G., Plautz, G., Boyce, F.M., Stanley, J.C., and Nabel, G.J. (1989). Recombinant gene expression *in vivo* within endothelial cells of the arterial wall. *Science*, **244**, (4910), 1342–4.
13. Blaese, R.M., Culver, K.W., Miller, A.D., Carter, C.S., Fleisher, T., Clerici, M., Shearer, G., Chang, L., Chiang, Y., and Tolstoshev, P. (1995).

T lymphocyte-directed gene therapy for ADA-SCID: initial trial results after 4 years. *Science*, **270**, (5235), 475–80.

14. Cavazzana-Calvo, M., Hacein-Bay, S., de Saint Basile, G., Gross, F., Yvon, E., Nusbaum, P., Selz, F., Hue, C., Certain, S., Casanova, J.-L., Bousso, P., Le Deist, F., and Fisher, A. (2000). Gene therapy of human severe combined immunodeficiency (SCID)-X1 disease. *Science*, **288**, (5466), 669–71.

15. Rosengart, T.K., Patel, S.R., and Crystal, R.G. (1999). Therapeutic angiogenesis: protein and gene therapy delivery strategies. *Journal of Cardiovascular Risk*, **6**, (1), 29–40.

16. Nabel, E.G., Yang, Z.-Y., San, H., Carr, D.P., and Nabel, G.J. (1997). Methods for liposome-mediated gene transfer to the arterial wall. In *Gene therapy protocols*, (ed. P.D. Robbins), pp. 127–33. Humana Press, Totowa.

17. Turunen, M.P., Hiltunen, M.O., Ruponen, M., Virkamaki, L., Szoka, F.C., Jr, Urtti, A., and Yla-Herttuala, S. (1999). Efficient adventitial gene delivery to rabbit carotid artery with cationic polymer-plasmid complexes. *Gene Therapy*, **6**, (1), 6–11.

18. Dunphy, E.J., Redman, R.A., Herweijer, H., and Cripe, T.P. (1999). Reciprocal enhancement of gene transfer by combinatorial adenovirus transduction and plasmid DNA transfection *in vitro* and *in vivo*. *Human Gene Therapy*, **10**, (14), 2407–17.

19. Muramatsu, T., Nakamura, A., and Park, H.M. (1998). *In vivo* electroporation: a powerful and convenient means of nonviral gene transfer to tissues of living animals (Review). *Bioorganic & Medicinal Chemistry Letters*, **1**, (1), 55–62.

20. Crystal, R.G. (1995). Transfer of genes to humans: early lessons and obstacles to success. *Science*, **270**, (5235), 404–10.

21. Narayanan, K., Williamson, R., Zhang, Y., Stewart, A.F., and Ioannou, P.A. (1999). Efficient and precise engineering of a 200 kb beta-globin human/bacterial artificial chromosome in E. coli DH10B using an inducible homologous recombination system. *Gene Therapy*, **6**, (3), 442–7.

22. Miller, A.D., Miller, D.G., Garcia, J.V., and Lynch, C.M. (1993). Use of retroviral vectors for gene transfer and expression. *Methods in Enzymology*, **217**, 581–99.

23. Cornetta, K., Morgan, R.A., and Anderson, W.F. (1991). Safety issues related to retroviral-mediated gene transfer in humans. *Human Gene Therapy*, **2**, (1), 5–14.

24. Rasko, J.E.J., Battini, J.-L., Gottschalk, R.J., Mazo, I., and Miller, A.D. (1999). The RD114/Simian type D retrovirus receptor is a neutral amino acid transporter. *Proceedings of the National Academy of Sciences of the United States of America*, **96**, (5), 2129–34.

25. Battini, J.-L., Rasko, J.E.J., and Miller, A.D. (1999). A Human cell-surface receptor for xenotropic and polytropic murine leukemia viruses: possible role in G protein-coupled signal transduction. *Proceedings of the National Academy of Sciences of the United States of America*, **96**, (4), 1385–90.

26. Bartz, S.R. and Vodicka, M.A. (1997). Production of high-titer human immunodeficiency virus type 1 pseudotyped with vesicular stomatitis virus glycoprotein. *Methods*, **12**, (4), 337–42.

27. Miller, D.G., Adam, M.A., and Miller, A.D. (1990). Gene transfer by retrovirus vectors occurs only in cells that are actively replicating at the time of infection. *Mol. Cell. Biol.*, **10**, 4239–42.

28. Lewis, P.F. and Emerman, M. (1994). Passage through mitosis is required for oncoretroviruses but not for the human immunodeficiency virus. *Journal of Virology*, **68**, (1), 510–6.

29. Zufferey, R., Dull, T., Mandel, R.J., Bukovsky, A., Quiroz, D., Naldini, L., and Trono, D. (1998). Self-inactivating lentivirus vector for safe and efficient *in vivo* gene delivery. *Journal of Virology*, **72**, (12), 9873–80.

30. Wang, Q. and Finer, M.H. (1996). Second-generation adenovirus vectors. *Nature Medicine*, **2**, (6), 714–6.

31. He, T.C., Zhou, S., da Costa, L.T., Yu, J., Kinzler, K.W., and Vogelstein, B. (1998). A simplified system for generating recombinant adenoviruses. *Proceedings of the National Academy of Sciences of the United States of America*, **95**, (5), 2509–14.

32. Bergelson, J.M., Cunningham, J.A., Droguett, G., Kurt-Jones, E.A., Krithivas, A., Hong, J.S., Horwitz, M.S., Crowell, R.L., and Finberg, R.W. (1997). Isolation of a common receptor for Coxsackie B viruses and adenoviruses 2 and 5. *Science*, **275**, (5304), 1320–3.

33. Van Belle, E., Maillard, L., Rivard, A., Fabre, J.E., Couffinhal, T., Kearney, M., Branellec, D., Feldman, L.J., Walsh, K., and Isner, J.M. (1998). Effects of poloxamer 407 on transfection time and percutaneous adenovirus-mediated gene transfer in native and stented vessels. *Human Gene Therapy*, **9**, (7), 1013–24.

34. Chen, H.H., Mack, L.M., Kelly, R., Ontell, M., Kochanek, S., and Clemens, P.R. (1997). Persistence in muscle of an adenoviral vector that lacks all viral genes. *Proceedings of the National Academy of Sciences of the United States of America*, **94**, (5), 1645–50.

35. Marshall, E. (1999). Gene Therapy Death Prompts Review of Adenovirus Vector. *Science*, **286**, 2244–5.

36. Berns, K.I. (1999). The Gordon Wilson Lecture. From basic virology to human gene therapy. *Transactions of the American Clinical & Climatological Association*, **110**, 75–85.

37. Nakai, H., Iwaki, Y., Kay, M.A., and Couto, L.B. (1999). Isolation of recombinant adeno-associated virus vector-cellular DNA junctions from mouse liver. *Journal of Virology*, **73**, (7), 5438–47.

38. Kaplitt, M.G., Xiao, X., Samulski, R.J., Li, J., Ojamaa, K., Klein, I.L., Makimura, H., Kaplitt, M.J., Strumpf, R.K., and Diethrich, E.B. (1996). Long-term gene transfer in porcine myocardium after coronary infusion of an adeno-associated virus vector. *Annals of Thoracic Surgery*, **62**, (6), 1669–76.

39. Koeberl, D.D., Bonham, L., Halbert, C.L., Allen, J.M., Birkebak, T., and Miller, A.D. (1999). Persistent, therapeutically relevant levels of human granulocyte colony-stimulating factor in mice after systemic delivery of adeno-associated virus vectors. *Human Gene Therapy*, **10**, (13), 2133–40.

40. Allen, J.M., Debelak, D.J., Reynolds, T.C., and Miller, A.D. (1997). Identification and elimination of replication-competent adeno-associated virus (AAV) that can arise by nonhomologous recombination during AAV vector production. *Journal of Virology*, **71**, (9), 6816–22.

41. Halbert, C.L., Standaert, T A., Wilson, C.B., and Miller, A.D. (1998). Successful readministration of adeno-associated virus vectors to the mouse lung requires transient immunosuppression during the initial exposure. *Journal of Virology*, **72**, (12), 9795–805.

42. Mesri, E.A., Federoff, H.J., and Brownlee, M. (1995). Expression of vascular endothelial growth factor from a defective herpes simplex virus type 1 amplicon vector induces angiogenesis in mice. *Circulation Research*, **76**, (2), 161–7.

43. Branch, A.D. and Klotman, P.E. (1998). Optimizing ribozymes for somatic cell gene therapy. *Experimental Nephrology*, **6**, (1), 78–83.

44. Macejak, D.G., Lin, H., Webb, S., Chase, J., Jensen, K., Jarvis, T.C., Leiden, J.M., and Couture, L. (1999). Adenovirus-mediated expression of a ribozyme to c-myb mRNA inhibits smooth muscle cell proliferation and neointima formation *in vivo*. *Journal of Virology*, **73**, (9), 7745–51.

45. Hanna, A.K., Fox, J.C., Neschis, D.G., Safford, S.D., Swain, J.L., and Golden, M. A. (1997). Antisense basic fibroblast growth factor gene transfer reduces neointimal thickening after arterial injury. *Journal of Vascular Surgery*, **25**, (2), 320–5.

46. Lemarchand, P., Jones, M., Yamada, I., Crystal, R.G. (1993). *In vivo* gene transfer and expression in normal uninjured blood vessels using replication-deficient recombinant adenovirus vectors. *Circulation. Research*, **72**, 1132–38.

47. Lee, S.W., Trapnell, B.C., Rade, J.J., Virmani, R., and Dichek D.A. (1993). *In vivo* adenoviral vector-mediated gene transfer into balloon-injured rat carotid arteries. *Circulation Research*, **73**, 797–807.

48. Steg, P.G., Feldman, L.J., Scoazec. J.-Y., Tahlil, O., Barry, J.J., Boulechfar, S., Ragot, T., Isner, J.M., Perricaudet, M. (1994). Arterial gene transfer to rabbit endothelial and smooth muscle cells using percutaneous delivery of an adenoviral vector. *Circulation*, **90**, 1648–56.

49. Qian, H., Neplioueva, V., Shetty, G.A., Channon, K.M., and George, S.E. (1999). Nitric oxide synthase gene therapy rapidly reduces adhesion molecule expression and inflammatory cell infiltration in carotid arteries of cholesterol-fed rabbits. *Circulation*, **99**, (23), 2979–82.

50. Willard, J.E., Landau, C., Glamann, B., Burns, D., Jessen, M.E., Pirwitz, M.J., Gerard, R.D., Meidell, R.S. (1994). Genetic modification of the vessel wall: comparison of surgical and catheter-based techniques for delivery of recombinant adenovirus. *Circulation*, **89**, 2190–7.

51. Dunn, P.F., Newman, K.D., Jones, M., Yamada, I., Shayani, V., Virmani, R., Dickek, D.A. (1996). Seeding of vascular grafts with genetically modified endothelial cells. *Circulation*, **93**, 1439–46.

52. Mann, M.J., Whittemore, A.D., Donaldson, M.C., Belkin, M., Conte, M.S., Polak, J.F., Orav E.J., Ehsan, A., Dell'Acqua, G., Dzau, V.J. (1999). Ex-vivo gene therapy of human vascular bypass grafts with E2F decoy: the PREVENT single-centre, randomised, controlled trial. *Lancet*, **354**, 1493–98.

53. von der Leyen, L.H., Gibbons, G.H., Morishita, R., Lewis, N.P., Zhang, L., Kaneda, Y., and Cooke, J.P. (1994). *In vivo* gene transfer to prevent hyperplasia after vascular injury: effect of overexpression of constitutive nitric oxide synthase. *FASEB Journal*, **8**, A802.

54. Kullo I.J., Mozes, G., Schwartz, R.S., Gloviczki, P., Crotty, T.B., Barber, D.A., Katusic, Z.S., O'Brien, T. (1997). Adventitial gene transfer of recombinant endothelial nitric oxide synthase to rabbit carotid arteries alters vascular reactivity. *Circulation*, **96**, 2254–61.

55. Buttrick, P.M., Kass, A., Kitsis, R.N., Kaplan, M.L., Leinwand, L.A. (1992). Behaviour of genes directly injected into the rat heart *in vivo*. *Circulation Research*, **70**, 193–8.

56. Schumacher, M.D., Pecher, M.D., von Specht, B.U., and Stegmann, T. (1998). Induction of neoangiogenesis in ischemic myocardium by human growth factors: first clinical results of a new treatment of coronary heart disease. *Circulation*, **97**, 645–50.

57. Rosengart, T.K., Yee, L.Y., Patel, S.R., Sanborn, T.A., Parikh, M., Bergman, G.W., Hachamovitch, R., Szulc, M., Kligfield, P.D., Okin, P.M., Hahn, R.T., Devereux, R.B., Post, M.R., Hackett, N.R., Foster, T., Grasso, T.M., Lesser, M.L., Isom, O.W., Crystal, R.G. (1999). Angiogenesis gene therapy: Phase I assessment of direct intramyocardial administration of an adenovirus vector expressing VEGF121 cDNA to individuals with clinically significant severe coronary artery disease. *Circulation*, 468–74.

58. Fromes, Y., Salmon, A., Wang X., Collin, H., Rouche, A., Hagege, A., Schwartz, K., Fiszman, M.Y. (1999). Gene delivery to the myocardium by intrapericardial injection. *Gene Therapy*, **6**, (4), 683–8.

59. Zhang, J.C., Woo, Y.J., Chen, J.A., Swain, J.L., and Sweeney, H.L. (1999). Efficient transmural cardiac gene transfer by intrapericardial injection in neonatal mice. *Journal of Molecular and Cellular Cardiology*, **31**, (4), 721–32.

60. Lundell, A., Kelly, A.B., Anderson, J., Marijianowski, M., Rade, J.J., Hanson, S.R., and Harker, L.A. (1999). Reduction in vascular lesion formation by hirudin secreted from retrovirus-transduced confluent endothelial cells on vascular grafts in baboons. *Circulation*, **100**, 2018–24.

61. Epstein, S.E., Siegall, C.B., Biro, S., Fu, Y.M., Fitzgerald, D., and Pastan, I. (1991). Cytoxic effects of a recombinant chimeric toxin on rapidly proliferating vascular smooth muscle cells. *Circulation*, **84**, 778–87.

62. Guzman, R.J., Hirschowitz, E.A., Brody, S.L., Crystal, R.G., Epstein, S.E., Finkel, T. (1994). *In vivo* suppression of injury-induced vascular smooth muscle cell accumulation using adenovirus-mediated transfer of herpes simplex thymidine kinase gene. *Proceedings of the National Academy of Sciences of the United States of America*, **91**, 10732–6.

63. Asahara, T., Bauters, C., Pastore, C., Kearney, M., Rossow, S., Bunting, S., Ferrara, N., Symes, J.F., and Isner, J.M. (1995). Local delivery of vascular endothelial growth factor accelerates reendothelialization and attenuates intimal hyperplasia in balloon-injured rat carotid artery. *Circulation*, **91**, 2793–801.

64. Callow, A.D., Choi, E.T., Trachtenberg, J.D., Stevens, S.L., Connolly, D.T., Rodi, C., and Ryan, U.S. (1994). Vascular permeability factor accelerates endothelial regrowth following balloon angioplasty. *Growth Factors*, **10**, 223–8.

65. Lee, J.S. and Feldman, A.M. (1998). Gene therapy for therapeutic myocardial angiogenesis: a promising synthesis of two emerging technologies. *Nature Medicine*, **4**, (6), 739–42.

66. Mack, C.A., Patel, S.R., Schwarz, E.A., Zanzonico, P., Hahn, R.T., Ilercil, A., Devereux, R.B., Goldsmith, S.J., Christian, T.F., Sanborn, T.A., Kovesdi, I., Hackett, N., Isom, O.W., Crystal, R.G., and Rosengart, T.K. (1998). 'Biologic bypass with the use of adenovirus-mediated gene transfer of the complementary deoxyribonucleic acid for vascular endothelial growth factor 121 improves myocardial perfusion and function in the ischemic porcine heart.' *Journal of Thoracic & Cardiovascular Surgery*, **115**, (1), 168–76; discussion 176–7.

67. Takeshita, S., Tsurumi, Y., Couffinahl, T., Asahara, T., Bauters, C., Symes, J., Ferrara, N., and Isner, J.M. (1996). Gene transfer of naked DNA encoding for three isoforms of vascular endothelial growth factor stimulates collateral development *in vivo*. *Laboratory Investigation*, **75**, (4), 487–501.

68. Mack, C.A., Magovern, C.J., Budenbender, K.T., Patel, S.R., Schwarz, E.A., Zanzonico, P., Ferris, B., Sanborn, T., Isom, P., Isom, O.W., Crystal, R.G., and Rosengart, T.K. (1998). Salvage angiogenesis induced by adenovirus-mediated gene

transfer of vascular endothelial growth factor protects against ischemic vascular occlusion. *Journal of Vascular Surgery*, **27**, (4), 699–709.

69. von der Leyen, H.E., Gibbons, G.H., Morishita, R., Lewis, N.P., Zhang, L., Nakajima, M., Kaneda, Y., Cooke, J.P., and Dzau, V.J. (1995). Gene therapy inhibiting neointimal vascular lesion: *in vivo* transfer of endothelial cell nitric oxide synthase gene. *Proceedings of the National Academy of Sciences of the United States of America*, **92**, (4), 1137–41.

70. Tio, R.A., Isner, J.M., and Walsh, K. (1998). Gene therapy to prevent restenosis, the Boston experience. *Seminars in Interventional Cardiology*, **3**, (3–4), 205–10.

71. Duckers, H.J. and Nabel, E.G. (1998). Gene therapy for coronary artery disease: The University of Michigan Program. *Seminars in Interventional Cardiology*, **3**, (3–4), 201–4.

72. Murakami, H., Yayama, K., Miao, R.Q., Wang, C., Chao, L., and Chao, J. (1999). Kallikrein gene delivery inhibits vascular smooth muscle cell growth and neointima formation in the rat artery after balloon angioplasty. *Hypertension*, **34**, (2), 164–70.

73. Neufeld, G., Cohen, T., Gengrinovitch, S., and Poltorak, Z. (1999). Vascular endothelial growth factor (VEGF) and its receptors. *FASEB Journal*, **13**, (1), 9–22.

74. Gerber, H.P., Hillan, K.J., Ryan, A.M., Kowalski, J., Keller, G.A., Rangell, L., Wright, B.D., Radtke, F., Aguet, M., and Ferrara, N. (1999). VEGF is required for growth and survival in neonatal mice. *Development*, **126**, (6), 1149–59.

75. Neufeld, G., Tessler, S., Gitay-Goren, H., Cohen, T., and Levi, B.Z. (1994). Vascular endothelial growth factor and its receptors. *Progress in Growth Factor Research*, **5**, (1), 89–97.

76. Suzuki, H., Seto, K., Shinoda, Y., Mori, M., Ishimura, Y., Suematsu, M., and Ishii, H. (1999). Paracrine upregulation of VEGF receptor mRNA in endothelial cells by hypoxia-exposed hep G2 cells. *American Journal of Physiology*, **276**, (1 Pt 1), G92–7.

77. Korpelainen, E.I. and Alitalo, K. (1998). Signaling angiogenesis and lymphangiogenesis. *Current Opinion in Cell Biology*, **10**, (2), 159–64.

78. Poltorak, Z., Cohen, T., Sivan, R., Kandelis, Y., Spira, G., Vlodavsky, I., Keshet, E., and Neufeld, G. (1997). VEGF145, a secreted vascular endothelial growth factor isoform that binds to extracellular matrix. *Journal of Biological Chemistry*, **272**, (11), 7151–8.

79. Isner, J.M., Pieczek, A., Schainfeld, R., Blair, R., Haley, L., Asahara, T., Rosenfield, K., Razvi, S., Walsh, K., and Symes, J.F. (1996). Clinical evidence of angiogenesis after arterial gene transfer of phVEGF165 in patient with ischaemic limb. *Lancet*, **348**, (9024), 370–4.

80. Baumgartner, I., Pieczek, A., Manor, O., Blair, R., Kearney, M., Walsh, K., and Isner, J.M. (1998). Constitutive expression of phVEGF165 after intramuscular gene transfer promotes collateral vessel development in patients with critical limb ischemia. *Circulation*, **97**, (12), 1114–23.

81. Slavin, J. (1995). Fibroblast growth factors: at the heart of angiogenesis. *Cell Biology International*, **19**, (5), 431–44.

82. Baffour, R., Berman, J., Garb, J.L., Rhee, S.W., Kaufman, J., and Friedmann, P. (1992). Enhanced angiogenesis and growth of collaterals by *in vivo* administration of recombinant basic fibroblast growth factor in a rabbit model of acute lower limb ischemia: dose-response effect of basic fibroblast growth factor. *Journal of Vascular Surgery*, **16**, (2), 181–91.

83. Jonca, F., Ortega, N., Gleizes, P.E., Bertrand, N., and Plouet, J. (1997). Cell release of bioactive fibroblast growth factor 2 by exon 6-encoded sequence of vascular endothelial growth factor. *Journal of Biological Chemistry*, **272**, (39), 24203–9.
84. Xu, X., Weinstein, M., Li, C., and Deng, C. (1999). 'Fibroblast growth factor receptors (FGFRs) and their roles in limb development.' *Cell & Tissue Research*, **296**, (1), 33–43.
85. Giordano, F.J., Ping, P., McKirnan, M.D., Nozaki, S., DeMaria, A.N., Dillmann, W.H., Mathieu-Costello, O., and Hammond, H.K. (1996). Intracoronary gene transfer of fibroblast growth factor-5 increases blood flow and contractile function in an ischemic region of the heart. *Nature Medicine*, **2**, (5), 534–9.

7 The endothelial cell in atherosclerosis

Paul Bannon
Cardiovascular Surgery, Royal Prince Alfred Hospital, Sydney, Australia

Natalie James
Micromedical Industries Ltd, Sydney, Australia

Wendy Jessup
The Heart Research Institute, Cell Biology Group, Camperdown, Sydney,
NSW 2050, Australia

7.1 Overview

7.1.1 Functions of the endothelium

Originally blood vessels were viewed as simple conduits, regulated at a distance by circulating hormones and the sympathetic nervous system. This view has now greatly expanded to recognize the central role of endothelial cells as active participants in vascular function. Endothelial cells form a continuous single-cell-thick layer lining blood vessels of the entire vascular tree, located between the circulating blood and the underlying vascular cells and tissues. As such, it is an important interface which controls the passage of molecules and cells and participates in signal transduction.

The anatomical position of the endothelium as a physical barrier between the blood and vascular smooth muscle has focused attention on its role as a regulator of normal vascular function and its dysfunctional role in certain pathologies, including atherosclerosis.[1] Endothelial cells actively maintain a non-thrombogenic surface to the blood, present a non-adherent surface for leukocytes, act as a permeability barrier regulating exchange and active transport of macromolecules and cells between blood and intima, and maintain the basement membrane collagen and proteoglycans and other matrix proteins upon which they rest. This latter can, in turn, regulate proliferative responses of vascular smooth muscle cells as well as endothelial barrier function. Endothelial cells also have the capacity to secrete a large range of growth factors, cytokines, chemokines (reviewed in Chapter 12), and enzymes which act on adjacent cells and the extracellular matrix. In addition, endothelial cells have been demonstrated to play a role in the maintenance of vascular tone by the production of small molecules such as nitric oxide, prostacyclin, and endothelin.[1] These functional properties define the endothelium as a distinct organ of the body. This realization has led to the suggestion that functional alterations and/or loss of integrity of the endothelium may be a crucial initial step in cardiovascular diseases of the heart, the cerebral circulation and the kidney;[2] see also Chapter 5.

7.1.2 The endothelium and atherogenesis

Two key events in atherogenesis appear to be the focal influx and accumulation of low-density lipoprotein (LDL) in the intima and the recruitment of blood monocytes, primarily at 'lesion-prone' sites.[3] These lesion-prone sites differ structurally and functionally from adjacent normal areas. They are characterized by increased permeability to plasma proteins such as albumin and LDL, and show increased intimal cholesterol accumulation in animals challenged with high dietary fat. The glycocalyx coat of the endothelium is said to be thinner at these points and spontaneous monocyte recruitment is greater. Endothelial cell turnover is also greater at these sites. The factors which predispose certain sites as lesion-prone are not completely understood, although factors such as shear stress and hypercholesterolemia appear to be implicated (see Section 7.2).

Circulating monocytes enter the intima by first attaching to the endothelial surface, a process that involves expression of specific adhesion molecules on the surface of both cell types, and then migration between adjacent endothelial cells. Cell adhesion molecule (CAM) expression is described in more detail in Section 7.3. The expression of CAMs is a highly regulated process that appears to be altered in atherosclerosis.

While both lipoprotein and monocyte penetration of the intima are recognized as early events, their relationship is still not clear. The movement of both lipoproteins and cells across the endothelium may occur simultaneously, driven by alterations in permeability at lesion-prone sites. Alternatively, endothelial expression of CAMs and chemotactic factors may occur secondarily to intimal LDL accumulation. Thus it has been suggested that intimal oxidation or other modifications to LDL can initiate the expression of inflammatory genes whose products mediate monocyte binding and chemotaxis into the intima.[4] Mechanisms that may contribute to lipoprotein retention and modification are further discussed in Section 7.4 and in Chapters 13 and 16 in this volume.

Subsequent to monocyte migration into the intima, several characteristic stages of lesion development ensue, such as massive cholesterol accumulation by macrophages, transforming them into foam cells[5] (see Chapter 9), smooth muscle migration into the intima and proliferation[6] (see Chapter 10), the generation of extracellular lipid deposits, fibrosis and calcification.[7] There is evidence for the participation of endothelial cells in several of these later events. Finally, rupture of the endothelium can lead either to generation of microthrombi or massive thrombosis, depending on the scale of the event.

Thus the endothelial cell is involved in all stages of atherosclerosis. In the present chapter we will consider several aspects of normal endothelial cell function and dysfunction. These include the function of the endothelium as a barrier to penetration of molecules and cells, its participation in intimal modification of lipoproteins, and its role in regulation of vascular tone. A discussion of endothelial-derived growth factors and cytokines is provided elsewhere in this volume.

In many respects the alterations to endothelial function which are associated with atherogenesis share features in common with those seen in inflammation,[8] which has led to the suggestion that atherosclerosis is a specific inflammatory condition. However, atherosclerosis differs from general inflammation in several respects. Unlike general inflammation, atherosclerosis is a disease restricted to large vessels, suggesting that features of their structure are important determinants of the disease process. The factors that predispose to atherosclerosis such as hypercholesterolemia and hypertension are not normally considered as inflammatory signals. Presumably only in the larger vessels does hypercholesterolemia generate enhanced cellular and intimal lipid accumulation or can hypertension produce sufficiently abnormal shear stresses to adversely affect endothelial function and promote atherogenesis. The effects of some of these atherosclerosis-specific stimuli are discussed in this chapter.

7.2 Relationships between blood flow and endothelial cell function

The endothelium is both responsive to and an important regulator of vascular tone. The mechanisms by which endothelial cells sense vascular tone appear to largely depend on the detection of shear stress at the cell/blood interface. Flow of blood across the endothelium generates shear stresses that have an important influence upon the susceptibility of the artery wall to atherogenesis. Steady laminar shear stress stimulates cellular responses that are essential for endothelial cell function and are atheroprotective. Conversely, dysfunctional activity of the endothelium, as indicated by increased permeability to and intimal retention of lipoproteins, increased adhesion molecule expression, secretion of chemotactic factors and growth factors, correlates with areas of the artery wall that are exposed to low mean shear stress, oscillatory flow and flow reversal.[9] Blood flow in arteries is generally laminar, but vessel geometry near branches, bifurcations and curved regions promotes flow separation and vortex formation. In these regions, shear stresses and pressures fluctuate greatly in magnitude and orientation. It is these sites which are also most susceptible to atherosclerotic lesion formation.

Blood flow regulates the internal diameter of the artery wall acutely through signalling of the relaxation/contraction of smooth muscle cells and chronically through the reorganization of vascular wall cellular and extracellular components. Both of these forms of response involve the endothelium functioning as a mechanically sensitive signal transduction interface between the blood and the artery wall. The mechanism by which flow signals are transduced within the vascular wall has received considerable attention, in part because of the correlations between regions of disturbed flow and atherosclerosis.[10, 11] This process involves cytoskeletal elements and mechanisms which translate the physical forces into biochemical signals at mechanotransduction sites;[12] for a review see Ref. 9. Multiple elements are proposed to be involved in mechanotransduction, including integrin–matrix interactions, ion channels, G proteins, and caveolae. The downstream events initiated in response to shear stress are similar to those in receptor-mediated signalling, involving G proteins, increases in intracellular Ca^{2+}, and altered gene expression.

There is a diverse range of endothelial cell responses that are affected by shear stress stimuli[12] and which are also implicated in atherogenic processes. Endothelial release of the vasodilator nitric oxide (NO) is exquisitely sensitive to shear stress stimuli; high shear stress can increase NO production to 13-fold the static control level.[13] Multiple signalling pathways are involved in fast NO responses to flow stimuli, while modulation of gene expression is responsible for chronic adaptations of eNOS expression.

Flow also exerts an influence upon the tissue renin–angiotensin system. Shear stress modulates vascular levels of angiotensin II and bradykinin essentially through an influence upon expression of angiotensin-converting enzyme (ACE). The endothelium is a source of ACE which transforms angiotensin I into angiotensin II and breaks down bradykinin into inactive products. Extended exposure to shear stress has been found to reduce endothelial ACE mRNA and protein expression *in vitro* and *in vivo*.[14] ACE-mediated degradation of bradykinin attenuates endothelial release of vasodilators, especially NO.[15, 16] Angiotensin II is a potent vasoconstrictor and has been found to have a range of influences upon vascular tissues which have pro-atherogenic actions. These include effects on the availability of NO,[16] activation of endothelin mRNA in endothelial cells,[15] and effects upon vascular remodelling, including smooth muscle cell migration, proliferation, apoptosis, and extracellular matrix elaboration.[17]

In addition, the secretion of chemotactic factors and expression of adhesion molecules, which modulate adhesion and migration of leukocytes and monocytes into the blood vessel wall, is also profoundly but not consistently affected by shear stress. Reduction of the shear stress in rabbit carotid arteries enhanced VCAM-1 expression by 30 times that of control vessels, while ICAM-1 expression was down-regulated in these areas of low shear stress.[18] Regions of concentrated monocyte adhesion correlated with the areas of low flow and high VCAM-1 expression, whereas arteries exposed to high shear stress had mild enhancement of VCAM-1 and ICAM-1 expression, but no monocyte adhesion. Other forms of mechanical stimuli, in particular pressure and cyclic stretch, also influence endothelial cell responses; however, specific shear stress regimes have shown the greatest correlations with atherogenic processes.

Blood flow also influences the surface concentrations of LDL to which the endothelium is exposed. In fact, the luminal surface concentration of LDL along the carotid artery has been found to be inversely proportional to the wall shear rate along the vessel wall, and increased LDL infiltration into the vessel wall, corresponded to regions of high surface LDL concentration.[19] Accumulation of LDL within the vascular wall correlated with regions of the vasculature where the endothelium is exposed to low shear stress and/or various types of disturbed flow.[20] The types of disturbed flow that have been associated with LDL accumulation within the vessel wall and plaque development are characterized by flow reversal (where the direction of flow is reversed during the normal pulsatile cycle), vortices in the flow stream, large gradients in the shear (within regions of relatively low shear), and other forms of departure from aligned unidirectional flow.[10, 21]

In summary, the blood flow milieu of the endothelium has a crucial influence upon the endothelial cell responses, and associated environmental factors, that are linked with atherogenesis. Cellular responses influenced directly by shear stress include the release of modulators of vessel tone, particularly vasodilatory NO, the secretion of chemotactic factors, and the expression of cell adhesion molecules. The exposure of the cells to high LDL concentrations, combined with enhanced LDL uptake into the vessel wall in these regions, characterized by low shear stresses and disturbed flow patterns, further contribute to the dysfunctional characteristics and atherogenic nature of the endothelium in these regions. In contrast, endothelium which experiences high shear stresses and laminar flow is comparatively protected from atherosclerosis.

7.3 Barrier function of the endothelium

7.3.1 Introduction

The endothelium provides a permeability barrier controlling transport of water and solutes, including macromolecules, into the intima. Production and maintenance of the basement membrane and extracellular matrix is also dependent on the endothelial cell. This meshwork of protein–glycosaminoglycan complexes (proteoglycans) bathed in a hydrated gel has vital roles in permeability as well as cell–matrix adhesion, cell–cell adhesion, cell migration, and lipoprotein retention. In addition, the endothelium normally presents a non-adherent surface for leukocytes. However, at sites of atherogenesis, endothelial cells alter to encourage adhesion and migration of circulating monocytes and neutrophils into the intima. This process, which is also a critical event in atherogenesis, involves the formation and expression of cell adhesion molecules (CAMs), adhesive cell-surface glycoproteins.[8]

7.3.2 Extracellular matrix

The extracellular space underlying endothelial cells is largely filled by an intricate network of macromolecules making up the extracellular matrix. The two main classes of macromolecules within the matrix are (i) fibrous proteins of two functional types – structural (collagen and elastin) and adhesive (fibronectin and laminin), and (ii) polysaccharide glycosaminoglycans (GAGs), which are usually covalently bound to protein in the form of proteoglycans. The collagen fibres give strength and organization to the matrix and elastin fibres confer resilience. Fibronectin and laminin also promote cell attachment. Proteoglycan molecules form a highly hydrated, gel-like ground substance in which the fibrous proteins are embedded.[22] GAG chains fill most of the extracellular space providing mechanical support to the tissues while still allowing the rapid diffusion of water-soluble molecules and the migration of cells. However, endothelial-derived proteoglycans may also bind and retain lipoproteins which penetrate into the intima, increasing their local concentration and likelihood of uptake by local macrophages.[23] There is evidence that the relative and absolute amounts of proteoglycans secreted by endothelial cells are altered at sites of atherosclerosis. This may influence lipoprotein retention and local cell function.

7.3.3 Permeability to macromolecules

The endothelium exerts some control over the transport of plasma proteins into the vessel wall. It allows free exchange of water but regulates the transport of solutes. Endothelial monolayers are selectively permeable with the rate of transport being inversely related to the size of the molecule, although additional factors such as charge may also affect transport rates. The mechanism(s) by which the endothelium controls solute transport remains the subject of speculation. Several routes have been proposed involving transcellular and/or paracellular pathways.

7.3.3.1 Transcellular transport

The movement of solutes across endothelial cells appears to occur largely by a vesicular transcytotic route involving 'plasmalemmal vesicles'. These have been directly observed in intact endothelium.[24–26] Recently it has been suggested that endothelial transcytosis may be mediated by specialized regions of the plasma membrane termed caveolae.[27] These are particularly richly expressed in endothelial cells and contain a characteristic spectrum of lipid and protein components. These domains of the plasma membrane are also suggested to be involved in mechanotransduction, signalling, cholesterol transport, and regulation of the activity of a number of cytosolic enzymes;[28] a more detailed description of these structures is provided in Section 7.5.1.1. Although initially considered to be static features of the cell surface, it is now clear that caveolae are also internalized[29] and that their activity can alter in response to various mediators.[27]

While such transport appears to occur by fluid-phase transcytosis, there is evidence for some selectivity in the rates of transport of various molecules based on size and charge. For example, albumin was transported seven-fold faster than LDL across pig aortic endothelium,[30] while cationized albumin was transported more rapidly than albumin.[31] The endothelial cell surface has a net negative charge, with microdomains of positive and negative charges across it.[32] The most abundant anionic plasma protein is albumin and its transport is suggested to occur by plasmalemmal vesicles expressing positively charged sites.[33] The high negative charge of interstitial macromolecules of the extracellular matrix, such as the glycosaminoglycans heparan, dermatan and chondroitin sulphate, may also provide a charge-dependent selectivity to the endothelium.[34] The basement membrane and the composition of the extracellular matrix are important determinants of endothelial permeability to macromolecules, with studies indicating that the interstitial matrix is capable of up to 14-fold reduction in the diffusion of albumin *in vivo*.[35] Certainly the extracellular matrix is an important player in cell–cell and cell–basement membrane contact and adhesion and, hence, cell shape which may in turn affect permeability. Exogenous unfractionated heparin increased vascular permeability in rabbits by releasing a protein–glycosaminoglycan complex, presumably from the vessel wall, though low molecular weight heparins and dermatan sulphate did not.[36]

Participation in the highly specific, polarized transport of proteins by receptor-mediated transcytosis that is well-developed in epithelial monolayers appears less important in endothelial cells. This may be related to the polarity of expression of cell-surface receptors which seems less extensive than in epithelia. Polarized transport across endothelial monolayers is also less consistently reported. Thus there are reports of both a 10-fold greater active transfer of albumin from the basolateral to the apical face of endothelial cells than in the reverse direction,[37] and of equal rates of transfer in both directions.[38] These studies were performed using endothelial cells grown on porous filters, where it is important to maintain a balanced hydrostatic pressure across the monolayer. Perhaps some of the inconsistencies in the literature may be attributed to differences in the degree to which this balance was achieved. Under such conditions, a four-fold greater rate of transfer of LDL from the basolateral to the apical surface was measured.[39]

7.3.3.2 Paracellular transport

The relative contributions of transcellular and paracellular pathways of macromolecular transport are debatable. It is likely that inflammatory mediators affect endothelial permeability through increased paracellular transport. Endothelial permeability to albumin can be increased by exposure to pro-inflammatory mediators such as thrombin, histamine, and IL-1. IL-1 increases vascular endothelial permeability to macromolecules including albumin,[40] perhaps by influencing the endothelial cytoskeleton and creating intercellular gaps. Thrombin is thought to lead to a rapid influx of calcium into the cell that then affects the alignment of actin fibres, thereby regulating cell shape. This may account for the appearance of inter-cellular gaps after thrombin exposure. After thrombin exposure, albumin transport increased more than that of IgG, a much larger molecule.[38, 41] This increased trans-port of the smaller molecule (albumin) may occur via transcytosis or it may be that intercellular gaps may restrict the movement of IgG.

The development of tight junctions between cells, reflected in transcellular elec-trical resistance, seems to be less extensive in endothelium compared to epithelial cell monolayers. Nevertheless, it is clear that at areas where the integrity of the endothelial barrier is compromised, excessive penetration of plasma macromol-ecules, particularly lipoproteins, is associated with increased propensity to athero-sclerosis. Thus paracellular transport is increased at sites of endothelial cell turnover,[21] which within large vessels correspond to lesion-prone sites.

7.3.4 Permeability to cells

The normal endothelium presents an effective block to the passage of leukocytes, by the presentation of both a non-adhesive surface and a physical barrier. Following various forms of stimulation, the endothelium becomes activated to express mol-ecules that promote monocyte margination and penetration. These include chemokines such as MCP-1, vasoactive compounds such as prostaglandins, platelet activating factor and chemotactic cytokines such as interleukin-8 as well as specif-ic cell adhesion molecules. The 'injury' may include mechanical factors such as trauma or altered shear, or responses to chemical signals including lipopolysaccharide and chemokines. Many of these responses have been detected at sites of atheroscle-rosis, where margination and endothelial transmigration is a very early and persistent feature in the development of atherosclerotic lesions.[42, 43]

Cell adhesion molecules are a heterogeneous group of molecules expressed on cell membranes taking part in cell–cell and cell–stromal interactions and involved in basic functions such as cell recognition, migration, and contact inhibition. Endothelial and leukocyte adhesion molecules have specific roles in the margina-tion of leukocytes, endothelial–leukocyte adhesion, and transendothelial migration. There are three major groups of CAMs based on structure and function: selectins, the immunoglobulin superfamily, and integrins. The three CAM families act co-operatively. Selectins slow the circulating leukocytes and keep them in contact with the endothelial cell, allowing the immunoglobulin superfamily and the integrins to form firm bonds between the endothelial cell and the leukocytes which permit tran-scytosis of leukocytes across the endothelium.

The selectins involved in leukocyte adhesion to endothelium are P-selectin and E-selectin (on endothelium), and L-selectin (on leukocytes); their function is to ini-tiate leukocyte attachment to endothelium by a process termed 'rolling', involving reversible and 'weak' binding of endothelial selectins to carbohydrate moieties on the leukocyte. E-selectin is expressed on endothelial cells, and its expression is induced by IL-1, TNF-α and endotoxin. Target cells for E-selectin include polymorphonuclear leukocytes, monocytes, eosinophils, and some tumour cells.[44] The immunoglobulin superfamily expressed in endothelial cells includes ICAM-1 and VCAM-1. Platelet endothelial cell adhesion molecule-1 (PECAM-1) is also expressed on endothelial cells and may play a role in diapedesis. ICAM-1 ligands are the leukocyte integrins LFA-1 and Mac-1. ICAM-1 is constitutively expressed on endothelial cells, but expression can be increased by exposure to IL-1, interferon-γ, TNF-α, and endotoxin. VCAM-1 basal expression is lower than ICAM-1 and it is upregulated by IL-1, TNF, endotoxin, and IL-4. The integrins act as leukocyte receptors for the immunoglobulin superfamily and matrix components. In endothelial–leukocyte interactions, the most important integrins are the β_2 group and VLA-4. The β_2 integrins are found exclusively on leukocytes. They all bind to endothelial ICAM-1.[45]

7.3.4.1 Regulation of CAM expression

CAM expression is generally upregulated by exposure of endothelial cells to inflammatory cytokines such as IL-1β and TNF-α. This is mediated through the activation of nuclear factor kappaB (NF-κB), a transcriptional regulator of several CAM genes, including ICAM-1 and VCAM-1. Nitric oxide synthesis[46] or L-arginine incubation[47] inhibits cytokine-induced CAM expression in endothelial cells. NO represses VCAM-1 gene transcription in part by inhibiting NF-κB.[46] The ability of NO to inhibit endothelial activation and monocyte adhesion may contribute to some of its anti-atherogenic and anti-inflammatory properties within the vessel wall. Similarly, high-density lipoprotein, well-known to protect against atherosclerosis, has been shown to inhibit cytokine-induced expression of VCAM-1, ICAM-1 and E-selectin in endothelial cells by an NF-κB-dependent mechanism (Ref. 48; see also Chapter 14 in this volume). In contrast, androgens stimulate VCAM-1 expression.[49]

7.3.4.2 CAM expression in atherosclerosis

There is ample evidence suggesting a major but complex role of CAMs in atherogenesis. Increased expression of VCAM-1 on the surface of aortic endothelial cells prior to the development of the fatty streak[50] and increased expression of ICAM-1 and E-selectin on the surface of arterial endothelium, especially over sites of lipid and macrophage accumulation,[51] have been reported. An atherogenic diet in rabbits leads to both an increased endothelial cell expression of VCAM-1 *in vivo* and a concomitant formation of macrophage-rich fatty streaks on the aortic intima.[52] The atherogenic signals which mediate CAM expression are not known, but may include altered availability of NO and/or HDL and also possibly altered generation of inflammatory mediators acting on the endothelium either directly or indirectly by the local production of cytokines by smooth muscle cells or macrophages. The result is upregulation of the expression of adhesion molecules such as E-Selectin, ICAM-1, and VCAM-1. Once the initial wave of leukocytes has been recruited, these in turn may secrete other endothelial activators and chemotactic factors to enhance the response.[53]

One mediator of increased leukocyte adhesion in atherosclerosis may be OxLDL, although reports are complex. Treatment of endothelial cells with OxLDL stimulates monocyte–endothelial adhesion and this effect may be mediated by changes in endothelial ICAM-1 or VCAM-1 expression.[54, 55] Lysophosphatidylcholine, a component of atherosclerotic plaque and present in OxLDL, was suggested to induce endothelial expression of ICAM-1 and VCAM-1.[56] However it appears that although both hyperlipidemic conditions and OxLDL induce monocyte–endothelial adhesion, this effect does not involve the endothelial cell adhesion molecules E-selectin, ICAM-1, or VCAM-1.[54, 57] OxLDL does enhance the effect of tumour necrosis factor (TNF) on endothelial cell expression of VCAM-1.[58] Monocytes exposed to OxLDL also secrete factors which stimulate the expression of E-selectin, ICAM-1, and VCAM-1.[59]

It is important to note that in the above work, heavily oxidized LDL was used. The presence of this material in atherosclerotic lesions is uncertain. Other studies have alternatively examined the effects of less heavily oxidized LDL, or 'minimally modified' LDL (MM-LDL).[55, 60, 61] This contains early oxidation products, but is generally more variable and less well-defined in composition. MM-LDL increases endothelial–monocyte adhesion, but has been shown to have no effect on E-selectin, ICAM-1 or VCAM-1 expression by endothelial cells, indicating that a separate mechanism may be involved.[60, 61]

As with all studies concerning the effects of oxidized LDL on cell function, its relevance to pathology cannot be established until more specific information is available on the types and amounts of oxidized products to which the endothelium is exposed.

7.4   Effects of the endothelial cell on intimal lipoprotein modification

The penetration of low-density lipoproteins into the intima and their uptake by local macrophages to generate foam cells are early and critical events in the development of atherosclerotic lesions. Since normal LDL uptake is tightly regulated in the macrophage, foam cell development appears to require the modification of LDL to a form which can be more extensively endocytosed, either by alternate receptors on the cell, by phagocytosis, or a combination of both (see Chapter 9). Such modifications include LDL oxidation, binding to extracellular GAGs and associated proteins such as lipoprotein lipase, and modifications to the particle which promote its aggregation and subsequent endocytosis. The endothelial cell is capable of participating in all of these processes.

7.4.1 Oxidation

Although LDL receptor-mediated uptake of LDL is tightly regulated to prevent excess cholesterol accumulation, several modifications to the lipoprotein particle can redirect it to alternative endocytic pathways in macrophages that are not sterol-regulated. Such alterations include changes to the surface charge (more negative) of the particle, increasing its affinity for scavenger receptors. One modification to LDL which converts it to a ligand for the scavenger receptor is oxidation. Endothelial cells are capable of stimulating LDL oxidation *in vitro*.[62, 63]

7.4.2 Arterial retention

Lipoprotein lipase (LPL), a key enzyme in lipoprotein triglyceride metabolism, has been shown to markedly increase LDL retention by sub-endothelial matrix, where it binds to heparan and dermatan sulphate.[64] Association of LDL to matrix may influence its endocytosis by macrophages and its susceptibility to oxidation.[65] LPL is secreted by macrophages.[66] Thus LPL released from sub-endothelial macrophages may associate with adjacent endothelial cells and matrix and act to tether LDL. This retained LDL is more likely to be a substrate for macrophage endocytosis, either directly or following oxidative modification.

7.4.3 Aggregation

Aggregation of LDL can lead to massive phagocytic uptake and accumulation of cholesterol by macrophages. Aggregation has been induced *in vitro* by mechanical means, and also by enzymatic lipolysis of LDL by phospholipase C[67] or by secretory sphinomyelinase (sSMase). The latter is an extracellular form of acid (lysosomal) SMase that is also active at neutral pH. SMase treatment of LDL *in vitro* leads to fusion of lipoprotein particles which are avidly endocytosed by macrophages and lead to foam cell formation. Aggregation induced by SMase also increases LDL binding to non-proteoglycan components of the EC matrix.[68] sSMase is strongly expressed in cultured endothelial cells and increased by inflammatory cytokines,[69] and a significant proportion is secreted basolaterally. Not surprisingly, expression of sSMase *in vivo* has been detected in atherosclerotic lesions but not in normal arteries.[70] Experiments with SMase transgenic and knockout mice have indicated that SMase may be atherogenic.[71]

7.4.4 HDL modification

Recently a novel lipase has been identified which is expressed in endothelial cells and which has some homology with lipoprotein lipase and hepatic lipase.[72]

Endothelial lipase (EL) has substantial phospholipase activity, but less triglyceride lipase activity. Over-expression of EL in mice reduced plasma concentrations of HDL cholesterol and its major protein apolipoprotein A-I. The endothelial expression, enzymatic profile and *in vivo* effects of EL suggest that it may have a role in lipoprotein metabolism and vascular biology. Further studies should indicate how this enzyme is regulated *in vivo*.

7.5 Influence of endothelial cells on vascular tone

The endothelium is an active participant in the relationship between blood flow and vascular health and disease, through its ability to release mediators which regulate vascular tone. These include both vasorelaxant and vasoconstrictive agents. There is evidence for altered production and/or availability of several of these agents in atherosclerosis.

7.5.1 Vasodilators

7.5.1.1 Nitric oxide

Nitric oxide is an important, dynamic mediator of cardiovascular function. It is released by endothelial cells in response to physical stimuli, especially shear stress due to blood flow, and circulating biochemical stimuli. It exerts a vasorelaxant effect on underlying smooth muscle cells (see Chapter 5). In addition, NO participates in a broad range of other processes that influence atherogenesis, including the modulation of endothelial lipoprotein uptake and metabolism, platelet aggregation, leukocyte binding to the endothelium, and inhibition of vascular smooth muscle cell growth. In fact, the atheroprotective effects attributed to steady shear stress appear to be mediated largely through NO. Consequently, it has been proposed that the beneficial anti-atherogenic effects of regular aerobic exercise may be mediated through shear stress-induced increases in NO secretion.[9, 73] Studies in eNOS knockout mice demonstrate that nitric oxide from eNOS regulates blood flow under physiological conditions.[74]

NO is synthesized in the endothelium through conversion of L-arginine to L-citrulline by the endothelial isoform of nitric oxide synthase (eNOS), an enzyme which is constitutively expressed and regulated by a range of interactions within the endothelial cell. In its resting, inactive state, eNOS is localized to the cytosolic face of the cell; most of the cellular complement of the enzyme is found within specialized signal-transducing regions of the plasma membrane called caveolae,[75] and some is associated with the Golgi membranes.[76] Caveolae are small invaginations (approximately 70 nm diameter) in the cell plasma membrane that are relatively enriched in specific glycosphingolipids and cholesterol. Endothelial cells have abundant caveolae, which are located primarily along the cell margin and the leading edge of the cell.[75] Caveolin is the principal structural caveolar protein and it is particularly important in the signalling functions of the caveolae. Recent investigations of caveolae and caveolar function reveal that they play an important, dynamic role in signal transduction by acting as organizing loci for signalling molecules. Signalling molecules that have been localized to caveolae include Src family kinases, nitric oxide synthases, a number of receptors (including those for epidermal growth factor, platelet-derived growth factor, B_2 bradykinin, endothelin, and HDL (SR-B1)), phospholipase Cγ, protein kinases, Ras, G protein subunits, adenylate cyclase, Mek1, and Erk2.[77] Targeting of eNOS to the plasmalemmal caveolae is dependent on myristoylation and palmitoylation of the enzyme.[75, 78]

eNOS is characteristically activated by Ca^{2+}-calmodulin binding, as a short-term response to receptor-dependent transient increases in intracellular Ca^{2+}. The interaction of caveolin with eNOS holds the enzyme in an inactive conformation, which may then be reversibly released by Ca^{2+}-calmodulin activation.[79] Reciprocal allosteric interactions between the caveolin scaffolding domain (residues 82–101 of caveolin) and Ca^{2+}-calmodulin control the activation of eNOS.[80] Intracellular calcium transients may be induced by shear stress or by circulating agents such as bradykinin or acetylcholine. In fact, the dramatic influence of caveolin on eNOS activation by calmodulin has been suggested as a potential mechanism for integration of the influences of hemodynamic shear stress and hormonally stimulated Ca^{2+} increases on eNOS activation.[79]

Increases in flow and pressure have been found to activate caveolar eNOS by causing dissociation of eNOS from inhibitory association with caveolin, leading to calmodulin-stimulated eNOS activation.[81] Consequently, caveolae have been proposed to act as flow-sensing organelles, since they display the capacity to rapidly transduce mechanical stimuli.[82] Caveolae have also been linked with signal transduction pathways associated with responses to flow including regulation of eNOS activity and Ras/MAPK.[81–83] The observation that caveolae-rich regions of endothelial cells are a site of initiation of intracellular Ca^{2+} waves[84] further implicates involvement of caveolae in flow signal transduction.

NO availability in atherosclerosis Reduced availability of NO contributes to the pathogenesis of atherosclerosis and a number of other disease states, including hypertension and diabetes. Hypercholesterolemia is also associated with impaired in vivo endothelium-dependent relaxation,[85] although NO synthesis is normal or increased. Since cholesterol can act in the regulation of caveolar function, by affecting both expression of caveolin and activity of caveolae,[86, 87] it is likely that elevated plasma cholesterol levels translate into increased endothelial cell cholesterol content and associated influences on NOS activity. Indeed, cholesterol loading was found to stimulate eNOS expression, association with caveolae and Ca^{2+}-dependent NO production.[88] Flow-stimulated activation of signal transduction pathways, including eNOS, and Ras/MAPK activation are also cholesterol dependent.[83] Thus the increased generation of NO in hypercholesterolemia may be explained by effects on caveolar function. The reduction in NO bioavailability in hypercholesterolemia, despite its increased formation, has been attributed to excessive consumption by superoxide radicals ($O^{2\bullet-}$), generated mainly by endothelial cells in response to the elevated cholesterol or the presence of lysophospholipids, or both of these.[89] Other sources of reactive oxygen species may also consume NO.[88]

Interest in the possible presence of oxidized low-density lipoprotein in atherosclerotic lesions has stimulated the study of its effects on NO generation. OxLDL treatment of vessels leads to inhibition of endothelial-dependent relaxation.[90] This effect may be mediated at several levels. Firstly, OxLDL has an immediate and dramatic effect upon eNOS localization and function,[91] inducing redistribution of eNOS from caveolae to an internal membrane compartment and inhibiting acetylcholine-mediated stimulation of eNOS. In parallel, co- and post-translational processing (myristoylation, palmitoylation, phosphorylation) of eNOS were unchanged, but caveolae were found to be depleted of cholesterol by OxLDL treatment. These researchers suggest that, by acting as an acceptor for cholesterol and depleting the caveolae of cholesterol, OxLDL causes dissociation of eNOS from caveolae, concomitant with inactivation of the enzyme. The effect of OxLDL on endothelial cell NO generation and on endothelial dysfunction could be reproduced by 7-ketocholesterol, the major oxysterol present in the particle.[92, 93] There is some evidence to indicate that oxysterols interfere with caveolar function.[86] Secondly, OxLDL may also scavenge NO directly.[94] Finally, exposure of endothelial cells to OxLDL reduced eNOS expression up to 56%, through a combination of transcriptional inhibition and post-transcriptional mRNA destabilization.[95] These studies indicate that OxLDL can reduce endothelial NO release through effects on eNOS activity and gene expression.

7.5.1.2 Prostacyclin

Prostacyclin is an additional biochemical mediator with a strong vasodilatory action. Both prostacyclin and thromboxane A_2 are derived from arachidonic acid in the vascular endothelium and in platelets. These labile metabolites antagonize each other: prostacyclin is a vasodilator, prevents platelet aggregation and dissipates preformed platelet clumps, whereas thromboxane constricts arteries and has a pro-aggregatory effect. Prostacyclin is in fact the most potent endogenous inhibitor of platelet aggregation so far discovered, and its release is an important element in the maintenance of circulatory hemostasis by the endothelium. Prostacyclin is produced by endothelial cells exposed to shear stress; release is greater under conditions of pulsatile flow than where the flow is steady[96] and synergistic interactions between prostacyclin and NO occur.[16] Prostacyclin acts through activation of adenylate cyclase, another signalling molecule with a caveolar localization, and leads to increased levels of cyclic adenosine monophosphate (cAMP). Prostacyclin synthesis by human aortic cells declines with age,[97] in diabetes mellitus and in atherosclerosis. Prostacyclin stimulates cholesteryl ester hydrolysis by smooth muscle cells, and high-density lipoprotein (HDL) augments this activity.[98] Further influences of prostacyclin and its analogues upon inhibition of proliferation of vascular smooth muscle have also been suggested.[99] These characteristics ndicate a number of mechanisms through which prostacyclin plays a role in atheroprotection.

7.5.2 Vasoconstrictors

The endothelium is a source of contracting factors such as endothelin-1, thromboxane A_2, and endoperoxides. Endothelin is the most significant of the contracting factors to be implicated in atherosclerosis.

7.5.2.1 Endothelin

The endothelins are a family of 21-amino acid peptides, which contain two intramolecular disulphide bonds and exist in a number of distinct isoforms (endothelin-1, endothelin-2, and endothelin-3). Endothelin-1 plays an important physiological role in the regulation of basal vascular tone and blood pressure in healthy humans, since it is the most potent vasopressor agent to be isolated. As discussed in Chapter 5, circulating endothelin-1 has been found to be enhanced in atherosclerosis and other conditions associated with endothelial dysfunction. Endothelin-1 is synthesized by endothelial cells and smooth muscle cells and it acts as a smooth muscle cell mitogen. Smooth muscle cells migrating into the intima of arteries in atherosclerosis have demonstrated upregulation of endothelin-1 expression, compared with regions of healthy vessel.[100] Treatment of endothelial cells with OxLDL induced endothelin expression and release from cultured cells and the endothelium of intact blood vessels.[101] Endothelin also stimulates the production of autocrine growth factors through enhancement of the expression of platelet-derived growth factor in smooth muscle cells. Thus endothelin release is enhanced during atherogenesis and it also contributes to atherogenic processes through a number of interactions, particularly those involving smooth muscle cells.

In summary, NO appears important for atheroprotection. Its release is stimulated by blood flow along the artery wall, through multiple signalling mechanisms. Recent studies indicate the involvement of plasmalemmal caveolae in the integration and transduction of hormonal and shear stress stimuli for NO release. Impaired endothelium-dependent relaxation, in parallel with reduced bioavailability of NO, is characteristic of atherosclerotic blood vessels and a range of interactions have been suggested to contribute to the reduced NO activity. These involve the influences of cholesterol and OxLDL upon caveolar signalling mechanisms, excess generation of superoxide radicals, and the inhibition of eNOS activity and expression by OxLDL. Prostacyclin has vasodilatory actions, interacts synergistically with NO, and is also implicated in atheroprotective processes. In contrast, endothelin is a potent vaso-constrictor and it is suggested that it plays a role in atherogenesis.

7.6 Summary

In recent years, our increased understanding of the role of the endothelium in vascular health and disease has transformed our view of its importance. Rather than presenting a simple mechanical barrier between blood and tissue, it is now clear that the endothelium is a dynamic, responsive and regulatory organ that plays an active and major role in vascular function. In the earliest detectable stages of atherosclerosis, alterations to endothelial permeability to both macromolecules and cells are found, suggesting that the endothelium is the initial site of dysfunction. As such it is a major target for therapeutic intervention.

References

1. Ross, R. (1993). The pathogenesis of atherosclerosis: a perspective for the 1990s. *Nature*, **362**, 801.
2. Luscher, T.F. and Tanner, F.C. (1993). Endothelial regulation of vascular tone and growth. *Am. J. Hypertens.*, **6**, 283S.
3. Schwartz, C.J., Valente, A.J., and Sprague, E.A. (1993). A modern view of atherogenesis. *Am. J. Cardiol.*, **71**, 9B.
4. Navab, M., Fogelman, A.M., Berliner, J.A., Territo, M.C., Demer, L.L., Frank, J.S., Watson, A.D., Edwards, P.A., and Lusis, A.J. (1995). Pathogenesis of atherosclerosis. *Am. J. Cardiol.*, **76**, 18C.
5. Gerrity, R.G. (1981). The role of the monocyte in atherogenesis: I. Transition of blood-borne monocytes into foam cells in fatty lesions. *Am. J. Pathol.*, **103**, 181.
6. Campbell, J.H. and Campbell, G.R. (1994). Cell biology of atherosclerosis. *J. Hypertens. Suppl.*, **12**.
7. Stary, H.C. *et al.* (1994). A definition of initial, fatty streak, and intermediate lesions of atherosclerosis. A report from the Committee on Vascular Lesions of the Council on Arteriosclerosis, American Heart Association. *Circulation*, **89**, 2462.
8. Munro, J.M. and Cotran, R.S. (1998). The pathogenesis of atherosclerosis: atherogenesis and inflammation. *Lab Invest.*, **58**, 249.
9. Traub, O. and Berk, B.C. (1998). Laminar shear stress: mechanisms by which endothelial cells transduce an atheroprotective force. *Arterioscler. Thromb. Vasc. Biol.*, **18**, 677.
10. Zarins, C.K., Giddens, D.P., Bharadvaj, B.K., Sottiurai, V.S., Mabon, R.F., and Glagov, S. (1983). Carotid bifurcation atherosclerosis. Quantitative correlation of plaque localization with flow velocity profiles and wall shear stress. *Circ. Res.*, **53**, 502.
11. Ku, D.N. and Giddens, D.P. (1983). Pulsatile flow in a model carotid bifurcation. *Arteriosclerosis*, **3**, 31.

12. Davies, P.F. (1995). Flow-mediated endothelial mechanotransduction. *Physiol. Rev.*, **75**, 519.

13. Corson, M., James, N., Latta, S., Nerem, R., Berk, B., and Harrison, D. (1996). Phosphorylation of endothelial nitric oxide synthase in response to fluid shear stress. *Circ. Res.*, **79**, 984.

14. Rieder, M.J., Carmona, R., Krieger, J.E., Pritchard, K.A., Jr, and Greene, A.S. (1997). Suppression of angiotensin-converting enzyme expression and activity by shear stress. *Circ. Res.*, **80**, 312.

15. Luscher, T.F. (1993). Angiotensin, ACE-inhibitors and endothelial control of vasomotor tone. *Basic Res. Cardiol.*, **88**, 15.

16. Mombouli, J.V. and Vanhoutte, P.M. (1999). Endothelial dysfunction: from physiology to therapy. *J. Mol. Cell. Cardiol.*, **31**, 61.

17. Gibbons, G.H. (1997). Vasculoprotective and cardioprotective mechanisms of angiotensin-converting enzyme inhibition: the homeostatic balance between angiotensin II and nitric oxide. *Clin. Cardiol.*, **20**, II.

18. Walpola, P., Gotlieb, A., Cybulski, M., and Langille, B. (1995). Expression of ICAM-1 and VCAM-1 and monocyte adherence in arteries exposed to altered shear stress. *Arterioscler. Thromb. Vasc. Biol.*, **15**, 2.

19. Deng, X., Marois, Y., How, T., Merhi, Y., King, M., Guidoin, R., and Karino, T. (1995). Luminal surface concentration of lipoprotein (LDL) and its effect on the wall uptake of cholesterol by canine carotid arteries. *J. Vasc. Surg.*, **21**, 135.

20. Berceli, S., Warty, V., Sheppeck, R., Mandarino, W., Tanksale, S., and HS, B. (1990). Hemodynamics and low density lipoprotein metabolism: rates of low density lipoprotein incorporation and degradation along medial and lateral walls of the rabbit aortoiliac bifurcation. *Arteriosclerosis*, **10**, 688.

21. Weinbaum, S. and Chien, S. (1993). Lipid transport aspects of atherogenesis. *J. Biomech. Eng.*, **115**, 602.

22. Borsum, T. (1991). Biochemical properties of vascular endothelial cells. *Virchows Arch. B. Cell. Pathol. Incl. Mol. Pathol.*, **60**, 279.

23. Vijayagopal, P., Srinivasan, S., Radhakrishnamurthy, B., and Berenson, G. (1992). Lipoprotein–proteoglycan complexes from atherosclerotic lesions promote cholesteryl ester accumulation in human monocytes/macrophages. *Arterioscler. Thromb.*, **12**, 237.

24. Willemsen, R., Wisselaar, H.A., and Vander Ploeg, A. (1990). Plasmalemmal vesicles are involved in transendothelial transport of albumin, lysosomal enzymes and mannose 6-phosphate receptor fragments in capillary endothelium. *Eur. J. Cell. Biol.*, **51**, 235.

25. Vasile, E., Antohe, F., Simionescu, M., and Simionescu, N. (1989). Transport pathways of beta-VLDL by aortic endothelium of normal and hypercholesterolemic rabbits. *Atherosclerosis*, **75**, 195.

26. Simionescu, M. and Simionescu, N. (1991). Endothelial transport of macromolecules: transcytosis and endocytosis. A look from cell biology. *Cell. Biol. Rev.*, **25**, 1.

27. Feng, Y., Venema, V.J., Venema, R.C., Tsai, N., Behzadian, M.A., and Caldwell, R.B. (1999). VEGF-induced permeability increase is mediated by caveolae. *Invest. Ophthalmol. Vis. Sci.*, **40**, 157.

28. Parton, R.G. (1996). Caveolae and caveolins. *Curr. Opin. Cell. Biol.*, **8**, 542.

29. Parton, R.G., Joggerst, B., and Simons, K. (1994). Regulated internalization of caveolae. *J. Cell. Biol.*, **127**, 1199.

30. Fry, D.L., Haupt, M.W., and Pap, J.M. (1992). Effect of endothelial integrity, transmural pressure, and time on the intimal-medial uptake of serum 125I-albumin and 125I-LDL in an *in vitro* porcine arterial organ-support system. *Arterioscler. Thromb.*, **12**, 1313.

31. Vorbrodt, A.W. and Trowbridge, R.S. (1991). Ultrastructural study of transcellular transport of native and cationized albumin in cultured sheep brain microvascular endothelium. *J. Neurocytol.*, **20**, 998.

32. Simionescu, N., Simionescu, M., and Palade, G.E. (1981). Differentiated microdomains on the luminal surface of the capillary endothelium. I. Preferential distribution of anionic sites. *J. Cell. Biol.*, **90**, 605.

33. Ghitescu, L., Fixman, A., Simionescu, M., and Simionescu, N. (1986). Specific binding sites for albumin restricted to plasmalemmal vesicles of continuous capillary endothelium: receptor-mediated transcytosis. *J. Cell. Biol.*, **102**, 1304.

34. Vernier, R.L., Klein, D.J., Sisson, S.P., Mahan, J.D., Oegema, T.R., and Brown, D.M. (1983). Heparan sulfate – rich anionic sites in the human glomerular basement membrane. Decreased concentration in congenital nephrotic syndrome. *N. Engl. J. Med.*, **309**, 1001.

35. Fox, J.R. and Wayland, H. (1979). Interstitial diffusion of macromolecules in the rat mesentery. *Microvasc. Res.*, **18**, 255.

36. Blajchman, M.A., Young, E., and Ofosu, F.A. (1989). Effects of unfractionated heparin, dermatan sulfate and low molecular weight heparin on vessel wall permeability in rabbits. *Ann. N. Y. Acad. Sci.*, **556**, 245.

37. Shasby, D.M. and Shasby, S.S. (1885). Active transendothelial transport of albumin. Interstitium to lumen. *Circ. Res.*, **57**, 903.

38. Malik, A.B., Lynch, J.J., and Cooper, J.A. (1989). Endothelial barrier function. *J. Invest. Dermatol.*, **93**, 62S.

39. Kim, M.J., Dawes, J., and Jessup, W. (1994). Transendothelial transport of modified low-density lipoproteins. *Atherosclerosis*, **108**, 5.

40. Campbell, W.N., Ding, X., and Goldblum, S.E. (1992). Interleukin-1 alpha and beta augment pulmonary artery transendothelial albumin flux *in vitro*. *Am. J. Physiol.*, **263**, L128.

41. Siflinger-Birnboim, A., Cooper, J.A., del Vecchio, P.J., Lum, H., and Malik, A.B. (1988). Selectivity of the endothelial monolayer: effects of increased permeability. *Microvasc. Res.*, **36**, 216.

42. Joris, I., Zand, T., Nunnari, J.J., Krolikowski, F.J., and Manjo, G. (1983). Studies on the pathogenesis of atherosclerosis. I. Adhesion and emigration of mononuclear cells in the aorta of hypercholesterolaemic rats. *Am. J. Pathol.*, **113**, 341.

43. Ross, R. and Agius, L. (1992). The process of atherogenesis-cellular and molecular interaction: from experimental animal models to humans. *Diabetologia*, **35** (Suppl. 2), S34.

44. McEver, R.P. (1994). Selectins. *Curr. Opin. Immunol.*, **6**, 75.

45. Springer, T.A. and Lasky, L.A. (1991). Cell adhesion. Sticky sugars for selectins. *Nature*, **349**, 196.

46. De Caterina, R., Libby, P., Peng, H.B., Thannickal, V.J., Rajavashisth, T.B., Gimbrone, M.J., Shin, W.S., and Liao, J.K. (1995). Nitric oxide decreases cytokine-induced endothelial activation. Nitric oxide selectively reduces endothelial expression of adhesion molecules and proinflammatory cytokines. *J. Clin. Invest.*, **96**, 60.

47. Adams, M., Jessup, W., Hailstones, D., and Celermajer, D. (1997). L-Arginine reduces human monocyte adhesion to vascular endothelium and endothelial cell expression of cell adhesion molecules. *Circulation*, **95**, 662.
48. Cockerill, G., Rye, K.-A., Clay, M., Vadas, M., Gamble, J., and Barter, P. (1995). High-density lipoproteins inhibit cytokine-induced expression of endothelial cell adhesion molecules. *Arterioscler. Thromb. Vasc. Biol.*, **15**, 1987.
49. McCrohon, J., Nakhla, S., Jessup, W., Handelsman, D., and Celermajer, D. (1999). Androgen exposure increases human monocyte adhesion to vascular endothelium and endothelial cell expression of VCAM-1. *Circulation*, **99**, 2317.
50. Cybulsky, M.I. and Gimbrone, M.A., Jr (1991). Endothelial expression of a mononuclear leukocyte adhesion molecule during atherogenesis. *Science*, **251**, 788.
51. Poston, R.N., Haskard, D.O., Coucher, J.R., Gall, N.P., and Johnson-Tidey, R.R. (1992). Expression of intercellular adhesion molecule-1 in atherosclerotic plaques. *Am. J. Pathol.*, **140**, 665.
52. Li, H., Cybulsky, M.I., Gimbrone, M.A., Jr, and Libby, P. (1993). An atherogenic diet rapidly induces VCAM-1, a cytokine-regulatable mononuclear leukocyte adhesion molecule, in rabbit aortic endothelium. *Arterioscler. Thromb.*, **13**, 197.
53. Bevilacqua, M.P., Nelson, R.M., Mannori, G., and Cecconi, O. (1994). Endothelial-leukocyte adhesion molecules in human disease. *Annu. Rev. Med.*, **45**, 361.
54. Jeng, J.R., Chang, C.H., Shieh, S.M., and Chiu, H.C. (1993). Oxidized low-density lipoprotein enhances monocyte-endothelial cell binding against shear-stress-induced detachment. *Biochim. Biophys. Acta.*, **1178**, 221.
55. Berliner, J.A., Territo, M.C., Sevanian, A., Ramin, S., Kim, J.A., Bamshad, B., Esterson, M., and Fogelman, A.M. (1990). Minimally modified low density lipoprotein stimulates monocyte endothelial interactions. *J. Clin. Invest.*, **85**, 1260.
56. Kume, N., Cybulsky, M.I., and Gimbrone, M.A., Jr (1992). Lysophosphatidylcholine, a component of atherogenic lipoproteins, induces mononuclear leukocyte adhesion molecules in cultured human and rabbit arterial endothelial cells. *J. Clin. Invest.*, **90**, 1138.
57. Bannon, P.G. Endothelial barrier function: assessment and preservation. Ph.D. Thesis, University of Sydney.
58. Khan, B.V., Parthasarathy, S.S., Alexander, R.W., and Medford, R.M. (1995). Modified low density lipoprotein and its constituents augment cytokine-activated vascular cell adhesion molecule-1 gene expression in human vascular endothelial cells. *J. Clin. Invest.*, **95**, 1262.
59. Frostegard, J., Wu, R., Haegerstrand, A., Patarroyo, M., Lefvert, A.K., and Nilsson, J. (1993). Mononuclear leukocytes exposed to oxidized low density lipoprotein secrete a factor that stimulates endothelial cells to express adhesion molecules. *Atherosclerosis*, **103**, 213.
60. Calderon, T.M., Factor, S.M., Hatcher, V.B., Berliner, J.A., and Berman, J.W. (1994). An endothelial cell adhesion protein for monocytes recognized by monoclonal antibody IG9. Expression *in vivo* in inflamed human vessels and atherosclerotic human and Watanabe rabbit vessels. *Lab. Invest.*, **70**, 836.
61. Kim, J.A., Territo, M.C., Wayner, E., Carlos, T.M., Parhami, F., Smith, C.W., Haberland, M.E., Fogelman, A.M., and Berliner, J.A. (1994). Partial characterization of leukocyte binding molecules on endothelial cells induced by minimally oxidized LDL. *Arterioscler. Thromb.*, **14**, 427.

62. van Hinsberg, V.W., Scheffer, M., Havekes, L., and Kempen, H.J. (1986). Role of endothelial cells and their products in the modification of low-density lipoproteins. *Biochim. Biophys. Acta*, **878**, 49.

63. Steinbrecher, U.P., Parthasarathy, S., Leake, D.S., Witztum, J.L., and Steinberg, D. (1984). Modification of low density lipoprotein by endothelial cells involves lipid peroxidation and degradation of low density lipoprotein phospholipids. *Proc. Natl. Acad. Sci. USA*, **81**, 3883.

64. Saxena, U., Ferguson, E., and Bisgaier, C.L. (1993). Apolipoprotein E modulates low density lipoprotein retention by lipoprotein lipase anchored to the subendothelial matrix. *J. Biol. Chem.*, **268**, 14812.

65. Hurt-Camejo, E., Camejo, G., Rosengren, B., Lopez, F., Ahlstrom, C., Fager, G., and Bondjers, G. (1992). Effect of arterial proteoglycans and gly-cosaminoglycans on low density lipoprotein oxidation and its uptake by human macrophages and arterial smooth muscle cells. *Arterioscler. Thromb.*, **12**, 569.

66. Chait, A., Iverius, P.-H., and Brunzell, J.D. (1982). Lipoprotein lipase secretion by human monocyte-derived macrophages. *J. Clin. Invest.*, **69**, 490.

67. Heinecke, J.W., Suits, A.G., Aviram, W, Chait, A. (1991). Phagoctytosis of lipase-aggregated low density lipoprotein promotes macrophage foam cell formation. *Arterioscler. Thromb.*, **11**, 1643.

68. Oorni, K., Hayala, J.K., Annila, A, Ala-Korpela, M, Kovanen, P.T. (1998). Sphingomyelinase induces aggregation and fusion, but phospholipase A2 only aggregation, of low density lipoprotein (LDL) particles. Two distinct mechanisms leading to increased binding strength of LDL to human aortic proteoglycans. *J. Biol. Chem.* **273**, 29127.

69. Marathe, S., Schissel, S.L., Yellin, M.J., Beatini, N., Mintzer, R., Williams, K.J., and Tabas, I. (1998). Human vascular endothelial cells are a rich and regulatable source of secretory sphingomyelinase. Implications for early atherogenesis and ceramide-mediated cell signaling. *J. Biol. Chem.*, **273**, 4081.

70. Marathe, S., Kuriakose, G., Williams, K.J., and Tabas, I. (1999). Sphingomyelinase, an enzyme implicated in atherogenesis, is present in atherosclerotic lesions and binds to specific components of the subendothelial extracellular matrix. *Arterioscler. Thromb. Vasc. Biol.*, **19**, 2648.

71. Wong, M.-L., Xie, B., Beatii, N., Phu, P,. Marathe, S., Johns, A., Hirsch, E., Williams, K., and Tabas, I. (1999). Sphingomyelinase transgenic and knockout mice: direct evidence that sphingomyelinase is atherogenic *in vivo*. *Circulation*, **100**, 1.

72. Jaye, M. *et al.* (1999). A novel endothelial-derived lipase that modulates HDL metabolism. *Nat. Genet.*, 424.

73. Sessa, W., Pritchard, K., Seyedi, N., Wang, J., and Hintze, T. (1994). Chronic exercise in dogs increases coronary vascular nitric oxide production and endothelial cell nitric oxide synthase gene expression. *Circ. Res.*, **74**, 349.

74. Mashimo, H. and Goyal, R.K. (1999). Lessons from genetically-engineered animal models. IV. Nitric oxide synthase gene knockout mice. *Am. J. Physiol.*, **277**, G745.

75. Garcia-Cardena, G., Oh, P., Liu, J., Schnitzer, J.E., and Sessa, W.C. (1996). Targeting of nitric oxide synthase to endothelial cell caveolae via palmitoylation: implications for nitric oxide signaling. *Proc. Natl. Acad. Sci. USA*, **93**, 6448.

76. Sessa, W.C., Garcia-Cardena, G., Liu, J., Keh, A., Pollock, J.S., Bradley, J., Thiru, S., Braverman, I.M., and Desai, K.M. (1995). The Golgi association of endothelial

nitric oxide synthase is necessary for the efficient synthesis of nitric oxide. *J. Biol. Chem.*, **270**, 17641.

77. Shaul, P.W. and Anderson, R.G. (1998). Role of plasmalemmal caveolae in signal transduction. *Am. J. Physiol.*, **275**, L843.

78. Shaul, P.W., Smart, E.J., Robinson, L.J., German, Z., Yuhanna, I.S., Ying, Y., Anderson, R.G., and Michel, T. (1996). Acylation targets endothelial nitric-oxide synthase to plasmalemmal caveolae. *J. Biol. Chem.*, **271**, 6518.

79. Michel, J.B., Feron, O., Sase, K., Prabhakar, P., and Michel, T. (1997).Caveolin versus calmodulin. Counterbalancing allosteric modulators of endothelial nitric oxide synthase. *J. Biol. Chem.*, **272**, 25907.

80. Michel, J.B., Feron, O., Sacks, D., and Michel, T. (1997). Reciprocal regulation of endothelial nitric-oxide synthase by Ca^{2+}-calmodulin and caveolin. *J. Biol. Chem.*, **272**, 15583.

81. Rizzo, V., Sung, A., Oh, P., and Schnitzer, J.E. (1998). Rapid mechanotransduction *in situ* at the luminal cell surface of vascular endothelium and its caveolae. *J. Biol. Chem.*, **273**, 26323.

82. Rizzo, V., McIntosh, D.P., Oh, P., and Schnitzer, J.E. (1998). *In situ* flow activates endothelial nitric oxide synthase in luminal caveolae of endothelium with rapid caveolin dissociation and calmodulin association. *J. Biol. Chem.*, **273**, 34724.

83. Park, H., Go, Y.M., St John, P.L., Maland, M.C., Lisanti, M.P., Abrahamson, D.R., and Jo, H. (1998). Plasma membrane cholesterol is a key molecule in shear stress-dependent activation of extracellular signal-regulated kinase. *J. Biol. Chem.*, **273**, 32304.

84. Isshiki, M., Ando, J., Korenaga, R., Kogo, H., Fujimoto, T., Fujita, T., and Kamiya, A. (1998). Endothelial Ca^{2+} waves preferentially originate at specific loci in caveolin-rich cell edges. *Proc. Natl. Acad. Sci. USA*, **95**, 5009.

85. Henry, P.D., Cabello, O.A., and Chen, C.H. (1995). Hypercholesterolemia and endothelial dysfunction. *Curr. Opin. Lipidol.*, **6**, 190.

86. Fielding, C.J., Bist, A., and Fielding, P.E. (1997). Caveolin mRNA levels are up-regulated by free cholesterol and down-regulated by oxysterols in fibroblast monolayers. *Proc. Natl. Acad. Sci. USA*, **94**, 3753.

87. Hailstones, D., Sleer, L.S., Parton, R.G., and Stanley, K.K. (1998). Regulation of caveolin and caveolae by cholesterol in MDCK cells. *J. Lipid. Res.*, **39**, 369.

88. Peterson, T.E., Poppa, V., Ueba, H., Wu, A., Yan, C., and Berk, B.C. (1999). Opposing effects of reactive oxygen species and cholesterol on endothelial nitric oxide synthase and endothelial cell caveolae. *Circ. Res.*, **85**, 29.

89. Pritchard, K.J., Groszek, L., Smalley, D.M., Sessa, W.C., Wu, M., Villalon, P., Wolin, M.S., and Stemerman, M.B. (1995). Native low-density lipoprotein increases endothelial cell nitric oxide synthase generation of superoxide anion. *Circ. Res.*, **77**, 510.

90. Plane, F., Bruckdorfer, K.R., Kerr, P., Steuer, A., and Jacobs, M. (1992). Oxidative modification of low-density lipoproteins and the inhibition of relaxations mediated by endothelium-derived nitric oxide in rabbit aorta. *Br. J. Pharmacol.*, **105**, 216.

91. Blair, A., Shaul, P.W., Yuhanna, I.S., Conrad, P.A., and Smart, E.J. (1999). Oxidized low density lipoprotein displaces endothelial nitric-oxide synthase (eNOS) from plasmalemmal caveolae and impairs eNOS activation. *J. Biol. Chem.*, **274**, 32512.

92. Deckert, V. *et al.* (1998). Inhibition by cholesterol oxides of NO release from human vascular endothelial cells. *Arterioscler. Thromb. Vasc. Biol.*, **18**, 1054.

93. Deckert, V., Persegol, L., Viens, L., Lizard, G., Athias, A., Lallemant, C., Gambert, P., and Lagrost, L. (1997). Inhibitors of arterial relaxation among components of human oxidized low-density lipoproteins: cholesterol derivatives oxidized in the 7-position are potent inhibitors of endothelium-dependent relaxation. *Circulation*, **95**, 723.

94. Chin, J.H., Azhar, S., and Hoffman, B.B. (1992). Inactivation of endothelial derived relaxing factor by oxidized lipoproteins. *J. Clin. Invest.*, **89**, 10.

95. Liao, J.K., Shin, W.S., Lee, W.Y., and Clark, S.L. (1995). Oxidized low-density lipoprotein decreases the expression of endothelial nitric oxide synthase. *J. Biol. Chem.*, **270**, 319.

96. Frangos, J.A., Eskin, S.G., McIntire, L.V., and Ives, C.L. (1985). Flow effects on prostacyclin production by cultured human endothelial cells. *Science*, **227**, 1477.

97. Tokunaga, O., Yamada, T., Fan, J., and Watanabe, T. (1991). Age-related decline in prostacyclin synthesis by human aortic endothelial cells. *Am. J. Pathol.*, **138**, 941.

98. Morishita, H., Yui, Y., Hattori, R., Aoyama, T., and Kawai, C. (1990). Increased hydrolysis of cholesteryl ester with prostacyclin is potentiated by high density lipoprotein through the prostacyclin stabilisation. *J. Clin. Invest.*, **86**, 1995.

99. Gryglewski, R. (1995). Interactions between endothelial mediators. *Pharmacol. Toxicol.*, **77**, 1.

100. Schiffrin, E.L., Intengan, H.D., Thibault, G., and Touyz, R.M. (1997). Clinical significance of endothelin in cardiovascular disease. *Curr. Opin. Cardiol.*, **12**, 354.

101. Boulanger, C.M., Tanner, F.C., Bea, M.L., Hahn, A.W., Werner, A., and Luscher, T.F. (1992). Oxidized low density lipoproteins induce mRNA expression and release of endothelin from human and porcine endothelium. *Circ. Res.*, **70**, 1191.

8 Oxidative stress and platelet function in atherosclerosis

John F. Keaney, Jr
Evans Memorial Department of Medicine and Whitaker Cardiovascular Institute, Boston University School of Medicine, 715 Albany Street, W507, Boston, Massachusetts, USA

Jane E. Freedman
Departments of Medicine and Pharmacology, Georgetown University Medical Center, Washington, DC 20007, USA

8.1 Introduction

Platelets are involved in all facets of atherosclerosis (Table 8.1). Beginning with the 'response to injury' hypothesis proposed by Ross *et al.*,[1] platelets, by virtue of their involvement in the healing response, have been implicated in atherogenesis. The role of platelets in atherosclerosis has since become more generalized and it is now recognized that platelets may support cellular cholesteryl ester accumulation as well as cellular proliferation. More recently, the role of platelets in atherosclerosis has focused on platelet activation, adhesion, and aggregation as critically important components of clinical atherosclerotic disease. These latter features of platelet function are associated with considerable oxidative stress in the form of reactive oxygen species production and lipid peroxidation. The purpose of this article will be to briefly review evidence implicating platelets in atherosclerosis and to discuss the role of oxidative stress and antioxidants in the modulation of platelet function.

8.2 Platelets in atherosclerotic lesion formation

The role of thrombosis in atherosclerosis dates back to the 'incrustation' theory of Rokitansky that suggested intimal thickening results from fibrin deposition in the arterial wall.[2] A competing contemporary hypothesis of atherosclerosis was offered by Virchow in 1858,[3] who proposed that atherosclerosis is the result of lipid

Table 8.1 Involvement of platelets in atherosclerosis

Stage	Mechanism	Reference
Atherogenesis	Smooth muscle cell proliferation	(11)
	Foam cell formation	(8)
Lesion progression	Adhesion to ruptured plaque	(15)
	Smooth muscle proliferation	(4)
Acute vascular syndromes	Adhesion and aggregation	(21)
	Vessel occlusion	(20)

transudation into the arterial wall followed by complexation with muco-polysaccharides. These two competing views of atherogenesis were combined in 1977 by Ross[1] in his characterization of atherosclerosis as a 'response to injury'. Since platelets had long been known to be critical for the healing process, their potential role in atherosclerosis captured considerable attention. This potential link between platelets and atherogenesis was strengthened by observations that rabbits rendered thrombocytopenic are resistant to atherosclerosis.[4] Early studies investigating the mechanism(s) responsible for platelet facilitation of atherosclerosis focused principally on foam cell formation and cell proliferation.

Cholesteryl ester accumulation within macrophages is a hallmark of early atherosclerotic lesions,[5] although the precise source of cholesterol for this observation was not initially clear. Although elevated levels of low-density lipoprotein (LDL) had been associated with atherosclerosis, co-incubation of cells with even very high levels of LDL did not result in the formation of cholesteryl ester-laden macrophages, so-called foam cells.[6] This latter point, combined with evidence that atherosclerotic lesion formation is facilitated by endovascular injury even in the absence of hypercholesterolemia,[7] suggested that some other source of cholesterol may play an important role in foam cell formation. Platelets represented one such source of cholesterol as they were known to liberate cholesterol during aggregation and to interact with the endothelium early after vascular injury. Furthermore, initial studies demonstrated that cultured aortic smooth muscle cells readily accept cholesterol from activated platelets.[8] With respect to foam cell formation, both cholesterol esterification and the accumulation of cholesteryl esters were increased in cultured macrophages exposed to activated platelets.[9] This formation of cholesteryl ester from platelet free cholesterol was not strictly dependent on elevated levels of blood cholesterol. Platelets from normocholesterolemic individuals also stimulated cholesteryl ester formation in monocytic cells,[10] although the extent of cholesteryl ester accumulation in monocytic cells was enhanced by elevated plasma cholesterol.[10]

The ability of platelets to stimulate atherosclerosis is not restricted to facilitating cholesteryl ester accumulation in foam cells. Platelets may adhere to injured endothelium, macrophages, and exposed collagen. Upon activation, platelets release their granule contents that contain, among other substances, platelet agonists such as thrombin and growth factors like platelet-derived growth factor (PDGF).[11] These compounds produce three important effects in the vascular wall that are germane to atherogenesis. The first effect involves stimulation of smooth muscle cell proliferation and inflammation from growth factors and cytokines. The second effect involves continued platelet activation and arterial inflammation through the conversion of arachidonic acid to thromboxane A_2 and leukotrienes, respectively. Finally, platelet activation and aggregation is associated with increased oxygen consumption[12] and lipid peroxidation.[13] Since lipid peroxidation in the form of LDL oxidation has been implicated in atherosclerotic lesion formation,[14] it is conceivable that oxidative stress imparted by aggregating platelets may further enhance foam cell formation and atherogenesis. Thus, platelet adhesion and activation in the arterial wall has a number of important implications for the progression of atherosclerosis.

Atherosclerosis is not a process of steady lipid accumulation and lesion progression. We now know that lesion progression is a dynamic process characterized by sequences of repeated atherosclerotic plaque rupture and healing. Typically, the developing atheroma is covered by a fibrous cap that shields the highly thrombogenic lesion from circulating blood components. Plaques are prone to periodic rupture that is characterized by platelet deposition, fibrin formation, and expansion of the atherosclerotic lesion during this healing process.[15] Acute vascular events are thought to result from an uncontrolled plaque rupture that produces total occlusion of the arterial lumen leading to organ ischemia manifested as either myocardial infarction or stroke. Evidence to support this notion is derived from observations that plaque rupture, platelet activation, and lesion progression are commonly found in autopsy studies of patients dying from ischemic heart disease.[15] Thus, in addition to participation in atherogenesis, platelets are an important feature of the acute clinical manifestations of atherosclerosis as well.

8.3   Platelets in the acute clinical manifestations of atherosclerosis

Since the angiographic severity of coronary stenoses is not an adequate predictor of acute coronary syndromes, it has become generally accepted that plaque rupture in mildly stenosed segments of coronary and carotid arteries leads to subsequent thrombus formation that is the precipitating event in acute vascular syndromes.[16, 17] A major feature of plaque rupture in acute arterial thrombosis is the exposure of the thrombogenic lipid core of the lesion to the circulating blood. Plaque rupture is facilitated by overlying endothelium that is often dysfunctional and cannot respond adequately to increased shear stress and blood flow.[18] As a consequence, shear may produce endothelial denudation or plaque fissuring, resulting in platelet adherence. This superficial layer of platelets provides the stimulus for thrombosis.

The importance of platelets in the precipitation of acute vascular events is supported by a wealth of clinical data indicating that inhibition of platelet stimulation prevents acute vascular events.[19] The acute event inciting cardiovascular disease is thrombus formation,[20] often precipitated by the activation of platelets by a ruptured atherosclerotic plaque.[15] Consistent with this idea, plasma platelet-derived thromboxane metabolites are increased in patients with acute coronary syndromes[21] and aspirin uniformly reduces the risk of thrombotic complications from atherosclerosis.[19]

8.4 Hyperlipidemia and platelet function

Since hyperlipidemia predisposes to clinical events from atherosclerosis, considerable effort has been directed at determining the effects of hyperlipidemia on platelet function. In experimental animals with diet-induced elevations of β-very low density lipoprotein (β-VLDL), platelet hypersensitivity in response to thromboxane and thrombin has been reported.[22] Consistent with these observations, circulating VLDL activates platelet adhesion through direct interaction with surface integrin receptors.[23] Support for this idea is derived from observations that individuals with Glanzmann's thrombasthenia, a disorder caused by lack of platelet $\alpha_2\beta_3$ (IIb/IIIa)-type integrin receptors, demonstrate significantly less VLDL-induced adhesion.

Human studies have focused more closely on the role of LDL in platelet hyperactivity. Patients with severe familial hypercholesterolemia are characterized by decreased platelet survival and platelet hyper-aggregation to ADP.[24] Profound reductions in LDL with apheresis produce prolonged platelet survival and a reduction in ADP-induced platelet aggregation, suggesting that LDL cholesterol does alter vascular homeostasis.[24] One possible mechanism for this effect of LDL on platelet aggregation is inhibition of the Na^+/H^+ antiport[25] or direct binding of LDL to the integrin via outside-in signalling.[26, 27] Thus, hyperlipidemia appears to sensitize platelets for activation, providing yet another mechanism for excess cardiovascular events in patients with hyperlipidemia.

8.5 Reactive oxygen species in platelet function

There is considerable evidence to suggest that oxidative stress is a normal consequence of platelet function. Platelet aggregation is associated with a burst of oxygen consumption,[28] lipid peroxidation,[13] and consumption of antioxidants.[29] Platelet superoxide production has received the most attention from an experimental point of view. Currently, there is not clear consensus on the source of superoxide or reactive oxygen species during platelet aggregation, and, as a consequence, intra- versus extracellular has not been addressed in any detail. With respect to the process of platelet aggregation, however, there is considerable data to suggest that oxidative stress in the form of either reactive oxygen species or lipid peroxidation tends to have a prothrombotic influence on the platelet aggregatory response. This relation between reactive oxygen species and platelet thrombosis is outlined in Table 8.2.

Superoxide has been shown to promote platelet aggregation. Isolated platelets incubated with a xanthine/xanthine oxidase system demonstrate aggregation and serotonin release.[30] This effect was inhibited by superoxide dismutase (SOD), but not catalase, implicating superoxide as the principal mediator of platelet activation. Cyclooxygenase did not appear to be important in this response as it was not inhibited by indomethacin or aspirin.[30] Likewise, lipid peroxidation did not seem to account for the effect of superoxide as malondialdehyde generation was not a major feature of platelet exposure to xanthine/xanthine oxidase. Thus, the precise mechanism for superoxide-mediated platelet stimulation remains unclear.

The implications of hydrogen peroxide for platelet function are less straightforward. In plasma[31] and buffer,[32] micromolar concentrations of hydrogen peroxide tend to inhibit platelet function. Higher concentrations of hydrogen peroxide (mM) have been shown to stimulate platelet aggregation through tyrosine phosphorylation of $\alpha_2\beta_3$ integrin receptors.[33] Obviously, an issue of paramount importance is the role of hydrogen peroxide in physiologic platelet function. In this regard, recent data indicate that H_2O_2 may be an important signal for platelet aggregation. In collagen-activated platelets, a burst of H_2O_2 production has been implicated in the activation of the arachidonic acid–phospholipase C pathway.[34] Current data suggests this effect is strictly related to collagen-induced aggregations (Table 8.3). Aggregation of platelets is also associated with the release of iron, which interacts with H_2O_2 leading to hydroxyl radical formation that may also contribute to collagen-induced platelet aggregation.[35]

Another platelet target of reactive oxygen species is the protein kinase C family of enzymes. Protein kinase C is an important component of thrombin-[36] and shear-mediated[37] platelet activation. Both H_2O_2[38] and lipid peroxidation by-products[39]

Table 8.2 Reactive oxygen species and platelet function

Reactive oxygen species	Effect on platelet function	Reference
Superoxide anion	Stimulation	(30, 87)
Hydrogen peroxide (μM)	Inhibition	(88)
Hydrogen peroxide (mM)	Stimulation	(31)
Hydroxyl radical	Stimulation	(35, 87)

Evidence derived from either direct addition of reactive oxygen species[30, 87, 88] or the use of specific enzymatic inhibitors such as SOD30 or catalase.[31]

Table 8. 3 Hydrogen peroxide and platelet aggregation

Agonist	Per cent aggregation at 5 min		
	Control	Catalase (250 U ml^{-1})	Catalase (1000 U ml^{-1})
Collagen (2 µg ml^{-1})	63.2 ± 9.4	31.0 ± 11.2*	6.1 ± 1.4*
ADP (2 µM)	52.7 ± 3.3	50.7 ± 4.7	44.0 ± 3.4
Thrombin (0.1 U ml^{-1})	92.0 ± 3.1	91.0 ± 2.6	84.2 ± 2.5

Data are mean ± SEM from five experiments and are taken from Ref. (34).
*$P < 0.05$ versus control.

have been shown to activate protein kinase C isoforms. Thus, protein kinase C stimulation represents one mechanism for platelet activation through oxidative stress.

8.6 Antioxidants and platelet function

Considering the evidence outlined above that reactive oxygen species and oxidative stress promote and participate in platelet activation, one might logically expect that platelet antioxidant status may have important implications for platelet function. In this context, there is *in vivo* evidence linking antioxidant status to platelet aggregatory responses. Selenium supplementation of normal individuals results in increased plasma glutathione peroxidase activity and prolongation of the bleeding time,[40] consistent with the notion that lipid hydroperoxides are, in part, responsible for platelet aggregation. This notion is supported by studies in selenium-deficient rats that indicate an enhancement of platelet aggregation in response to arachidonic acid.[41] Similarly, human platelet antioxidant content is found to be deficient in a number of clinical settings, including diabetes, aging, and tobacco use, and these abnormalities are associated with enhanced platelet activation responses.[42–44] In aged individuals, arachidonic acid oxidation is enhanced, in part, due to reduced glutathione peroxidase activity.[43] Tobacco use induces the formation of lipid hydroperoxides resulting in platelet hyperactivity,[44] and this abnormality can be reversed with antioxidant treatment. In coronary artery disease patients, impaired enzymatic antioxidant defenses in platelets are associated with enhanced platelet aggregation and evidence of increased lipid peroxidation.[45]

There is also considerable evidence linking platelet vitamin E status to aggregation responses. Higashi and Kikuchi[46] initially demonstrated that α-tocopherol inhibits the aggregation of platelets using hydrogen peroxide as the stimulus. Subsequent studies by Steiner and Anastasi[47] demonstrated that α-tocopherol also inhibited platelet aggregation in response to epinephrine, collagen, and ATP.

Considerable effort has been directed at identifying the mechanism responsible for vitamin E inhibition of platelet aggregation. Since platelet aggregation results in the formation of lipid hydroperoxides that is limited by α-tocopherol,[13] one potential mechanism involved the inhibition of lipid peroxidation that had been linked to cyclooxygenase function.[48] A number of studies examining the effect of α-tocopherol on the activity of cyclooxygenase produced conflicting results.[49–51] Given that the antioxidant activity of α-tocopherol is not required for the inhibition of platelet aggregation,[52, 53] the inhibition of lipid peroxidation seems an unlikely mechanism.

Until recently, the only mechanism attributed to α-tocopherol was the inhibition of pseudopodia formation upon platelet contact with fibrinogen-coated glass slides.[54] More recent studies have begun to shed light on this issue. Platelet aggregation by phorbol ester is particularly sensitive to α-tocopherol, whereas ADP-induced platelet aggregation is rather insensitive.[55] Since phorbol ester-induced platelet aggregation is sensitive to protein kinase C (PKC), whereas ADP-induced aggregation is not,[56] PKC has been implicated as a target for the effect of vitamin E. This suspicion has been borne out by experimental evidence indicating that α-tocopherol supplementation of normal volunteers potently inhibits *ex vivo* platelet aggregation in response to PKC-dependent agonists (Fig. 8.1),[55] a finding that has recently been confirmed in megakaryoblastic cells as well.[57]

The notion that vitamin E inhibits platelets in a PKC-dependent manner is consistent with available clinical data. Early trials of oral α-tocopherol supplementation failed to demonstrate any significant effect on platelet aggregation,[47, 58] because they employed ADP, epinephrine, and collagen at concentrations that largely effect platelet aggregation in a PKC-independent manner.[56] The activity of α-tocopherol to inhibit shear-induced platelet pseudopodia formation[54] is also consistent with a PKC-mediated mechanism. Shear stress evokes PKC stimulation in platelets[37] and platelet adherence to fibrinogen is strictly dependent upon PKC activity.[59] Thus, α-tocopherol inhibition of platelet PKC stimulation reconciles a number of previous observations.

It is important to realize that oxidative stress and platelet antioxidant status may also modulate platelet function in an indirect manner. In this context, one germane feature of platelet activation is the production of nitric oxide (NO). This autocrine and paracrine messenger is exquisitely sensitive to oxidative stress and antioxidant status and, as such, represents one potential target for the effects of antioxidants on platelet function.[60]

8.7 Oxidative stress and platelet-derived NO bioactivity

The control of platelet adhesion to the endothelium is a complex process controlled in part by a number of endothelial paracrine factors including prostacyclin and NO.

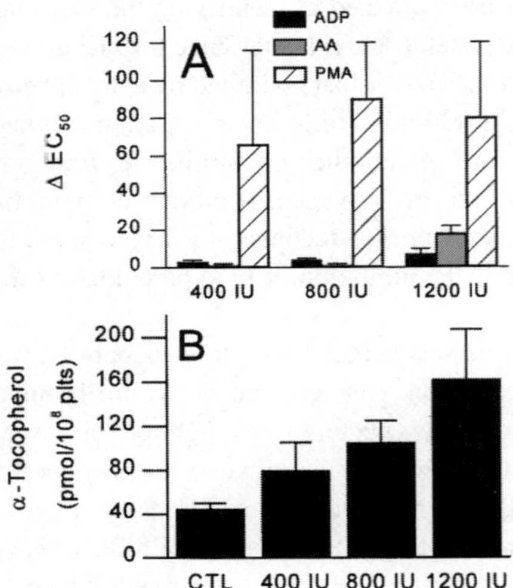

Fig. 8.1 (A) Gel-filtered platelets were prepared from subjects treated with the indicated dose of α-tocopherol for 2 weeks and aggregation was tested with ADP (0.15–6.1 μM), arachidonic acid (AA; 10–100 μnM), and FMA (1.6–1600 nM). Data represent the shift in the EC_{50} of the dose-response curve for each agonist. (B) Platelet α-tocopherol content was determined by HPLC with electrochemical detection. $n = 5$ individuals in each treatment arm and the effect of α-tocopherol was significant ($P < 0.05$) in shifting the EC_{50} for aggregation due to AA and PMA. Data are derived from Ref. (55).

With respect to the latter, NO inhibits the activation of platelets[61, 62] and is known to prevent thrombosis.[63] Nitric oxide has been shown to provide significant feedback regulatory effects on platelet activation including inhibiting the surface expression of integrins such as P-selectin and $\alpha_2 \beta_3$.[64] Similarly, NO limits intracellular calcium, thereby preventing calcium-induced activation of platelets. Finally, NO has been shown to stimulate the cyclic GMP-dependent phosphorylation of the thromboxane receptor leading to inhibition of platelet activation.[65]

In the vascular wall, NO is not derived exclusively from endothelial cells. A constitutive NO synthase (cNOS) has been identified in megakaryoblastic cells and in platelets.[66] *In vitro* studies have demonstrated that platelet aggregation is mildly sensitive to platelet-derived NO. Inhibitors of NOS enhance platelet aggregation, whereas L-arginine, the substrate for cNOS, modestly inhibits platelet aggregation.[67] Consistent with these *in vitro* observations, inhibition of NOS with the L-arginine analogue L-N^G-monomethyl arginine (l-NMMA) produces a reduction in bleeding time and enhanced aggregation of platelets to agonist stimulation.[68] Available evidence suggests that the magnitude of platelet NO release is comparable to that of endothelial cells when size considerations are controlled.[69]

Although platelet aggregation itself is only modestly sensitive to platelet-derived NO, the recruitment of platelets to a developing thrombus is highly sensitive to platelet-derived NO.[70] The clinical significance of this observation is highlighted by the fact that activated platelets from patients with acute coronary syndromes produce less NO than patients with inactive coronary artery disease.[71] Therefore, platelet-derived NO appears to have an important role in vascular homeostasis and may be one important mechanism for the limitation of thrombus formation.

8.7.1 Superoxide and platelet-derived NO

Among sources of oxidative stress in the vasculature, superoxide has the most direct relevance for NO bioactivity. Nitric oxide and superoxide combine readily in a diffusion-limited reaction to produce peroxynitrite,[72, 73] a potent oxidant that has considerably less bioactivity than NO itself.[74] In endothelial cells, it is known that ambient superoxide levels are an important component of NO bioactivity.[75] In platelets, superoxide is produced during aggregation response[76, 77] and likely has an important impact on platelet-derived NO bioactivity. A logical extension of this argument would indicate that SOD should have important implications for platelet-derived NO bioactivity, and this contention is supported by available evidence. In canine coronary arteries, endothelial injury and vascular constriction produce cyclic flow variations that are known to be mediated by platelet aggregation.[78] In this model, a combined infusion of SOD and catalase reduces *in vivo* platelet activity.[79] Conversely, infusion of an O_2^-/H_2O_2 generating system enhances the cyclic flow variations consistent with increased *in vivo* platelet activation. Using an analogous rabbit carotid artery model, Meng *et al.*[80] abrogated platelet-mediated thrombosis with intravenous infusion of SOD. In the same study, inhibition of NOS with L-NMMA decreased platelet cGMP levels and enhanced platelet-mediated cyclic flow variations.[80] Thus, it appears that modifying the balance between NO and superoxide in favour of the former results in reduced platelet-mediated cyclic flow variations.

8.7.2 Glutathione peroxidase and platelet-derived NO

Reactive oxygen species produced during platelet aggregation lead to the formation of lipid hydroperoxides.[13] Intact cells metabolize such oxidized lipids using the glutathione peroxidase enzyme system, a selenocysteine-containing family of proteins that reduce lipid hydroperoxides to their corresponding alcohols using glutathione as a source of electrons. Glutathione peroxidase activity has important implications for NO bioactivity as the decomposition of lipid hydroperoxides is associated with the production of lipid peroxyl radicals that have been shown to react directly with NO and impair its bioactivity.[81]

In light of such observations, it is not surprising that the activity of glutathione peroxidase has been examined with respect to platelet NO bioactivity. Glutathione peroxidase has been shown to potentiate the inhibitory effect of NO on platelet aggregation through the reduction of lipid hydroperoxides.[82] It also appears that a lack of glutathione peroxidase activity has important implications for homeostasis. A congenital lack of plasma glutathione peroxidase activity is associated with spontaneous platelet aggregation and thrombotic strokes in childhood.[83] Thus, glutathione peroxidase activity appears important in maintaining normal NO-mediated platelet homeostasis.

8.7.3 Vitamin E and platelet-derived NO

The implications of platelet α-tocopherol for platelet function are not limited solely to the aggregation response. For example, platelets contain NOS, which results in the production of NO late during the aggregation response that has important implications for vascular homeostasis.[70, 84] Platelets derived from mice lacking the eNOS gene exhibit bleeding times that are 43% shorter than wild-type mice and release no measurable NO[85] (see Fig. 8.2). Platelets loaded with physiologically relevant levels of vitamin E demonstrate increased platelet NO production during aggregation (unpublished observations). Similar results can be produced through direct protein kinase C inhibition with chelerythrine, perhaps through the limitation of superoxide elaborated during aggregation. The precise mechanisms that regulate the relative platelet production of superoxide and nitric oxide remain unknown.

It is likely that inhibition of protein kinase C stimulation by α-tocopherol will have important implications in developing platelets. Protein kinase C inhibition in endothelial cells results in upregulation of eNOS transcription[86] and an increase in endothelium-derived NO. A similar effect in megakaryocytes would explain

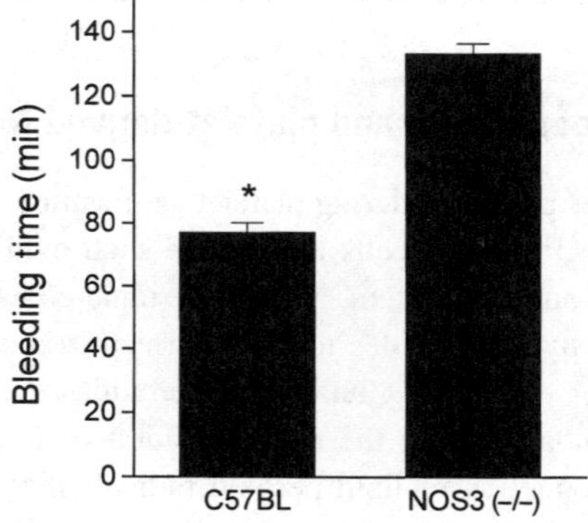

Fig. 8.2 Measurement of bleeding times in wild-type (C57BL) and eNOS-deficient (NOS3 −/−) mice. Data represent mean ± SEM for $n = 24$ and are taken from Ref. (85) (unpublished observation). *$P < 0.001$ versus wild-type.

observations that platelet ˙NO production correlates with plasma α-tocopherol levels in patients undergoing coronary angiography ($R = 0.5$; $P < 0.01$). The impact of atherogenic risk factors such as hypercholesterolemia, diabetes, and hypertension on platelet-derived NO remains unknown, but may have important implications for the clinical manifestations of atherosclerotic vascular disease.

8.8 Conclusions

In summary, platelets are involved in all facets of atherosclerotic vascular disease from atherogenesis to the clinical manifestations of end organ ischemia. Clinical trial data indicate that inhibition of platelet aggregation with aspirin or antagonists to the platelet fibrinogen receptor result in reduced clinical events in patients with atherosclerosis. We are only now beginning to understand the role of redox-sensitive events in the control of platelet function. In particular, platelet-derived NO appears important in the control of platelet recruitment and may have an important impact on the vascular tendency toward thrombosis in the setting of atherosclerosis. Further investigation will be needed to clarify the role of platelet-derived NO and its deficiency in the clinical manifestations of atherosclerosis.

Acknowledgements

Dr Keaney is supported by grants from the National Institutes of Health (HL59346, HL55854, HL03195, and HL58976) and is the recipient of an Established Investigator Award from the American Heart Association. Dr Freedman is supported by NIH grants (HL61795, AG08226) and a Grant-in-Aid from the Mid-Atlantic Affiliate of the American Heart Association, and is the recipient of an NIH Clinical Investigator Development Award (HL03556).

References

1. Ross, R., Glomset, J., and Harker, L. (1977). Response to injury and atherogenesis. *Am. J. Pathol.*, **86**, 675.
2. Rokitansky, C. (1852). In *A manual of pathological anatomy*, p. 261. Sydenham Society, London.
3. Virchow, R. (1989). Cellular pathology. As based upon physiological and pathological history LXVI – Atheromatous effection of arteries. *Nutr. Rev.*, **47**, 23.
4. Freidman, R.J., Stemerman, M.B., Moore, S., Gauldie, J., Gent, M., Tiell, M.L., and Spaet, T. (1977). The effect of thrombocytopenia on experimental arteriosclerotic lesion formation in rabbits. *J. Clin Invest.*, **16**, 1191.

5. Stary, H.C. (1983). Evolution of atherosclerotic plaques in the coronary arteries of young adults. *Arteriosclerosis*, **3**, 417A.

6. Goldstein, J.L. and Brown, M.S. (1977). The low-density lipoprotein pathway and its relation to atherosclerosis. *Ann. Rev. Biochem.*, **46**, 897.

7. Moore, S. (1973). Thromboatherosclerosis in normal lipemic rabbits. *Lab. Invest.*, **29**, 478.

8. Kruth, H. (1985). Platelet-mediated cholesterol accumulation in cultured aortic smooth muscle cells. *Science*, **227**, 1243.

9. Curtiss, L.K., Black, A.S., Takagi, I., and Plow, E.F. (1987). A new mechanism for foam-cell generation in atherosclerotic lesions. *J. Clin. Invest.*, **80**, 367.

10. Mendelsohn, M.E. and Loscalzo, J. (1988). Role of platelets in cholesteryl ester formation by U-937 cells. *J. Clin. Invest.*, **81**, 62.

11. Ross, R., Glomset, J., Kariya, B., and Harker, L. (1974). A platelet-dependent serum factor that stimulates the proliferation of arterial smooth muscle cells. *Proc. Natl. Acad. Sci. USA*, **71**, 1207.

12. Murer, E.H. (1968). Release reaction and energy metabolism in blood platelets with special reference to the burst in oxygen uptake. *Biochim. Biophys. Acta.*, **162**, 320.

13. Okuma, M., Steiner, M., and Bladini, M.G. (1971). Studies on lipid peroxides in platelets. II. Effect of aggregating agents and platelet antibody. *J. Lab. Clin. Med.*, **77**, 728.

14. Steinberg, D. (1997). Low density lipoprotein oxidation and its pathobiological significance. *J. Biol. Chem.*, **272**, 20963.

15. Davies, M.J. and Thomas, A.C. (1985). Plaque fissuring: the cause of acute myocardial infarction, sudden ischemic death, and crescendo angina. *Br. Heart J.*, **53**, 363.

16. Ambrose, J.A. (1992). Plaque disruption and the acute coronary syndromes of unstable angina and myocardial infarction: if the substrate is similar, why is the clinical presentation different? *J. Am. Coll. Cardiol.*, **19**, 1653.

17. Ambrose, J. A., Winters, S. L., Stern, A., Eng, A., Teichholz, L. E., Gorlin, R., and Fuster, V. (1985). Angiographic morphology and the pathogenesis of unstable angina pectoris. *J. Am. Coll. Cardiol.*, **5**, 609.

18. Levine, G.N., Keaney, J.F., Jr, and Vita, J.A. (1995). Cholesterol reduction in cardiovascular disease. Clinical benefits and possible mechanisms. *N. Engl. J. Med.*, **332**, 512.

19. Antiplatelet Trialists Collaboration (1994). Collaborative overview of randomized trials antiplatelet therapy: I. Prevention of death, myocardial infarction, and stroke by prolonged antiplatelet therapy in various categories of patients. *Br. Med. J.*, **308**, 81.

20. DeWood, M.A., Spores, J., Notske, R., Mouser, L.T., Burroughs, R., Golden, M.S., and Lang, H.T. (1980). Prevalence of total coronary occlusion during the early hours of transmural myocardial infarction. *N. Engl. J. Med.*, **303**, 897.

21. Fitzgerald, D.J., Roy, L., Catella, F., and FitzGerald, G.A. (1986). Platelet activation in unstable coronary disease. *N. Engl. J. Med.*, **315**, 983.

22. Gross, P., Rand, M.L., Barrow, D.V., and Packham, M. (1991). Platelet hypersensitivity in cholesterol-fed rabbits: enhancement of thromboxane A2-dependent and thrombin-induced, thromboxane A2-independent platelet responses. *Atherosclerosis*, **88**, 77.

23. Kowalska, M.A., Tuszynski, G.P., and Capuzzi, D. (1990). Plasma lipoproteins mediate platelet adhesion. *Biochem. Biophys. Res. Commun.*, **172**, 118.

24. Sinzinger, H., Pirich, C., Bednar, J., and O'Grady, J. (1996). *Ex-vivo* and *in-vivo* platelet function in patients with severe hypercholesterolemia undergoing LDL-apheresis. *Thromb. Res.*, **82**, 291.

25. Nofer, J.R., Tepel, M., Kehrel, B., Wierwille, S., Walter, M., Seedorf, U., Zidek, W., and Assmann, G. (1997). Low-density lipoproteins inhibit the Na^+/H^+ antiport in human platelets. A novel mechanism enhancing platelet activity in hypercholesterolemia. *Circulation*, **95**, 1370.

26. Hackeng, C.M., Huigsloot, M., Pladet, M.W., Nieuwenhuis, H.K., van Rijn, H.J., and Akkerman, J.W. (1999). Low-density lipoprotein enhances platelet secretion via integrin-alphaIIb beta3-mediated signaling. *Arterioscler. Thromb. Vasc. Biol.*, **19**, 239.

27. Hackeng, C.M., Pladet, M.W., Akkerman, J.W., and van Rijn, H.J. (1999). Low-density lipoprotein phosphorylates the focal adhesion-associated kinase P125 (FAK) in human platelets independent of integrin alphaIIb beta3. *J. Biol. Chem.*, **274**, 384.

28. Bressler, N. M., Broekman, M. J., and Marcus, A. J. (1979). Concurrent studies of oxygen consumption and aggregation in stimulated human platelets. *Blood.*, **53**, 167.

29. Chan, A.C., Tran, K., Raynor, T., Ganz, P.R., and Chow, C.K. (1991). Regeneration of vitamin E in human platelets. *J. Biol. Chem.*, **266**, 17290.

30. Handin, R.I., Karabin, R., and Boxer, G.J. (1977). Enhancement of platelet function by superoxide anion. *J. Clin. Invest.*, **59**, 959.

31. Rodvien, R., Lindon, J., and Levine, P. (1976). Physiology and ultrastructure of the blood platelet following exposure to hydrogen peroxide. *Br. J. Haematol.*, **33**, 19.

32. Naseem, K.M., Chirico, S., Mohammadi, B., and Bruckdorfer, K.R. (1996). The synergism of hydrogen peroxide with plasma S-nitrosothiols in the inhibition of platelet activation. *Biochem. J.*, **318** , 759.

33. Irani, K., Pham, Y., Coleman, L.D., Roos, C., Cooke, G.E., Miodovnik, A., Karim, D., Wilhide, C.C., Bray, P.F., and Goldschmidt-Clermont, P.J. (1998). Priming of platelet $\alpha_{IIb}\beta_{III}$ by oxidants is associated with tyrosine phosphorylation of β_3. *Arterioscler. Thromb. Vasc. Biol.*, **18**, 1698.

34. Pignatelli, P., Pulcinelli, F.M., Lenti, L., Gazzaniga, P.P., and Violi, F. (1998). Hydrogen peroxide is involved in collagen-induced platelet activation. *Blood*, **91**, 484.

35. Pratico, D., Pasin, M., Barry, O.P., Ghiselli, A., Sabatino, G., Iuliano, L., FitzGerald, G.A., and Violi, F. (1999). Iron-dependent human platelet activation and hydroxyl radical formation: involvement of protein kinase C. *Circulation*, **99**, 3118.

36. Baldassare, J.J., Henderson, P., Burns, D., Loomis, C., and Fischer, G.J. (1992). Translocation of protein kinase C isoenzymes in thrombin-stimulated human platelets – correlation with 1,2 diacylglycerol levels. *J. Biol. Chem.*, **267**, 15585.

37. Kroll, M.H., Hellums, J.D., Guo, Z., Durante, W., Razdan, D., Hrbolich, J.K., and Schafer, A.I. (1993). Protein kinase C is activated in platelets subjected to pathological shear stress. *J. Biol. Chem.*, **268**, 3520.

38. Konishi, H., Tanaka, M., Takemura, Y., Matsuzaki, H., Ono, Y., Kikkawa, U., and Nishizuka, Y. (1997). Activation of protein kinase C by tyrosine phosphorylation in response to H_2O_2. *Proc. Natl. Acad. Sci. USA*, **94**, 11233.

39. Sugiyama, S., Kugiyama, K., Ohgushi, M., Fujimoto, K., and Yasue, H. (1994). Lysophosphatidylcholine in oxidized low-density lipoprotein increases endothelial susceptibility to polymorphonuclear leukocyte-induced endothelial dysfunction in porcine coronary arteries: role of protein kinase C. *Circ. Res.*, **74**, 565.

40. Schiavon, R., Freeman, G.E., Guidi, G.C., Perona, G., Zatti, M., and Kakkar, V.V. (1984). Selenium enhances prostacyclin production by cultured endothelial cells: possible explanation for increased bleeding times in volunteers taking selenium as a dietary supplement. *Thromb. Res.*, **34**, 389.

41. Schoene, N., Morris, V., and Levander,O. (1986). Altered arachidonic acid metabolism in platelets and aortas from selenium-deficient rats. *Nutr. Res.*, **6**, 75.

42. Johnson, M., Ramey, E., and Ramwell, P. (1975). Sex and age differences in human platelet aggregation. *Nature*, **253**, 355.

43. Vericel, E., Rey, C., Calzada, C., Haond, P., Capuy, P., and Lagarde, M. (1992). Age-related changes in arachidonic acid peroxidation and glutathione-peroxidase activity in human platelets. *Prostaglandins*, **43**, 75.

44. Blache, D. (1995). Involvement of hydrogen and lipid peroxidase in acute tobacco smoking-induced platelet hyperactivity. *Am. J. Physiol.*, **268**, H679.

45. Buczynski, A.B., Wachowicz, B., Dedziora-Kornatowska, K., Tkaczewski, W., and Kedziora, J. (1993). Changes in antioxidant enzyme activities, aggregability, and malonyldialdehyde concentrations in blood platelets from patients with coronary heart disease. *Atherosclerosis*, **100**, 223.

46. Higashi, O. and Kikuchi, Y. (1974). Effects of vitamin E on the aggregation and the lipid peroxidation of platelets exposed to hydrogen peroxide. *Tohoku J. Exp. Med.*, **112**, 271.

47. Steiner, M. and Anastasi, J. (1976). Vitamin E: an inhibitor of the platelet release reaction. *J. Clin. Invest.*, **57**, 732.

48. Hemler, M.E. and Lands, W.E.M. (1980). Evidence for a peroxide-initiated free radical mechanism of prostaglandin biosynthesis. *J. Biol. Chem.*, **225**, 6253.

49. Kockman, V., Vericel, E., Croset, M., and Lagade, M. (1988). Vitamin E fails to alter the aggregation and oxygenated metabolism of arachidonic acid in normal human platelets. *Prostaglandins*, **36**, 607.

50. Ali, M., Gudbranson, C.G., and McDonald, J.W. (1980). Inhibition of human platelet cyclooxygenase by alpha-tocopherol. *Prostaglandins & Medicine*, **4**, 79.

51. Mower, R. and Steiner, M. (1983). Biochemical interaction of arachidonic acid and vitamin E in human platelets. *Prostaglandins. Leukotrienes. Med.*, **10**, 389.

52. Cox, A.C., Rao, G.H.R., Gerrard, J.M., and White, J.G. (1980). The influence of α-tocopherol quinone on platelet structure, function and biochemistry. *Blood*, **55**, 907.

53. Mower, R. and Steiner, M. (1982). Synthetic byproducts of tocopherol oxidation as inhibitors of platelet function. *Prostaglandins*, **24**, 137.

54. Jandak, J., Steiner, M., and Richardson, P.D. (1989). Alpha-tocopherol, an effective inhibitor of platelet adhesion. *Blood*, **73**, 141.

55. Freedman, J.E., Farhat, J.H., Loscalzo, J., and Keaney, J.F., Jr (1996).
α-Tocopherol inhibits aggregation of human platelets by a protein kinase
C-dependent mechanism. *Circulation*, **94**, 2434.

56. Packham, M.A., Livne, A.A., Ruben, D.H., and Rand, M.L. (1993). Activation of
phospholipase C and protein kinase C has little involvement in ADP-induced
primary aggregation of human platelets: effects of diacylglycerols, the
diacylglycerol kinase inhibitor R59022, staurosporine, and okadaic acid. *Biochem.
J.*, **290**, 849.

57. Steiner, M., Li, W., Ciaramella, J.M., Anagnostou, A., and Sigounas, G.
(1997). dl-alpha-tocopherol, a potent inhibitor of phorbol ester induced shape
change of erythro- and megakaryoblastic leukemia cells. *J. Cell. Physiol.*,
172, 351.

58. Steiner, M. (1983). Effect of α-tocopherol administration on platelet function in
man. *Thromb. Haemost.*, **49**, 73.

59. Haimovich, B., Kaneshiki, N., and Ji, P. (1996). Protein kinase C regulates
tyrosine phosphorylation of pp125FAK in platelets adherent to fibrinogen. *Blood*,
87, 152.

60. Huang, A. and Keaney, J.F., Jr (2000). Antioxidants and endothelium-derived nitric
oxide action In *Nitric oxide and the cardiovascular system*. (ed. J. Loscalzo and
J.A. Vita), Humana, Totowa, in press.

61. Stamler, J., Mendelsohn, M.E., Amarante, P., Smick, D., Andon, N., Davies, P.F.,
Cooke, J.P., and Loscalzo, J. (1989). *N*-Acetylcysteine potentiates platelet
inhibition by endothelium-derived relaxing factor. *Circ. Res.*, **65**, 789.

62. Cooke, J.P., Stamler, J., Andon, N., Davies, P. F., McKinley, G., and Loscalzo, J.
(1990). Flow stimulates endothelial cells to release nitrovasodilator that is
potentiated by reduced thiol. *Am. J. Physiol.*, **259**, H804.

63. Shultz, P. and Raij, L. (1992). Endogenously synthesized nitric oxide prevents
endotoxin-induced glomerular thrombosis. *J. Clin. Invest.*, **90**, 1718.

64. Michelson, A.D., Benoit, S.E., Furman, M.I., Breckwoldt, W.L., Rohter, M.J.,
Barnard, M.R., and Loscalzo, J. (1996). Effects of endothelium-derived
relaxing factor/nitric oxide on platelet surface glycoproteins. *Am. J. Physiol.*, **39**,
H1640.

65. Wang, G.R., Zhu, Y., Halushka, P.V., Lincoln, T.M., and Mendelsohn, M.E. (1998).
Mechanism of platelet inhibition by nitric oxide: *in vivo* phosphorylation of
thromboxane receptor by cyclic GMP-dependent protein kinase. *Proc. Natl. Acad.
Sci. USA*, **95**, 4888.

66. Sase, K. and Michel, T. (1995). Expression of constitutive endothelial nitric oxide
synthase in human blood platelets. *Life Sci.*, **57**, 2049.

67. Chen, L.Y. and Mehta, J.L. (1996). Variable effects of *L*-arginine–nitric oxide
pathway in human neutrophils and platelets may relate to different nitric oxide
synthase isoforms. *J. Pharmacol. Exp. Ther.*, **267**, 253.

68. Bodzenta-Lukaszyke, A., Gabryelewicz, A., Lukaszyke, A., Bielawiec, M.,
Konturek, J., and Domschke, W. (1994). Nitric oxide synthase inhibition and
platelet function. *Thromb. Res.*, **75**.

69. Zhou, Q., Hellermann, G.R., and Solomonson, L.P. (1995). Nitric oxide release
from resting human platelets. *Thromb. Res.*, **77**, 87.

70. Freedman, J.E., Loscalzo, J., Barnard, M.R., Alpert, C., Keaney, J.F., Jr, and
Michelson, A.D. (1997). Nitric oxide released from activated platelets inhibits
platelet recruitment. *J. Clin. Invest.*, **100**, 350.

71. Freedman, J.E., Ting, B., Hankin, B., Loscalzo, J., Keaney, J.F., Jr, and Vita, J.A. (1998). Impaired platelet production of nitric oxide predicts the presence of acute coronary syndromes. *Circulation.*, **98**, 1481.

72. Saran, M., Michel, C., and Bors, W. (1990). Reaction of NO with O^{2-}. Implications for the action of endothelium-derived relaxing factor (A2RF). *Free. Radic. Res. Commun.*, **10**, 221.

73. Kissner, R., Nauser, T., Bugnon, P., Lye, P.G., and Koppenol, W.H. (1997). Formation and properties of peroxynitrite as studied by laser flash photolysis, high-pressure stopped-flow technique, and pulse radiolysis. *Chemical Res. Toxicol.*, **10**, 1285.

74. Tarpey, M.M., Beckman, J.S., Ischiropoulos, H., Gore, J.Z., and Brock, T.A. (1995). Peroxynitrite stimulates vascular smooth muscle cell cyclic GMP synthesis. *FEBS Lett.*, **364**, 314.

75. Mugge, A., Elwell, J.K., Peterson, T.E., and Harrison, D.G. (1991). Release of intact endothelium-derived relaxing factor depends on endothelial superoxide dismutase activity. *Am. J. Physiol.*, **260**, C219–25.

76. Marcus, A.J., Silk, S.T., Safier, L.B., and Ullman, H.L. (1977). Superoxide production and reducing activity in human platelets. *J. Clin. Invest.*, **59**, 149.

77. Freedman, J.E. and Keaney, J.F., Jr, (1999). NO and superoxide detection in human platelets. *Methods Enzymol.*, **301**, 61.

78. Folts, J.D., Crowell, E.B., and Rowe, G.G. (1976). Platelet aggregation in partially obstructed vessels and their elimination with aspirin. *Circulation*, **54**, 365.

79. Ikeda, H., Koga, Y., Od, T., Kuwano, K., Nakayama, I., Ueno, T., Toshima, H., Michael, L., and Entman, M. (1994). Free oxygen radicals contribute to platelet aggregation and cyclic flow variations in stenosed and endothelium-injured canine coronary arteries. *J. Am. Coll. Cardiol.*, **24**, 1749.

80. Meng, Y.Y., Trachtenburg, J., Ryan, U.S., and Abendschein, D.R. (1995). Potentiation of endotenous nitric oxide with superoxide dismutase inhibits platelet-mediated thrombosis in injured and stenotic arteries. *J. Am. Coll. Cardiol.*, **25**, 269.

81. Rubbo, H., Parthasarathy, S., Barnes, S., Kirk, M., Kalyanaraman, B., and Freeman, B.A. (1995). Nitric oxide inhibition of lipoxygenase-dependent liposome and low-density lipoprotein oxidation: termination of radical chain propagation reactions and formation of nitrogen-containing oxidized lipid derivatives. *Arch. Biochem. Biophys.*, **324**, 15.

82. Freedman, J.E., Frei, B., Welch, G.N., and Loscalzo, J. (1995). Glutathione peroxidase potentiates the inhibition of platelet function by S-nitrosothiols. *J. Clin. Invest.*, **96**, 394.

83. Freedman, J.E., Loscalzo, J., Benoit, S.E., Valeri, C.R., Barnard, M.R., and Michelson, A.D. (1996). Decreased platelet inhibition by nitric oxide in two brothers with a history of arterial thrombosis. *J. Clin. Invest.*, **97**, 979.

84. Malinski, T., Radomski, M.W., Taha, Z., and Moncada, S. (1993). Direct electrochemical measurement of nitric oxide released from human platelets. *Biochem. Biophys. Res. Commun.*, **194**, 960.

85. Freedman, J.E., Sauter, R., Battinelli, E.M., Ault, K., Knowles, C., Huang, P.L., and Loscalzo, J. (1999). Deficient platelet-derived nitric oxide and

enhanced hemostasis in mice lacking the NOSIII gene. *Circ. Res.*, **84**, 1416.

86. Ohara, Y., Sayegh, H.S., Yamin, J.J., and Harrison, D.G. (1995). Regulation of endothelial constitutive nitric oxide synthase by protein kinase C. *Hypertension*, **25**, 415.

87. Leo, R., Pratico, D., Iuliano, L., Pulcinelli, F.M., Ghiselli, A., Pignatelli, P., Colavita, A.R., FitzGerald, G.A., and Violi, F. (1997). Platelet activation by superoxide anion and hydroxyl radicals intrinsically generated by platelets that had undergone anoxia and then reoxygenated. *Circulation*, **95**, 885.

88. Stuart, M.J. and Holmsen, H. (1977). Hydrogen peroxide, an inhibitor of platelet function: effect on adenine nucleotide metabolism, and the release reaction. *Am. J. Hematol.*, **2**, 53.

9 Macrophage lipid metabolism and atherosclerosis

Leonard Kritharides
Clinical Research Group, Heart Research Institute,
Camperdown, Sydney, NSW 2050,
Department of Cadiology, Concord Hospital, University of Sydney,
NSW 2139, Australia

Wendy Jessup
Cell Biology Group, Heart Reasearch Institute, Camperdown, Sydney, NSW 2050,
Australia

9.1 Introduction

Atherosclerotic lesion development is preceded by intimal penetration of blood-borne mononuclear cells which subsequently undergo transformation into macrophage foam cells in fatty streak lesions.[1] The macrophage is an early and persistent feature of atherosclerosis and is therefore likely to be pivotal in the disease development. Indeed, the importance of the macrophage in atherosclerosis was revealed in the osteopetrotic (op) mouse (lacking macrophage colony-stimulating factor (M-CSF) and thus severely deficient in monocytes) in which a profound inhibition of atherosclerosis development was observed.[2]

The role of the macrophage in atherogenesis is likely to be multifactorial, including influences on the levels of inflammatory mediators such as cytokines and extracellular matrix-degrading enzymes and on generation and uptake of modified forms of low-density lipoprotein (LDL) which may promote atherogenesis. There is evidence that lipid loading of macrophages, a characteristic and unusual feature of atherosclerotic lesions, profoundly affects their physiology, often in a manner likely to promote lesion development.

In the present review we will consider the nature of lipid deposition in foam cell macrophages, its likely origins, effects and reversal, and their impact on atherosclerosis.

9.2 Lipid and morphological characteristics of foam cell macrophages in plaque

9.2.1 Ultrastructural morphology of macrophage foam cells in advanced human lesions and in animal models

As described in detail elsewhere in this book, lipid-rich foam cell macrophages are characteristic of fatty streak lesions, and are present in fibroatheroma and advanced plaque.[3] Macrophages without lipid droplets can also be found in plaque, particularly near the arterial lumen, whereas foam cell macrophages are more typically observed deeper in the intima, in lateral margins of the necrotic core and at the lateral edges of plaque.[3] Foam cells of macrophage origin, as defined by morphology or immunological phenotype, may have lipid droplets stored in both cytosolic and in lysosomal compartments. The sources and types of lipid in these compartments may be quite distinct, and differentiation of their respective sources and mechanisms of accumulation may have important implications for understanding atherogenesis.

Foam cells (predominantly macrophages) from rabbit atheroma have been separated according to their flotation density, which is determined by their lipid content.[4] Low density cells ($1.02 < d < 1.07$) contain 4 times more free cholesterol, 17 times more esterified cholesterol, and 10-fold higher concentrations of lysosomal enzymes β-glucosaminidase, β-galactosidase and neutral glucosidase than high-density cells. It is likely that other lysosomal enzymes such as proteinases would also be elevated. Low density foam cells contain two morphologically distinct types of lipid inclusions, which comprise 50% of the total cell cytoplasmic volume: (i) clear droplets without surrounding membranes; and (ii) polymorphic structures, limited by a membrane containing intravacuolar debris, and variably containing myelin-like whorls, reticulations, and crystalline inclusions.[4] High-density cell populations also contain more lipid inclusions than seen in comparable cells derived from normal aortae, but lipid accumulation in these cells is more variable and less marked than that seen in low-density cells.

9.2.1.1 Lipid accumulation in cytosolic and lysosomal compartments

^{125}I-LDL adminstered intravenously to rabbits fed a high cholesterol diet is rapidly accumulated and degraded by macrophages within fatty streak lesions and in atheromatous plaque.[5] Such accumulation is not dependent on the presence of LDL receptor as it also occurs in Watanabe rabbits. After endocytosis and lysosomal hydrolysis of LDL proteins and lipids, lipoprotein-derived cholesterol which is not incorporated into the plasma membrane or other cell membranes is esterified by acetyl CoA cholesteryl acyl transferase (ACAT) and stored in the cytosol as cholesteryl ester (CE) (see Section 9.4).

CE thus generated forms lipid droplets which are not surrounded by a membrane bilayer. Morphologically similar cytosolic droplets are formed during triglyceride synthesis by cells. CE-derived droplets show birefringence under polarized light consistent with the liquid crystalline state, and are present in foam cell macrophages in primates and rabbits.[6] The physical state of lipid droplets in relation to phase transition contributes to the rate of intracellular CE hydrolysis. Triglyceride (TG)-rich droplets may be distinguished from those containing only CE by polarized light, but are indistinguishable by microscopy using stains such as oil-red O or light microscopic morphology alone. This can confuse when primary cultures of TG-rich monocyte-derived human macrophages are used *in vitro*.[7] The distinction is biochemically important because foam cells from atherosclerotic plaque are characterized by the mass accumulation of larger quantities of CE than TG[8] (discussed further below).

Within weeks of initiation of a high fat diet, lesions in rabbits, primates, pigeons, and mice all contain foam cells with prominent cytosolic lipid accumulation. These droplets are surrounded by a single membrane[6] or no membrane, but are not surrounded by a true membrane bilayer. Macrophage foam cells contain numerous dense granules typical of lysosomes.[6] In high-density cell populations, 70% of intracellular lipid droplets are cytosolic, whereas in low-density cells, 40–50% are cytosolic and more than half are lysosomal.[4] This implies that an alteration in the distribution of cell lipids occurs with progressive cell lipid accumulation.

The relative abundance of cytosolic and lysosomal lipid droplets (see below) appears to vary within different regions in atherosclerotic plaques. Using a combination of acid phosphatase cytochemistry and three-dimensional microscopy, stage-dependent changes in morphology in macrophage foam cells were identified within White Carneau pigeons.[9] Foam cells from early lesions, and the foam cells located on the edge of advanced lesions, principally contain cytoplasmic lipid (72% of cell lipid). Foam cell macrophages located in the centre of lesions contain most lipid (59%) in secondary lysosome-like structures.[9] Lysosomal lipid structures vary from small discrete spheres to multilobed complex lysosomes with acid phosphatase staining the periphery and multiple lipid droplets within, suggesting a process of progressive lysosome–lysosome fusion. Support for sequential cellular lipid redistribution has been obtained by varying the duration of cholesterol feeding. After 5 weeks of cholesterol feeding, 74% of foam cell lipid is in the form of cytoplasmic droplets, whereas after 10 weeks, 73% of foam cell lipid is lysosomal.[10] These changes correspond to a doubling in the number of secondary lysosomes per cell, a doubling of the size of each lysosome, and the appearance of complex multichambered structures between 5 and 10 weeks.[10] The number of cytoplasmic droplets does not increase between 5 and 10 weeks feeding, suggesting that maximal accumulation of cytoplasmic lipid is achieved first, with later accumulation of lipid occurring within lysosomes.

Cytosolic CE are cell-synthesized (either from plasma membrane-derived cholesterol, endogenously synthesized cholesterol, or lysosomal cholesterol). The factors which influence accumulation of CE in foam cells may thus contribute to the maintenance of the excess cellular cholesterol characteristic of atherosclerosis, and cellular studies of their regulation are described below. The accumulation of lipid in the lysosomal compartment may indicate a primary or secondary perturbation of lysosomal function, the progressive accumulation of lipoprotein-derived or other material which is resistant to lysosomal hydrolysis, or the cellular generation of new lipid structures following accumulation of lipoprotein cholesterol. The increased abundance of lysosomal lipids in advanced lesions relative to that in fatty streaks may indicate important differences in lipoprotein modification and in factors regulating lysosomal lipid metabolism in fatty streaks and advanced plaques, such as differential accumulation of oxidized lipids (see below).

9.2.1.2 Macrophages and the necrotic core

Implicit in the differential intracellular distribution of lipid with different macrophage location and the duration of cholesterol feeding, is the notion that foam cell macrophages within atherosclerotic lesions may be of varying ages. This has been investigated by labelling dividing cells in the White Carneau pigeon with thymidine.[11] Lesion edges contain six-fold the number of adherent monocytes as middle regions of lesions, and macrophages in shoulder regions appear to derive from more recently infiltrated monocytes than centrally located macrophages. Interpretation of these data needs to be qualified by the potential of regional inflammatory stimuli to induce proliferation and thymidine uptake in older cells, but remain interesting. As the shoulder region is typically the site of inflammation, erosion and rupture in lipid-rich plaques, the biology of recently infiltrated monocyte-derived foam cells may be particularly important for the evolution of acute inflammatory cardiac complications. In the core region of plaque, macrophage foam cells may show signs of cell injury, cell death and degrees of disintegration, and contribute to the high level of expression of tissue factor in this region.[3] This may be important, as the necrotic core contributes greatly to the thrombotic burden of plaque[12] and suggests that older foam cell macrophages may contribute to the thrombotic complications of atherosclerosis. Consequently, recently infiltrated monocyte macrophages at the lesion edge, and older necrotic macrophages in the core may differentially contribute to the inflammatory and thrombotic complications of atherosclerosis. The respective accumulation of cytosolic and lysosomal lipids of foam cells in these sites may contribute to these differential effects.

Cholesterol crystals are typical of the necrotic core. As indicated above, the necrotic core is associated with older foam cell macrophages which show signs of cell death and disintegration. The prospect of a common aetiology of both core formation and cell toxicity has generated *in vitro* studies which suggest that cellular accumulation of cytotoxic oxidized lipids, including oxysterols, could contribute to cytotoxicity within the necrotic core.[13, 14] As atherosclerosis progresses, ultrastructural studies in rabbits show transition from predominantly cytosolic to predominantly lysosomal lipid accumulation, and the appearance of cholesterol crystals within the lysosomal compartment.[15]

In vitro models have demonstrated that cell toxicity can follow the intracellular accumulation of free cholesterol in both lysosomal and cytosolic compartments.[16, 17] In both cases, crystal formation requires the local accumulation of free cholesterol in concentrations exceeding its aqueous solubility and the absence of the opportunity for esterification to CE. In the lysosomal compartment, hydrolysis of CE to free cholesterol (FC) faster than the transport of cholesterol out of the lysosome may mediate cholesterol accumulation and subsequent crystal formation. In the case of cytosolic crystals, defective movement of cholesterol to ACAT, defects in ACAT, or rates of cytosolic CE hydrolysis in excess of the rates of cellular esterification and cellular cholesterol efflux may mediate FC crystallization.[18] The accumulation of FC stimulates cell phospholipid synthesis by inducing phosphocholine cytidyltransferase.[18] This may serve as a protective mechanism against cholesterol-induced cytotoxicity by maintaining optimal FC to phospholipid (PL) ratios in cell membranes, and can be associated with the formation of intracellular whorls,[18] commonly observed within lysosomes of authentic foam cells. It should be recalled, as described in Chapter 13, that there is evidence to suggest that lipid structures typical of the core of human atherosclerotic plaques can occur without evident foam cells, suggesting the potential for cell-free aetiology.[19]

9.2.2 Lipid analysis of *ex vivo* foam cell macrophages and cell-rich plaques

The relevance of *in vitro* studies of foam cell macrophages requires that models adequately reflect the lipid composition of foam cells within atherosclerotic tissue. Surprisingly few studies have systematically investigated the lipid composition of human *ex vivo* foam cell macrophages; consequently, much of our current understanding is based upon animal studies and whole tissue analyses.

9.2.2.1 Whole intima metabolic studies

There is evidence of active cellular lipid metabolism in the atherosclerotic arterial wall. Aortic intima from cholesterol-fed pigs incorporated greater amounts of ^{14}C-oleic acid into CE, PL, and TG fractions than did aortic intima from normally fed pigs.[20] Active incorporation indicates the presence of functioning cellular synthetic enzymes in the atherosclerotic intima, and implies the intracellular incorporation of oleate into these lipids by fat-filled foam cells. In normally fed pigs, incorporation of ^{14}C-labelled oleate into intimal PL was 4-fold that of TG, and 100-fold that of CE. After cholesterol feeding, incorporation of ^{14}C-labelled oleate into intimal PL was 3-fold that of TG and only 3.5-fold that of CE. Thus, as a result of cholesterol-feeding, incorporation of FFA into CE increased (to 20% of total label incorporated) far more dramatically than incorporation into either PL or TG. However, after cholesterol feeding, both TG and PL remained metabolically more active than CE. This may indicate either greater synthesis and mass of TG and PL than of CE, or a greater turnover of these pools independent of the mass accumulated. It is difficult to draw confident conclusions regarding the relative mass of the distinct lipid pools within foam cell macrophages as these data are derived from whole intima.

9.2.2.2 Lipid accumulation within foam cell-rich plaques and *ex vivo* foam cells

CE is the predominant lipid fraction in all plaque types, representing up to 52% of plaque lipid, and CE content correlates positively with macrophage area at the centre and edge of all plaque types. The number of macrophages in human atherosclerotic plaques is positively associated with the CE content of plaques.[21] Additionally, the shoulder region of plaques is associated with a greater ratio of esterified to free cholesterol which parallels the increased cellular macrophage number in this site.[21] Phospholipids represent 14% of plaque lipid, and TG 9% of plaque lipid. As TGs are lower at the edge of disrupted plaques than at the centre, this suggests that macrophage foam cells do not contain large quantities of this lipid. In contrast, PL and CE are greater at the edges of disrupted plaques than at the centre, consistent with a foam cell macrophage contribution to both these lipids. Atherosclerotic arterial tissues contain increased cellular phospholipid, especially phosphatidylcholine, and to a lesser extent also sphingomyelin, and most of the increase in phospholipid appears to be intracellular.[18]

Foam cell macrophages have been isolated from rabbit atheroma using monoclonal antibodies to cell surface antigens and magnetic microspheres. These studies identified that the major lipid classes CE, FC, TG, PL and FFA respectively comprise 42–63%, 13–17%, 2–3%, 19–35% and 2–3% of cell lipid. CE varies from 2–5-fold the amount of FC (mass) in foam cell macrophages.[22] In a further study of cells isolated from human atherosclerotic intima, 40% of lesion mononuclear cells were recovered as CD14 positive, CD14 positive cells were predominantly foam cells on the basis of oil-red O staining, and these cells contained 408 ± 349 μg CE/mg cell protein, 133 ± 103 μg TG/mg cell protein, and 333 ± 200 μg FC/mg cell protein.[8] Although sampling variation between and within plaque samples may be significant, these data indicate that the FC to CE ratio in human foam cells is much closer to 1:1 than suggested by data from rabbit atheroma. These data also suggest that TGs represent a much larger proportion of total human foam cell lipid than anticipated from rabbit foam cell data or total human plaque data. The latter may indicate that lipoprotein lipids inevitably overestimate the contribution of CE in whole plaque lipid analysis. In a separate earlier study with less defined foam cell population, cells isolated from human atherosclerotic lesions had more TG, C, and CE compared with whole tissue lipids,[23] consistent with the above data. The major contribution to excess cellular lipid accumulation in cells from atherosclerotic lesions was made by CE.[23]

None of these observations address the intracellular distribution of CE, in particular the lysosomal or cytoplasmic location, nor whether the CEs are lipoprotein-derived or cell ACAT-esterified.

9.2.2.3 CE oxidation products, and their relationship to *in vitro* OxLDL and OxLDL-derived foam cells

Numerous studies have identified oxidized lipids in plaque.[24] Epitopes of oxidized LDL are co-localized with macrophages and occur intracellularly, thus differing from the extracellular location of LDL epitopes.[25] This may indicate either selective intracellular accumulation of extracellularly oxidized lipoproteins, or intracellular, perhaps lysosomal, oxidation of lipoproteins after endocytosis. Although many lipoprotein modifications may contribute to foam cell formation, the bioactivity and potential toxicity of oxidized lipids may particularly contribute to perturbed lysosomal function. As lysosomal acid lipase is pivotal in the metabolism of LDL lipids in the lysosome, perturbed lysosomal clearance probably involves inhibition of this enzyme.[26] A number of properties of substrates are known to affect lysosomal lipase-mediated degradation and these include the physical state of endocytosed material and the degree of polyunsaturation of the CE acyl chain.[27] Ceroid and lipofuscin are lipid–protein complexes which are insoluble in organic solvent, are identified within the lysosomes of foam cell macrophages in atherosclerotic plaque, and can be generated by uptake of *in vitro* oxidized lipoproteins.[28, 29] Chemical characterization of intracellular oxidized lipids is described in more detail elsewhere in this chapter (Section 9.4.7).

9.3 Interactions between macrophage lipid accumulation, inflammation and proliferation

9.3.1 *In vivo* and *in vitro* evidence for relationships between lipid accumulation and release of inflammatory mediators

The foam cell macrophage is an early and persistent feature of atherosclerosis and is therefore likely to play a key role in the development and progression of the disease. This may involve production of inflammatory mediators and effectors such as cytokines and extracellular matrix-degrading enzymes, which are considered active agents in atherosclerosis.

  The expression of several cytokines co-localized with macrophages in human atherosclerotic plaques is consistent with this hypothesis (reviewed in Ref. (30)). The cytokines include PDGF-B, TGF-β, TNF-α, IL-1, IL-6, IL-8, and M-CSF. Elevated expression of several of these cytokines is also detected in macrophages isolated from human atheroma. Macrophages are a source of a number of enzymes implicated in ECM degradation, including the matrix metalloproteinases (MMP), components of the plasmin activation system, lysosomal proteases such as cathepsin S, heparanases and sulphatases. Several studies have demonstrated elevated expression and production of several MMP, plasmin, and cathepsin S in lesion macrophages.

To determine the importance of lipid loading in lesion macrophage expression of inflammatory mediators, the effects of *in vitro* cholesterol loading of cultured macrophages have been studied. Cholesterol loading stimulates the release of some, but not all, cytokines and matrix-degrading enzymes (reviewed in Ref. (30)). Both stimulatory and inhibitory effects of OxLDL (and oxidized lipid components of it) on macrophage cytokine expression have been reported (for reviews, see, e.g. Refs (14, 30)). It should be noted that in most such studies, little attention has been given to the amounts of oxidation products which are introduced into the cells and how these compare with those present in lesion foam cells *in vivo*. This is an important consideration, since these compounds are generally present in foam cells in small amounts, while at high concentrations they are often toxic.

9.3.2 Lipoproteins and macrophage proliferation

The macrophages which accumulate in atherosclerotic lesions are derived from blood-borne monocytes that have been recruited into the intima. Local proliferation of macrophages within the vessel wall could also augment the numbers of these cells in atherosclerotic lesions. Studies of human atherosclerotic lesions, using either *in situ* tritiated thymidine incorporation or expression of proliferating cell nuclear antigen (PCNA) to determine the extent of proliferation, have shown that the rates of cell division are higher in atherosclerotic than in normal vessels. Moreover, macrophages are the predominant proliferating cell type, even in lesions derived mainly from smooth muscle cells. Proliferation was enhanced in lipid-rich lesions[31] and in focal collections of lipid-laden cells of monocytic appearance.[32]

Lipids have been recognized as mild stimuli of macrophage proliferation for some time.[33] This is strongly enhanced when such lipids are subjected to peroxidation before exposure to the cells.[34] Perhaps not surprisingly, OxLDL is also mitogenic for human and murine macrophages, while comparable sterol loading induced by AcLDL is less active[35] or inactive.[36] However, while the activity of OxLDL is established, the active component(s) and mechanism of action are still uncertain. Thus it was suggested that lysophosphatidylcholine (lyso-PC) is the active component of OxLDL, since generation of lyso-PC in acetylated LDL by treatment with PLA_2 increased its mitogenic activity. However, lyso-PC alone is not mitogenic, and removal of lyso-PC from OxLDL does not affect its influence on macrophage proliferation.[36] Perhaps aggregation induced by PLA_2 which would greatly accelerate its phagocytic uptake by macrophages, may explain this anomaly. Recently, modifications to apoB during LDL oxidation have been suggested to account for much of the growth-promoting activity of OxLDL.[36]

It has been suggested that OxLDL may induce GM-CSF secretion, inducing autocrine growth stimulation. However, macrophages from op/op GM-CSF -/- double knockout mice (deficient in both CSF-1 and GM-CSF) responded normally to growth induction by OxLDL,[35] indicating that this is not the major mechanism for growth induction. The downstream signalling pathways involved in OxLDL-mediated growth induction are still not clear, with suggestions that either phosphatidylinositol 3-kinase or protein kinase C are involved. OxLDL and CSF-1 act synergistically to stimulate macrophage proliferation,[35] indicating that they may operate through different effector mechanisms.

9.3.3 PPARα and PPARγ

Peroxisome proliferator-activated receptors (PPARs) are ligand-activated transcription factors belonging to the nuclear receptor superfamily. Three forms (α, δ/β, and γ) have been identified, each separate gene products and with distinct tissue distributions and patterns of expression. PPARs have roles in lipid, glucose and energy homeostasis, and in regulation of differentiation. Their ligands are predominantly lipids.

PPARγ, predominantly expressed in adipose tissue, adrenal gland and spleen, is also present in cells of the monocyte-macrophage lineage, where the level of expression is affected by several factors including the state of cell differentiation or activation.[37] Interestingly, a number of studies have shown that PPARγ expression in macrophages is also increased on exposure to oxidized LDL but not native or acetylated LDL.

Upon ligand activation, PPARγ regulates the expression of genes involved in adipocyte differentiation and glucose homeostasis. PPARγ also forms heterodimers with the retinoid X receptor (RXR) on a number of regulatory elements involved in adipocyte lipid metabolism. In monocyte/macrophages, activation of PPARγ inhibits the expression of inducible nitric oxide synthase (IFN-γ-induced), gelatinase B and scavenger receptor A (PMA-induced) genes, and the expression of several pro-inflammatory cytokines such as TNF-α, IL-6 and IL-1 induced by PMA but not by lipopolysaccharide. PPARγ activation can also lead to macrophage apoptosis in some circumstances. Some of these effects appear to be mediated, at least partly, through antagonism of transcription factors AP-1, STAT and/or NF-κB.[38]. In combination with agonists of RXR, activation of PPARγ causes induction of differentiation markers CD11b, CD18 and CD14.

'Natural' ligands for PPARγ include several lipids such as prostaglandin J2 (15d-PGJ2; $K_d = 2\,\mu M$) and certain polyunsaturated fatty acids such as linoleic acid ($K_d = 10–20\,\mu M$). The antidiabetic thiazolidinedione (TZD) drugs such as troglitazone are highly specific and potent agonists for PPARγ ($K_d = 30–700\,nM$), which has been very useful experimentally in determining the importance of these receptors. OxLDL also activates PPARγ.[39] In an attempt to determine the active component(s) of OxLDL, a systematic examination of a range of oxidized lipids likely to be present in OxLDL indicated that oxidized forms of linoleic and arachidonic acids, 9-HODE, 13-HODE and 15-HETE, were efficient stimuli of PPARγ-mediated reporter gene transcription, while the major oxysterols were inactive. In fact, the active oxidation products are present at relatively low levels in the heavily oxidized LDL preparations used in such studies, but may be generated endogenously by 12- and 15-lipoxygenases expressed in macrophages. The protection against atherosclerosis in rabbits expressing transgenic macrophage-specific 15-lipoxygenase[40] might be related to anti-inflammatory effects of its products acting as PPARγ ligands.

The interest in OxLDL-mediated induction of both PPARγ expression and activation has been largely stimulated by indirect evidence of its existence in atherosclerotic lesions and of its known biological properties. OxLDL has affinity for macrophage scavenger receptors SR-A and SR-B (CD36), which mediate uptake of OxLDL and consequent lipid accumulation, producing cells with some features of foam cells (see Sections 9.4 and 9.5). OxLDL has been shown to induce expression of SR-A and CD36 through a PPARγ-dependent mechanism, amplified in the presence of RXR agonists.[39] The existence of a positive feedback loop has been proposed, in which endocytosed OxLDL, through activation of PPARγ, further enhances expression of receptors for its own uptake. The likelihood of such a mechanism operating *in vivo* depends on the generation and uptake of heavily oxidized LDL by macrophages. As discussed in Section 9.2, and elsewhere in this volume, the evidence for this is not compelling at present.

PPARγ appears to be strongly expressed in atherosclerotic lesions, to a large degree associated with lesion macrophages.[37] Is this likely to affect the rate of lesion formation? If PPARγ ligands can suppress monocyte activation and the secretion of cytokines and matrix-degrading enzymes, then its expression should exert an anti-atherogenic influence. On the other hand, should uptake of OxLDL be the driving force in macrophage foam cell development, PPARγ-dependent upregulation of OxLDL receptors could promote foam cell formation. At the present time, TZD drugs are widely used to treat patients with non-insulin-dependent diabetes. Preliminary data from humans taking troglitazone show a reduction in atherosclerosis. While troglitazone lowers blood lipids and therefore would be expected to reduce cardiovascular risk, these results at least indicate that the beneficial effects of some PPAR agonists outweigh potentially negative outcomes *in vivo*. Whether the use of such agents more widely as anti-atherosclerotic or anti-inflammatory agents is appropriate remains to be determined.

PPARα is also expressed in macrophages,[37] and is reported to be present in atherosclerotic lesions, but not normal vessels, at levels generally more abundant than PPARγ. Like PPARα, its ligands include fatty acids and derivatives, including leukotriene B4 and 8-HETE. Pharmacologically it is stimulated by fibrates but not TZG drugs. PPARα activation increases hepatic uptake and esterification of free fatty acids, skeletal and cardiac fatty acid uptake and oxidation, stimulation of lipoprotein lipase, over-expression of apoA-I and apoA-II and inhibition of apoCII expression. Its effects in macrophages are less well understood than those of PPARγ. For example, LTB4 stimulated iNOS expression while a synthetic agonist (WY-14,643) inhibited it. It is not known if PPARα activation affects macrophage uptake and the metabolism of fatty acids. Studies in human and animal models have shown that fibrates limit the progression of atherosclerosis, but it is not possible to determine which of their many effects are most important features of their activity.

9.4 Normal cellular processing of lipoprotein lipids

9.4.1 Introduction

Cellular cholesterol homeostasis is a delicate balance between uptake and endogenous synthesis on the one hand, and its metabolism and export on the other. Although cholesterol is necessary for many cellular functions, excess accumulation of free sterol must be prevented to avoid potentially toxic perturbation of membrane function and the formation of insoluble crystals. Figure 9.1 illustrates the major pathways that contribute to cholesterol homeostasis in the macrophage. As in all peripheral cells, cholesterol can be acquired either by uptake of plasma lipoproteins, principally low-density lipoprotein (native LDL), or by endogenous synthesis. Receptor-mediated uptake of LDL involves binding to the lipoprotein to cell-surface expressed receptors, internalization into endosomes, their fusion with lysosomes followed by complete degradation of both the protein and lipid components of the lipoprotein.

9.4.2 Lysosomal lipolysis

Lysosomal acid lipase (LAL) plays an important role in cellular metabolism of endocytosed neutral lipids. It catalyses the intralysosomal hydrolysis of cholesteryl esters and triglycerides from endocytosed lipoproteins and generates free cholesterol, mono- and di-glycerides and free fatty acids, which are transported into the cytosol for oxidation or further metabolism. Normally this enzymic activity is

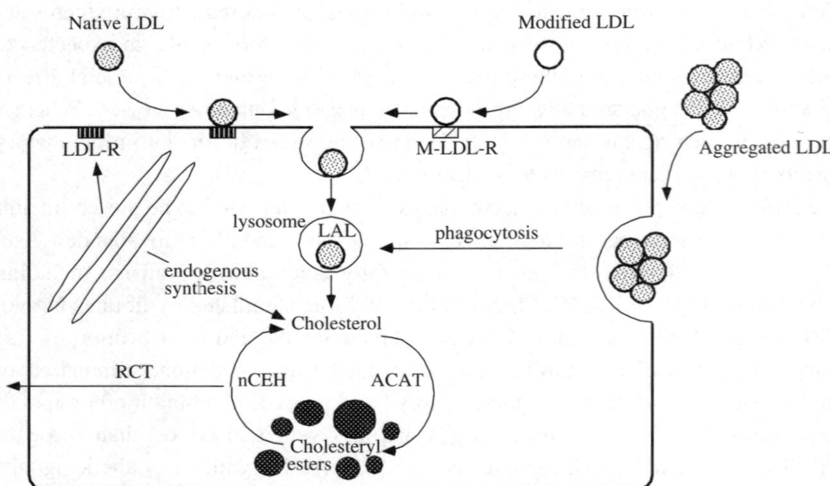

Fig. 9.1 Pathways of macrophage sterol metabolism. (Abbreviations: LAL, lysosomal acid lipase; nCEH, neutral cholesteryl ester hydrolase; ACAT, acyl CoA cholesterol acyltransferase; LDL, low-density lipoprotein; LDL-R, low-density lipoprotein receptor; M-LDL-R, modified LDL receptors; RCT, reverse cholesterol transport.)

present in excess, and its substrates are efficiently degraded. The importance of this enzyme is demonstrated in Wolman's disease, an inherited deficiency of LAL which results in massive intralysosomal accumulation of undegraded substrates and is lethal in early infancy.[41] A partial deficiency produces a less severe pathology (cholesteryl ester storage disease) and survival to adulthood. LAL is expressed in human monocytes and macrophages and its activity is increased approximately 10-fold during differentiation. Activity is regulated transcriptionally under control of AP-2 and Sp1. The enzyme is partially inactivated following endocytosis of peroxidized acetylated LDL by macrophages,[42] which may be of relevance in atherosclerotic intima, in which similar oxidation products are present. An important point frequently overlooked is the presence of substantial lysosomal lipid deposits in foam cells, particularly in advanced lesions (see Section 9.2). This indicates either that the rate of lysosomal degradation is insufficient to match uptake and/or some intralysosomal material is inherently resistant to degradation. Uptake of oxidized LDL may contribute to both of these mechanisms; inactivation of lysosomal lipases by lightly oxidized LDL,[42] while both protein[43] and lipid[44] components of heavily oxidized LDL are resistant to lysosomal proteolysis and accumulate in macrophage lysosomes (see Section 9.4.7).

9.4.3 Normal regulation of cellular cholesterol status

The cell has a sensitive mechanism for monitoring and maintaining cholesterol status. This involves membrane-bound transcription factors called sterol regulatory element binding proteins (SREBPs) that activate the genes controlling cholesterol and fatty acid synthesis and LDL receptor (LDL-R) expression.[45] SREBPs are proteins that are bound to the membranes of the endoplasmic reticulum and nuclear envelope in a hairpin orientation, with the N-terminal and C-terminal domains projecting into the cytosol. The N-terminal domain is the transcription factor region. In sterol-depleted cells this is released from the membranes by two sequential proteolytic cleavages, mediated by specific proteases (Site-1 protease (S1P) and Site-2 protease (S2P)). Cleavage by S1P is dependent on the tight association of SREBP with another membrane-bound regulatory protein, SREBP cleavage activating protein (SCAP). Within cells, SREBP and SCAP are found in a tight complex which is necessary for S1P cleavage to occur. SCAP is also the target for sterol suppression of its activity, losing its activity when sterols accumulate within cells. A discrete 160 aa region of the protein has been identified as the sterol-sensitive domain, but whether its sterol-sensing mechanism depends on direct interaction with sterols or indirect interaction with other proteins is not presently known. Interestingly, Lange *et al.*[46] have recently provided evidence of a non-linear relationship between endoplasmic reticulum cholesterol and cell cholesterol status which may provide a direct influence on sterol-sensing endoplasmic reticulum proteins. Above a certain threshold level, endoplasmic reticulum cholesterol levels rose much more rapidly than total cell cholesterol, effectively doubling for every 10% rise in cell cholesterol. 25-Hydroxycholesterol (25-OH) alters the shape of this curve, shifting and linearizing it. This is consistent with the effects of 25-OH cholesterol on SREBP-dependent gene expression.

9.4.4 Mechanisms for foam cell cholesterol accumulation

In most cells these cholesterol-sensitive mechanisms closely control the uptake and accumulation of cholesterol. Additionally, most cells are also able to export excess cholesterol to extracellular acceptors by a process of reverse cholesterol transport (RCT; see Section 9.5). In macrophages the export pathway may be particularly important, since these scavenging phagocytic cells frequently endocytose large amounts of cholesterol-containing materials such as dead or damaged cells. Nevertheless, macrophage cholesterol balance is normally well controlled. The development of the atherosclerotic foam cell phenotype is one outstanding exception to this general observation. In this case cholesterol accumulation clearly exceeds export, although whether dysfunction in uptake, export or both is rate-limiting is not known.

Most consideration has been given to routes for macrophage uptake of lipoproteins which bypass the cholesterol-sensitive LDL-R. Several alternative routes involving different receptors, phagocytosis, or a combination of both, have been identified. Central to all of these uptake routes is the assumption that a proportion of intimal LDL becomes either chemically or physically modified, such that it becomes a ligand for one or more of the alternative pathways. This concept was originally proposed to explain how macrophages accumulated excess LDL cholesterol despite the close control of LDL-R activity. A scavenger receptor was first demonstrated (now identified as SR-A) which bound large amounts of chemically modified (acetylated) LDL, converting macrophages into cholesteryl ester droplet-filled cells with a morphology similar to that of atherosclerotic foam cells.[47] This receptor also mediates uptake of oxidized LDL. This is of interest because of evidence for LDL oxidation in the intima. Subsequently a number of other macrophage receptors for oxidized LDL have been identified, as well as a range of other non-oxidative mechanisms of varying physiological plausibility (Table 9.1). These latter mechanisms mainly depend on aggregation of LDL particles induced by physical association with extracellular molecules or partial hydrolysis by a range of enzymes, or a combination of both. Aggregated LDLs provide a mechanism for massive lipoprotein uptake by phagocytosis and are particularly attractive candidates for stimulation of foam cell formation *in vivo*. Large particles which appear to derive from LDL have been detected in the intima.

9.4.5 Macrophage cholesteryl ester synthesis (ACAT)

In the face of abnormal cholesterol accumulation by macrophages, the esterification of excess cholesterol by the enzyme ACAT provides a mechanism for sterol sequestration in a relatively inert form. ACAT is active at the cytoplasmic face of the

Table 9.1 Routes for macrophage uptake of modified LDL

Modification	Uptake route
Acetylation	SR-A
Oxidation	SR-A
	CD36
	macrosialin/CD68
	LOX-1
LPL	LDL-R
Phospholipase D	LDL-R
Ig-binding	Fc receptor, phago
Aggregation induced by:	Phagocytosis
Proteoglycan association	
Phospholipase C	
Mechanical	
Sphingomyelinase degradation	
Mast cell granules	
Platelets	

endoplasmic reticulum and its products accumulate in the cytoplasm as lipid droplets. Human ACAT cDNA was first identified through expression cloning in 1993[48] and was termed ACAT-1. The importance of this enzyme in macrophage cholesteryl ester synthesis was clearly shown when macrophages from ACAT-1 $-/-$ mice were found to have a markedly reduced ability to synthesize cholesteryl esters. Immunodepletion experiments confirmed its major role in macrophage cholesterol esterification. Subsequently a second form of ACAT, which has 44% homology with ACAT-1, has been cloned,[49] termed ACAT-2. This is expressed mainly in liver and intestine, but does not appear of importance in macrophage cholesteryl ester synthesis. ACAT-1 is strongly expressed in human aortic atherosclerotic lesions, predominantly within macrophages[50] and also in macrophages of several non-atherosclerotic tissues.[51]

ACAT-1 expression increases up to 10-fold during differentiation of human monocyte/macrophages, occurring relatively early during differentiation (within the first 2 days). While the activity of the enzyme is affected by cholesterol levels in the cell, increasing once levels reach a certain threshold above basal, ACAT-1 protein levels do not change with cholesterol loading.[51] It seems that cholesterol levels affect ACAT-1 activity rather than expression. This is in agreement with the effects of cholesterol on the activity of purified ACAT-1 reconstituted into PL vesicles, where the increasing cholesterol content of the vesicles stimulated ACAT in sigmoid manner, indicating likely allosteric regulation. The mechanism of cholesterol sensing is not yet identified, and could depend either on a cholesterol-binding region of the protein or the effects of the sterol on membrane, since ACAT has several transmembrane domains. The threshold relationship between changes in endoplasmic reticulum and cell cholesterol changes may explain how ACAT activity is regulated in response to sterol loading.[46] Recently it has been shown that the subcellular location of ACAT alters upon cholesterol loading, shifting from being mainly in tubular rough endoplasmic reticulum in basal cells to approximately 30–40% in small-sized vesicles containing endoplasmic reticulum marker in cholesterol-loaded cells.[51] How this relates to changes in ACAT activity is not yet known.

9.4.6 Macrophage cholesteryl ester hydrolysis (nCEH)

Neutral cholesteryl esterase (nCEH) is responsible for the mobilization of cytoplasmic cholesteryl ester, an essential step in the process of cholesterol efflux from foam cell to extracellular acceptors (see Section 9.5), in which cholesterol export occurs principally at the expense of cholesteryl ester stores. Generally it is considered that ester hydrolysis can be rate-limiting, and that significant species differences in rates of cholesterol export from macrophage foam cells are related to their levels of nCEH activity.[52, 53]

It is suggested that in macrophages at least part of this activity is contributed by hormone-sensitive lipase (HSL). HSL is an enzyme of relatively broad specificity, having the ability to hydrolyse tri-, di- and mono-acylglycerols as well as cholesteryl esters and small water-soluble substrates. It is strongly expressed in adipose tissue, where it is responsible for lipolysis of triglycerides. The enzyme is expressed in murine and human macrophages, although the level of expression in human macrophages is 40-fold less than in adipose cells.[54, 55] On the other hand, evidence which favours a role for HSL are the removal of most nCEH activity by immuno-depletion of HSL from macrophage homogenates and increased cholesteryl ester hydrolysis in RAW 264.7 macrophages over-expressing HSL. HSL activity depends on cAMP-dependent phosphorylation; in macrophages, exposure to dibutyryl cAMP increased both HSL phosphorylation and nCEH activity. Interestingly, over-expression of HSL in mice did not attenuate but rather increased atherosclerosis;[56] the opposite would be predicted if cholesterol clearance from foam cells is limited by HSL activity.

However, other nCEH enzymes exist in mammalian systems. A secretory bile salt-stimulated cholesteryl esterase activity expressed in liver and pancreas has also been detected in human macrophages, although this is not considered likely to represent the cytosolic activity. Recently a separate neutral cholesteryl esterase enzyme has been identified in liver[57] which, like HSL, is regulated post-transcriptionally by phosphorylation. The expression of this nCEH is sterol-sensitive with a site in the promoter binding to SREBP-2. This enzyme is also expressed in human macrophages (Death, Brown and Jessup, unpublished).

Thus, at present, the identity of the enzyme(s) responsible for cytoplasmic ester hydrolysis in macrophages is not certain. While a body of evidence supports HSL as the active enzyme, this may not be the exclusive activity. It will be of considerable interest to compare the expression of all the above enzymes in macrophage foam cells of different species and their regulation in response to cholesterol loading and efflux.

9.4.7 Metabolism of oxidized lipoproteins

Chemical support for accumulation of oxidized lipids comes from demonstration of malondialdehyde-lysine and 4-hydroxynonenal-lysine adduct epitopes in lipid-laden macrophages in animal atherosclerosis and in patients with lysosomal dysfunction in Wolman's disease.[29] apoB and some oxidized lipids of OxLDL have both been shown to be resistant to lysosomal hydrolysis and accumulate in macrophage lysosomes.[43, 44, 58] Chemically characterized products of oxidation of cholesteryl linoleate, such as cholesteryl ester aldehydes, are resistant to lysosomal hydrolysis in macrophages, complex with proteins and are present in atherosclerotic lesions of humans.[59] Products of mild lipoprotein oxidation may inhibit lysosomal hydrolysis of unoxidized CE, and this may relate to direct inhibitor effects of hydroperoxides on the activity of cholesteryl esterases.[42] Intracellular oxidation of LDL would be expected to be enhanced by endosomal/lysosomal acidity. Uptake of oxidized LDL, and intralysosomal oxidation may also perturb lysosomal function by lysosomal enzyme relocation. Human THP-1 cells appear particularly susceptible to such effects of OxLDL.[60]

A number of secondary consequences may arise from accumulation of OxLDL in lysosomes, such as hypertrophy of the golgi apparatus and the trans-golgi network and loss of lysosomal integrity. Endocytic and secretory activities of macrophages have been altered by loading cells with OxLDL[61] and a number of lysosomal enzymes may be secreted in response to frustrated phagocytosis.

In summary, these data suggest that mild chemical modification of LDL may secondarily perturb lysosomal function, and that more advanced oxidation products may be themselves resistant to lysosomal hydrolysis.

9.5 Cholesterol efflux from macrophages

9.5.1 Introduction

Low plasma HDL concentrations confer an increased risk of atherosclerotic coronary events, elevation of HDL by pharmacological means decreases this risk and this appears attributable at least in part to a direct athero-protective effect of HDL. That HDL has anti-atherogenic properties is inferred from diverse animal studies. These include demonstration that apoA-I-transgenic apoE-KO mice develop less atherosclerosis than apoE-KO controls,[62] and that late transfection of LDLr-KO mice with human apoA-I gene induces regression of pre-formed fatty streaks.[63] The histological changes associated with regression, i.e. decreased cellularity and increased fibrosis, are also seen with regression following the lowering of LDL cholesterol. This supports the notion that cholesterol clearance is common to both low LDL-related and high HDL-related regression. In the case of HDL, cholesterol clearance is believed to occur by the pathway of reverse cholesterol transport (RCT) where cholesterol is directly and/or indirectly delivered to the liver for eventual biliary clearance.[64] Any component of this putative RCT pathway could be perturbed and could contribute to the failed clearance of cell cholesterol from lesions *in vivo*.

Non-steroidogenic peripheral cells such as macrophages cannot catabolize cholesterol once accumulated, but must retain or release it to an extracellular acceptor such as HDL or related apolipoprotein and phospholipid constituents. This initial cholesterol removal, or cholesterol efflux, is considered the first step in the RCT pathway, and, as the point of removal of cholesterol from cholesterol-rich foam cells in the artery wall, is potentially a limiting factor for the regression of atherosclerosis. This aspect of RCT is the particular focus of this chapter.

One metabolic route for the spontaneous elimination of cholesterol from macrophages is the metabolism of cholesterol to 27-hydroxycholesterol (27-OH).[65] Sterol 27-hydroxylase is a widely distributed mitochondrial cytochrome P450 enzyme which is expressed at particularly high activity in human macrophages and co-localizes with macrophages in human carotid lesions. Its product, 27-OH, is one of the most abundant oxysterols in human atherosclerotic lesions and macrophage-derived foam cells; the enzyme can hydroxylate the same methyl group three times in sequence to eventually form 3β-hydroxy-5-cholestenoic acid. Both 27-OH and 3β-hydroxy-5-cholestenoic acid are synthesized by cultured macrophages in response to cholesterol loading. It has been suggested that the polarity of 3β-hydroxy-5-cholestenoic acid relative to cholesterol allows its export even in the absence of an extracellular acceptor. This route may make a significant contribution to cholesterol export in tissues where HDL concentrations are relatively low, and there is some evidence to support this hypothesis; for example, patients with cerebrotendinous xanthomatosis (who lack this enzyme) die prematurely from atherosclerotic disorders, despite having normal circulating levels of cholesterol.

Many different molecules and particles can act as cholesterol acceptors. These include whole HDL, its constituent apolipoproteins such as apoA-I, phospholipid liposomes, reconstituted HDL particles, and synthetic molecules such as cyclodextrins. *In vivo*, a range of HDL and HDL-related particles are likely to be involved in cholesterol efflux, the most important of these for initial efflux being small apoA-I-only lipid-free or lipid-poor particles. Small apoE-containing particles may also contribute to initial cholesterol efflux. Physicochemical properties and receptor–ligand interactions are operative in cholesterol efflux, the extent of each probably depending on both cell and acceptor investigated and local concentrations of respective acceptors in arterial tissues. Factors regulating the efflux of cholesterol have been elucidated in some detail and there have been significant recent developments in this field, some of which are considered briefly below.

9.5.2 Cellular factors and cholesterol efflux

9.5.2.1 Plasma membrane and general physicochemical aspects of cholesterol efflux

Most simply, movement of cholesterol from the plasma membrane to a phospholipid-containing acceptor can be considered as analogous to movement of cholesterol between small unilamellar membrane vesicles containing pure phospholipid and cholesterol.[66] Transmembrane cholesterol movement (flip-flop) is generally much faster than desorption, indicating desorption is likely to be limiting. Molecules which disrupt phospholipid packing in the bilayer, such as bile salts or phospholipids with polyunsaturated acyl chains, enhance the desorption of phospholipids and cholesterol. Conversely, increasing acyl chain saturation of membrane PC decreases cholesterol efflux by increasing packing density.[67] The efflux of cholesterol is proportional to the concentration of cholesterol in the donor particles, and inversely related to the cholesterol concentration of the donor. At high acceptor/donor ratios, the rate of exchange is primarily dependent on the rate of desorption of the cholesterol from the donor membrane. The most important physical parameters of the plasma membrane are the FC content, the degree of unsaturation of PL, and the PC/SM ratio.[66]

In such systems, kinetic data indicates movement of cholesterol from donor membrane and into the unstirred water layer before being received by the acceptor vesicle. This aqueous diffusion model is supported by various data[66] and implies that direct acceptor–cell contact is not mechanistically important. This may not apply to apolipoprotein-mediated efflux where surface activity mediated by membrane penetration by amphipathic helices is important, to efflux mediated by cyclodextrins, and certainly cannot comprehensively model receptor–ligand interactions. All these efflux processes appear to be important in human macrophages.

Cholesterol flux is bidirectional. Net efflux of cholesterol requires greater efflux from donor to acceptor than influx from acceptor to donor, and is determined by the initial concentration of free cholesterol in donor and acceptor particles and the ratio of rate constants for efflux and influx. Thus net efflux from macrophages is most marked after cholesterol enrichment as after foam cell formation.[68] Because of bidirectional flux, the chemical cholesterol gradient between cell and acceptor is a major determinant of net efflux achieved by diverse initiating mechanisms, including receptors such as SRB1, apolipoprotein-mediated efflux or phospholipid vesicles.

In vitro studies have shown that there is significant variation between cells in their efficiency of cholesterol efflux to phospholipid-containing acceptors.[64] For example, FU5Ah hepatoma cells are very fast ($t_{1/2}$ 2 h), and smooth muscle cells are relatively slow ($t_{1/2}$ 24 h). Macrophages tend to demonstrate efflux intermediate between these two extremes, although human macrophages appear to be especially slow.[69] Any cell cholesterol (FC or CE) accumulated in lysosomal, cytosolic or other intracellular pools will also limit cholesterol efflux; slow intracellular processing of exogenous cholesterol appears to be particularly relevant in human macrophages. However, it is clear that the plasma membrane itself demonstrates significant heterogeneities in efflux which contribute to the cell–cell variation.[70]

Variations of plasma membrane-dependent cholesterol efflux between cells can be described kinetically in terms of overall rate (half-times of efflux) and in terms of the relative sizes of kinetically fast and slow pools.[66] For example, the relative size of slow pools appears to be especially great in human macrophages which demonstrate very slow overall cholesterol efflux.[52, 69] Apparent differences between cells in their rates of efflux of cholesterol to physiological and/or phospholipid-containing acceptors largely resolve on exposure to high concentrations of physicochemical acceptors of very high efficiency such as cyclodextrins.[64] These observations imply that efflux-resistant cells such as human monocyte-derived macrophages potentially stand to show maximal increment when exposed to very efficient synthetic agents such as cyclodextrins.

The relationship between kinetic and structural heterogeneities has been the subject of intense investigation. Even simple phospholipid vesicles demonstrate kinetic heterogeneity with fast- and slow-exchanging pools of cholesterol,[71] and phospholipid–cholesterol monolayers demonstrate regional lateral phase separation of phospholipid–cholesterol clusters from free phospholipid.[72] Physiological concentrations of cholesterol and sphingomyelin (SM) spontaneously form detergent-insoluble lipid phases in model membranes[73] which may be analogous to SM-rich domains in plasma membranes (see below). Additional structural variation specific to cell plasma membranes may be attributable to plasma membrane proteins including receptors for cholesterol acceptors, heterogeneous cholesterol distribution between inner and outer leaflets of the plasma membrane, and the presence of specific cholesterol-enriched structures such as caveolae and other SM-enriched domains (known as detergent-resistant DIGs or rafts).

9.5.2.2 Phospholipid composition of cell membranes – importance of sphingomyelin

Although PC is the predominant PL in cell membranes, and is the main PL released during cholesterol efflux, there appears to be a major role for cell SM in cholesterol homeostasis. As indicated above, SM and cholesterol spontaneously generate detergent-insoluble domains in model membranes,[73] and thus may be particularly important in caveolar and DIG-related cholesterol flux. SM is predominantly located in the exofacial leaflet, with minimal translocation between membrane leaflets, and up to 90% of cell SM is present in the plasma membrane.[74] The concentration of SM plays a major role in determining cholesterol solubilization in the plasma membrane. Increases in cell and plasma membrane SM result in increased cholesterol synthesis and decreased cholesterol esterification, and decreases in SM achieved by treatment of cells with SMase result in translocation of cholesterol from the plasma membrane to intracellular compartments and esterification by

ACAT.[74] The same does not occur with selective degradation of plasma membrane PC, suggesting SM has a particularly important role to play in cholesterol homeostasis. This may relate to the $1:1$ molar solubilization of cholesterol by PC and the $2:1$ solubilization of cholesterol by SM. Increased SM increases the packing density of cholesterol in membranes, and this increases the affinity of binding of cholesterol and decreases partitioning into the aqueous layer as compared with membranes containing only PC.[67] As oxysterols such as 25-OH cholesterol increase SM synthesis, SM may contribute to the effect of these lipids on cell cholesterol metabolism and efflux.

9.5.2.3 Caveolae and DIGs

Cell plasma membranes contain glycolipid-rich and FC-rich domains. Some of these are in the form of caveolae which are clathrin-free cell surface invaginations containing a 4–5-fold enrichment in cholesterol to protein ratio over surrounding plasma membrane, and some are in the form of lateral detergent-insoluble glycolipid-rich rafts (DIGs). Caveolae are abundant in adipocytes, SMC, EC and fibroblasts, but deficient in lymphocytes and macrophages, although recent studies indicate caveolin in human THP-1 macrophages.[75] Caveolae have been implicated as the source of cholesterol removed from fast pools to the primary acceptors of cell cholesterol preβ-HDL, and are upregulated by free cholesterol and downregulated by oxysterols.[76] Caveolin is a 22 kDa integral membrane protein associated with caveolae which is important for the structural integrity of caveolae and, in its soluble cytoplasmic form, the bidirectional delivery of cholesterol between caveolae and the ER. Redirection of caveolin away from the plasma membrane by oxidation of plasma membrane cholesterol with cholesterol oxidase does not necessarily alter the number or morphology of caveolae, indicating a degree of separation of the biology of caveolin and caveolae. Its expression, and the expression of caveolae are modulated by cell cholesterol – upregulated after accumulation of free cholesterol and downregulated by removal of cell cholesterol or suppression of cell cholesterol synthesis. In addition, the incorporation of caveolin into lipid membranes is cholesterol-dependent.[77] The paucity of caveolae in macrophage foam cells complicates the applicability of these structures to initial cholesterol efflux from atherosclerotic lesions, but this role could be taken by non-caveolar DIGs.

Caveolae may be especially important for secondary effects of cholesterol accumulation and efflux on cell signalling. As caveolae are associated with important and diverse proteins such as CD36, cytoplasmic signalling molecules, multiple GPI-linked proteins and cytoskeletal elements, modulation of caveolin expression by cholesterol flux will have major implications for cell function.[78] For example, caveolae appear to connect plasma membrane cholesterol levels to activation of the MAP kinase pathway and cell division.[79] The secondary effects of cholesterol depletion are also relevant in cells which do not express caveolae but have DIGs.

More recently, caveolae have been linked with the selective uptake of CE from HDL via the class B type I scavenger receptor SRB1, which is localized to caveolae.[80] Here, caveolae act as a site for reversible transient deposition of CE in the plasma membrane, receiving CE from HDL but allowing subsequent CE removal by HDL. Thus, in addition to roles in transcytosis, potocytosis and cholesterol trafficking in peripheral cells, caveolae appear to have a major role in the net clearance of cholesterol by the RCT pathway, as SRB1 appears to play a major role in this process.

9.5.2.4 Receptors SRB1 and ABC-1

Receptor-mediated processes have long been implicated in cholesterol efflux from cells, but, until recently, characterization of functionally important receptors has been problematic. Two important proteins have been identified as having important roles in different aspects of the RCT pathway.

SRB1 was identified in 1996 and has since been established as fundamental to HDL metabolism.[64] It mediates the selective uptake of CE from HDL by liver and adrenal cells, and is thus important for cholesterol delivery for hormone synthesis and for biliary secretion of cholesterol via the RCT pathway. Over-expression of hepatic SRB1 increases biliary secretion of cholesterol and decreases plasma HDL, LDL and VLDL cholesterol levels, but is associated with diminished development of atherosclerosis in heterozygous LDL receptor-deficient mice. Conversely, SRB1/apoE double knockout mice developed much more atherosclerosis than apoE-KO mice, and this correlated with a diminished excretion of biliary cholesterol.[81] Whether the effect is solely due to the clearance of CE by the liver or due to the roles of SRB1 in cholesterol efflux from peripheral cells is unclear.

SRB1 also markedly alters the traffic of lipids into and out of peripheral cells such as macrophages. By generating a hydrophobic channel, selective uptake of CE occurs from HDL, but, in addition, bidirectional flux of free cholesterol is facilitated, thus accelerating cholesterol efflux down a chemical gradient.[64] Expression of SRB1 in CHO cells achieves increases in cholesterol efflux and, to some extent, natural differences in cholesterol efflux between cell lines correlate with SRB1 expression.[82] The binding characteristics of this receptor are complex. SRB1 binds a wide range of apolipoproteins and lipoprotein particles, yet does not mediate apolipoprotein-mediated efflux but strongly enhances efflux to phospholipids.[82] The fact that simple binding of HDL to a closely related receptor CD36 does not enhance cholesterol efflux indicates that simple tethering of acceptors to the plasma membrane alone may not mediate its enhanced cholesterol efflux.[83] It is more likely that a combination of specific hydrophobic channel formation and/or plasma membrane organization secondary to SRB1 may be more important. In this respect, SRB1 may function as a means of partially overcoming limited partitioning of plasma membrane cholesterol across the water layer before coming into contact with phospholipid acceptors.

Tangier disease (TD) is a rare condition associated with absence of HDL, tonsillar enlargement, peripheral neuropathy and, at least in some cases, premature atherosclerosis.[84] In TD, HDL undergoes rapid catabolism, and fibroblasts from patients with Tangier disease show normal cholesterol efflux to phospholipid-containing acceptors but severely impaired efflux of cholesterol and phospholipid to delipidated apoA-I.[85, 86] The ability of apoA-I to remove cell cholesterol from peripheral cells, including macrophages, is closely linked to its ability to mediate phospholipid efflux (see below). This suggests that the inability of apoA-I in TD to acquire cell phospholipid is linked to both accelerated HDL clearance and to impaired apoA-I-mediated cholesterol efflux, and that phospholipid efflux is regulated by a specific protein which is subject to mutations in TD.

Recently the ATP binding cassette transporter 1 (ABC-1) protein has been identified as mutated in families with TD.[87] Importantly, with respect to macrophage biology, its expression is induced during differentiation of monocytes into macrophages and is upregulated by cholesterol enrichment.[88] As apolipoprotein-mediated cholesterol efflux is dependent upon the cholesterol enrichment of cells, this significantly supports the biological importance of ABC-1 expression in apoA-I-mediated cholesterol efflux. Other studies have previously indicated that apolipoprotein-mediated cholesterol efflux may be mediated by specific pathways, some of which relate to cell surface proteins that can be upregulated by cAMP.[89] Whether these are all related to ABC-1 expression will require further investigation. ABC-1 has many complex biological activities, including roles in multidrug resistance, transmembrane translocation of phosphatidylcholine, and presentation of endogenous antigens by HLA class I molecules. Consequently, the *in vivo* effects of deficiency and over-expression in animal models may be complex. To date, atherosclerosis *in vivo* deletion and upregulation studies have not been completed with ABC-1 and these studies are awaited with interest.

The combination of SRB1 and ABC-1 appears to mediate two poles of the RCT pathway. Both are involved in initial cholesterol efflux from macrophages: SRB1 for phospholipid-containing acceptors, and ABC-1 for apolipoprotein acceptors. SRB1 is also important in terminal uptake of HDL lipids by the liver. In combination with strategies to enhance the physicochemical gradient of cholesterol efflux, future genetic manipulation of these receptors may provide the prospect of therapy for established atherosclerosis.

9.5.3 Properties of cholesterol acceptors

In plasma, the smallest HDL particle, pre β-HDL, achieves a disproportionate initial cholesterol efflux relative to larger more abundant HDL particles.[90] This may relate both to the advantage of smaller acceptors over larger acceptors in inducing cholesterol efflux, and to specific properties of apolipoprotein-mediated efflux. Studies in phospholipid vesicle systems and in cell culture agree that smaller acceptor molecules are more efficient mediators of cholesterol efflux than larger acceptors.[91] The frequency of collision of smaller acceptors with cholesterol molecules in the aqueous layer is greater than that of larger acceptors for any given total mass of acceptor, and is due to the collision frequency being proportional to the total particle concentration in the medium.[66] However, acceptors achieving optimal rates of initial cholesterol efflux may achieve less net efflux over longer incubations than larger acceptors with greater capacity for binding cholesterol such as large phospholipid vesicles.

Initial studies using pure apolipoproteins incubated with smooth muscle cells and fibroblasts indicated that pure apolipoproteins were poor acceptors of cell cholesterol, and were much less effective than whole HDL. Subsequent studies using cholesterol-enriched murine macrophages showed that apolipoproteins A-I, A-II, E, and A-IV are all effective acceptors of cell cholesterol.[92] Cellular cholesterol enrichment enhances both cholesterol and phospholipid efflux from macrophages,[68] and increases apolipoprotein binding to lipid bilayers. Major properties of apolipoproteins which enhance cholesterol efflux are related to their surface activity, such as their content of hydrophobic segments, their overall flexibility, and their content of amphipathic helices.[93] That apolipoproteins function at least to some extent by their insertion between phospholipid molecules in the plasma membrane and subsequently microsolubilizing plasma membrane phospholipid has been supported in studies using apolipoprotein mutants. Other studies indicate that apolipoprotein-mediated efflux, as distinct from phospholipid vesicle-mediated efflux, is an energy-requiring active cellular process. In view of the proposed importance of ABC-1 in apolipoprotein-mediated efflux from peripheral cells such as macrophages, it is possible that ABC-1 may alter the phospholipid milieu in macrophage plasma membrane, which then facilitates the solubilization effects of apolipoproteins.

Among the smallest and most efficient acceptors of cell cholesterol from foam cell macrophages are small cyclical sugar particles known as cyclodextrins.[94, 95] As well as optimizing cholesterol efflux by enhancing collision frequency, the reduced activation energy for efflux to these particles has been interpreted as indicating direct contact with the cell surface and solubilization of cholesterol without desorption into the aqueous layer. In combination with phospholipid vesicles, cyclodextrins achieve massive synergistic cholesterol efflux, consistent with their acting as low-affinity but high-efficiency shuttles that remove cholesterol from the cell and deliver it to high-affinity but low-efficiency vesicles.[96] A similar process of transfer of foam cell cholesterol from smaller HDL particles to phospholipid vesicles has been reported.[97] This suggests that the apparent transfer of cell cholesterol from pre β-HDL to larger HDL species could involve remodelling as originally envisaged, or could involve transfer from smaller to larger particles as a more general process in cholesterol efflux. Our own unpublished data suggests that for primary human foam cell macrophages, synergistic efflux between cyclodextrins and phospholipid vesicles overcomes deficient cholesterol efflux far more effectively than any physiological acceptor (Liu, S., Dean, R.T., Jessup W., and Kritharides, L., unpublished).

9.5.4 Protein secretion in response to efflux: apoE

Although the anti-atherogenic effects of HDL have been largely attributed to its ability to deplete foam cell macrophages of excess cholesterol, it is known to exert many potential anti-atherogenic effects during atherogenesis. These include the inhibition of adhesion molecule expression by endothelial cells (Chapter 14) and the modulation of second messenger systems. In the artery wall, apoE expression by foam cell macrophages is well described, and is probably antiatherogenic.[98]

We and others have recently established that apolipoprotein A-I stimulates the secretion of apoE from human and murine foam cell macrophages in a time- and concentration-dependent manner.[99, 100] Importantly, because the secretion of apoE removes apoE from sites of intracellular degradation, the net accumulation of apoE in the artery wall may be markedly enhanced by this process. Although phospholipid vesicles also promote apoE secretion, cholesterol efflux induced by cyclodextrins does not, and apoA-I and phospholipid vesicles interact suggesting that these two essential components of HDL stimulate apoE secretion by complementary mechanisms.

It is quite possible that apolipoproteins will mediate a range of secondary biology effects which are variably related to the efflux of cell cholesterol. Some of these may relate to their effects on proteins located in cholesterol-rich domains, such as GPI-anchored proteins, and others to their induction of second messenger systems. It appears that the anti-atherogenic effects of these molecules may become increasingly complex, but may reveal specific processes which cannot be modulated by simple synthetic mediators of cholesterol efflux.

9.5.5 Resistance of human macrophages to cholesterol efflux

A number of studies have shown that human monocyte-derived macrophages (HMDM) are particularly resistant to cholesterol efflux induced by physiological acceptors such as HDL and apoA-I. Elucidating the components of intracellular lipid metabolism or plasma membrane cholesterol efflux which are limiting, may be of great importance in applying *in vitro* cholesterol data to *in vivo* results in humans. As human macrophages appear to be more susceptible to lysosomal accumulation lipids than non-human macrophages, and both esterify and hydrolyse CE more slowly than murine macrophages, intracellular processes may certainly be limiting.[52, 69] However, the ability to induce massive efflux with physiological depletion of cholesteryl ester by using combinations of cyclodextrins with phospholipid vesicles suggests that the plasma membrane is a key limiting factor in cholesterol efflux from HMDM. Whether caveolar expression, expression of receptors such as SRB1 and ABC-1, or phospholipid microdomains are mechanistically involved in this obstruction will be the subject of intense future investigation.

9.6 Conclusions

Macrophages play a key role in the development and progression of atherosclerotic lesions. It is the macrophage- and lipid-rich plaques rather than those with the greatest degree of stenosis that are most prone to rupture and thus to precipitate clinical events. Further, lipid lowering progressively reduces lesion macrophage numbers and improves plaque stability. Such observations indicate that treatments which prevent macrophage accumulation and lipid uptake and/or enhance their clearance are likely to improve health outcomes.

References

1. Gerrity, R.G. (1981). The role of the monocyte in atherogenesis: I. Transition of blood-borne monocytes into foam cells in fatty lesions. *Am. J. Pathol.*, **103**, 181.
2. Smith, J.D., Trogan, E., Ginsberg, M., Grigaux, C., Tian, J., and Miyata, M. (1995). Decreased atherosclerosis in mice deficient in both macrophage colony-stimulating factor (op) and apolipoprotein E. *Proc. Natl. Acad. Sci. USA*, **92**, 8264.
3. Stary, H.C. *et al.* (1995). A definition of advanced types of atherosclerotic lesions and a histological classification of atherosclerosis. A report from the Committee on Vascular Lesions of the Council on Arteriosclerosis, American Heart Association. *Arterioscler. Thromb. Vasc. Biol.*, **15**, 1512.
4. Shio, H., Haley, N.J., and Fowler, S. (1978). Characterization of lipid-laden aortic cells from cholesterol-fed rabbits II. Morphometric analysis of lipid-filled lysosomes and lipid droplets in aortic cell populations. *Lab. Invest.*, **39**, 390.
5. Rosenfeld, M.E., Carew, T.E., von Hodenberg, E., Pittman, R.C., Ross, R., and Steinberg, D. (1992). Autoradiographic analysis of the distribution of 125I-tyramine-cellobiose-LDL in atherosclerotic lesions of the WHHL rabbit. *Arterioscler. Thromb.*, **12**, 985.
6. Schaffner, T., Taylor, K., Bartucci, E.J., Fischer, D.K., Beeson, J.H., Glagov, S., and Wissler, R.W. (1980). Arterial foam cells with distinctive immunomorphologic and histochemical features of macrophages. *Am. J. Pathol*, **100**, 57.
7. Garner, B., Baoutina, A., Dean, R.T., and Jessup, W. (1997). Regulation of serum-induced lipid accumulation in human monocyte-derived macrophages by interferon-γ. Correlations with apolipoprotein E production, lipoprotein lipase activity and LDL receptor-related protein expression. *Atherosclerosis*, **128**, 47.
8. Mattsson, L., Johansson, H., Ottosson, M., Bondjers, G., and Wiklund, O. (1993). Expression of lipoprotein lipase mRNA and secretion in macrophages isolated from human atherosclerotic aorta. *J. Clin. Invest.*, **92**, 1759.

9. Lewis, J.C., Taylor, R.G., and Ohta, K. (1988). Lysosomal alterations during coronary atherosclerosis in the pigeon: correlative cytochemical and three-dimensional HVEM/IVEM observations. *Exp. Mol. Pathol.*, **48**, 103.

10. Jerome, W.G. and Lewis, J.C. (1987). Early atherogenesis in the White Carneau pigeon. III. Lipid accumulation in nascent foam cells. *Am. J. Pathol.*, **128**, 253.

11. Jerome, W.G., and Lewis, J.C. (1997). Cellular Dynamics in early atherosclerotic lesion progression in White Carneau pigeons. Spatial and temporal analysis of monocyte and smooth muscle invasion of the intima. *Arterioscler. Thromb. Vasc. Biol.*, **17**, 654.

12. Fernandez-Ortiz, A., Badimon, J.J., Falk, E., Fuster, V., Meyer, B., Mailhac, A., Weng, D., Shah, P.K., and Badimon, L. (1994). Characterization of the relative thrombogenicity of atherosclerotic plaque components: implications for consequences of plaque rupture. *J. Am. Coll. Cardiol.*, **23**, 1562.

13. Guyton, J.R., Black, B.L., and Seidel, C.L. (1990). Focal toxicity of oxysterols in vascular smooth muscle cell culture. *Am. J. Path.*, **137**, 425.

14. Brown, A. and Jessup, W. (1999). Oxysterols and atherosclerosis. *Atherosclerosis*, **142**, 1.

15. Lupu, F., Danaricu, I., and Siionescu, N. (1987). Development of intracellular lipid deposits in the lipid-laden cells of atherosclerotic lesions. *Atherosclerosis*, **67**, 127.

16. Tangirala, R.K., Jerome, W.G., Jones, N.L., Small, D.M., Johnson, W.J., Glick, J.M., Mahlberg, F.H., and Rothblat, G.H. (1994). Formation of cholesterol monohydrate crystals in macrophage-derived foam cells. *J. Lipid Res.*, **35**, 93.

17. Kellner-Weibel, G., Yancey, P.G., Jerome, W.G., Walser, T., Mason, R.P., Phillips, M.C., and Rothblat, G.H. (1999). Crystallization of free cholesterol in model macrophage foam cells. *Arterioscler. Thromb. Vasc. Biol.*, **19**, 1891.

18. Tabas, I. (1997). Phospholipid metabolism in cholesterol-loaded macrophages. *Curr. Opin. Lipidol.*, **8**, 263.

19. Guyton, J.R. and Klemp, K.F. (1989). The lipid-rich core region of human atherosclerotic fibrous plaques. Prevalence of small lipid droplets and vesicles by electron microscopy. *Am. J. Pathol.*, **134**, 705.

20. Day, A.J., Bell, F.P., and Schwartz, C.J. (1974). Lipid metabolism in focal areas of normal-fed and cholesterol-fed pig aortas. *Exp. Mol. Pathol.*, **21**, 179.

21. Felton, C.V., Crook, D., Davies, M.J., and Oliver, M.F. (1997). Relation of plaque lipid composition and morphology to the stability of human aortic plaques. *Arterioscler. Thromb. Vasc. Biol.*, **17**, 1337.

22. Mattsson, L., Bondjers, G., and Wiklund, O. (1991). Isolation of cell populations from arterial tissue using monoclonal antibodies and magnetic microspheres. *Atherosclerosis*, **89**, 25.

23. Mukhin, D.N., Orekhov, A.N., Andreeva, E.R., Schindeler, E.M., and Smirnov, V.N. (1991). Lipids in cells of atherosclerotic and uninvolved human aorta. III. Lipid distribution in intimal sublayers. *Exp. Mol. Pathol.*, **54**, 22.

24. Suarna, C., Dean, R.T., May, J., and Stocker, R. (1995). Human atherosclerotic plaque contains both oxidized lipids and relatively large amounts of alpha-tocopherol and ascorbate. *Arterioscler. Thromb. Vasc. Biol.*, **15**, 1616.

25. Sugiyama, N., Marcovina, S., Gown, A.M., Sefetel, H., Joffe, B., and Chait, A. (1992). Immunohistochemical distribution of lipoprotein epitopes in xanthomata from patients with familial hypercholesterolemia. *Am. J. Path.*, **141**, 99.

26. Goldstein, J.L., Dana, S.E., Faust, J.R., Beaudet, A.L., and Brown, M.S. (1975). Role of lysosomal acid lipase in the metabolism of plasma low density lipoprotein. Observations in cultured fibroblasts from a patient with cholesteryl ester storage disease. *J. Biol. Chem.*, **250**, 8487.

27. Lusa, S. and Somerharju, P. (1998). Degradation of low density lipoprotein cholesterol esters by lysosomal lipase *in vitro*. Effect of core physical state and basis of species selectivity. *Biochim. Biophys. Acta*, **1389**, 112.

28. Mitchinson, M.J. (1982). Insoluble lipids in human atherosclerotic plaques. *Atherosclerosis*, **45**, 11.

29. Hoff, H.F. and Hoppe, G. (1995). Structure of cholesterol-containing particles accumulating in atherosclerotic lesions and the mechanisms of their derivation. *Curr. Opin. Lipidol.*, **6**, 317.

30. van Reyk, D.M. and Jessup, W. (1999). The macrophage in atherosclerosis: modulation of cell function by sterols. *J. Leukoc. Biol.*, **66**, 557.

31. Orekhov, A., Andreeva, E., Mikhailova, I., and Gordon, D. (1998). Cell proliferation in normal and atherosclerotic human aorta: proliferative splash in lipid-rich lesions. *Atherosclerosis*, **139**, 41.

32. Villaschi, S. and Spagnoli, L. (1983). Autoradiographic and ultrastructural studies on human fibro-atheromatous plaque. *Atherosclerosis*, **48**, 95.

33. Yui, S. and Yamazaki, M. (1986). Induction of macrophage growth by lipids. *J. Immunol.*, **136**, 1334–8.

34. Yui, S. and Yamazaki, M. (1990). Augmentation of macrophage growth-stimulating activity of lipids by their peroxidation. *J. Immunol.*, **144**, 1466–71.

35. Hamilton, J.A., Myers, D., Jessup, W., Cochrane, F., Byrne, R., Whitty, G., and Moss, S. (1999). Oxidized LDL can induce macrophage survival, DNA synthesis and enhanced proliferature response to CSF-1 and GM-GSF. *Aterioscler. Thromb. Vasc. Biol.*, **19**, 98.

36. Martens, J.S., Lougheed, M., Gomez-Munoz, A., and Steinbrecher, U.P. (1999). A modification of apolipoprotein B accounts for most of the induction of macrophage growth by oxidized low density lipoprotein. *J. Biol. Chem.*, **274**, 10903.

37. Ricote, M. *et al.* (1998). Expression of the peroxisome proliferator-activated receptor γ(PPARγ) in human atherosclerosis and regulation in macrophages by colony stimulating factors and oxidized low-density lipoprotein. *Proc. Natl. Acad. Sci. USA*, **95**, 7614.

38. Ricote, M., AC, L., Willson, T., Kelly, C., and Glass, C. (1998). The peroxisome proliferator-activated receptor-gamma is a negative regulator of macrophage activation. *Nature*, **391**, 79.

39. Nagy, L., Tontonoz, P., Alvarez, J., Chen, H., and Evans, R. (1998). Oxidized LDL regulates macrophage gene expression through ligand activation of PPARγ. *Cell*, **93**, 229.

40. Shen, J., Herderick, E., Cornhill, J., Zsigmond, E., Kim, H.-S., Kuhn, H., Guevara, N., and Chan, L. (1996). Macrophage-mediated 15-lipoxygenase expression protects against atherosclerosis development. *J. Clin. Invest.*, 98.

41. Anderson, R.A., Byrum, R.S., Coates, P.M., and Sando, G.N. (1994). Mutations at the lysosomal acid cholesteryl ester hydrolase gene locus in Wolman disease. *Proc. Natl. Acad. Sci. USA*, **91**, 2718.

42. Kritharides, L., Upston, J., Jessup, W., and Dean, R.T. (1998). Accumulation and metabolism of cholesteryl linoleate hydroperoxide and hydroxide by macrophages. *J. Lipid Res.*, **39**, 2394.

43. Mander, E.L., Dean, R.T., Stanley, K.K., and Jessup, W. (1994). Apolipoprotein B of oxidized LDL accumulates in the lysosomes of macrophages. *Biochim. Biophys. Acta*, **1212**, 80.

44. Brown, A., EL, M., IC, G., Kritharides, L., Dean, R., and Jessup, W. (2000). Cholesterol and oxysterol metabolism and subcellular distribution in macrophage foam cells: accumulation of oxidized esters in lysosomes. *J. Lipid. Res.*, in press.

45. Brown, M. and JL, G. (1999). A proteolytic pathway that controls the cholesterol content of membranes, cells and blood. *Proc. Natl. Acad. Sci. USA*, **96**, 11041.

46. Lange, J., Ye, J., Rigney, M., and Steck, T. (1999). Regulation of endoplasmic reticulum cholesterol by plasma membrane cholesterol. *J. Lipid. Res.*, **40**, 2264.

47. Goldstein, J.L., Ho, Y.K., Basu, S.K., and Brown, M.S. (1979). Binding site on macrophages that mediates the uptake and degradation of acetylated low density lipoprotein, producing massive cholesterol deposition. *Proc. Natl. Acad. Sci. USA*, **76**, 333.

48. Chang, C.C., Huh, H.Y., Cadigan, K.M., and Chang, T.Y. (1993). Molecular cloning and functional expression of human acyl-coenzyme A: cholesterol acyltransferase cDNA in mutant Chinese hamster ovary cells. *J. Biol. Chem.*, **268**, 20747.

49. Cases, S. *et al.* (1998). ACAT-2, a second mammalian acyl-CoA: cholesterol acyltransferase. Its cloning, expression and characterization. *J. Biol. Chem.*, **273**, 26755.

50. Miyazaki, A., Sakashita, N., Lee, O., Takahashi, K., Horiuchi, S., Hakamata, H., Morganelli, P.M., Chang, C.C., and Chang, T.Y. (1998). Expression of ACAT-1 protein in human atherosclerotic lesions and cultured human monocytes-macrophages. *Arterioscler. Thromb. Vasc. Biol.*, **18**, 1568.

51. Sakashita, N., Miyazaki, A., Takeya, M., Horiuchi, S., Chang, C.C., Chang, T.Y., and Takahashi, K. (2000). Localization of human acyl-coenzyme A: cholesterol acyltransferase-1 (ACAT-1) in macrophages and in various tissues. *Am. J. Pathol.*, **156**, 227.

52. Graham, A., Angell, A.D., Jepson, C.A., Yeaman, S.J., and Hassall, D.G. (1996). Impaired mobilisation of cholesterol from stored cholesteryl esters in human (THP-1) macrophages. *Atherosclerosis*, **120**, 135.

53. Yancey, P.G. and St. Clair, R.W. (1994). Mechanism of the defect in cholesteryl ester clearance from macrophages of atherosclerosis-susceptible White Carneau pigeons. *J. Lipid. Res.*, **35**, 2114.

54. Reue, K., Cohen, R., and Schotz, M. (1997). Evidence for hormone-sensitive lipase mRNA expression in human monocyte-macrophages. *Arterioscler. Thromb. Vasc. Biol.*, **17**, 3428.

55. Li, F. and Hui, D. (1997). Modified low-density lipoprotein enhances the secretion of bile salt-stimulated cholesterol esterase by human monocyte-macrophages. Species-specific differences in macrophage cholesteryl ester hydrolase. *J. Biol. Chem.*, **272**, 28666.

56. Escary, J.L., Choy, H.A., Reue, K., Wang, X.P., Castellani, L.W., Glass, C.K., Lusis, A.J., and Schotz, M.C. (1999). Paradoxical effect on atherosclerosis of hormone-sensitive lipase overexpression in macrophages. *J. Lipid. Res.*, **40**, 397.

57. Ghosh, S., Mallonee, D.H., Hylemon, P.B., and Grogan, W.M. (1995). Molecular cloning and expression of rat hepatic neutral cholesteryl ester hydrolase. *Biochim. Biophys. Acta*, **1259**, 305.

58. Jessup, W., Mander, E.L., and Dean, R.T. (1992). The intracellular storage and turnover of apolipoprotein B of oxidized LDL in macrophages. *Biochim. Biophys. Acta*, **1126**, 167.

59. Hoppe, G., Ravandi, A., Herrera, D., Kuksis, A., and Hoff, H.F. (1997). Oxidation products of cholesteryl linoleate are resistant to hydrolysis in macrophages, form complexes with proteins, and are present in human atherosclerotic lesions. *J. Lipid. Res.*, **38**, 1347.

60. Yancey, P.G. and Jerome, W.G. (1998). Lysosomal sequestration of free and esterified cholesterol from oxidized low density lipoprotein in macrophages of different species. *J. Lipid Res.*, **39**, 1349.

61. Bolton, E.J., Jessup, W., Stanley, K.K., and Dean, R.T. (1997). Loading with oxidised low density lipoprotein alters endocytic and secretory activities of murine macrophages. *Biochim. Biophys. Acta*, **1356**, 12.

62. Paszty, C., Maeda, N., Verstuyft, J., and Rubin, E.M. (1994). Apolipoprotein AI transgene corrects apolipoprotein E deficiency-induced atherosclerosis in mice. *J. Clin. Invest.*, **94**, 899.

63. Tangirala, R., Tsukamoto, K., Chun, S.H., Usher, D., Pure, E., and Rader, D.J. (1999). Regression of atherosclerosis induced by liver-directed gene transfer of apolipoprotein A-I in mice. *Circulation*, **100**, 1816.

64. Rothblat, G.H., de la Llera-Moya, M., Atger, V., Kellner-Weibel, G., Williams, D.L., and Phillips, M.C. (1999). Cell cholesterol efflux: integration of old and new observations provides new insights. *J. Lipid Res.*, **40**, 781.

65. Bjorkhem, I., Andersson, O., Diczfalusy, U., Sevastik, B., Xiu, R.-J., Duan, C., and Lund, E. (1994). Atherosclerosis and sterol 27-hydroxylase: evidence for a role of this enzyme in elimination of cholesterol from human macrophages. *Proc. Natl. Acad. Sci. USA*, **91**, 8582.

66. Phillips, M.C., Johnson, W.J., and Rothblat, G.H. (1987). Mechanisms and consequences of cholesterol exchange and transfer. *Biochim. Biophys. Acta*, **906**, 223.

67. Lund-Katz, S., Laboda, H.M., McLean, L.R., and Phillips, M.C. (1988). Influence of molecular packing and phospholipid type on rates of cholesterol exchange. *Biochemistry*, **27**, 3416.

68. Yancey, P.G., Bielicki, J.K., Johnson, W.J., Lund-Katz, S., Palgunachari, M.N., Anatharamaiah, G.M., Segrest, J.P., Phillips, M.C., and Rothblat, G.H. (1995). Efflux of cellular cholesterol and phospholipid to lipid-free apolipoproteins and class A amphipathic peptides. *Biochemistry*, **34**, 7955.

69. Kritharides, L., Christian, A., Stoudt, G., Morel, D., and Rothblat, G.H. (1998). Cholesterol metabolism and efflux in human THP-1 macrophages. *Arterioscler. Thromb. Vasc. Biol.*, **18**, 1589.

70. Mahlberg, F.H. and Rothblat, G.H. (1992). Cellular cholesterol efflux. Role of cell membrane kinetic pools and interaction with apolipoproteins AI, AII, and Cs. *J. Biol. Chem.*, **267**, 4541.

71. Bar, L.K., Barenholz, Y., and Thompson, T.E. (1986). Fraction of cholesterol undergoing spontaneous exchange between small unilamellar phosphatidylcholine vesicles. *Biochemistry*, **25**, 6701.

72. McLean, L.R. and Phillips, M.C. (1982). Cholesterol desorption from clusters of phosphatidylcholine and cholesterol in unilamellar vesicle bilayers during lipid transfer or exchange. *Biochemistry*, **21**, 4053.

73. Ahmed, S.N., Brown, D.A., and London, E. (1997). On the origin of sphingolipid/cholesterol-rich detergent-insoluble cell membranes: physiological concentrations of cholesterol and sphingolipid induce formation of a detergent-insoluble, liquid-ordered lipid phase in model membranes. *Biochemistry*, **36**, 10944.

74. Slotte, J.P. (1997). Cholesterol–sphingomyelin interactions in cells – effects on lipid metabolism In *Subcellular biochemistry Volume 28: Cholesterol: its functions and metabolism in biology and medicine*, (ed. R. Bittman), p. 277. Plenum Press, New York.

75. Matveev, S., van der Westhuyzen, D., and Smart, E. (1999). Co-expression of scavenger receptor-B1 and caveolin-1 is associated with enhanced selective cholesteryl ester uptake in THP-1 macrophages. *J. Lipid Res.*, **40**, 1647.

76. Fielding, C.J. and Fielding, P.E. (1997). Intracellular cholesterol transport. *J. Lipid Res.*, **38,** 1503.

77. Li, S., Song, K.S., and Lisanti, M.P. (1996). Expression and characterization of recombinant caveolin. *J. Biol. Chem.*, **271**, 568.

78. Lisanti, M.P., Scherer, P.E., Vidugiriene, J., Tang, Z., Hermanowski-Vosatka, A., Tu, Y.H., Cook, R.F., and Sargiacomo, M. (1994). Characterization of caveolin-rich membrane domains isolated from an endothelial-rich source: implications for human disease. *J. Cell Biol.*, **126**, 111.

79. Furuchi, T. and Anderson, R.G.W. (1998). Cholesterol depletion of caveolae cause hyperactivation of extracellular signal-related kinase (ERK). *J. Biol. Chem.*, **273**, 21099.

80. Graft, G.A., Connell, P.M., van der Westhuyzen, D.R., and Smart, E.J. (1999). The class B, type I scavenger receptor promotes the selective uptake of high density lipoprotein cholesterol ethers into caveolae. *J. Biol. Chem.*, **274**, 12043.

81. Trigatti, B. *et al.* (1999). Influence of the high density lipoprotein receptor SR-BI on reproductive and cardiovascular pathophysiology. *Proc. Natl. Acad. Sci. USA*, **96**, 9322.

82. Jian, B., de la Llera-Moya, M., Ji, Y., Wang, N., Phillips, M.C., Swaney, J.B., Tall, A.R., and Rothblat, G.H. (1998). Scavenger receptor class B type I as a mediator of cellular cholesterol efflux to lipoproteins and phospholipid acceptors. *J. Biol. Chem.*, **273**, 5599.

83. de la Llera-Moya, M., Rothblat, G.H., Connelly, M.A., Kellner-Weibel, G., Sakr, S.W., Phillips, M.C., and Williams, D.L. (1999). Scavenger receptor BI (SR-BI) mediates free cholesterol flux independently of HDL tethering to the cell surface. *J. Lipid Res.*, **40**, 575.

84. Young, S.G. and Fielding, C.J. (1999). The ABCs of cholesterol efflux. *Nat. Genet.*, **22**, 316.

85. Oram, J.F. and Yokoyama, S. (1996). Apolipoprotein-mediated removal of cellular cholesterol and phospholipids. *J. Lipid Res.*, **37**, 2473.

86. Remaley, A.T., Schumacher, U.K., Stonik, J.A., Farsi, B.D., Nazih, H., and Brewer, H.B. (1997). Decreased reverse cholesterol transport from Tangier disease fibroblasts. *Arterioscler. Thromb. Vasc. Biol.*, **17**, 1813.

87. Brooks-Wilson, A. *et al.* (1999). Mutations in ABC1 in Tangier disease and familial high-density lipoprotein deficiency. *Nat. Genet.*, **22**, 336.

88. Langmann, T., Klucken, J., Reil, M., Liebisch, G., Luciani, M.F., Chimini, G., Kaminski, W.E. and Schmitz, G. (1999). Molecular cloning of the human ATP-binding cassette transporter 1(hABC1): evidence of sterol-dependent regulation in macrophages. *Biochem. Biophys. Res. Commun.*, **257**, 29.

89. Smith, J.D., Miyata, M., Ginsberg, M., Grigaux, C., Shmookler, E., and Plump, A.S. (1996). Cyclic AMP induces apolipoprotein E binding activity and promotes cholesterol efflux from a macrophage cell line to apolipoprotein acceptors. *J. Biol. Chem.*, **271**, 30647.

90. Castro, G.R. and Fielding, C.J. (1998). Early incorporation of cell-derived cholesterol into pre-β migrating high-density lipoprotein. *Biochemistry*, **1988**, 25.

91. Letizia, J. and Phillips, M. (1991). Effects of apolipoprotein on the kinetics of cholesterol exchange. *Biochemistry*, **30**, 866.

92. Hara, H. and Yokoyama, S. (1991). Interaction of free apolipoproteins with macrophages. *J. Biol. Chem.*, **266**, 3080.

93. Phillips, M.C. (1992). Interactions of apolipoproteins at interfaces. In *Structure and function of apolipoproteins* (ed. M. Rosseneu), p.186. CRC press, Ann Arbor.

94. Yancey, P.G., Rodrigueza, W.V., Kilsdonk, E.P.C., Stoudt, G.W., Johnson, W.J., Phillips, M.C., and Rothblat, G.H. (1996). Cellular cholesterol efflux mediated by cyclodextrins. *J. Biol. Chem.*, **271**, 16026.

95. Kritharides, L., Kus, M., Brown, A.J., Jessup, W., and Dean, R.T. (1996). Hydroxypropyl-beta-cyclodextrin-mediated efflux of 7-ketocholesterol from macrophage foam cells. *J. Biol. Chem.*, **271**, 27450.

96. Atger, V.M., de la Llera Moya, M., Stoudt, G.W., Rodrigueza, W.V., Phillips, M.C., and Rothblat, G.H. (1997). Cyclodextrins as catalysts for the removal of cholesterol from macrophage foam cells. *J. Clin. Invest.*, **99**, 773.

97. Rodrigueza, W.V., Williams, K.J., Rothblat, G.H., and Phillips, M.C. (1997). Remodeling and shuttling. Mechanisms for the synergistic effects between different acceptor particles in the mobilization of cellular cholesterol. *Arterioscler. Thromb. Vasc. Biol.*, **17**, 383.

98. Linton, M., Atkinson, J., and Fazio, S. (1995). Prevention of atherosclerosis in apolipoprotein E-deficient mice by bone marrow transplantation. *Science*, **267**, 1034.

99. Rees, D., Sloane, T., Jessup, W., Dean, R.T., and Kritharides, L. (1999). Apolipoprotein A-I stimulates secretion of apolipoprotein E by foam cell macrophages. *J. Biol. Chem.*, **274**, 27925.

100. Bielicki, J.K., McCall, M.R., and Forte, T.M. (1999). Apolipoprotein A-I promotes cholesterol release and apolipoprotein E recruitment from THP-1 macrophage-like foam cells. *J. Lipid Res.*, **40**, 85.

10 Smooth muscle cells and the connective tissue matrix of the intima

Gordon R. Campbell, John A. Bingley, Ian P. Hayward, and Julie H. Campbell
Centre for Research in Vascular Biology, Department of Anatomical Sciences, University of Queensland, Brisbane, Qld 4072, Australia

10.1 Introduction

Diffuse intimal thickenings in large and medium-sized arteries are sites for the development of atherosclerosis. For the first one to two decades in most humans, the thickening is composed of smooth muscle cells completely surrounded by extracellular matrix and covered by an endothelium. Occasionally, isolated macrophages are observed beneath this endothelium layer. The extracellular matrix provides two primary functions for cells: (i) it maintains the spatial relationships between cells; and (ii) maintains their differentiated state. In this context, it is intriguing to consider the fact that most of the extracellular matrix in the human intimal thickening is synthesized by the smooth muscle cells themselves. The consequences of this relationship mean that if under pathological circumstances there is a change in the function of either the smooth muscle cells or the extracellular matrix, then one could expect a concomitant change in the other component.

This article explores the relationship between these two components of the intima, particularly with regard to factors involved in atherogenesis. This scenario is further complicated by the fact that smooth muscle cells of the artery wall are heterogeneous. Some of these heterogeneous populations and their properties which could exacerbate the atherosclerosis process are explored.

10.2 Evolution of the atherosclerotic plaque

In recent years, our understanding of the development of the atherosclerotic plaque has been greatly increased both by the work of Stary,[1] who studied sequential sections of coronary arteries from 691 human subjects aged between birth and 39 years, and the Pathological Determinants of Atherosclerosis in Youth (PDAY) study,[2] which examined specific designated sites in arteries of over 2000 youths aged 15–24 years. From these studies, vital information has been provided on how the normal adaptive intimal thickening (also called diffuse intimal thickening, intimal cushion, smooth muscle mats) accumulates lipid to become the fatty streak, some of which (type IIa lesion or progression-prone type II) become intermediate lesions (type II lesions or pre-atheroma) and then atheroma (type III lesions or fibrous plaques).[1]

The importance of smooth muscle cells in the development of the atherosclerotic plaque is highlighted by the 1995 Report from the Committee on Vascular Lesions of the Council on Atherosclerosis, American Heart Association,[3] which states: 'Functional modulation and numerical increases in intimal smooth muscle cells stand out among cellular changes in advanced human lesions. The smooth muscle cells are in part resident intimal cells that preceded the lesions and in part their progeny that arose as a response to various stimuli. The stimuli include lipid accumulation, disruption of intimal structure, damage to intimal cells and matrix, and deposits of platelets and fibrinogen, all of which may activate resident cells to produce mitogenic factors.'

Three features of the immediate lesion need to be highlighted:

(1) Four different types of intermediate lesions have been categorized, and these lesions are not always rich in macrophages.

(2) Smooth muscle cells play a major role in the accumulation of lipid and in the further development of the atherosclerotic plaque.

(3) The smooth muscle cells present in the intermediate lesion express a range of phenotypes from the 'myofilament-rich (contractile) to the myofilament-poor but relatively rich in rough surfaced endoplasmic reticulum (synthetic)'.[1]

10.2.1 Functional modulation of smooth muscle

As well as maintaining tension (tone) via contraction or relaxation, the vascular smooth muscle cell (SMC) is responsible for vessel integrity by proliferation and synthesis of the extracellular matrix.[4] A number of cell types retain multifunctional abilities. Indeed, various cells can reversibly alter their character or phenotype when the environment changes. Modulations in cell phenotype may be due to cell interactions, changes in the extracellular matrix, altered functional demands, or in response to other signals such as hormones.[5] At one end of the spectrum of SMC phenotypes is the cell whose function is almost exclusively one of contraction ('contractile state'). The cytoplasm of the contractile-state cell contains numerous myofilament bundles, i.e. a high volume fraction of myofilaments ($V_{\mathrm{v\,myo}}$). Organelles, such as rough endoplasmic reticulum, Golgi and free ribosomes, are few in number and located in the perinuclear region. The majority of cells in the media of normal adult arteries are toward this state. At the opposite end of the SMC phenotypic spectrum is the cell whose function is almost exclusively synthesis. This cell has been descriptively termed as being in a 'synthetic' state. Similar to other cell types actively engaged in the production of extracellular matrix such as fibroblasts, the cytoplasm of synthetic-state cells contains few filament bundles but large amounts of rough endoplasmic reticulum, Golgi, and free ribosomes (i.e. a high volume of organelles and a low $V_{v\,myo}$). SMCs in the media of foetal arteries express this synthetic-state phenotype. Thus, the normal development of an artery (in line with the required functional changes) involves the SMCs altering from a synthetic state (foetal) to a contractile state (adult). Reversal of this process (phenotype modulation) occurs in situations such as arterial repair.[5]

10.2.2 How could a change in smooth muscle function contribute to the development of the atherosclerotic plaque?

Three important events in the development of the atherosclerotic lesion include (a) proliferation of SMC, (b) synthesis of large amounts of extracellular matrix by the SMC, and (c) accumulation of lipid within SMC and macrophages to form foam cells.[6] Studies of SMCs in culture indicate that a change in phenotype is probably necessary for all three events to occur.

(a) Smooth muscle phenotypic change and proliferation   When aortic SMCs from the adult pig and monkey are grown in primary culture only between 3 and 5% of SMCs, in the first 6 or 7 days, incorporate ^{3}H-thymidine into their nuclei during a 4 h pulse-labelling period in the presence of 5% whole blood serum. On day 7 or 8 (the day after phenotypic modulation to the synthetic state is first evident morphologically), the number of labelled nuclei increases to between 10 and 20%. The percentage of cells incorporating ^{3}H-thymidine into DNA then increases linearly with time until day 14, when between 40 and 60% of cell nuclei are labelled during the 4 h pulse-labelling period. The cells begin to proliferate logarithmically, in 5% whole blood serum, about 24 h after they show an increased incorporation of ^{3}H-thymidine.[7] Similar changes have been shown with rabbit SMCs, but occurring a few days earlier. Here changes in $V_{v\,myo}$ and α-smooth muscle actin messenger RNA have been correlated as a per cent of total actin messenger RNA prior to the onset of proliferation.[8] In addition, Thyberg and Sjölund[9] studied isolated vascular SMCs from 8-month-old rat aortic arteries, and found that by day 3 structural modification had occurred, and on day 5 the cells began to proliferate logarithmically.

Generally, no significant increase has been found in cell number while vascular SMCs are in the contractile state. It must be remembered, however, that the phenotypes expressed by vascular SMCs cover a continuous spectrum. The cells are often not fully contractile or fully synthetic, but somewhere between these two states. Thus, some cells, while still contractile, are also capable of undergoing division.[10]

(b) Smooth muscle phenotypic change and extracellular matrix production
Synthetic-state aortic SMCs under basal conditions synthesize between 25- and 46-fold the amount of collagen as contractile-state cells, measured by the incorporation of ^{3}H-proline into hydroxyproline.[11] The amount of non-collagen protein synthesized doubles under the same conditions. Synthetic-state cells also synthesize five-fold the contractile-state level of glycosaminoglycans, as measured by the incorporation of inorganic ^{3}H-acetate.[12] One of the major increases is in the synthesis of chondroitin sulphate, which is three-fold greater than that produced by the contractile cells. These increases are not related to cell proliferation because in these experiments the synthetic-state cells were maintained in a quiescent-growth state.

(c) Smooth muscle phenotypic change and cellular lipid accumulation Smooth muscle cells in the 'contractile' (high $V_{v\,myo}$) phenotype in culture bind, internalize and degrade very little β-VLDL; however, upon phenotypic modulation to a 'synthetic' (low $V_{v\,myo}$) state, there is a dramatic increase.[13] As the synthetic- state cells approach senescence through repeated cell divisions, the amount of cholesterol accumulated increases by up to seven-fold. Phenotypic modulation to the synthetic state in culture is also associated with a decrease in high affinity binding sites for native HDL.[14] Only about 60% of the β-VLDL is bound to the classic LDL or B/E receptor, with the remainder bound to a form of scavenger receptor. Indeed, two cDNAs that encode type I and type II scavenger receptors have been isolated from rabbit SMC and are homologous to macrophage scavenger receptors. The amount of β-VLDL bound to the scavenger receptors of synthetic-state cells in culture is further increased by up to 30-fold following oxidation of the lipoprotein by endothelial cells or macrophages.[15] Macrophages also decrease the activity of both neutral and acid cholesteryl ester hydrolase in synthetic state smooth muscle.[16] This leads to massive accumulation of lipid in the cells such that they become indistinguishable from the 'foam' cells of smooth muscle origin which can be isolated from atherosclerotic plaques.

Scavenger receptors are upregulated on SMC in neointimal thickenings of balloon-injured rabbit arteries, with or without cholesterol feeding, and can be upregulated in passaged SMC by TNF-α and IFN-α, PDGF-BB and phorbol ester, making them more macrophage-like.[17, 18] Meitus-Snyder *et al.*[19] have more recently shown that scavenger receptors on SMC are regulated by protein kinase C, and suggest a role for reactive oxygen species in upregulation of these receptors in atherogenesis.

10.2.3 Extracellular matrix of the intima

The major components of the extracellular matrix of the intima of the artery wall are:[20]

(a) Collagens The collagens are a family of proteins of at least 19 genetically distinct types classified by differences in their amino acid composition and the

proportion of the molecule forming the triple helix. Six collagen types, I, III, IV, V, VI and VIII, have been identified in blood vessels. The predominant types are I and III, which comprise up to 80–90% of the total artery wall collagens.

There are minor amounts of other collagens which provide important properties to blood vessels. Types IV and VIII collagens are in endothelial cell basement membranes and surround smooth muscle cells. Also present in small amounts is type V collagen. This collagen is found in association with type I collagen and may participate in the formation of collagen heteropolymers.

Ninety per cent of the total protein in the atherosclerotic plaque consists of collagens, predominantly types I and III. Fibrillar collagens and other members of the collagen family, types IV and V, play important roles in vessel wall homeostasis. Type VIII has recently been shown to be unregulated in the rabbit aorta following a cholesterol diet and balloon injury.[21]

(b) Elastic fibres   Elastic fibres are complex structures which include a hydrophobic 70 kDa elastin protein associated with hydrophilic glycoproteins and enzymes responsible for the internal cross-linking of elastin peptides.

(c) Proteoglycans   Proteoglycans are protein polysaccharides which share a common structural feature of one or more glycosaminoglycan (GAG) chains covalently attached to a core glycoprotein backbone. The GAG chains are linear polymers of repeating disaccharides which contain a hexosamine and either a carboxylate or a sulphate ester or both.

The major types of proteoglycans identified in blood vessels include large (~1000 kDa) chondroitin sulphate proteoglycan (CSPG) such as versican,[22] small (~120–300 kDa) leucine-rich dermatan sulphate proteoglycan (DSPG) such as decorin and biglycan, keratin sulphate proteoglycan (KSPG) such as lumican, and heparan sulphate proteoglycan (HSPG) such as perlecan and other basement membrane proteoglycans.[23] All these proteoglycans associate with different components of the vascular ECM. Versican, the most abundant proteoglycan of the intima, is a large member of the aggrecan family. It has a core protein of 263 kDa and 15–20 GAG chains made of ≈70% chondroitin-6-sulphate, 20% chondroitin-4-sulphate and 10% dermatan sulphate, and is defined as a CSPG.

Versican is a major CSPG produced by smooth muscle cells which binds to hyaluronan, a high molecular weight polysaccharide. Increased synthesis of hyaluronan is associated with smooth muscle cell proliferation and its receptor CD44 is also upregulated at this time. Both hyaluronan and versican are major components of the primary atherosclerotic plaque. Recent studies have shown that hyaluronan/versican form matrixes around smooth muscle cells facilitating their migration and mitosis.

Two small proteoglycans, decorin and biglycan, with core proteins of 36 kDa and two to three GAG chains of dermatan sulphate, also exist in the intimal interstitium and are involved in the organization of collagen fibres. These proteoglycans are mostly products of smooth muscle cells of the media. Decorin, a small leucine-rich interstitial proteoglycan, interacts with both LDL and collagen type I. Studies on atherosclerotic plaques in Japanese quails demonstrate decorin in the core of the plaque with collagen type I on the plaque surface.[24] The basement membrane of the endothelium contains perlecan, with a large multidomain core protein (≈ 450 kDa) and with three heparan sulphate chains that interact with collagen type IV, vitronectin, and laminin. Altered expressions of perlecan have been identified in proliferating smooth muscle cells and have an association with the severity of atherosclerotic lesions.[25] Syndecan is an additional proteoglycan associated with the cell membrane of most cells, and it contains three to five chondroitin sulphate and heparan sulphate GAG chains attached to 30 kDa core proteins. The main function of syndecan appears to be the accretion at the cell surface of growth factors, cytokines, lipases, and homeostasis factors that have GAG-binding segments.

Proteoglycans are also present on the surface of vascular cells.[25] These proteoglycans may be inserted directly into vascular cell membranes via hydrophobic sequences in the core proteins, as is the case for the heparan sulphate proteoglycans of syndecans. Other cell surface heparan sulphate proteoglycans associate with vascular cell membranes by phosphatidylinositol linkages and comprise a separate family of membrane proteoglycans termed glypicans.

Of particular relevance to atherosclerosis and smooth muscle cells are the heparan sulphates. These are one of a group of unbranching carbohydrate polymers, GAGs, characterized by repeats of a basic disaccharide unit which consist of glucuronic acid (GlcA) and either *N*-acetyl glucosamine (Glcan) (in the case of heparan sulphate or heparin) or *N*-acetyl galactosamine (in the cases of chondroitin, keratan, and dermatan sulphates). This simple structure undergoes several modifications, notably sulphations, which produce substantial heterogeneity within the final GAG molecule. Heparin and heparan sulphates are qualitatively similar, differing structurally predominantly in the degree to which they are sulphated. Heparin tends to be more heavily N-sulphated relative to heparan sulphate: the N-sulphate ratio being $> 2/1$ for heparin, and roughly equimolar in heparan sulphate. In either case, the final GAG is anionic (negatively charged in solution). Indeed, heparin is more negatively charged than any other polyelectrolyte within mammalian tissues.[26]

Heparan sulphate and other GAGs, as a result of their net negative charge, have the ability to bind to positively charged peptides and proteins,[27] although this does not necessarily equate to physiological function.[26] The binding to heparin of a number of unrelated proteins has uncovered common consensus sequences for heparin binding within the peptide structure involving basic and hydrophobic amino acids. These sequences can be found within thrombin, lipoprotein lipase, apolipoprotein B, fibronectin, vitronectin, heparin cofactor II, lipocortin, protein C inhibitor, several viral coat proteins,[28] anti-thrombin III[27] and heparin-binding growth factors – fibroblast growth factor, hepatocyte growth factor[29] and epidermal growth factor.[30]

While the arrangement of basic amino acids may confer on these peptides the ability to bind GAGs generally, highly specific recognition is created by unique patterns of sulphation with the GAG chain. This specific binding is necessary for the targeting and modulation of proteoglycans for regulatory processes.[28] Minimal functional sequences have been identified for the binding by heparan sulphate of lipoprotein lipase,[31] bFGF, and for heparin binding of anti-thrombin III.[32] Interestingly, the minimum anti-thrombin III binding sequence is the same minimum pentasaccharide which is anti-proliferative for smooth muscle cell growth, even though anti-coagulation and anti-proliferation are independent of each other in the heparin molecule.[32]

Deliberate alterations of the heparan sulphate and heparin chains can be achieved chemically, allowing examination of the importance of particular structural features. Desulphation of the heparin molecule, particularly the N-site of N-acetyl glucosamine, completely removes both the anti-coagulant and anti-proliferative functions,[32] while returning an acetyl group to that site returns the inhibitory but not the anti-coagulant activity of full-length heparin. The anti-proliferative potency of heparin resides in the 2-O-sulphate-rich oligosaccharide segments of the heparan sulphate molecule, with sulphate-poor regions having no inhibitory action. These sulphate-rich domains are in fact more inhibitory than the native heparan sulphate molecule. Using nitrous acid to partially digest heparins, Castellot *et al.*[33] showed that a 12-saccharide residue glycosaminoglycan holds similar anti-proliferative activity to the full molecule, that two- and four-residue glycosaminoglycans were unable to inhibit smooth muscle cell proliferation, and that the intermediate-sized oligosaccharides possessed intermediate inhibitory activity. Deliberate oversulphation increases the anti-proliferative action of heparin fragments that are too small to be active otherwise.[34] Cingi *et al.*[35] found that larger heparan sulphate molecules did not necessarily hold more inhibitory potential, with molecules from 6800 to 10 700 kDa proving more inhibitory than a 25 500 kDa heparan sulphate. To determine the functional limits of size and sulphation of heparin further, Choay *et al.*[36] produced GAG polymers from monosaccharide residues, allowing examination of the effect of discrete residue sequence on activity. The smallest molecule that can inhibit smooth muscle cell proliferation is a pentasaccharide. If the 3-O position in this molecule is unsuccessful, however, the molecule loses all its inhibitory action.

(d) Glycoproteins The principal glycoproteins in arterial tissue synthesized by the vascular cells include fibronectin, laminin, thrombospondin, tenascin, and osteopontin.[20]

Fibronectins Fibronectins are a family of glycoproteins present in blood plasma and in the ECM of most tissues. They consist of two similar peptide chains of ~220 kDa held together at one end by disulphide bonds. Fibronectins are a principal attachment protein for vascular cells and serve as a substrate for the migration of cells in the artery wall during development and remodelling.

Laminin Laminin is an ~800 kDa trimeric glycoprotein present in endothelial and smooth muscle cell basement membranes. Like fibronectin, this molecule consists of sub-units of polypeptides which contain multiple binding sites for cell surface receptors as well as other ECM molecules. Laminin interacts with arterial smooth muscle cells through more than one receptor. This interaction in part maintains these cells in a contractile phenotype, unlike fibronectin, which promotes modulation of arterial smooth muscle cells to a synthetic phenotype.[37]

Thrombospondin   Thrombospondin is a 450 kDa trimeric glycoprotein which consists of three identical 150 kDa chains joined together by disulphide linkages. The glycoprotein exists in more than one form and may constitute a family of related proteins generated by alternative splicing, although more than one thrombospondin gene exists. Thrombospondin is present in different vascular layers but is elevated in intimal thickenings in human diseased arteries.[38]

Tenascin   Tenascin is a glycoprotein only transiently present in vascular ECM. It is synthesized by both vascular smooth muscle and endothelial cells and regulated in part by factors such as PDGF, TGF-β1, and angiotensin II. It can influence vascular cell adhesion. It is present in the vascular ECM at early stages of embryonic vasculogenesis, and is present in increased levels in the neointima following experimental balloon de-endothelialization[39] and in macrophage-rich human coronary atherosclerotic plaque.[40]

Osteopontin   Osteopontin is an acidic, highly phosphorylated glycoprotein first identified in bone, but subsequently found in a variety of tissues including blood vessels. Osteopontin is small compared to most ECM molecules and has an average molecular mass ranging from 44 to 85 kDa. Osteopontin is not present in the ECM of normal blood vessels but appears in the neointima following experimental balloon angioplasty and is present in human atherosclerotic plaques.[41]

10.2.4   Changes in the extracellular matrix in cell culture, injury and atherosclerosis

Camejo *et al.*[42] have shown that cultured proliferating smooth muscle cells produce chondroitin-sulphate-rich proteoglycans of a differing size and composition from those produced by quiescent cells. Measured with a gel mobility shift assay, the apparent affinity constant of LDL for proteoglycans from proliferating cells is three-fold higher than that for proteoglycans from quiescent cells. Proteoglycans produced by intimal smooth muscle cells from injured vessels (synthetic state) also differ from those produced by smooth muscle cells in the media (contractile state) both in physical and chemical properties.[43]

By 4 weeks after balloon injury in the rabbit and rat models, the smooth muscle cell proliferation within the arterial media is reduced to the very low pre-injury rate, about 0.06%.[44] This is accompanied by a reversion of phenotype to the 'contractile' state characterized by a high $V_{v\,myo}$. Despite the cessation of cellular proliferation, there continues to be an accumulation of extracellular matrix within the intima, predominantly around the more quiescent smooth muscle cells away from the lumen, until finally the matrix represents 50–60% of the neointimal area. Collagen, elastin, and proteoglycan production increase from the first week through to at least the fourth week after arterial injury, indicating that while the more chronic changes in the healing artery are associated with matrix production, there is accumulation of matrix from the early timepoints.

Within normal blood vessels and in culture, smooth muscle cells produce predominantly CSPGs and DSPGs, with HSPGs accounting for less than 10% of the total proteoglycans; proteoglycans in turn make up only 2–3% of the dry weight of the vessel.[45] Proteoglycans in the normal rabbit aorta decrease in concentration from the intima to the outer media.[46] One large CSPG laid down by smooth muscle cells in the extracellular matrix has seemingly little association with other major components of the ECM such as collagen, unlike smaller DSPG. A second CSPG has been identified on the cell surface, where its likely role in cell–matrix interactions is uncertain.[47] Interestingly, CSPG and DSPG, which bind low-density lipoprotein avidly in the presence of lipoprotein lipase, are found in higher concentrations in arteries prone to atherosclerosis, such as the abdominal aorta.[48] Endothelial cells, by contrast, produce mainly HSPGs and DSPGs, which are found either tightly bound to elastin or closely attached to the cell surface.[47] Cultured bovine arterial smooth muscle cells similarly produce cell-associated and soluble HSPG.

The types of HSPG produced by smooth muscle cells from the arterial wall in culture alter with disease. Smooth muscle cells from atherosclerosis-susceptible pigeons show a 50% decrease, and differing charge densities, in their cell-associated HSPG compared with smooth muscle cells from atherosclerosis-resistant pigeons.[49] Rat aortic smooth muscle cell transmembrane proteoglycans syndecan and fibroglycan are independently regulated within the injury response. Syndecan-1 mRNA is increased in situations that stimulate smooth muscle cell proliferation (e.g. PDGF stimulation) and decreased in situations of inhibition (e.g. increased cell density), while fibroglycan mRNA is decreased in response to proliferative stimuli, and either increased or unaffected by inhibitory stimuli. Such changes suggest that these transmembrane HSPGs fulfil individual roles in cellular control. Downregulation of fibroglycan may occur following an injury stimulus, for example, and may play an important role in allowing cell migration.[50]

In vivo, diseased arteries also show alteration of the various proteoglycan types.[51] The general increase in GAG production seen when smooth muscle cells modulate from a quiescent to a proliferative phenotype is similar to changes which occur within a neointima and within atherosclerotic lesions.[12] The smooth muscle cells of the neointima are responsible for the abundance of proteoglycans surrounding them. Li *et al.*[45] have shown that smooth muscle cells from injured aortic explants *ex vivo* incorporate ^{35}S sulphate into proteoglycan synthesis significantly faster than cells from uninjured aortae, whether or not the aorta had been re-endothelialized. The neointima produced following re-endothelialization of rabbit aortae shows an absolute increase in total proteoglycan production, and also a relative increase in chondroitin sulphate, little change to heparan sulphate, and an absolute decrease in dermatan sulphate.[52] Similarly, in human atherosclerotic plaque there is an increase in the amount of chondroitin sulphate relative to heparan; the relative decrease in heparan sulphate is thought to be coincident with loss of control over smooth muscle proliferation with the plaque.[53] The proteoglycans from neointimal smooth muscle cells have an increased capacity to bind low-density lipoprotein, a feature which may help explain lipid accumulation in atherosclerosis.[54]

Re-endothelialization significantly increases proteoglycan accumulation compared with those areas of neointima which remain de-endothelialized.[55] ^{35}S incorporation in the intima after air-drying and balloon injury is maximal at 3 days in the rabbit, suggesting increased turnover. From 6 days on, the incorporation in ballooned vessels is similar to that in uninjured vessels, whereas air-drying caused a slightly more prolonged increase in ^{35}S incorporation. Similar changes occur in the degree of ruthenium red staining at 3 and 6 days, as measured morphometrically under electron microscopy.[52]

Nikkari *et al.*[56] examined the alterations in matrix- and cell-associated proteoglycans and proteins following arterial injury in the rat carotid model. Northern analysis showed an increase in RNA expression of syndecans 1 and 4 within 2 days after injury, sustained through to 4 weeks after injury. Perlecan, biglycan, and versican all increased after the first week and remain elevated. Similarly, collagen and elastin increase from 1 week after injury, with collagen mRNA expression remaining elevated to 4 weeks while elastin levels returned towards normal. These changes were mirrored in *in situ* hybridization findings.

10.2.5 Extracellular matrix and sequestration of lipoprotein in the intima

A number of studies suggest that particular proteoglycans selectively cause the accumulation of specific types of lipoproteins in the intima[57] and these proteoglycans may be due to a particular activity of the smooth muscle cells. For instance, Tao *et al.*[58] have shown that proliferating cultured vascular smooth cells produce a proteoglycan sub-fraction which has a very high affinity for LDL. In pre-lesional sites that are susceptible to atherogenesis, two common features have been reported: an altered proteoglycan structure and increased lipoprotein retention.[59]

It has been suggested that this process of retention of lipoprotein by proteoglycans is the first step in a sequence of events which lead to further modification of the lipoproteins such as oxidation and aggregation.[60] This retention by proteoglycans and consequential modification of lipoprotein form the basis of the 'Response-to-Retention hypothesis of early atherogenesis'.[61] The authors of this hypothesis point out how proteoglycan-bound LDL *in vitro* forms aggregates and vesicular structures which resemble material seen *in vitro* and these LDL–proteoglycan complexes are susceptible to oxidation. Finally, they highlight how aggregated LDL or LDL modified by proteoglycans is avidly taken up by macrophages and smooth muscle cells to form foam cells. For example, Tirziu *et al.*[62] have recently shown that incubation of LDL with chondroitin-6-sulphate results in modified lipoprotein. These lipoproteins are mitogenic for smooth muscle cells, induce a significant migratory stimulus, and accumulate within the cells as lipid droplets. Not only do chondroitin sulphate-rich proteoglycans bind LDL but they also bind Lp(a).[63]

10.2.6 What maintains the phenotype of smooth muscle cells?

Studies in our laboratory implicate cell–matrix interactions in control of smooth muscle phenotype. In primary cell culture, a crude extract of GAGs from the aortic intima plus inner media maintains sparsely seeded smooth muscle with a high $V_{v\,myo}$. Treatment of this aortic extract with heparinase from *Flavobacterium heparinum* destroys the active factor, indicating that GAGs of the heparan sulphate species are responsible; and indeed, addition of the closely related GAG heparin maintains sparsely seeded smooth muscle with a high $V_{v\,myo}$.[64] A similar effect has been found with pentosan polysulphate, a semi-synthetic sulphated polysaccharide. Smooth muscle cells in primary culture can also be maintained with a high $V_{v\,myo}$ by seeding the freshly dispersed cells at confluent density or by placing sparsely seeded cells with a spatially separated feeder layer of confluent high $V_{v\,myo}$ smooth muscle cells or confluent endothelial cells.[65] Both cell types are known to produce large amounts of an anti-proliferative heparan sulphate species which is readily incorporated into their own basal lamina. It thus appears that the presence of heparin-like GAG is an important determinant of the phenotype that smooth muscle expresses and that high levels of this GAG species maintain the cells in a high $V_{v\,myo}$ state. Any factor which removes this substance (e.g. enzyme dispersion, washing and dilution through sparse seeding) may therefore induce the cells to undergo a change in phenotype to a low $V_{v\,myo}$ state. Smooth muscle cells which have been seeded sparsely onto a layer of type IV collagen or basement membrane Matrigel (a solubilized extract of EHS tumour basement membrane containing collagen type IV, laminin, heparan sulphate proteoglycans, and entactin) do not undergo a change in phenotype, nor do isolated smooth muscle cells which have been completely embedded in a gel of collagen type I.[65] Thus replacement of certain components of the basal lamina, or providing a microenvironment in which the basal lamina can be rapidly reconstituted, encourages maintenance of the high $V_{v\,myo}$ state.

Thus, it appears that the presence of heparan sulphate is an important determinant of the phenotype expressed by the smooth muscle, and high levels of this GAG species maintain the cells in a high $V_{v\,myo}$ state. Any factor that removes this substance (or prevents its synthesis) may therefore induce the cells to undergo a change in phenotype to a low $V_{v\,myo}$ state.

10.2.7 What could cause the change in smooth muscle cells during atherogenesis?

Peritoneal macrophages grown in culture with a confluent monolayer of high $V_{v\,myo}$ smooth muscle cells induce a significant decrease in $V_{v\,myo}$ after 3 days as compared with both the freshly isolated smooth muscle cells and those grown for 3 days in the absence of macrophages.[66] Since invasion of the artery wall by monocytes (which subsequently become macrophages) is one of the earliest cellular events in experimental atherogenesis, it may be through the influence of these cells that smooth muscle phenotypic change is induced in the initial stages of the disease. We demonstrated that macrophages completely degrade heparan sulphate-rich matrix to low molecular weight fragments and that the sub-cellular site of enzyme activity is the lysosome. Furthermore, when a confluent monolayer of living smooth muscle cells is labelled with ^{35}S, macrophages both degrade the labelled HSPG on the smooth muscle surface and induce a change in smooth muscle phenotypic expression, indicating that removal of heparan sulphate from the pericellular compartment of smooth muscle cells may be the trigger that initiates smooth muscle phenotypic change.

From the above data, we have hypothesized the following scenario for atherogenesis. Monocyte-macrophages enter the vessel wall in response to hyperlipidaemia and release HSPG from the surface of smooth muscle cells by the action of secreted proteases, or proteases present on the plasma membrane. The proteoglycans are phagocytosed by the macrophages and the heparan sulphate chains are completely degraded by heparanases in the lysosomes. This temporarily removes all heparan sulphate from the surface of the smooth muscle cells such that there is none available for internalization and degradation by the smooth muscle cells themselves until they can synthesize more.[67] This somehow (mechanism unknown) initiates modulation of the smooth muscle cells phenotype and, under specific conditions in the human artery wall (such as the presence of elevated levels of β-VLDL or LDL and growth factors), promotes the development of atherosclerosis.

10.2.8 Matrix metalloproteinases and atherogenesis

Matrix metalloproteinases (MMPs) have been implicated in the remodelling of the extracellular matrix and the penetration of both normal and tumour cells through tissue barriers. A role for vascular smooth muscle cell migration has been convincingly demonstrated by the over-expression of TIMPs (tissue inhibitors of MMPs) in the vascular wall *in vivo* and *in vitro*.

MMP over-expression enhances smooth muscle cell migration and attenuates the response to arterial injury.[68] Increased expression of MMPs has been observed in atherosclerotic plaques, where it has been suggested they could lead to plaque rupture.[69] T-lymphocytes may play a role in this process.[70]

It has been suggested that proteolytic activity residing in the extracellular matrix and/or expressed by invading cells may be required to produce a more accessible substrate for the heparanase enzymes.[67] MMPs are good candidates for this facilitation of heparanase activity. Matrix metalloproteinases have been divided into three sub-classes: collagenases which specifically degrade connective tissue collagens (e.g. interstitial collagenase or MMP1); gelatinases which degrade basement membrane collagens as well as denatured collagen or gelatin (e.g. MMP2, MMP9); and the stromelysin sub-class of MMPs which include MMP3.[71] Matrix metalloproteinases, including MMP2 and MMP9, are upregulated following balloon catheter de-endothelialization of rat, pig or rabbit arteries.[72] Human atherosclerotic plaques display increased expression of MMP1, 2, 3 and 9 in smooth muscle cells and macrophages,[73] and abdominal aortic aneurysms have been shown to express activated forms of MMP2, 3 and 9, localized particularly to infiltrating macrophages.[74] A recent study of MMP expression in abdominal aortic aneurysms demonstrated increased expression of MMP2 in small aneurysms and increased expression of MMP9 in larger aneurysms.[75] Furthermore, it has been reported that inhibition of MMPs prevents DNA synthesis by smooth muscle cells *in vitro*.[76]

Recently we have demonstrated[77] that the MMP inhibitor BB94 inhibits the macrophage modulation of phenotypic change and DNA synthesis in smooth muscle cells freshly dispersed from rabbit aorta, which we previously showed was mediated by macrophage heparanase.[67] Heparanase activity was upregulated with MMP activity in rabbit carotid arteries soon after balloon catheter de-endothelialization, and both are present in human end-stage complex lesions from coronary arteries, carotid endarterectomies and abdominal aortic aneurysms. It is possible therefore that MMPs are involved together with heparanase in the promotion of phenotypic modulation of smooth muscle cells *in vivo*, in the formation of myointimal thickening and in complex atherosclerotic lesions.

10.2.9 Smooth muscle heterogeneity and atherosclerosis

It has been suggested that the accumulation of synthetic-state smooth muscle cells in the intima may not simply be the result of a phenotypic modulation of mature smooth muscle cells migrating from the media, but is the result of an expansion of a phenotypically unique subset of vascular cells.[78, 79] This hypothesis is largely based on cell culture studies, although sub-populations have been reported *in vivo*.[80]

Smooth muscle cells passaged from rat neointima 2 weeks after balloon injury differ from cells cultured from the normal media. These cells are epitheloid, express the PDGF-B gene and secrete large amounts of PDGF-BB, but have little or no PDGF α-receptor mRNA. These changes are similar to medial smooth muscle cells multipassaged from very young or pup rats. However, they are unlike those from the arterial media of adult rats, which express abundant PDGF α-receptor mRNA, but little PDGF-B mRNA and little or no PDGF-BB. Consequently, they are called 'pup-intimal' smooth muscle cells. Despite the large differences in the PDGF-B mRNA level of these passaged 'stem' cells compared with adult medial smooth muscle cells, Majesky *et al.* found only similar low levels of PDGF-B transcripts in intact aorta from newborn and adult rats. Indeed, primary cultures of neointimal smooth muscle cells from newborn rats had low levels of PDGF-B mRNA similar to those in the intact artery, suggesting that the earlier observation could represent induction of gene expression *in vitro*.

Majesky *et al.*[81] constructed and screened a smooth muscle cDNA library for molecular markers of the 'pup-intimal' (pi or π) smooth muscle cell phenotype. Two cDNA clones were identified and elevated levels of the two mRNAs were maintained in cultures of neointimal (but not medial) smooth muscle cells. The cDNA clones encoded rat tropoelastin and α_1 procollagen (type I). Further screening of Majesky's smooth muscle cDNA library by the use of a subtracted probe enriched in π-specific sequences identified a third extracellular matrix protein, osteopontin.[82] Osteopontin is found in calcified tissues and is also synthesized by bone marrow stromal cells with a four-fold increase in levels of production in the presence of PDGF. It is interesting to note that all transformed cells thus far examined, regardless of origin, synthesize significantly higher levels of osteopontin than their untransformed counterparts,[83] and osteonectin and osteopontin are found in human breast cancer. Like transformed cells, π cells have a mechanism for growth that does not require either exogenous PDGF or fibroblast growth factor (FGF).

Osteopontin expression has been shown to be associated with smooth muscle migration and/or proliferation both *in vivo*[84] and *in vitro*.[85] A number of studies have demonstrated osteopontin in atheroma,[86] where it appears to be associated with macrophages or smooth muscle-derived foam cells. Macrophages involved in repair of myocardial necrosis also express osteopontin.[87] Thus synthesis of osteopontin by cells could be a generalized response in the reaction to tissue injury.

More recent studies have shown distinct patterns of gene expression between the 'pup' and 'adult' phenotypes[79, 77] as well as a differential cholesterol ester accumulation.[89]

As pointed out by Jackson,[90] one potential problem with cultured cells is that they are usually the progeny of a small proportion of cells in the tissue of origin. He cites his own study where after enzymatic dispersion of the rat aorta, only 9% of the cells originally in the tissue survived in primary culture.

10.2.10   Subpopulations of smooth muscle cells within the artery wall

In a recent review on the origin and heterogeneity of smooth muscle cell, Gittenberger de Groot *et al.*[91] state that 'in disease it remains to be proven whether there is really a phenotypic modulation of one cell type into another, or whether the stem cell population provide for the required phenotype'.

Certainly in the artery wall there are heterogeneous populations of cells, a few examples being:

(a) Smooth muscle cells of ectodermal origin   During arterial development, medial smooth muscle cells are recruited by endothelium from the surrounding tissue. Therefore, while the majority of arterial medial smooth muscle cells are mesodermal in origin, some are of ectodermal origin.[91, 92] Some vessels, in fact, have a mixture of cells of mesodermal and ectodermal origin. Several culture studies have found that ectodermal and mesodermal smooth muscle cells from arteries differ dramatically. Mesodermal cells express about 10 times more α-smooth muscle actin and tropoelastin and 15 times more *c-jun* than neural crest derived smooth muscle cell. Protein kinase C activation and intracellular levels of 1, 2-diacylglycerol are increased after stimulation of ectodermal smooth muscle cells, whereas mesodermal smooth muscle cells show no evidence of either response. Majesky and Topouzis[92] found that smooth muscle cells of different embryological origin differ in their growth responses to TGF-β, in spite of them having the same number of receptors on their cell surface, suggesting a difference in signal transduction pathways.

(b) Langhans cells   In 1886, Theodor Langhans[93] described a population of stellate cells in the intima just beneath the endothelium of normal and arteriosclerotic arteries. These cells have been isolated by alcoholic alkaline dissociation of fixed specimens, allowing quantitation of cell population in normal and atherosclerotic aortae. There is a high positive correlation between increase in the number of stellate shaped smooth muscle cells within the intima and atherosclerosis. Similar cells have been reported in myointimal thickenings produced in rabbit aorta by de-endothelialization.[94]

(c) Pseudoendothelium   Smooth muscle cells can form a 'pseudoendothelium' which lines the luminal surface of vessels after removal of large areas of endothelium.[95] The cell body is embedded within the neointima and contains large amounts of rough endoplasmic reticulum and free ribosomes; filament bundles are few in number and localized toward the luminal surface of the vessel. Like true endothelium, these luminal smooth muscle cells are non-thrombogenic; With time, they produce increasing amounts of prostacyclin (PGI_2) such that by 35–70 days after de-endothelialization they produce an amount similar to that of endothelium of the normal vessel. Larrue *et al.*[96] suggest that this activation of PGI_2 production by smooth muscle cells is associated with their phenotypic expression. The luminal smooth muscle cells lack Weibel-Palade bodies, have loose cell-to-cell contacts, do not stain with antibodies to factor VIII, and do not exclude horseradish peroxidase, Evans blue, or ferritin. Also, they do not produce endothelium-derived relaxing factor.[97]

Areas of long-term injured arteries which stain 'white' and 'blue' with Evans blue therefore represent areas covered with endothelium and smooth muscle, respectively.

(d) Class II MHC antigen expression   It was once thought that class II antigens of the major histocompatibility complex (MHC) were only expressed by cells of the immune system. These surface antigens participate in the presentation of foreign antigens to T-cells, a function normally confined to macrophages. While there is no expression of class II antigens in normal arteries, smooth muscle cells involved in atherosclerotic plaques do express HLA-DR, HLA-DQ, and the invariant γ-chain.[98]

Smooth muscle cells enzymatically isolated from human atherosclerotic plaques can still express HLA-DR when grown in culture for up to 1 week. DR > DP ≫ DQ expression can be induced, in that order, in smooth muscle cells by growing them in the presence of recombinant IFN-γ (γ interferon). Tumour necrosis factor α (TNF-α) enhances IFN-γ-induced expression of HLA-DR.[99] This suggests that DR expression in smooth muscle cells of atheroma could be due to IFN-γ release from other cells within the plaque, such as T-lymphocytes. Smooth muscle cell phenotypic change can be stimulated by IFN-γ released from T-lymphocytes.[100]

(e) Polyploid smooth muscle cells   The observation of increased DNA synthesis in the aorta of the spontaneous hypertensive rat (SHR) with no change in cell number, led to the conclusion that polyploidy (DNA replication without cell division) occurs in medial smooth muscle cell.[101] Polyploidy is usually associated with smooth muscle cells of large elastic arteries, with the incidence increasing with age and accentuated by hypertension. Polyploidy incidence is low in small blood vessels of the SHR like the caudal artery and mesenteric arterioles.[102]

Polyploid nuclei, both mononucleate and multinucleate, have been observed in the normal and hypertensive media and intima of human elastic arteries, in fatty streaks and atherosclerotic plaques. Up to 20% of the medial and intimal smooth muscle cell population is polyploid in the human aortic wall in hypertensives compared with 10% in normotensives.[103]

Tetraploid (4N) smooth muscle cell from adult rat aorta, when isolated and cultured, can undergo DNA synthesis without division, resulting in octaploidy (8N). Studies of the proliferative capacities of these cells show that their ability to undergo division, and the ultimate cell densities achieved, decreases as the ploidy increases. Passaged smooth muscle cell can develop polyploidy in culture, with some rat strains having a greater propensity to polyploidy induction than others (WKY > SHR > Fischer and Sprague-Dawley).[104] Gordon *et al.*[105] have described a high propensity for polyploidy induction in cultured neonatal rat smooth muscle cell in comparison to cells from the adult aorta.

10.3 Conclusion

The cellular unit of the artery wall is the smooth muscle cell plus its immediate ECM. Any change in one part of this unit influences the other, which in turn impacts on the first. This is particularly true under pathological conditions such as atherogenesis, where smooth muscle cells and extracellular matrix both play crucial and multiple roles in the aetiology of the disease.

References

1. Stary, H.C. (1996). The histological classification of atherosclerotic lesions in human coronary arteries. In *Atherosclerosis and coronary artery disease* (ed. V. Fuster, R. Ross, E.J. Topal), p. 463. Lippincott-Raven, Philadelphia.
2. Wissler, R.W., Hiltscher, L., and Oinuma, T. (1996). The lesions of atherosclerosis in the young. From fatty streaks to intermediate lesions. In *Atherosclerosis and coronary artery disease* (ed. V. Fuster, R. Ross, and E.J. Topal), p. 475. Lippincott-Raven, Philadelphia.
3. Stary, H.C., Chandler, A.B., Dinsmore, R.E., Fuster, V., Glagor, S., Insull, W., Jr, Rosenfeld, M.E., Schwartz, C.J., Wagner, W.D., and Wissler, R.W. (1995). A definition of advanced types of atherosclerotic lesions and a histological classification of atherosclerosis. *Circulation*, **92**, 1355.
4. Wissler, R.W. (1968). The arterial medial cell, smooth muscle or multifunctional mesenchyme. *J, Atheroscler. Res.*, **8**, 201.
5. Campbell, G.R., Chamley-Campbell, J.H., and Burnstock, G. (1981). Differentiation and phenotypic modulation of arterial smooth muscle cells. In *Structure and function of circulation* (ed. C.J. Schwartz, N.T. Werthessen, and S. Wolf), Vol. 3, p. 357. Plenum Press, Orlando, Florida.
6. Ross, R. (1999). Atherosclerosis – an inflammatory disease. *New. Engl. J. Med.*, **340**, 115.
7. Chamley-Campbell, J.H., Campbell, G.R., and Ross, R. (1981). Phenotype-dependent response of cultured aortic smooth muscle to serum mitogens. *J. Cell. Biol.*, **89**, 379.
8. Campbell, J.H., Kocher, O., Skalli, O., Gabbiani, G., and Campbell, G.R. (1989). Cytodifferentiation and expression of smooth muscle actin mRNA and protein during primary culture of aortic smooth muscle cells. *Arteriosclerosis*, **9**, 633.
9. Sjölund, M., Madsen, K., von der Mark, K., and Thyberg, J. (1986). Phenotype modulation in primary cultures of smooth muscle cells from rat aorta. *Differentiation*, **32**, 173.
10. Chamley, J.H. and Campbell, G.R. (1974). Mitosis of contractile smooth muscle cells in tissue culture. *Exp. Cell. Res.*, **84**, 105.
11. Ang, A.H., Tachas, G., Campbell, J.H., Bateman, J.F., and Campbell, G.R. (1990). Collagen synthesis by cultured rabbit aortic smooth muscle cells: alteration with phenotype. *Biochem. J.*, **265**, 461.

12. Merrilees, M.J., Campbell, J.H., Spanidis, E., and Campbell, G.R. (1990). Glycosaminoglycan synthesis by smooth muscle cells of differing phenotype and their response to endothelial cell conditioned medium. *Atherosclerosis*, **81**, 245.

13. Campbell, J.H., Reardon, M.F., Koudounas, S., Popadynec, L., Nestel, P.J., and Campbell, G.R. (1986). Influence of smooth muscle phenotype on lipid accumulation. In *Atherosclerosis*, *VII*, p. 399.

14. Dusserre, E., Bourdillon, M.C., Pulcini, T., and Berthezene, F. (1994). Decrease in high density lipoprotein binding sites is associated with decrease in intracellular cholesterol effect in dedifferentiated aortic smooth muscle cells. *Biochim. Biophys. Acta*, **1212**, 235.

15. Horrigan, S., Campbell, J.H., and Campbell, G.R. (1988). Effect of endothelium on β-VLDL metabolism by cultured smooth muscle cells of differing phenotype. *Atherosclerosis*, **71**, 57.

16. Rennick, R.E., Campbell, J.H., and Campbell, G.R. (1994). Macrophages enhance binding of β-VLDL to cultured smooth muscle cells and their accumulation of cholesterol esters. *Heart and Vessels*, **9**, 19.

17. Pitas, R.E., Friera, A., McGuire, J., and Dejager, S. (1992). Further characterization of the acetyl LDL (scavenger) receptor expressed by rabbit smooth muscle cells and fibroblasts. *Arterioscl. Thromb.*, **12**, 1235.

18. Li, H., Freeman, M.W., and Libby, P. (1995). Regulation of smooth muscle scavenger receptor expression *in vivo* by atherogenic diets and *in vitro* by cytokines. *J. Clin. Invest.*, **95**, 122.

19. Meitus-Snyder, M., Friera, A., Glass, C.K., and Pitas, R.E. (1997). Regulation of scavenger receptor expression in smooth muscle cells by protein kinase C. *Arterioscler. Thromb. Vasc. Bio.*, **17**, 969.

20. Wight, T.N. (1996). The vascular extracellular matrix. In *Atherosclerosis and coronary artery disease* (ed. V. Fuster, R. Ross, and E.J. Topol), p. 421. Lippincott-Raven Philadelphia.

21. Plenz, G., Dorszewski, A., Breithardt, G., and Robenek, K. (1999). Expression of Type VIII collagen after cholesterol diet and injury in the rabbit model of atherosclerosis. *Arterioscler. Thromb. Vasc. Biol.*, **19**, 1201.

22. Camejo, G., Hurt-Camejo, E., Olsson, U., Bondjers, G. (1993). Proteoglycans and lipoproteins in atherosclerosis. *Curr. Opin. Lipidol.*, **4**, 385.

23. Murdock, A.D. and Iozzo, R. (1993). Perlecan: the multidomain heparan sulfate proteoglycan basement membrane and extracellular matrix, *Virchows Arch. [A] Pathol. Anat.*, **423**, 237.

24. Jarrold, B.B., Bacon, W.L., and Velleman, S.G. (1999). Expression and localization of the proteoglycan decorin during the progression of cholesterol induced atherosclerosis in Japanese quail: implications for interaction with collagen type I and lipoproteins. *Atherosclerosis*, **146**, 299.

25. Evanko, S.P., Angello, J.C., and Wight, T.N. (1999). Formation of hyaluronan- and versican-rich pericellular matrix is required for proliferation and migration of vascular smooth muscle cells. *Arterioscler. Thromb. Vasc. Biol.*, **19**, 1004.

26. Kjellen, L. and Lindahl, U. (1991). Proteoglycans: structures and interactions. *Ann. Rev. Biochem.*, **60**, 443.

27. Jackson, R.L., Busch, S.J., and Cardin, A.D. (1991). Glycosaminoglycans: molecular properties, protein interactions, and role in physiological functions. *Physiol. Rev.*, **71**, 481.

28. Templeton, D.M. (1992). Proteoglycans in cell regulation. *Critical Reviews in Clinical Laboratory Sciences*, **29**, 141.

29. Mizuno, K., Inoue, H., Hagiya, M., Shimizu, S., Nose, T., Shimohiagashi, Y., and Nakamura, T. (1994). Hairpin loop and second Kringle domain are essential sites for heparin binding and biological activity of hepatocyte growth factor. *J. Biol. Chem.*, **269**, 1131.

30. Thompson, S.A., Higashiyama, S., Wood, K., Pollit, N.S., Damm, D., McEnroe, G., Garrick, B., Ashton, N., Lau, K., Hancock, N., Klagbrun, M., and Abraham, J.A. (1994). Characterization of sequence within heparin-binding EGF-like factor that mediates interaction with heparin. *J. Biol. Chem.*, **269**, 2541.

31. Parthasarathy, N., Goldberg, I.J., Sivaram, P., Mulloy, B., Flory, D.M., and Wagner, W.D. (1994). Oligosaccharide sequences of endothelial cell surface heparan sulfate proteoglycan with affinity for lipoprotein lipase. *J. Biol. Chem.*, **269**, 22391.

32. Karnovsky, M.J., Wright, T.C., Castellot, J.J., Choay, J., Lormeau, J.-C., and Petitou, M. (1989). Heparin, heparan sulfate, smooth muscle cells, and atherosclerosis. In *Lane DA*. (ed. U. Lindahl), Heparin, Arnold, London.

33. Castellot, J.J., Choay, J., Lormeau, J.C., Petitou, M., Sache, E., and Karnovsky, M.J. (1986). Structural determinants of the capacity of heparin to inhibit the proliferation of vascular smooth muscle cells. II. Evidence for a pentasaccharide sequence that contains a 3-O-sulfate group. *J. Cell. Biol.*, **102**, 1979.

34. Castellot, J.J., Wright, T.C., and Karnovsky, M.J. (1987). Regulation of vascular smooth cell growth by heparin and heparan sulfates. *Sem. Thromb. Hemostas.*, **13**, 489.

35. Cingi, M.R., Tiozzo, R., Croce, M.A. (1993). Heparan sulphate fractions: pharmacological properties and effect on smooth muscle cell proliferation *in vitro*. *Int. J. Tiss. Reac.*, **XV**, 77.

36. Choay, J., Petitou, M., Lormeau, J.-C., Sinay, P., Casu, B., and Gatti, G. (1983). Structure–function relationship in heparin: a synthetic pentamer with high affinity for antithrombin III and eliciting high anti-factor Xa. *Biochem. Biophys. Res. Comm.*, **116**, 492.

37. Hedin, V. (1989). Extracellular matrix components in the regulation of arterial smooth muscle cell phenotype and growth. Ph.D. thesis, Karolinska Institue, Stockholm.

38. Wight, T.N., Raugi, G.J., Mumby, S.M., and Bornstein, P. (1984). Light microscopic immunolocation of thrombospondin secretion in human tissues. *J. Histochem. Cytyochem.*, **33**, 280.

39. Hedin, U., Holm, J., and Hansson, G.K. (1991). Induction of tenascin in rat arterial injury, relationship to altered smooth muscle phenotype. *Am. J. Pathol.*, **139**, 649.

40. Wallner, K., Shah, P.K., Fishbein, M.C., Forrester, J.S., Kaul, S., and Sharifi, B.G. (1999). Tenascin-C is expressed in macrophage-rich human atherosclerotic plaque. *Circulation*, **99**, 1284.

41. O'Brien, E.R., Garvin, M.R., Stewart, D.K., Hinohara, R., Simpson, J.B., Schwartz, S.M., and Giachelli, C.M. (1994). Osteopontin is synthesized by macrophage smooth muscle and endothelial cells in primary and restenotic human coronary atherosclerotic plaques. *Atheroscl. Thromb.*, **14**, 1648.

42. Camejo, G., Fagar, G., Rosengren, R., Hurt-Camejo, E., and Bondjers, G. (1993). Binding of low density lipoproteins by proteoglycans synthesized by proliferating and quiescent human arterial smooth muscle cells. *J. Biol. Chem.*, **268**, 14131.

43. Alavi, M.Z. and Moore, S. (1987). Proteoglycan composition of rabbit arterial wall under conditions of experimentally induced atherosclerosis. *Atherosclerosis*, **63**, 65.

44. Clowes, A.W. and Clowes, M.M. (1986). Kinetics of cellular proliferation after arterial injury IV: heparin inhibits rat smooth muscle mitogenesis and migration. *Circ. Res.*, **58**, 839.

45. Li, Z., Alavi, M., Wasty, F., Galis, Z., and Moore, S. (1993). Proteoglycan synthesis by the neointimal smooth muscle cells cultured from rabbit aortic explants following de-endothelialisation. *Pathobiol.*, **61**, 89.

46. Massaro, T.A. and Glatz, C.E. (1979). Distribution of glycosaminoglycans in consecutive layers of the rabbit aorta. *Artery*, **5**, 1.

47. Wight, T.N. (1989). Cell biology of arterial proteoglycans. *Arterioscler. Thromb. Vasc. Biol.*, **9**, 1.

48. Cardoso, L.E.M. and Mourao, P.A.S. (1994). Glycosaminoglycan fractions from human arteries presenting diverse susceptibilities to atherosclerosis have different binding affinities to plasma LDL. *Arterioscler. Thromb. Vasc. Biol.*, **1**, 115.

49. Edwards, I.J. and Wagner, W.D. (1992). Cell surface heparan sulfate proteoglycan and chondroitin sulfate proteoglycan of arterial smooth muscle cells. *Am. J. Path.*, **140**, 193.

50. Cizmeci-Smith, G., Stahl, R.C., Showalter, L.J., and Carey, D.J. (1993). Differential expression of transmembrane proteoglycan in vascular smooth muscle cells. *J. Biol. Chem.*, **268**, 18740.

51. Evanko, S.P., Raines, E.W., Ross, R., Gold, L.I., and Wight, T.N. (1998). Proteoglycan distribution in lesions of atherosclerosis depends on lesion severity, structural characteristics, and the proximity of platelet-derived growth factor and transforming growth factor-beta. *Amer. J. Pathol.*, **152**, 533.

52. Richardson, M. and Hatton, M.W.C. (1993). Transient morphological and biochemical alterations of arterial proteoglycan during early wound healing. *Exp. Molec. Pathol.*, **58**, 77.

53. Hollman, J., Schmidt, A., von Bassewitz, D.-B., and Buddecke, E. (1989). Relationship of sulfated glycosaminoglycans and cholesterol content in normal and atherosclerotic human aorta. *Arterioscler. Thromb. Vasc. Biol.*, **9**, 154.

54. Srinivasan, S.R., Xu, J.-H., Vijayagopol, P., Radhakrishnamurthy, B., and Berenson, G.S. (1993). Injury to the arterial wall of rabbits produces proteoglycan variants with enhances low-density lipoprotein-binding property. *Biochem. Biophys. Acta*, **1168**, 158.

55. Ismail, N.A., Wasty, F., Moore, S., and Alavi, M.Z. (1997). Injury-induced alterations in newly synthesized sulphated proteoglycans from rabbit arterial neointima covered by regenerated endothelium. *Int. J. Exp. Pathol.*, **78**, 71.

56. Nikkari, S.T., Jarvelainen, H.T., Wight, T.N., Ferguson, M., Clowes, A.W. (1994). Smooth muscle cell expression of extracellular matrix genes after arterial injury. *Am. J. Path.*, **144**, 1348.

57. Camejo, G., Hurt-Camejo, E., Wiklund, O., and Bondjers, G. (1998). Association of apo B lipoproteins with arterial proteoglycans: pathological significance and molecular basis. *Atherosclerosis*, **139**, 205.

58. Tao, Z., Smart, F.W., Figueroa, J.E., Glancy, D.L., and Vijayagopal, P. (1997). Elevated expression of proteoglycans in proliferating smooth muscle cells. *Atherosclerosis*, **135**, 171.

59. Hollman, J.A., Schmidt, A., and von Bassewitz, E. (1989). Relationship of sulfated glycosaminoglycans and cholesterol content in normal and atherosclerotic human aortas. *Arteriosclerosis*, **9**, 154.

60. Hurt-Camejo, E., Olsson, U., Wiklund, O., Bondjers, G., and Camejo, G. (1997). Cellular consequences of the association of the Apo-B lipoproteins with

proteoglycans. Potential contribution to atherogenesis. *Arterioscler. Thromb. Vasc. Biol.*, **17**, 1011.

61. Williams, K.J. and Tabas, I. (1995). The response to retention hypothesis of early atherogenesis. *Arterioscler. Thromb. Vasc. Biol.*, **15**, 551.

62. Tirziu, D., Jinga, V.V., Serban, G., and Simionescu, M., (1999). The effects of low density lipoproteins modified by incubation with chondroitin 6-sulphate on human aortic smooth muscle cells. *Atherosclerosis*, **147**, 155.

63. Lundstam, U., Hurt-Camejo, E., Olsson, G., Sartipy, P., Camejo, G., and Wiklund, O. (1999). Proteoglycans contribution to association of Lp(a) and LDL with smooth muscle cell extracellular matrix. *Arterioscler. Thromb. Vasc. Biol.*, **19**, 1162.

64. Chamley-Campbell, J.H. and Campbell, G.R. (1981). What controls smooth muscle phenotype? *Atherosclerosis*, **40**, 347.

65. Stadler, E., Campbell, J.H., and Campbell, G.R. (1989). Do cultured vascular smooth muscle cells resemble those of the artery wall? If not, why not? *J. Cardiovas. Pharmacol.*, **14**, 51.

66. Rennick, R.E., Campbell, J.H., and Campbell, G.R. (1988). Vascular smooth muscle phenotype and growth behaviour can be influenced by macrophages *in vitro*. *Atherosclerosis*, **71**, 35.

67. Campbell, J.H., Rennick, R.E., Kalevitch, S.G., and Campbell, G.R. (1992). Heparan sulfate-degrading enzymes induce modulation of smooth muscle phenotype. *Exp. Cell. Res.*, **200**, 156.

68. Mason, D.P., Kenagy, R.D., Hasenstab, D., Bowen-Pope, D.F., Seifert, R.A., Coats, S., Hawkins, S.M., and Clowes, A.W. (1999). Matrix metalloproteinase-9 overexpression enhances vascular smooth muscle cell migration and alters remodeling in the injured rat carotid artery. *Circ. Res.*, **85**, 1179.

69. Lee, R.T. and Libby, P. (1997). The unstable atheroma. *Arterioscler. Thromb. Vasc. Biol.*, **17**, 1859.

70. Schonbeck, V., Mach, F., Sukhova, G.K., Murphy, C., Bonnefoy, J.-Y., Fabunmi, R.P., and Libby, P. (1997). Regulation of matrix metalloproteinase expression in human vascular smooth muscle cells by T lymphocytes. A role for CD40 signaling in plaque rupture? *Circ. Res.*, **81**, 448.

71. Matrisian, L.M. (1990). Metalloproteinases and their inhibitors in matrix remodelling. *Trends. Genet.*, **6**, 121.

72. Southgate, K.M., Fisher, M., and Banning, A.P., *et al.* (1996). Upregulation of basement membrane-degrading metalloproteinase secretion after balloon injury of pig arteries. *Circ. Res.*, **79**, 1177.

73. Li, Z., Li, L., Zielke, H.R., *et al.* (1996). Increased expression of 72-kd type IV collagenase (MMP-2) in human aortic atherosclerotic lesions *Am. J. Pathol.*, **148**, 121.

74. Newman, K.M., Ogata, Y., Malon, A.M., *et al.* (1994). Identification of matrix metalloproteinases 3 (stromelysin-1) and 9 (gelatinase B) in abdominal aortic aneurysm. *Arterioscler. Thromb.*, **14**, 1315.

75. Freestone, T., Turner, R.J., Coady, A., *et al.* (1995). Inflammation and matrix metalloproteinases in the enlarging aortic aneurysms. *Arterioscler. Thromb. Vasc. Biol.*, **15**, 1145.

76. Newby, A.C., Southgate, K.M., and Davies, L. (1994). Extracellular matrix degrading metalloproteinases in the pathogenesis of arteriosclerosis. *Basic Res. Cardiol.*, **89**, 59.

77. Fitzgerald, M., Hayward, I.P., Thomas, A.C., Campbell, G.R., and Campbell, J.H. (1999). Matrix metalloproteinases can facilitate the heparanase-induced promotion of phenotypic change in vascular smooth muscle cells. *Atherosclerosis*, **145**, 97.

78. Schwartz, S.M., de Blois, D., and O'Brien, E.R.M. (1995). The intima. Soil for atherosclerosis and restenosis. *Circ. Res.*, **77**, 445.

79. Frid, M.G., Aldashev, A.A., Nemenoff, R.A., Higashito, R., Westcott, J.Y., and Stenmark, K.R. (1999). Subendothelial cells from normal bovine arteries exhibit autonomous growth and constitutively activated extracellular signalling. *Arterioscler. Thromb. Vasc. Biol.*, **19**, 2884.

80. Sartore, S., Chiavegato, A., Franch, R., Faggin, E., and Pauletto, P. (1997). Myosin gene expression and cell phenotypes in vascular smooth muscle during development, in experimental models, and in vascular disease. *Arterioscler. Thromb. Vasc. Biol.*, **17**, 1210.

81. Majesky, M.W., Giachelli, C.M., and Schwartz, S.M. (1992). Rat carotid neointimal smooth muscle cells re-express a developmentally regulated phenotype during repair of arterial injury. *Circ. Res.*, **71**, 759.

82. Giachelli, C.M., Bae, N., Lombardi, D., Majesky, M., and Schwartz, S.M. (1991). Molecular cloning and characterization of 2B7, a rat mRNA which distinguishes smooth muscle cell phenotypes *in vitro* and is identical to osteopontin (secreted phosphoprotein 1,2aR). *Biochem. Biophys. Res. Commun.*, **177**, 867.

83. Craig, A.M., Nemir, M., Mukherjee, B.B., Chambers, A.F., and Denhart, D.T. (1998). Identification of the major phosphoprotein secreted by many rodent cell lines as 2 arlosteopontin: enhanced expression in H-ras-transformed 3T3 cells. *Biochem. Biophys. Res. Commun.*, **157**, 166.

84. Giachelli, C.M., Bae, N., Almeida, M., Denhardt, D.T., Alpers, C.E., and Schwartz, S.M. (1993). Osteopontin is elevated during neointima formation in rat arteries and is a novel component of human atherosclerotic plaques. *J. Clin. Invest.*, **92**, 1686.

85. Gadeau, A.P., Campan, M., Millet, D., Candresse, T., and Desgranges, C. (1993). Osteopontin overexpression is associated with arterial smooth muscle cell proliferation *in vitro*. *Arterioscler. Thromb.*, **13**, 120.

86. Giachelli, C.M., Schwartz, S.M., and Liaw, L. (1995). Molecular and cellular biology of osteopontin, potential role in cardiovascular disease. *Trends. Cardiovasc. Med.*, **5**, 88.

87. Murray, C.E., Giachelli, C.M., Schwartz, S.M., and Vracko, R. (1994). Macrophages express osteopontin during repair of myocardial necrosis. *Amer. J. Pathol.*, **145**, 1450.

88. Adams, L.D., Lemire, J.M., and Schwartz, S.M. (1999). A systematic analysis of 40 random genes in cultured vascular smooth muscle subtypes reveals a heterogeneity of gene expression and identifies the tight junction gene zonula occludens 2 as a marker of epitheloid 'pup' smooth muscle cells and a participant in carotid neointimal formation. *Arterioscler. Thromb. Vasc. Biol.*, **19**, 2600.

89. Llorente-Cortes, V., Martinez-Gonzalez, J., and Badimon, L. (1999). Differential cholesteryl ester accumulation in two human vascular smooth cell subpopulations exposed to aggregrated LDL: effect of PDGF-stimulation and HMG-CoA reductase inhibition. *Atherosclerosis*, **144**, 335.

90. Jackson, C.L. (1995). Pharmacology of smooth cell proliferation. In *The vascular smooth muscle cell* (ed. S.M. Schwartz and R.P. Mccham), p. 297. Academic Press, New York.

91. Gittenberger de Groot, A.C., DeRuiter, M.C., Berghoff, and Poelmann, G. (1999). Smooth muscle cell origin and its relation to heterogeneity in development and disease. *Arterioscler. Thromb. Vasc. Biol.*, **19**, 1589.

92. Majesky, M.W. and Topouzis, S. (1995). Smooth muscle lineage diversity and atherosclerosis. In *Atherosclerosis X* (ed. F.P. Woodford, J. Davignon, and A. Sniderman). Elsevier Science BV NY, 56.

93. Langhans, T.H. (1886). Beiträge zur normalen und pathologischen Anatomie der Arterien. *Arch. Pathol. Anat. Physiol. Klin. Med.*, **36**, 187.

94. Rekhter, M.D., Andreeva, E.R., Andrianova, I.V., Moronov, A.A, and Orekhov, A.N. (1992). Stellate cells of aortic intima: I. Human and rabbit. *Tissue Cell*, **24**, 689.

95. Campbell, G.R. and Campbell, J.H. (1985). Smooth muscle phenotypic changes in arterial wall homeostasis. Implications for the pathogenesis of atherosclerosis. *Exp. Mol. Pathol.*, **42**, 139.

96. Larrue, J., Daret, D., Demond-Henri, J., Allieres, C., and Bricaud, H. (1984). Prostacyclin synthesis by proliferative smooth muscle cells. A kinetic *in vivo* and *in vitro* study. *Atherosclerosis*, **50**, 63.

97. Cocks, T.M., Manderson, J., Mosse, P.R.L., Campbell, G.R., and Angus, J.A. (1987). Development of a large fibromuscular intimal thickening does not impair endothelium-dependent relaxation in the rabbit carotid artery. *Blood Vessels*, **24**, 192.

98. Jonasson, L., Holm, J., Skalli, O., Gabbiani, G., and Hansson, G.K. (1985). Expression of class II transplantation antigen on vascular smooth muscle cells in human atherosclerosis. *J. Clin. Invest.*, **76**, 125.

99. Stemme, S., Fager, G., and Hansson, G.K. (1990). MHC class II antigen expression in human vascular smooth muscle cells is induced by interferon-gamma and modulated by tumor necrosis factor and lymphotoxin. *Immunology*, **69**, 243.

100. Rolfe, B.E., Campbell, J.H., Smith, N.J., Cheong, M.W., and Campbell, G.R. (1995). T lymphocytes affect smooth muscle cell phenotype and proliferation. *Arterioscler. Thromb. Vasc. Biol.*, **15**, 1204.

101. Owens, G.K., Rabinovitch, P.S., Schwartz, S.M. (1981). Smooth muscle cell hypertrophy versus hyperplasia in hypertension. *Proc. Natl. Acad. Sci. USA*, **78**, 7759.

102. Black, M.J., Adams, M.A., Bobik, A., Campbell, J.H., and Campbell, G.R. (1988). Vascular smooth muscle polyploidy in the development and regression of hypertension. *Clin. Exp. Pharmacol.*, **15**, 345.

103. Printseva, O.Y. and Tjurmin, A.V. (1992). Proliferative response of smooth muscle cells in hypertension. *Amer. J. Hypertension.*, **5**, 1185.

104. Rosen, E.M., Goldberg, I.D., Shapiro, H.M., Levenson, S.E., Halpin, P.A., and Faraggi, D. (1986). Strain and site dependence of polyploidization of cultured rat smooth muscle. *J. Cell. Physiol.*, **128**, 337.

105. Gordon, D., Mohai, L.G., and Schwartz, S.M. (1986). Induction of polyploidy in cultures of neonatal rat aortic smooth muscle cells. *Circ. Res.*, **59**, 633.

11 Lymphocytes in atherogenesis

Alan Daugherty
Gill Heart Institute, University of Kentucky, L543 Kentucky Clinic, Lexington,
KY 40536-0284, USA

Göran K. Hansson
Center for Molecular Medicine, Karolinska Institute, Karolinska Hospital,
S-17176 Stockholm, Sweden

11.1 Introduction

Atherosclerosis is increasingly described as an inflammatory disease.[1] Many descriptions of the inflammatory nature of the disease process have focused on the role of innate immunity through infiltration of macrophages. However, while it has been known for many years that lymphocytes are present within lesions, only recently has there been a wider appreciation for a role of adaptive immunity in the atherogenic process. This recognition is largely based on the presence of activated lymphocytes within lesions that penetrate the vasculature at all stages of the atherogenic process.

In this brief and selective review, we will initially provide a background on the nature of lymphocyte infiltration in atherosclerosis and define features that have been mimicked in animal studies. Subsequently, we will review studies in which lymphocyte function has been regulated by a number of techniques in order to define an effect of adaptive immunity on development of atherosclerotic lesions. Finally, there is a highly selective list to illustrate the complex array of effects that can be exerted on the disease process by cytokines secreted during lymphocyte activation.

11.2 Overview of lymphocytes

Lymphocytes display considerable diversity. The major classes of B and T lymphocytes, their differentiation markers, and examples of the spectrum of cytokines

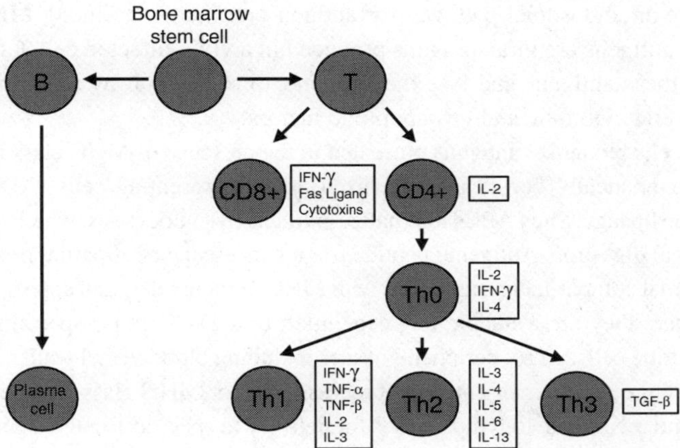

Fig. 11.1 Major classification of lymphocytes, their major role, markers, and selected cytokines secreted by specific populations.

released on activation are summarized in Fig. 11.1. The major subclassification of lymphocytes are B and T cells that function in adaptive immunity, and natural killer (NK) cells that participate in innate immune responses. B and T cells account for humoral and cellular adaptive immune responses, respectively, but there is significant cross-talk between the two systems of cells.

The B cell is activated when a specific antigen ligates its antigen receptor, which is an immunoglobulin inserted into the plasma membrane. Binding of antigen induces a process of differentiation into a B lymphoblast (or activated B cell), which produces large amounts of IgM antibodies. Continued or repeated antigenic stimulation leads to differentiation of the activated B cell into a plasma cell, which secretes copious amounts of immunoglobulins. These carry the same antigenic specificity as the original B cell, but are usually of the IgG type. The latter phenomenon is called isotype switching and depends on T cell signals (T cell help).

T lymphocytes are differentiated in the thymus into two major subclasses, CD8 + and CD4 +, that may be distinguished by the presence of the respective glycoproteins. CD8 serves as a co-receptor molecule for major histocompatibility complex (MHC) class I molecules, while CD4 serves as a co-receptor molecule of MHC class II molecules. The activation of T cells depends not only on the mere presence of its cognate antigen but on its binding, in fragmented form, to MHC proteins on antigen-presenting cells.

MHC class I dependent antigens are synthesized by antigen-presenting cells. Peptide fragments of a nascent protein antigen associate with MHC class I proteins in the Golgi compartment. The peptide–MHC-I complex is transported to the cell surface, where it can be recognized by T cells that simultaneously carry CD8 molecules that ligate the MHC class I protein on the antigen-presenting cell and

T cell antigen receptors (TCRs) that can recognize the peptide. The MHC-I pathway is therefore an endosomal pathway for antigen presentation. Among MHC class I dependent antigens are viral antigens produced in a virus-infected cell. CD8 + cells recognize these antigens and lyse the antigen-expressing cell by a machinery containing several cytotoxic and pro-apoptotic factors.

CD4 + cells recognize antigens presented in the presence of MHC class II proteins. This characteristically occurs in specialized antigen-presenting cells (APCs) of the macrophage lineage. Such APCs internalize antigens by endocytosis, which is followed by lysosomal digestion. Antigenic peptide fragments escape after partial proteolysis in pro-lysosomal compartments, associate with MHC-II molecules, and appear on the cell surface where they are available for recognition by CD4 + cells. Specialized APCs, called dendritic cells, patrol peripheral tissues including blood vessel walls, where they internalize large numbers of antigens, process them for MHC class II dependent presentation and recirculate to lymph nodes for delivery to specific T cells. Upon cytokine stimulation, dendritic cells 'freeze' their peptide–MHC-II complexes in high density on their surfaces, which leads to efficient activation of antigen-specific T cells.

Undifferentiated 'naive' precursor CD4 + T cells proliferate upon primary activation and produce interleukin-2 (IL-2), but have a limited effector spectrum. Upon reactivation, they differentiate into Th1 or Th2 effector cells, depending on the cytokine signals received by the cell during activation. As an initial step in this pathway, the reactivated T cells may express a large variety of cytokines and are referred to as Th0 cells. Th1 differentiation is promoted by the cytokine IL-12 that is produced by macrophages and NK cells in inflammatory lesions and lymphoid tissues. Th1 cells are considered to be pro-inflammatory cells that function to activate macrophages and activate the defence against intracellular microorganisms. The major cytokine secretions on activation of Th1 cells are IL-2, interferon-γ and tumour necrosis factors (both TNF-α and TNF-β, also termed lymphotoxin).

Conversely, Th2 cells are considered to be anti-inflammatory and have a major function in assisting in the activation of B lymphocytes to generate antibodies. Their development is stimulated by IL-4 produced by surrounding T cells, eosinophils, and mast cells. Activated Th2 cells secrete a myriad of cytokines including IL-4, -5, -6, -10, and -13. The overall effect of these activities is to stimulate allergic responses and the defence against parasites and extracellular microorganisms.

A third type of Th cell, Th3, has recently been described. Its major secretory product is transforming growth factor-β (TGF-β) that also promotes its development in an autocrine manner. Th3 cell differentiation characteristically occurs among mucosal lymphocytes and after oral immunization. Since TGF-β is anti-inflammatory and inhibits activation of T cells and macrophages, it is possible that TGF-β-secreting Th3 cells account for the suppressor cell activity that is often observed in parallel with immune activation.

All T cells depend on immunoglobulin-like antigen receptors (TCRs) for recognition of antigen–MHC complexes. TCRs develop through somatic rearrangement during T blast differentiation in a process that is analogous to the immunoglobulin gene rearrangement during B blast development. Among many different variable (V) gene segments present in the genome, one is randomly selected in each cell and fused to diversity (D), joining (J) and constant (C) segments. Together, these V-D-J-C gene segments form a functioning TCR gene that can be transcribed and translated. An α and a β chain gene product formed by this mechanism associate to form the dimeric TCR, which contains complementarity-determining regions used for binding of antigenic peptides. The conformation and, therefore, the sequence of the TCR determines the specificity of the T cell.

Only a small proportion of all TCRs formed by the stochastic process of somatic rearrangement is used for antigen recognition in the adult individual. This is due to the process of clonal selection, which first eliminates all T cells carrying TCRs that are unable to bind MHC. In a second step, T cells that express TCRs that bind MHC–self-antigen complexes with high affinity are killed. This leaves approximately 1–2% of all T cells surviving and able to leave the thymus to fight antigens throughout the body. Such T cells have the capacity to bind self-MHC molecules that have bound 'foreign' antigenic peptides. In theory, all autoreactive T cells should have been eliminated in the thymus.

More than 95% of all T cells use TCR$\alpha\beta$ antigen receptors, recognize MHC–peptide complexes, and are selected through thymic education. However, a small proportion of T cells express another type of TCR, a $\gamma\delta$ dimer that is structurally related to, but functionally different from, TCR$\alpha\beta$. $\gamma\delta$ T cells do not interact with MHC proteins. Instead, they bind to CD1 proteins expressed on the surface of antigen-presenting cells. Importantly, antigens bound to CD1 are often complex lipids rather than oligopeptides. $\gamma\delta$ T cells do not undergo thymic education but appear to mature in the bone marrow and mucosal tissues. In addition to $\gamma\delta$ T cells, certain $\alpha\beta$ T cells are also MHC-independent and CD1-restricted. These include the NK-T cells that express a few specific types of TCR$\alpha\beta$ molecules that recognize antigen–CD1 complexes and exhibit cytotoxic NK-like activity upon activation.

To summarize, the effector mechanisms of adaptive immunity include antibodies, cytotoxic activity, and cytokines regulating immunity and inflammation. All these actions are initiated when antigen receptors on B and T cells encounter their cognate antigens. In order to evoke immune effector responses, initial antigen recognition has to be followed by a series of events involving cytokines, cell surface receptors, and signalling pathways on interacting T cells, B cells, macrophages and other immune cells. The precise type and order of such interactions determine the effector mechanism and, hence, the type of immune response that is mounted upon each encounter of antigen.

11.3   Effect of humoral immunity on atherogenesis

Deposits of immunoglobulins have been detected in human atherosclerotic lesions.[2] These antibodies may largely be derived from the circulating blood, but some of them could emerge from B cells and plasma cells of lesions and the periadventitial connective tissue.[3] An implication that this accumulation of immunoglobulins is related to the disease process is derived from the presence of activated complement complexes in atherosclerotic lesions. The classic pathway of complement activation is initiated by binding of antibody to the C1 component of this multiprotein system. This initiates a cascade that results in the formation of a C5b-9 terminal complex. This end product of the cascade has been detected in atherosclerotic lesions from both humans[4] and animals.[5] Although it does not cause a massive lytic response in lesions, it may provoke many other atherogenic responses, such as stimulating the secretion of monocyte chemoattractant protein-1 (MCP-1) and IL-8.[6] A causal role of complement in atherogenesis has been implicated by the marked reduction in extent of lesions formed in complement C6 deficiency cholesterol-fed rabbits.[7]

A major focus of humoral immunity is on antibodies to oxidized forms of LDL. IgG that recognizes both malondialdehyde and copper modified forms of LDL has been detected in atherosclerotic lesions.[8] Furthermore, titres of IgG autoantibodies to oxidized LDL have also been detected in serum. These titres have been implicated as a diagnostic tool that is positively correlated to the severity of human atherosclerosis in both the carotid and coronary arteries.[9, 10] However, a lack of correlations has also been reported.[11] Part of the disagreement between studies may be related to the lack of standardization of assays for autoantibodies to oxidized lipoproteins.[12] The inconsistencies may also be in part due to the complexity of oxidized forms of LDL which may contain a variable amount of many antigenic lipid and protein moieties.

To determine whether autoantibodies against oxidized LDL are causal in atherogenesis, titres have been increased in Watanabe heritable hyperlipidemic rabbits,[13] cholesterol-fed rabbits,[14] and LDL receptor-deficient mice[15] by immunization. Although the earlier human studies showed that autoantibody titres were positively correlated with the severity of atherosclerosis, each of these studies demonstrated that increased titres lead to decreased severity of disease. Overall, these studies demonstrate that humoral responses to modified forms of LDL may have profound effects on the atherogenic process, although this is a complex interaction that needs to be defined further.

11.4 Presence of T lymphocytes in atherosclerotic lesions

11.4.1 Studies in human tissues

Substantial numbers of T lymphocytes have been detected in specific regions of human atherosclerotic lesions.[16] T lymphocytes may precede the entry of mononuclear cells in the earliest stages of lesion formation. Furthermore, their numbers may be greater than that of macrophages.[17, 18] As lesions progress, T lymphocytes have a distinct regional distribution (Fig. 11.2). They are present in numbers comparable to the numbers of macrophages in the shoulder and fibrous cap regions of more advanced lesions, making up approximately 20% of the total cell population. T lymphocytes represent a lesser number of the cells in the lipid core, but still account for nearly 10%.[16] Sites of coronary thrombosis, whether caused by erosion or rupture, are closely associated with lesion regions that are composed of activated T lymphocytes.[19] There is a potential for these cells to be recruited and activated as a consequence of the thrombotic event, although the location of this cell type within lesions is consistent with their presence prior to the thrombotic event. Thus, a role of T lymphocytes in atherogenesis may be inferred by their presence in lesions from the initiation phase to the final stage that causes acute clinical events.

Several investigators have reported on the phenotype of T lymphocytes within atherosclerotic lesions. CD4/CD8 ratios vary widely between studies, but most investigators find a predominance of CD4+ cells in advanced lesions.[16, 20, 21] Resident lymphocytes have also been characterized by their expression of TCRαβ and TCRγδ antigen receptors. In one study a predominance of αβ chains was demonstrated, as would be expected since ~95% of peripheral lymphocytes in blood express this form of the T cell receptor.[22] However, a relative enrichment of γδ T cells in lesions compared with blood has been reported.[23] It remains to be determined whether such cells recognize lipid antigens derived from the plaque.

The mere presence of T lymphocytes does not necessarily imply that these cells are performing an adaptive immune response, since lymphocytes could be attracted to lesions by the combination of adhesion molecules and chemoattractants that are

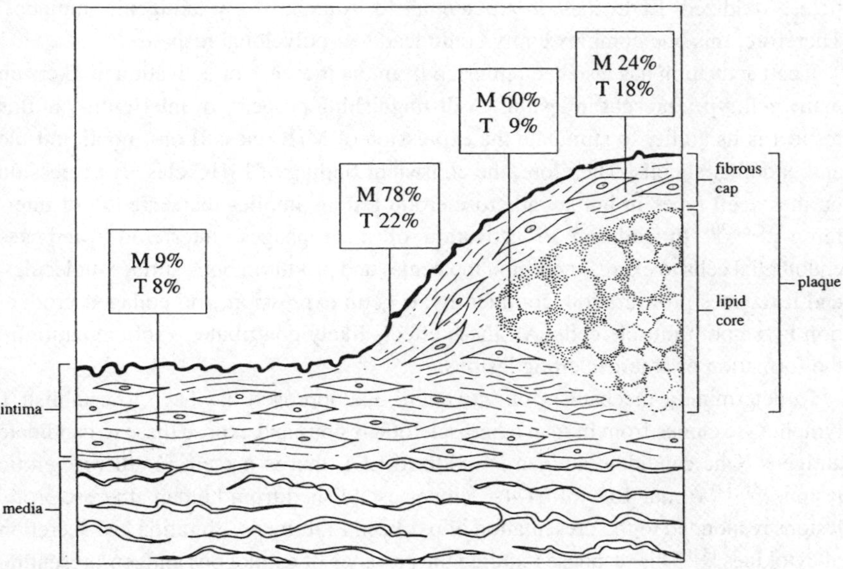

Fig. 11.2 The distribution of T lymphocytes and monocyte-derived macrophages in specific regions of advanced human atherosclerotic lesions.

expressed in diseased tissue. For example, the combined expression of vascular cell adhesion molecule-1 (VCAM-1) and MCP-1 would be sufficient to explain the presence of lymphocytes in lesions. To determine whether T lymphocytes are involved in an immunological response, immunocytochemical analysis has been performed to detect activation markers. The activation markers MHC class II, very late activation antigen-1, and the cytokine, interferon-γ, are present on many cells. Interleukin-2 receptors (CD25) are also present on a smaller number of cells.[24, 25] Consistent with the presence of CD25, T lymphocyte proliferation occurs in human lesions.[26] Additional evidence that T lymphocytes are activated within atherosclerotic lesions is implied by the close association of this cell type with macrophages. T lymphocytes are frequently directly apposed to macrophages in lesions and have been shown to be linked by specialized membrane contacts that are consistent with a functional interaction.[27, 28]

A further method of determining whether T cells are activated by specific mechanisms within lesions is to define potential antigens present in atherosclerosis. The gene reorganization that provides the basis of the specificity of TCRs enables determination of whether a population of cells is mono- or polyclonal. Studies of advanced human lesions have been consistent with heterogeneous TCR gene rearrangement, thus indicative of a polyclonal population.[22, 28] However, early lesions in an experimental murine model show a restricted TCRαβ heterogeneity and the presence of oligoclonal T cells.[29] This points to activation and clonal expansion of a small set of T cells that probably recognize specific antigens of the lesions. However, as mentioned previously, one of the major proposed antigenic particles, oxidized LDL, has the potential to contain many antigenic moieties. Therefore, this one complex entity could lead to a polyclonal response.

T cell activation has also been inferred from the presence of activation markers on many cell types present in lesions. A distinguishing property of interferon-γ in this respect is its ability to stimulate the expression of MHC class II on smooth muscle and endothelial cells. Therefore, the consistent finding of MHC class II expression on these cell types in human atherosclerotic lesions implies the secretion of interferon-γ.[16, 30] In addition to activation of macrophages, interferon-γ activates endothelial cells to express adhesion molecules and pro-thrombotic surface molecules, and it reduces proliferation, contractility, α-actin expression, and collagen production by smooth muscle cells. All these effects likely contribute to inflammation in the formation of atherosclerotic lesions.

To determine a mechanism of activation, one approach has been to establish T lymphocyte clones from human atherosclerotic lesions and expose these to candidate antigens. One candidate antigen is oxidized LDL that is present in atherosclerotic lesions.[31, 32] About 15% of CD4+ clones established from human atherosclerotic lesions responded to the presentation of oxidized LDL by proliferation and secretion of cytokines.[33] This response required the presence of autologous antigen-presenting cells, which is consistent with the presence of committed CD4+ cells. In the presence of both antigen and antigen-presenting cells, all the positive clones responded to oxidized LDL with secretion of interferon-γ, while only 25% secreted interleukin-4. In contrast, oxidized LDL was unable to generate a proliferative response to any clones generated from peripheral blood. Therefore, at least part of the activation of T lymphocytes in lesions is attributable to the presence of oxidized LDL. Similar data are also available that a population of T lymphocytes in rabbit lesions will proliferate in response to heat shock protein 65.[34] Currently, oxidized LDL and heat shock protein 60/65 are on a short list of potential antigens that are thought to promote T lymphocyte responses within atherosclerotic lesions.

Other emerging antigens include those derived from viruses and bacteria.[35] Viruses of the herpes family have been detected in atherosclerotic lesions by several investigators. Cytomegalovirus (a member of the herpes family) has been linked to transplant arteriosclerosis, and Marek's disease virus (another herpes virus) accelerates arteriosclerosis in cholesterol-fed chickens. Finally, *Chlamydia pneumoniae* has been linked to atherosclerosis in seroepidemiological studies and has been isolated from human atherosclerotic tissue.[36] It remains to be determined to what extent the proposed pro-atherogenic effects of these microbes are due to immune responses or to direct effects of the microorganisms on the vascular cells.

Further insight on lymphocyte properties within lesions may be gleaned from the spectrum of cytokines that have been secreted in the diseased tissue. As noted earlier, activation of lymphocytes can lead to an array of cytokines being secreted, and the type of cytokine secreted can provide an insight into the differentiation status of the cell. Two of the principal cytokines that have been studied are interferon-γ and interleukin-4, as indicative of the presence of Th1 and Th2 cells, respectively. Interferon-γ has been the most consistently detected cytokine, both at the protein and mRNA levels.[24, 37] However, NK cells can also secrete substantial interferon-γ, although there is currently no indication that this cell type is present in atherosclerotic lesions. Finally, recent data suggest that macrophages may secrete interferon-γ under certain circumstances.[38] Interleukin-4 has been detected in only a small number of lesions.[39, 40] Again, any detection of interleukin-4 would have to take into account its secretion by other cell types, especially mast cells that have been detected in human lesions.[41] In conclusion, while many questions remain, currently available data on human atherosclerotic plaques point to a predominance of Th1 cytokines. This would suggest that cellular immune responses in lesions promote a macrophage-activating inflammation resembling delayed-type hypersensitivity reactions.

The ability to detect cytokines, either at the mRNA or protein level, is difficult in human atherosclerotic lesions. Acquisition of human vascular tissue in a fresh state to permit isolation of mRNA is only permissible under specific conditions such as carotid endarterectomy. However, only lesions that are in advanced disease stages can be obtained by such procedures. There are insurmountable barriers to the routine acquisition of early or intermediate lesions from humans to permit reliable detection of mRNA or protein. This problem is compounded for lymphocyte-derived cytokines, since they are usually secreted in small discrete areas termed immunological synapses. In a chronic inflammatory response such as atherosclerosis, cytokines exert powerful actions when secreted in masses that may not be detectable by currently available technology. Therefore, while positive results implicate a specific lymphocyte reaction, negative results do not negate the involvement of a specific cytokine in the human disease process.

11.4.2 Studies in animal tissues

Much of our information on the cellular sequence of events in atherogenesis has been derived from animal models of the disease. While few of these studies have included lymphocytes, the inclusion of immunocytochemical procedures to quantify and characterize this cell type is becoming increasingly common. One model that has been used in a large number of studies is rabbits fed a cholesterol-enriched diet. Despite this being a simple dietary stimulus to the formation of atherosclerotic lesions, lymphocytes are present in lesions from these animals, albeit in relatively small numbers.[42–44]

Contemporary atherosclerosis studies are more commonly performed in mice. Early atherosclerosis studies used C57BL/6 mice fed a diet enriched in cholesterol, cholate, and saturated fat. These mice develop small lesions that are restricted to the aortic root. The lesions have a simple morphology of macrophages and no T lymphocytes. There is now an increased use of mice that have been genetically manipulated to generate pronounced lesions either spontaneously or in response to modified diets. The most commonly used are LDL receptor $-/-$ and apoE $-/-$, both of which develop lesions in many regions of the vasculature. Lesions in LDL receptor $-/-$ mice are composed predominantly of lipid-laden macrophages, while those present in apoE $-/-$ mice progress to a more complex morphology.[45–47] Lesions from both these strains contain T lymphocytes, which are predominantly CD4 $+$.[48–50] In lesions from apoE $-/-$ mice, these cells are clustered in either the luminal aspect of lesions,[48, 51] or in other regions consistent with a local proliferative response.[49] Studies in both mice[48] and rats[52] have demonstrated that the presence of T lymphocytes relative to macrophages is greatest at the earlier stages of the disease evolution.

While there is solid evidence for the presence of T lymphocytes at all stages of the atherogenic process, B lymphocytes have not been consistently detected, despite the large amounts of immunoglobulins present within lesions. However, expressions of immunoglobulin genes have been detected by differential hybridization techniques in lesions from Watanabe heritable hyperlipidemic rabbits. This study also performed electron microscopic analysis of lesions and was able to demonstrate the presence of plasma cells.[53] B lymphocytes have also been observed in lesions from apoE-deficient mice by immunocytochemical detection of CD22.[54]

There have been more limited studies to define the presence of cytokines in lesions from animal models. However, the limited information derived from lesions of apoE $-/-$ mice is consistent with findings in human tissue, with interferon-γ being present in all atherosclerotic arteries that were examined. Conversely, interleukin-4 was only detected in lesions of apoE$-/-$ that were fed a modified diet that provoked a severely hyperlipidemic response.[55]

Overall, there is abundant evidence that lymphocytes are present in the atherosclerotic lesions of many animal models of atherosclerosis and particularly in genetically manipulated forms of mice. Therefore, this infers that regulation of lymphocyte function in these models should modulate formation of atherosclerotic lesions if this cell type has an active role in the disease process.

11.5 Effects of regulation of T lymphocyte function on atherogenesis

The presence of activated lymphocytes at each stage of the human lesion formation provides compelling evidence for a role of this cell type in the disease process. However, defining the nature of that role will provide a formidable obstacle. There are limited circumstances in humans in which lymphocyte function is modified so that the effect on atherosclerosis can be demonstrated as a secondary endpoint.

Some potential insight into the role of lymphocytes in the development of atherosclerosis in humans can be assimilated from individuals who are engaged in a course of immunosuppressive therapy. These are predominantly transplant groups of patients. Unfortunately, these patients are seldom on monotherapy, and there is the confounding variable of the presence of the transplanted organ. In the case of heart transplants, the donated organ is well known to have greatly increased vascular disease. However, the pathology of this disease is not the same as atherosclerotic lesions present in non-transplant tissues. Some information could be obtained by quantifying and characterizing atherosclerosis in non-transplanted areas of immunosuppressed patients, although no systematic study has been performed.

Individuals with AIDS may also provide an understanding into the role of lymphocytes in atherosclerosis. Indeed, there are some indications that patients with AIDS have markedly increased severity of atherosclerosis.[56] While prolonging the interval between HIV infection and clinical AIDS, the recently introduced HAART combination therapy may lead to increased cardiovascular disease. However, the complexity of the disease in HIV-infected individuals, the frequent occurrence of associated complications, and the fact that current therapy elevates risk factors for atherosclerosis provides a barrier to clear extrapolation of the specific effects of immune suppression on atherogenesis.

Given the barriers to obtaining information on immunosuppression in humans, definition of the role of lymphocytes in atherogenesis will have to be garnered from animal experimentation. There have been multiple ways of addressing this problem which generally involves the inhibition of lymphocyte function. Discussed below are the results obtained from several different modes of manipulating the immune system.

11.5.1 Antibody ablation approaches

T lymphocyte subpopulations can be markedly ablated by administration of antibodies to specific cell markers. The first experiments were performed in a rat model of vascular injury.[57] T cell ablation with a cytolytic antibody increased restenosis lesions, implying an inhibitory role in the proliferative response of smooth muscle cells. However, the post-injury lesion lacks a significant inflammatory component and therefore differs from the early atherosclerotic lesion that is dominated by inflammatory cells.

When antibody ablation was used to decrease T cells in C57BL/6 mice fed a high-fat diet,[58] ablation of CD4+ cells dramatically reduced the formation of atherosclerosis in the aortic root. CD8+ cell ablation also produced a modest, but statistically significant, decrease in atherosclerosis. Furthermore, this effect could have been related to this treatment promoting a decrease in plasma cholesterol by an undefined mechanism. The combined ablation of CD4+ and CD8+ cells did not have an additive effect. The magnitude of this effect of ablation is surprising given the lack of T lymphocytes that have been demonstrated in this model.

Deletion of specific lymphocyte populations with antibodies has not been performed in genetically altered mice that exhibit larger and more complex lesions. However, a recent study has used this approach to inhibit CD40 signalling in LDL receptor-deficient mice. Signalling through this molecule is related both to humoral and cellular adaptive immunity following interaction with the CD40 ligand. The CD40 ligand is a non-soluble effector molecule that reacts with CD40 that is present on many of the cell types present in lesions. In support of a role of this molecule in atherosclerosis, anti-CD40 ligand antibody administration into LDL receptor-deficient mice reduced both the size and lipid content of lesions.[59] This ablation was also associated with a decreased presence of macrophages and T lymphocytes in lesions. The CD40 ligand was previously considered to be specific for lymphocytes, although the functional molecule has recently been detected on endothelial cells, smooth muscle cells, and macrophages.[60] Therefore, at present the inhibition of the CD40 ligand cannot be taken as definitive evidence of the role of T lymphocytes.

An alternative approach is to modulate immune activity by administration of large doses of polyclonal immunoglobulins. In addition to transfer of specific antibodies, this can lead to inhibition of cellular immune responses. Such therapy is used successfully to prevent coronary inflammatory disease and myocardial infarction in children with Kawasaki disease. Nicoletti *et al.*[61] found that a few injections of polyclonal immunoglobulins reduced atherosclerotic lesions in apoE-deficient mice by ~50%. This supports the notion that atherosclerosis is an inflammatory disease that can be controlled by immunotherapy.

11.5.2 Pharmacological approaches

There has been a relatively limited number of studies in which conventional pharmacological agents have been used to suppress the adaptive immune response, which in part is due to the limited armory of immunosuppressive drugs. Cyclosporin A has been the most widely used drug for immunosuppressive studies in atherosclerosis. The first study was performed in C57BL/6 mice fed a diet enriched in cholesterol, cholate, and saturated fats. While the administration of cyclosporin A was associated with an increase in the extent of atherosclerosis in the aortic root, there was also a profound increase in plasma cholesterol concentrations that negates definition of a mechanism based specifically on lymphocyte inhibition.[62] Two further studies have determined the effect of cyclosporin A on the development of atherosclerotic lesions in cholesterol-fed rabbits. However, conflicting results were obtained with the drug either inhibiting[63] or promoting[42] the extent of the disease. Cyclosporin A had no effect on plasma lipid concentrations in either study. Roselaar *et al.*[42] administered cyclosporin A initially in a daily loading dose followed by a maintenance dose to titre blood concentrations between 100 and 200 ng ml^{-1}. This scheme provided effective immunosuppression as defined by the maintenance of patency of an allogeneic skin graft. Conversely, Drew *et al.*[63] administered cyclosporin A on a daily basis that led to blood concentrations at the terminal bleed of over 600 mg ml^{-1}. Despite these high blood concentrations, there was no demonstration of overt toxicity, as defined by plasma creatinine concentrations. The difference in the results may be related to the spectrum of activities that cyclosporin A can exert on lymphocyte function. Smaller doses are more selective on Th1 T lymphocyte subtypes, while higher doses act on a broader spectrum of lymphocyte subtypes and may also inhibit smooth muscle proliferation.[64]

A more recent study has administered FK506 to cholesterol-fed rabbits.[66] FK506 is a newer immunosuppresive agent that is structurally distinct from cyclosporin A. It has a similar pharmacological profile in reducing the production of the secretion of IL-2, IL-3, and interferon-γ, but with much greater efficacy.[66] FK506 produced a dose-dependent increase in the extent of atherosclerotic lesions covering the aortic intima. The mechanism of this effect of FK506 was attributed to the promotion of cholesterol esterification in macrophages.

While pharmacological approaches have provided an appreciation that lymphocytes have a role in the atherogenic process, the limited number of drugs have poorly defined mechanisms of action. Therefore, there is limited usefulness of this approach for providing mechanistic insight into defining a specific function of lymphocytes in the atherogenic process.

11.5.3 Genetic approaches

Probably the most useful approach to define mechanisms by which lymphocytes influence the atherogenic process is the use of mice with specific genetic manipulations of the immune system.

Initial investigations used mice strains in which the development of atherosclerosis required the feeding of a diet enriched in saturated fat, cholate, and cholesterol. One of the earliest studies was unable to demonstrate changes in the severity of aortic root lesion formation in mice with severe combined immune deficiency, athymic nude mice, or MHC class II deficiency.[67] However, since T lymphocytes have not been detected in lesions of this model, it is unclear that this is a suitable model to study the effects of lymphocyte deficiency. Despite this lack of appropriate cellularity, deficiency of MHC class I led to a three-fold increase in lesion size. A subsequent study also quantified the development of atherosclerosis in nude (nu/nu) C57BL/6 mice fed a modified diet.[58] In this study, there was a pronounced decrease in atherosclerosis in the aortic root with lesion size in homozygous nude (nu/nu) mice, being only 10% of that obtained in the heterozygotes (nu/+). The reason for the disparity between these studies is not apparent. In agreement with the study of Emeson et al.,[58] a further study was performed in rats that had the same nude phenotype that demonstrated euthymic (rnu/+) rats developed larger lesions than rnu/rnu rats. Furthermore, these lesions contained a greater number of macrophages that were more engorged with lipid.[68] In contrast, genetic T cell deficiency (rnu/rnu) led to enhanced proliferative post-injury lesions in rats.[57] The discrepancy in results between smooth muscle-dominated post-injury lesions and macrophage-dominated fatty streak lesions clearly indicates that the mechanisms driving the growth of these two different pathologies are completely different.

More recent studies have used compound genetically manipulated mice in a specific immune deficiency bred into mice of an atherosclerosis-susceptible background. The most commonly used of these atherosclerosis-susceptible mice are the apoE-deficient strain.[69, 70] Unlike the lesions formed in C57BL/6 mice, several studies have demonstrated the presence of T lymphocytes in lesions of these mice.[48–50]

Two studies have compared the effects of total lymphocyte deficiency in the development of lesions in apoE −/− mice.[51, 71] These studies used mice that were deficient in genes of the recombinant activation gene cascade. This deficiency leads to an inability to perform VDJ rearrangement, thus negating the ability to produce mature B or T lymphocytes. In both studies, there was no effect of total lymphocyte deficiency on the development of atherosclerosis in mice fed a diet enriched in saturated fat that promoted a dramatic hypercholesterolemic response (with serum cholesterol concentrations of approximatately $1000–2000\,\mathrm{mg\,dl^{-1}}$, or $25–50\,\mathrm{mmol\,l^{-1}}$). In contrast, mice maintained on a normal laboratory diet had a 43% reduction in the severity of atherosclerotic lesions in the aortic root. Therefore, these studies indicate that lymphocytes modulate atherogenesis in apoE −/− mice but this regulatory effect is overwhelmed by excessive hypercholesterolemia.

Several recent studies have defined the effects of a specific cytokine or cell surface molecule on atherogenesis. One example in the lymphocyte field is interferon-γ receptor-deficient animals that were bred into an apoE-deficient background.[50] Interferon-γ receptor-deficient mice had a decrease in atherosclerosis, thus providing evidence for a pro-atherogenic role of interferon-γ. In addition to quantitative changes, lesions also had decreased lipid deposition, reduced cellularity, and increased collagen content. This effect may not have been accounted for completely by local immune reactions within the vessel wall, since compound deficient mice also had an increase in potentially protective phospholipid/apoA-IV particles. Furthermore, this study is potentially compromised by some variation in the genetic background of the single, compared to the compound, deficient mice. Nevertheless, it provides the first indication that a specific lymphocyte-derived cytokine may influence the atherogenic process.

The importance of CD40 signalling was recently confirmed when CD40 ligand (CD154)-deficient mice were crossed with apoE-deficient animals.[72] The compound mutant mice showed a significant reduction in atherosclerotic lesions compared to apoE single-knockout mice. Interestingly, the effect of CD40 signalling could be ascribed to a role in the progression rather than the initiation of lesions.

It is likely that there will be an increasing number of studies performed on compound deficient animals. The specificity of the immune deficiency that is produced by genetic manipulation is a distinct advantage of this approach. However, this approach may also highlight some of the contradictions that are likely to arise in this field. Different mouse strains have long been known to provide differing responses to stimuli that activate adaptive immune responses. For example, the commonly used strain in atherosclerotic studies, C57BL/6, is known to be more oriented toward Th1 responses, while BALB/c mice are more oriented toward Th2 responses. Therefore, in addition to the many potentially confounding factors in quantifying atherosclerosis in mice (reviewed in Ref. (73)) careful attention must be applied to strain in data interpretation.

11.6 Potential effects of lymphocyte-derived cytokines on atherogenesis

Activated lymphocytes secrete a considerable repertoire of cytokines that have the potential to influence every stage of the atherogenic process by an influence on all the cell types involved in the disease process. Some of these effects are highlighted in Table 11.1, which shows there is the potential for lymphocyte-derived cytokines to exert a wide array of mechanisms. The table focuses on selected effects of interferon-γ and interleukin-4 and -13 as examples of the cytokines that define the distinction between Th1 or Th2 lymphocyte subpopulations. This illustrates that many of these potential effects of lymphocyte-derived cytokines can be influenced by antagonistic mechanisms of these selected cytokines. Therefore, this emphasizes the need to determine the specific nature of the lymphocyte populations that infiltrate lesions.

11.7 Conclusions

An important role of lymphocytes in the development of atherosclerosis is heavily implied by the large number of activated lymphocytes that are present at all stages

Table 11.1 Selected effects of cytokines released from activated lymphocytes on the mechanisms that have been implicated in the atherogenic process

Potential mechanism influencing atherogenesis	Cytokines that increase the process	Cytokines that decrease the process
LDL oxidation		Interferon-γ[74]
15-Lipoxygenase expression	Interleukin-4[75] Interleukin-13[76]	Interferon-γ[75]
Class A scavenger receptor expression	Interleukin-4[77]	Interferon-γ[78]
CD36 expression	Interleukin-4[79]	
Nitric oxide production	Interferon-γ[80]	Interleukin-4[81] Interleukin-13[81]
VCAM-1 expression	Interleukin-4[82]	
Smooth muscle cell growth		Interferon-γ[83]

of lesion development. Based on these data, it would seem that lymphocytes are an integral component of the disease process and that adaptive immunity to either self or foreign antigens is important in the pathogenesis of atherosclerosis. At the present time, the complexities of both the atherogenic process and the lymphocytic system have confounded attempts to define the specific role of adaptive immunity in the development of lesions. However, the increased availability of mice with specific immune deficiencies is likely to lead to a great understanding of the experimental disease process that will hopefully be extrapolated to the human disease.

References

1. Ross, R. (1999). Mechanisms of disease – atherosclerosis – an inflammatory disease. *N. Engl. J. Med.*, **340**, 115–26.
2. Hollander, W., Colombo, M.A., Kirkpatrick, B., and Paddock, J. (1979). Soluble proteins in the human atherosclerotic plaque. With spectral reference to immunoglobulins, C3-complement component, alpha 1-antitrypsin and alpha 2-macroglobulin. *Atherosclerosis*, **34**, 391–405.
3. Parums, D.V., Chadwick, D.R., and Mitchinson, M.J. (1986). The localisation of immunoglobulin in chronic periaortitis. *Atherosclerosis*, **61**, 117–23.
4. Vlaicu, R., Niculescu, F., Rus, H.G., and Cristea, A. (1985). Immunohistochemical localization of the terminal C5b-9 complement complex in human aortic fibrous plaque. *Atherosclerosis*, **57**, 163–77.
5. Seifert, P.S., Hugo, F., Hansson, G.K., and Bhakdi, S. (1989). Prelesional complement activation in experimental atherosclerosis. Terminal C5b-9

complement deposition coincides with cholesterol accumulation in the aortic intima of hypercholesterolemic rabbits. *Lab. Invest.*, **60**, 747–54.

6. Torzewski, J., Bowyer, D.E., Waltenberger, J., and Fitzsimmons, C. (1997). Processes in atherogenesis: complement activation. *Atherosclerosis*, **132**, 131–8.

7. Schmiedt, W., Kinscherf, R., Deigner, H.P., Kamencic, H., Nauen, O., Kilo, J., Oelert, H., Metz, J., and Bhakdi, S. (1998). Complement C6 deficiency protects against diet-induced atherosclerosis in rabbits. *Arterio. Thromb. Vasc. Biol.*, **18**, 1790–5.

8. Ylä-Herttuala, S., Palinski, W., Butler, S.W., Picard, S., Steinberg, D., and Witztum, J.L. (1994). Rabbit and human atherosclerotic lesions contain IgG that recognizes epitopes of oxidized LDL. *Arterio. Thromb.*, **14**, 32–40.

9. Salonen, J.T., Yla-Herttuala, S., Yamamoto, R., Butler, S., Korpela, H., Salonen, R., Nyyssonen, K., Palinski, W., and Witztum, J.L. (1992). Autoantibody against oxidised LDL and progression of carotid atherosclerosis. *Lancet*, **339**, 883–7.

10. Eber, B., Schumacher, M., Tatzber, F., Kaufmann, P., Luha, O., Esterbauer, H., and Klein, W. (1994). Autoantibodies to oxidized low density lipoproteins in restenosis following coronary angioplasty. *Cardiology*, **84**, 310–5.

11. Hulthe, J., Wikstrand, J., Lidell, A., Wendelhag, I., Hansson, G.K., and Wiklund, O. (1998). Antibody titers against oxidized LDL are not elevated in patients with familial hypercholesterolemia. *Arterio. Thromb. Vasc. Biol.*, **18**, 1203–11.

12. Craig, W.Y. (1995). Autoantibodies against oxidized low density lipoprotein: a review of clinical findings and assay methodology. *J. Clin. Lab. Anal.*, **9**, 70–4.

13. Palinski, W., Miller, E., and Witztum, J.L. (1995). Immunization of low density lipoprotein (LDL) receptor-deficient rabbits with homologous malondialdehyde-modified LDL reduces atherogenesis. *Proc. Natl. Acad. Sci. USA*, **92**, 821–5.

14. Ameli, S., HultgardhNilsson, A., Regnstrom, J., Calara, F., Yano, J., Cercek, B., Shah, P.K., and Nilsson, J. (1996). Effect of immunization with homologous LDL and oxidized LDL on early atherosclerosis in hypercholesterolemic rabbits. *Arterio. Thromb. Vasc. Biol.*, **16**, 1074–9.

15. Freigang, S., Horkko, S., Miller, E., Witztum, J.L., and Palinski, W. (1998). Immunization of LDL receptor-deficient mice with homologous malondialdehyde-modified and native LDL reduces progression of atherosclerosis by mechanisms other than induction of high titers of antibodies to oxidative neoepitopes. *Arterio. Thromb. Vasc. Biol.*, **18**, 1972–82.

16. Hansson, G.K., Jonasson, L., Lojsthed, B., Stemme, S., Kocher, O., and Gabbiani, G. (1988). Localization of T lymphocytes and macrophages in fibrous and complicated human atherosclerotic plaques. *Atherosclerosis*, **72**, 135–41.

17. Shimokama, T., Haraoka, S., and Watanabe, T. (1995). Morphological fate and sequelae of human atherosclerosis: evaluation of immune mechanisms in atherogenesis through immunohistological and ultrastructural analysis. *Pathol. Internat.*, **45**, 801–14.

18. Xu, Q.B., Oberhuber, G., Gruschwitz, M., and Wick, G. (1990). Immunology of atherosclerosis – cellular composition and major histocompatibility complex class-II antigen expression in aortic intima, fatty streaks, and atherosclerotic plaques in young and aged human specimens. *Clin. Immunol. Immunopathol.*, **56**, 344–59.

19. van der Wal, A.C., Koch, K.T., Piek, J.J., de Boer, O.J., and Becker, A.E. (1997). Inflammation in atherosclerotic plaques: a clinically crucial event. *Fibrinolysis Proteolysis*, **11**, 125–8.

20. Van der Wal, A.C., Das, P.K., Van de Berg, D.B., Van der Loos, C.M., and Becker, A.E. (1989). Atherosclerotic lesions in humans. *In situ* immunophenotypic analysis suggesting an immune mediated response. *Lab. Invest.*, **61**, 166–70.

21. Emeson, E.E. and Robertson A.L., Jr (1988). T lymphocytes in aortic and coronary intimas. Their potential role in atherogenesis. *Am. J. Pathol.*, **130**, 369–76.

22. Swanson, S.J., Rosenzweig, A., Seidman, J.G., and Libby, P. (1994). Diversity of T-cell antigen receptor V beta gene utilization in advanced human atheroma. *Arterio. Thromb.*, **14**, 1210–14.

23. Kleindienst, R., Xu, Q.B., Willeit, J., Waldenberger, F.R., Weimann, S., and Wick, G. (1993). Immunology of atherosclerosis – demonstration of heat shock protein-60 expression and T-lymphocytes bearing alpha/beta or gamma/delta receptor in human atherosclerotic lesions. *Am. J. Pathol.*, **142**, 1927–37.

24. Hansson, G. K., Holm, J., and Jonasson, L. (1989). Detection of activated T lymphocytes in the human atherosclerotic plaque. *Am. J. Pathol.*, **135**, 169–75.

25. Stemme, S., Patarroyo, M., and Hansson, G.K. (1992). Adhesion of activated T lymphocytes to vascular smooth muscle cells and dermal fibroblasts is mediated by beta1- and beta2-integrins. *Sc. J. Immunol.*, **36**, 233–42.

26. Gordon, D. and Rekhter, M.D. (1997). The growth of human atherosclerosis: cell proliferation and collagen synthesis. *Cardiovasc. Pathol.*, **6**, 103–16.

27. van der Wal, A.C., Dingemans, K.P., Weerman, M.V., Das, P.K., and Becker, A.E. (1994). Specialized membrane contacts between immunocompetent cells in human atherosclerotic plaques. *Cardiovasc. Pathol.*, **3**, 81–5.

28. Stemme, S., Rymo, L., and Hansson, G.K. (1991). Polyclonal origin of lymphocytes-T in human atherosclerotic plaques. *Lab. Invest.*, **65**, 654–60.

29. Paulsson, G., Zhou, X., Törnquist, E., and Hansson, G.K. (1999). Oligoclonal T cell expansions in atherosclerotic lesions of apoE-deficient mice. *Arterioscl. Thromb. Vasc. Biol.*, in press.

30. Jonasson, L., Holm, J., Skalli, O., Gabbiani, G., and Hansson, G.K. (1985). Expression of class II transplantation antigen on vascular smooth muscle cells in human atherosclerosis. *J. Clin. Invest.*, **76**, 125–31.

31. Daugherty, A., Zweifel, B.S., Sobel, B.E., and Schonfeld, G. (1988). Isolation of low density lipoprotein from atherosclerotic vascular tissue of Watanabe heritable hyperlipidemic rabbits. *Arteriosclerosis*, **8**, 768–77.

32. Palinski, W., Rosenfeld, M.E., Yla-Herttuala, S., Gurtner, G.C., Socher, S.S., Butler, S.W., Parthasarathy, S., Carew, T.E., Steinberg, D., and Witztum, J.L. (1989). Low density lipoprotein undergoes oxidative modification *in vivo. Proc. Natl. Acad. Sci. USA*, **86**, 1372–6.

33. Stemme, S., Faber, B., Holm, J., Wiklund, O., Witztum, J.L., and Hansson, G.K. (1995). T lymphocytes from human atherosclerotic plaques recognize oxidized low density lipoprotein. *Proc. Natl. Acad. Sci. USA*, **92**, 3893–7.

34. Xu, Q.B., Kleindienst, R., Waitz, W., Dietrich, H., and Wick, G. (1993). Increased expression of heat shock protein-65 coincides with a population of infiltrating T-lymphocytes in atherosclerotic lesions of rabbits specifi-cally responding to heat shock protein-65. *J. Clin. Invest.*, **91**, 2693–702.

35. Epstein, S.E., Zhou, Y.F., and Zhu, J.H. (1999). Infection and atherosclerosis – emerging mechanistic paradigms. *Circulation*, **100**, E20–8.

36. Saikku, P. (1997). *Chlamydia pneumoniae* and atherosclerosis – an update. *Scand. J. Infect. Dis. Suppl.*, **104**, 53–6.

37. Uyemura, K., Demer, L.L., Castle, S.C., Jullien, D., Berliner, J.A., Gately, M.K., Warrier, R.R., Pham, N., Fogelman, A.M., and Modlin, R.L. (1996). Cross-regulatory roles of interleukin (IL)-12 and IL-10 in atherosclerosis. *J. Clin. Invest.*, **97**, 2130–8.

38. Wang, J., Wakeham, J., Harkness, R., and Xing, Z. (1999). Macrophages are a significant source of type 1 cytokines during mycobacterial infection. *J. Clin. Invest.*, **103**, 1023–9.

39. De Boer, O.J., VanderWal, A.C., Verhagen, C.E., and Becker, A.E. (1999). Cytokine secretion profiles of cloned T cells from human aortic atherosclerotic plaques. *J. Pathol.*, **188**, 174–9.

40. Frostegard, J., Ulfgren, A.K., Nyberg, P., Hedin, U., Swedenborg, J., Andersson, U., and Hansson, G.K. (1999). Cytokine expression in advanced human atherosclerotic plaques: dominance of pro-inflammatory (Th1) and macrophage-stimulating cytokines. *Atherosclerosis*, **145**, 33–43.

41. Kaartinen, M., Penttila, A., and Kovanen, P.T. (1994). Accumulation of activated mast cells in the shoulder region of human coronary atheroma, the predilection site of atheromatous rupture. *Circulation*, **90**, 1669–78.

42. Roselaar, S.E., Schonfeld, G., and Daugherty, A. (1995). Enhanced development of atherosclerosis in cholesterol-fed rabbits by suppression of cell-mediated immunity. *J. Clin. Invest.*, **96**, 1389–94.

43. Hansson, G.K., Seifert, P.S., Olsson, G., and Bondjers, G. (1991). Immunohisto-chemical detection of macrophages and lymphocytes-T in atherosclerotic lesions of cholesterol-fed rabbits. *Arterio. Thromb.*, **11**, 745–50.

44. Drew, A.F. and Tipping, P.G. (1995). T helper cell infiltration and foam cell proliferation are early events in the development of atherosclerosis in cholesterol-fed rabbits. *Arterio. Thromb. Vasc. Biol.*, **15**, 1563–8.

45. Ishibashi, S., Goldstein, J.L., Brown, M.S., Herz, J., and Burns, D.K. (1994). Massive xanthomatosis and atherosclerosis in cholesterol-fed low density lipoprotein receptor-negative mice. *J. Clin. Invest.*, **93**, 1885–93.

46. Nakashima, Y., Plump, A.S., Raines, E.W., Breslow, J.L., and Ross, R. (1994). ApoE-deficient mice develop lesions of all phases of atherosclerosis throughout the arterial tree. *Arterio. Thromb.*, **14**, 133–40.

47. Reddick, R.L., Zhang, S.H., and Maeda, N. (1994). Atherosclerosis in mice lacking apo E – evaluation of lesional development and progression. *Arterio. Thromb.*, **14**, 141–7.

48. Roselaar, S.E., Kakkanathu, P.X., and Daugherty, A. (1996). Lymphocyte populations in atherosclerotic lesions of ApoE −/− and LDL receptor −/− mice – decreasing density with disease progression. *Arterio. Thromb. Vasc. Biol.*, **16**, 1013–18.

49. Zhou, X.H., Stemme, S., and Hansson, G.K. (1996). Evidence for a local immune response in atherosclerosis: CD4(+) T cells infiltrate lesions of apolipoprotein-E-deficient mice. *Am. J. Pathol.*, **149**, 359–66.

50. Gupta, S., Pablo, A.M., Jiang, X.C., Wang, N., Tall, A.R., and Schindler, C. (1997). IFN-gamma potentiates atherosclerosis in apoE knock-out mice. *J. Clin. Invest.*, **99**, 2752–61.

51. Dansky, H.M., Charlton, S.A., Harper, M.M., and Smith, J.D. (1997). T and B lymphocytes play a minor role in atherosclerotic plaque formation in the apolipoprotein E-deficient mouse. *Proc. Natl. Acad. Sci. USA*, **94**, 4642–6.

52. Haraoka, S., Shimokama, T., and Watanabe, T. (1995). Participation of T lymphocytes in atherogenesis: sequential and quantitative observation of aortic lesions of rats with diet-induced hypercholesterolaemia using *en face* double immunostaining. *Virchows Archiv.*, **426**, 307–15.

53. Sohma, Y., Sasano, H., Shiga, R., Saeki, S., Suzuki, T., Nagura, H., Nose, M., and Yamamoto, T. (1995). Accumulation of plasma cells in atherosclerotic lesions of Watanabe heritable hyperlipidemic rabbits. *Proc. Natl. Acad. Sci. USA*, **92**, 4937–41.

54. Zhou, X. and Hansson, G.K. (1999). Detection of B cells and proinflammatory cytokines in atherosclerotic plaques of hypercholesterolaemic apolipoprotein E knockout mice. *Scand. J. Immunol.*, **50**, 25–30.

55. Zhou, X.H., Paulsson, G., Stemme, S., and Hansson, G.K. (1998). Hypercholesterolemia is associated with a T helper (Th) 1/Th2 switch of the autoimmune response in atherosclerotic apo E-knockout mice. *J. Clin. Invest.*, **101**, 1717–25.

56. Paton, P., Tabib, A., Loire, R., and Tete, R. (1993). Coronary artery lesions and human immunodeficiency virus infection. *Res. Virolo.*, **144**, 225–31.

57. Hansson, G.K., Holm, J., Holm, S., Fotev, Z., Hedrich, H.J., and Fingerle, J. (1991). Lymphocytes-T inhibit the vascular response to injury. *Proc. Natl. Acad. Sci. USA*, **88**, 10530–4.

58. Emeson, E.E., Shen, M.L., Bell, C.G.H., and Qureshi, A. (1996). Inhibition of atherosclerosis in CD4 T-cell-ablated and nude (nu/nu) C57BL/6 hyperlipidemic mice. *Am. J. Pathol.*, **149**, 675–85.

59. Mach, F., Schonbeck, U., Sukhova, G.K., Atkinson, E., and Libby, P. (1998). Reduction of atherosclerosis in mice by inhibition of CD40 signalling. *Nature*, **394**, 200–3.

60. Mach, F., Schonbeck, U., Sukhova, G.K., Bourcier, T., Bonnefoy, J.Y., Pober, J.S., and Libby, P. (1997). Functional CD40 ligand is expressed on human vascular endothelial cells, smooth muscle cells, and macrophages: implications for CD40–CD40 ligand signaling in atherosclerosis. *Proc. Natl. Acad. Sci. USA*, **94**, 1931–6.

61. Nicoletti, A., Kaveri, S., Caligiuri, G., Bariety, J., and Hansson, G.K. (1998). Immunoglobulin treatment reduces atherosclerosis in apo E knockout mice. *J. Clin. Invest.*, **102**, 910–18.

62. Emeson, E.E. and Shen, M.L. (1993). Accelerated atherosclerosis in hyperlipidemic C57BL/6 mice treated with cyclosporin-A. *Am. J. Pathol.*, **142**, 1906–15.

63. Drew, A.F. and Tipping, P.G. (1995). Cyclosporine treatment reduces early atherosclerosis in the cholesterol-fed rabbit. *Atherosclerosis*, **116**, 181–9.

64. Thyberg, J. and Hansson, G.K. (1991). Cyclosporine A inhibits induction of DNA synthesis by PDGF and other peptide mitogens in cultured rat aortic smooth muscle cells and dermal fibroblasts. *Growth Factors*, **4**, 209–19.

65. Matsumoto, T., Saito, E., Watanabe, H., Fujioka, T., Yamada, T., Takahashi, Y., Ueno, T., Tochihara, T., and Kanmatsuse, K. (1998). Influence of FK506 on experimental atherosclerosis in cholesterol-fed rabbits. *Atherosclerosis*, **139**, 95–106.

66. Kino, T. and Goto, T. (1993). Discovery of FK-506 and update. *Ann. NY Acad. Sci.*, **685**, 13–21.

67. Fyfe, A.I., Qiao, J.H., and Lusis, A.J. (1994). Immune-deficient mice develop typical atherosclerotic fatty streaks when fed an atherogenic diet. *J. Clin. Invest.*, **94**, 2516–20.

68. Haraoka, S., Shimokama, T., and Watanabe, T. (1997). Role of T lymphocytes in the pathogenesis of atherosclerosis: animal studies using athymic nude rats. *Ann. NY Acad. Sci.*, **811**, 515–18.

69. Piedrahita, J.A., Zhang, S.H., Hagaman, J.R., Oliver, P.M., and Maeda, N. (1992). Generation of mice carrying a mutant apolipoprotein-E gene inactivated by gene targeting in embryonic stem cells. *Proc. Natl. Acad. Sci. USA*, **89**, 4471–5.

70. Plump, A.S., Smith, J.D., Hayek, T., Aaltosetala, K., Walsh, A., Verstuyft, J.G., Rubin, E.M., and Breslow, J.L. (1992). Severe hypercholesterolemia and atherosclerosis in apolipoprotein-E-deficient mice created by homologous recombination in ES cells. *Cell*, **71**, 343–53.

71. Daugherty, A., Pure, E., Delfel-Butteiger, D., Chen, S., Leferovich, J., Roselaar, S.E., and Rader, D.J. (1997). The effects of total lymphocyte deficiency on the extent of atherosclerosis in apolipoprotein E $-/-$ mice. *J. Clin. Invest.*, **100**, 1575–80.

72. Lutgens, E., Gorelik L., Daemen, M.J., de Muinck, E.D., Grewal, I.S., Koteliansky, V.E., and Flavell, R.A. (1999). Requirement for CD154 in the progression of atherosclerosis. *Nat. Med.*, **5**, 1313–16.

73. Daugherty, A. and Whitman, S.C. (2000). Mouse models of atherosclerosis. In The transgenic mouse: methods and protocols (ed. J. van Deursen and M. Hofker). The Humana Press Inc.

74. Christen, S., Thomas, S.R., Garner, B., and Stocker, R. (1994). Inhibition by interferon-gamma of human mononuclear cell-mediated low density lipoprotein oxidation – participation of tryptophan metabolism along the kynurenine pathway. *J. Clin. Invest.*, **93**, 2149–58.

75. Conrad, D.J., Kuhn, H., Mulkins, M., Highland, E., and Sigal, E. (1992). Specific inflammatory cytokines regulate the expression of human monocyte 15-lipoxygenase. *Proc. Natl. Acad. Sci. USA*, **89**, 217–21.

76. Nassar, G.M., Morrow, J.D., Roberts, L.J., Lakkis, F.G., and Badr, K.F. (1994). Induction of 15-lipoxygenase by interleukin-13 in human blood monocytes. *J. Biol. Chem.*, **269**, 27631–4.

77. Cornicelli, J.A., Rateri, D.L., Butteiger, D., Welch, K., and Daugherty, A. (2000). Interleukin-4 augments acetylated LDL induced cholesterol esterification in macrophages. *J. Lipid Res.*, **41**, 376–83.

78. Geng, Y.J. and Hansson, G.K. (1992). Interferon-γ inhibits scavenger receptor expression and foam cell formation in human monocyte-derived macrophages. *J. Clin. Invest.*, **89**, 1322–30.

79. Yesner, L.M., Huh, H.Y., Pearce, S.F., and Silverstein, R.L. (1996). Regulation of monocyte CD36 and thrombospondin-1 expression by soluble mediators. *Arterio. Thromb. Vasc. Biol.*, **16**, 1019–25.

80. Lorsbach, R.B., Murphy, W.J., Lowenstein, C.J., Snyder, S.H., and Russell, S.W. (1993). Expression of the nitric oxide synthase gene in mouse macrophages activated for tumor cell killing. Molecular basis for the synergy between interferon-gamma and lipopolysaccharide. *J. Biol. Chem.*, **268**, 1908–13.

81. Taub, D.D. and Cox, G.W. (1995). Murine Th1 and Th2 cell clones differentially regulate macrophage nitric oxide production. *J. Leuk. Biol.*, **58**, 80–9.

82. Barks, J.L., McQuillan, J.J., and Iademarco, M.F. (1997). TNF-alpha and IL-4 synergistically increase vascular cell adhesion molecule-1 expression in cultured vascular smooth muscle cells. *J. Immunol.*, **159**, 4532–8.

83. Hansson, G.K. and Holm, J. (1991). Interferon-gamma inhibits arterial stenosis after injury. *Circulation*, **84**, 1266–72.

12 Growth factors and cytokines in atherogenesis

Levon M. Khachigian
Centre for Thrombosis and Vascular Research, School of Pathology,
The University of New South Wales, Sydney, NSW 2052, Australia

Carolyn L. Geczy
Cytokine Research Unit, School of Pathology, The University of New South Wales,
Sydney, NSW 2052, Australia

Atherosclerosis is an inflammatory-fibroproliferative process that typically evolves over decades and involves the local action of cytokines and growth factors, among other molecules, in the vessel wall.[1] These classes of regulatory molecules are produced by endothelial cells, smooth muscle cells (SMCs), mast cells, as well as by lymphocytes (see Chapter 11) and monocytes recruited from the circulation in early atherogenesis. These cells may be activated either by direct cell–cell contact or by a wide array of factors including modified low-density lipoprotein (LDL), advanced glycosylation end products (AGE), lipid mediators, infectious agents, oxidants, growth factors, and cytokines.

One of the earliest detectable histological features of atherosclerosis is leukocyte adhesion to the endothelium. Monocytes and lymphocytes accumulate in the subendothelial space, where monocytes ingest lipid and become foam cells. SMCs respond by migrating from the medial compartment of the artery wall to the intima, where they proliferate and elaborate the extracellular matrix. These cellular responses are mediated by the local action of growth factors and cytokines. Cytokines and growth factors relevant to atherosclerosis are summarized in Table 12.1.

12.1 Cytokines in the atheroma

A relationship is emerging between levels of systemic leukocyte activation and risk factors for atherosclerotic vascular disease. For example, in subjects with

Table 12.1 Cytokines and growth factors in atherosclerotic lesions

Cytokines
Interleukins (IL-1, IL-4, IL-5, IL-6, IL-8, IL-10, IL-11, IL-12, IL-15)
Colony-stimulating factors (G-CSF, M-CSF, GM-CSF)
Chemokines (MCP-1, IL-8, RANTES, Gro-α)
Tumour necrosis factor
Interferon-γ

Growth factors
Fibroblast growth factors (FGF-1, FGF-2)
Colony-stimulating factors (GM-CSF, M-CSF)
Insulin-like growth factor-I
Platelet-derived growth factors (PDGF-A, PDGF-B)
Transforming growth factors
Vascular endothelial growth factor / vascular permeability factor

asymptomatic early atherosclerosis, plasma levels of neutrophil activation markers and TNF, as a marker of monocyte/macrophage activation, correlate with age and blood pressure, and are higher in smokers and hypertensives.[2] Cytokines have been linked to many pathophysiological events in the vessel wall, as summarized by Libby.[3] Cytokines contribute to adhesion molecule expression, monocyte recruitment and SMC migration, activation of pro-coagulant (Ref. (4) and Chapter 18) and fibrinolytic factors, nitric oxide synthases, growth factors, and apoptosis. Macrophages activated by cytokines can produce growth factors which, in turn, stimulate migratory and mitogenic events in the developing lesion. Some cytokines can modulate expression of their own genes, and the regulators of these genes.

The pro-inflammatory cytokines, tumour necrosis factor TNF and interleukin-1 (IL-1) are key mediators initiating vascular responses. They have similar activities and affect numerous processes, including stimulation of adhesion molecule expression, leukocyte migration via induction of chemoattractant cytokines of the chemokine families, platelet activation and thrombogenesis, SMC proliferation, hemorrhagic necrosis and release of vasoactive agents.[5] These cytokines are thought to initiate and promote the maturation of atheromatous plaque. SMCs in atherosclerotic lesions express more TNF than endothelial cells or macrophages, and administration of soluble TNF receptors to heart allografts in rabbits reduces development of graft coronary disease.[6] Moreover, balloon catheter injury of the rat femoral artery results in medial expression of TNF in SMCs which remains elevated until SMCs migrate into the intima 7 days after injury. The roles of TNF and IL-1 in fatty streak formation in apolipoprotein E-deficient mice were investigated using an IL-1 receptor antagonist and TNF binding protein. Blocking TNF seemed to be active in female mice but not in males, whereas IL-1 receptor antagonist was as effective or more effective than estradiol in both sexes, indicating a crucial role for IL-1 in the early atherosclerotic process in this model.[7]

Although activated macrophages express IL-1, the cell source expressing this cytokine in vascular cells has only recently more clearly delineated. IL-1 is not secreted by classical secretory pathways and is converted to the active form by a cysteine protease, IL-1-converting enzyme (ICE).[8] However, activated vascular SMCs only elaborate IL-1 activity when endothelial cells contact their surface. Stimulation through interactions of CD40–CD40L results in the release of biologically active IL-1, indicating processing of the native IL-1β precursor induced by the ligand via ICE activation.[9]

IL-6 is a multifunctional cytokine expressed in atherosclerotic plaque. Vascular cells produce large amounts of IL-6 in response to IL-1[10] and angiotensin II.[11] IL-1 also induces IL-6 in endothelial cells. IL-6 regulates levels of acute phase reactants by the liver, including production of fibrinogen, a risk factor associated with atherosclerosis. It may be involved in antibody production by B-lymphocytes.[12]

Disruption of advanced atherosclerotic plaques with associated thrombus is responsible for the majority of acute coronary syndromes. Plaque instability is related to the degree of inflammation. Inflammatory cells within the plaque produce cytokines that inhibit collagen production by SMCs and increase the production of metalloproteinases which degrade the extracellular matrix in the fibrous cap.[13] However, other enzymes generated as a result of cell activation (see below) and by localized pro-coagulants could also contribute to this process.

12.2 Monocyte recruitment in atherosclerosis

Sub-endothelial accumulation of macrophage-derived foam cells is a hallmark of atherosclerosis, and monocyte adhesion and infiltration is an early event in its pathogenesis.[1, 14] Mechanisms of leukocyte recruitment to sites of inflammation, including rolling and adhesion, the involvement of chemokines and other chemoattractants, are well accepted. Monocyte recruitment involves binding to leukocyte adhesion receptors on the endothelial surface such as ICAM-1 and VCAM-1.[3, 15] ICAM-1 expression is upregulated by IL-1 and TNF, and VCAM-1 on endothelium and SMC by TNF and IL-4.[16] The level of expression of CAMs can determine susceptibility to the formation of fatty streaks. Using C57BL/6 mice fed a high-fat diet, 50–75% reduction in fatty streaks was observed in mice with homozygous mutations for ICAM-1, P-selectin, CD18, both ICAM-1 and CD18, or both ICAM-1 and P-selectin.[17]

Once adherent to the endothelium, monocytes enter the intima by directed migration. The C-C chemokine which plays a key role in monocyte recruitment is MCP-1, which is expressed in atheromatous lesions of rabbits, primates and man[18] and expression correlates with lesion formation in murine models.[19, 20] It is produced by endothelial cells, SMCs and monocyte/macrophages activated by cytokines, including TNF and IL-1, by minimally modified LDL,[21] thrombin[22] and by fibroblasts, and SMC stimulated with PDGF. MCP-1-positive endothelial cells are prominent in diffuse intimal thickening and fatty streaks but weak in atheromatous lesions. Sub-endothelial macrophages express high levels of MCP-1 in the fatty streak and atherosclerotic plaque.[20] Macrophages from mice lacking the MCP-1 receptor, CCR2, fail to migrate in response to MCP-1 and, when crossed with apo E-null mice, lesion formation is reduced. These studies suggest that MCP-1 induction by minimally oxidized LDL in early atherosclerotic lesions may be an important link between hyperlipidemia and fatty streak formation.

12.3 Do mast cells contribute to atherogenesis?

Although still speculative, mast cells may play an important role in coronary events[23] by maintaining a local inflammatory response. They accumulate in the shoulders of human atheroma and their presence in areas of excessive degradation of the extracellular matrix in other inflammatory diseases[24, 25] suggests a role in the weakening of the fibrous cap and subsequent erosion. Following activation, mast cells release large amounts of stored TNF, heparin, serotonin and histamine.[26] Seratonin causes gap formation and histamine rapidly induces P-selectin expression by mobilizing Wiebel-Palade bodies in blood vessels, thereby altering the adhesive properties of the endothelium.[27, 28] They also contain or, as a result of activation, express a number of cytokine genes, including interferon-γ, IL-1, IL-4, IL-6, GM-CSF, TGF-β, a number of chemokines including MCP-1, lipid mediators including PGD2, PAF and leukotrienes, FGF-2 and the cysteine proteases tryptase and chymase.[29, 30] Moreover, mast cell granule remnants non-specifically bind LDL[23] which can be phagocytosed by macrophages to form foam cells. Stimulation of mast cell recruitment by IL-8,[31] TGF-β[32] or stem cell factor (SCF) [33, 34] produced by activated macrophages,[35] and of mast cell activation by SCF, activated complement components (C5a, C3b), bacterial products including LPS and by the chemokines MCP-1 and MIP-1 could propagate pathogenesis. Moreover, mast cell heparin could also contribute to the directed migration and enhanced activity of chemokines,[36, 37] and localization of heparin-binding growth factors such as FGF-2[38] may be important in the development of tissue injury in particular areas of the vessel wall and in the conversion of an asymptomatic plaque to a symptomatic state. Our earlier studies showed that localized DTH responses induced with antigen together with heparin profoundly enhances skin test responses.[39] Moreover, mast cell tryptases and chymases may also contribute to leukocyte recruitment[40] and matrix degradation; tryptase activates pro-MMP-3 (pro-stromelysin) and chymase activates interstitial collagenase (pro-MMP-1).[23, 41] Although there are relatively few studies to date, it is obvious that mast cells could potentially regulate atherogenesis, and their unique ability to secrete heparin may play a role in localization of lesions.

12.4 Cytokines with anti-inflammatory activity

In addition to soluble receptors (e.g. for TNF) and receptor antagonists which can block activation by these cytokines, pro-inflammatory cytokine genes are regulated by anti-inflammatory cytokines including IL-4, IL-10, IL-13 and TGF-β. IL-4, IL-10 and IL-13 are products of activated Th2 lymphocytes, although activated macrophages also produce IL-10 and mast cells produce IL-4, IL-10, IL-13[42] and TGF-β.[43] IL-4, IL-10 and IL-13 downregulate IL-1, IL-6 and TNF induction in macrophages although these cytokines can have different effects on various cell types. For example, IL-13, but not IL-10, significantly enhanced IL-8 and MCP-1 release from SMC in response to IL-1 or TNF.[44] IL-10 inhibits cytokine generation from mast cells.[45] The expression and role of anti-inflammatory cytokines in human atherosclerosis is not well understood. IL-10 is expressed by macrophages and SMCs in most lesions studied and is associated with decreased levels of inducible nitric oxide synthase expression, suggesting that it may modulate inflammation in plaque and protect against excessive cell death.[46] Because IL-4 is apparently not a major component of atherosclerotic lesions, IL-10 may be the prime regulator of immune-mediated injury. In addition to the effects of TGF-β described below, it can downmodulate endothelial cell adhesion[47] and upregulate expression of IL-1 receptor antagonist by SMC,[48] suggesting a role in regulating early events of atherogenesis. Cytokines are also regulated by numerous other pathways including cell–cell adhesion, generation of prostaglandins and other lipid mediators, changes in cyclic nucleotide levels and by certain hormones, all of which are likely to contribute to the extent of tissue injury.

12.5 S100 proteins and atherogenesis

S100 is a multigene family of calcium binding proteins of 19 members which are implicated in important intra- and extracellular processes.[49] *Extracellular functions* indicate that this class of highly conserved proteins have cytokine-like activities including regulation of cell migration, adhesion[50] and activation of plasminogen.[51] The human 'myeloid specific' S100 proteins S100A8, S100A9 and S100A12 (A8, A9 and A12) occur, together with most other S100 genes, within a cluster on chromosome 1q21 and are constitutive in neutrophils. Their hallmark is expression at sites of chronic inflammation including rheumatoid arthritis, cystic fibrosis, Crohn's disease, ulcerative colitis, allergic dermatitis and infection,[49, 52–54] and in atherosclerotic plaque, but not normal vessels, in cells opposing the endothelium (our unpublished data). The A8/A9 complex, known as calprotectin,[53] is antimicrobial. Murine A8 (mA8) is chemotactic for monocytes and PMN although the human homologue is inactive.[55, 56]

Human A12 is chemotactic for monocytes and was isolated from bovine lung, and is a ligand for the receptor for advanced glycosylation end products (RAGE).[57]

Ligation of RAGE on EC, monocytes and lymphocytes by A12 induces key inflammatory genes including IL-1 and VCAM-1. Moreover, delayed-type hypersensitivity (DTH) responses in mice are reduced by anti-A12[57] and A12 is chemotactic for leukocytes *in vivo* (our studies and Ref. (57)). This represents a novel paradigm in inflammation and suggests that A12 and RAGE have roles in chronic inflammatory conditions and may be particularly important in complications of diabetes and associated atherosclerosis.

We have extensively characterized mA8 (initially called CP-10[58–60]). It occurs in lesions associated with infection[61] and acute inflammation,[62] DTH responses and atherosclerosis (unpublished), and elicits an infiltrate with cellular composition reminiscent of a DTH response over 4–48 h *in vivo*.[63, 64] The gene is upregulated by specific cytokines in macrophages (interferon-γ, IL-1, TNF), endothelial cells (IL-1) and fibroblasts (FGFs and PDGF), and is regulated by changes in calcium flux (Xu and Geczy, submitted), and IL-4 and IL-13 suppress activation (in preparation). It is located in trophoblasts in the ectoplacental cone, and gene deletion is lethal; embryos implant normally and then resorb.[65] A8 expression in the cells listed may be related to a positive migration and mechanisms are currently under investigation.

Macrophages elicited intraperitoneally by mA8 exhibit a 'pro-atherogenic phenotype'.[64] Apart from the coordinated upregulation of CD11b/CD18, LFA-1 and VLA-4 integrins are not altered, confirming these as newly recruited cells. Levels of scavenger receptor are markedly higher than those expressed on macrophages elicited by other agents including macrophage colony-stimulating factor-1 (CSF-1). A8-elicited macrophages accumulate high levels of cholesterol esters, profiles of which are similar to those from cells freshly isolated from human atherosclerotic plaque, in response to acetylated low-density lipoprotein *in vitro* and *in vivo*.[64] Scavenger receptor activity increases with macrophage maturation and differentiation and contributes to potential foam cell development.[66, 67] Moreover, mA8-elicited Mac express high levels of TNF, Fc receptors, and are strongly phagocytic (Lau and Geczy, unpublished), additional parameters of an activated phenotype. To date, there have been few similar studies with other factors that regulate monocyte migration, and our observations suggest a role for A8 in the modulation of monocyte function in atheroma.

12.6 Growth factors in atherosclerotic lesions

Several lines of evidence have strongly implicated the involvement of growth factors in the migratory and/or proliferative events associated with neointima formation. Barrett and Benditt[68] showed that platelet-derived growth factor (PDGF)-A and PDGF-B mRNA are present in human carotid atherosclerotic plaques. By reprobing the blots with cell-specific markers and performing regression analysis, PDGF-A was localized to SMCs, whereas PDGF-B was associated mainly with macrophages.[68] PDGF-B was also found to be associated with endothelial cells.[69] Later studies showed that both PDGF-B and the PDGF β-receptor were significantly expressed in early atherosclerotic lesions and in all stages of lesion development.[70] SMCs and macrophages in the lesion undergoing migration and proliferation expressed both the ligand and the receptor.[71] PDGF-B and transforming growth factor-β (TGF-β1) were detected in macrophages in intermediate and advanced lesions,[70, 72, 73] whereas PDGF-A, TGF-β2 and TGF-β3 were observed in both SMCs and macrophages.[72, 74] PDGF-A and PDGF-B are expressed in human coronary arteries following percutaneous transluminal coronary angioplasty.[75] Cells positive for PDGF mRNA expression after angioplasty are macrophages, SMCs and endothelial cells, at sites of neovascularization.[75] Findings from investigations *in vitro* suggest that PDGF expression in the developing early atherosclerotic lesion may be under the influence of cytokines. For example, PDGF-like mitogenic activity is secreted from endothelial cells exposed to TNF.[76] PDGF-A expression, like PDGF-B, is induced by TNF, IL-1 and IL-6.[77]

Several other growth factors have been detected in developing atherosclerotic lesions. Basic and acidic fibroblast growth factor (FGF-2 and FGF-1, respectively) are expressed in early, simple and advanced atherosclerotic plaques principally in macrophages and SMCs.[78] FGF-2 and FGF-R1 immunoreactivity was associated with adventitial microvasculature in virtually all lesions examined.[78] Additionally, Brogi *et al.*[79] found that FGF-1 expression in human atheroma was associated with plaque microvessels and macrophages. Vascular endothelial growth factor (VEGF) is a specific mitogen and chemoattractant for endothelial cells that can also promote vascular permeability and monocyte migration. VEGF, structur-ally related to PDGF, was not significantly detectable in normal human arterial segments, but was expressed by endothelial cells, macrophages and smooth muscle cells in human atherosclerotic lesions.[80] VEGF, as well as its two receptors (flt-1 and flk-1), were associated with accumulated macrophages and microvascular endothelial cells in occlusive lesions with intense neovascularization.[80] Plenz *et al.*[81] determined that GM-CSF is basally expressed by sub-populations of intimal SMCs in undiseased human arteries and is markedly upregulated in atherosclerotic lesions. Donnelly *et al.*[82] found that foam cells of Watanabe heritable hyperlipidemic rabbits produce M-CSF. Interestingly, M-CSF expression increased upon treatment of the rabbits with recombinant human M-CSF and was coincident with reduced cholesterol content in the cells.[82] Finally, Grant *et al.*[83] could not detect IGF-I or its receptor in SMCs of normal coronary arteries, but found that IGF-I was expressed primarily in SMCs in both primary human atherosclerotic and restenotic plaques. Morphometric analysis revealed that the IGF-I receptor was present in greater levels in atherosclerotic plaques than restenotic lesions.[83] These studies suggest roles for multiple growth factors in monocyte/macrophage activation, differentiation and proliferation in the initiation and progression of atherosclerotic lesions.

12.7 Lessons in growth factor function from animal models

More direct evidence for involvement of growth factors in SMC migration and/or proliferation in the neointima has been derived from animal models of vascular injury. Balloon inflation triggers platelet adherence, followed several days later by SMC migration to the intima and neointima formation within 14 days.[84] PDGF-A and PDGF-B are expressed at low or undetectable levels in the uninjured rat vessel wall.[85] Balloon injury results in a dramatic increase in PDGF-A expression within 6 h, whereas PDGF-B levels are unaffected.[85] However, *in situ* hybridization of *en face* prepared balloon-injured arteries revealed that a sub-population of SMCs actually express PDGF-B.[86] PDGF α- and β-receptors are basally expressed in the uninjured rat vessel wall.[85] Injury results in a modest induction in the expression of the PDGF β-receptor after 2 weeks,[85] whereas PDGF α-receptor expression appears to decrease.[85] Following catheter-based endothelial injury in rat aorta, PDGF-A, PDGF-B and the PDGF α-receptor are rapidly inducibly expressed by endothelium at the wound edge.[87, 88] Intriguingly, PDGF-A is expressed by SMCs as these cells migrate toward the intima.[89]

Numerous studies suggest that PDGF plays a critical role in the response to vascular injury. Administration of polyclonal antibodies recognizing all three isoforms of PDGF after injury to rat arteries inhibits intimal thickening without altering the thymidine labelling index,[90] suggesting that PDGF regulates SMC migration after injury but not proliferation. Similarly, infusion of recombinant PDGF-BB after injury stimulates SMC migration to the intima and neointima formation.[91] Antisense oligonucleotides targeting the PDGF β-receptor inhibit neointima formation in the rat after balloon injury.[92] Introduction of a PDGF-B expression vector into porcine iliofemoral arteries results in intimal hyperplasia,[93] an effect involving SMC migration, proliferation and synthesis of extracellular matrix.[94] Hart *et al.*[95] recently used a monoclonal antibody targeting the PDGF β-receptor to inhibit intimal hyperplasia after injury to saphenous arteries of baboons.

TGF-β1[96] as well as TGF-β2, TGF-β3 and TGF-β receptors[97] are induced in the rat artery following balloon injury. Infusion of recombinant TGF-β1 into rats with a pre-existing neointima resulted in increased SMC DNA synthesis.[96] Expression of recombinant TGF-β1 in porcine arteries led to significant extracellular matrix deposition as well as intimal and medial hyperplasia.[93] Injection of TGF-β1 into rabbits after balloon injury increased neointimal thickening and matrix synthesis.[98] Similarly, FGF-2 administration to rats after balloon injury increased DNA synthesis and size of the intimal lesion.[99, 100] Moreover, infusion of neutralizing antibodies to FGF-2 after injury blocked SMC replication.[99, 101] VEGF is induced in the artery wall of baboons[102] and pigs[103] following balloon injury. In rats, Couper *et al.*[104] did not observe VEGF or flk-1 following balloon injury, but did detect high levels of flt-1 mRNA and protein in neointimal SMCs. VEGF was induced in medial SMCs following *ex vivo* balloon angioplasty of rat thoracic aorta explanted segments.[105] Localized delivery of VEGF after balloon injury in rats stimulates re-endothelialization and reduces intimal hyperplasia.[106] Similarly, transfer of phVEGF165 into rabbit arteries after balloon injury promotes re-endothelialization, reduced thrombogenicity and inhibition of neointimal thickening.[107] Collectively, these findings demonstrate that growth factors play a critical role in the development of intimal lesions.

12.8 Egr-1 as a mediator of growth factor production in the vessel wall

The promoter regions of many growth factor genes, including FGF-2, TGF-β1, PDGF-A and PDGF-B, contain binding sites for the transcription factor and immediate-early gene product, early growth response factor-1 (Egr-1).[108] Egr-1 is rapidly and transiently induced in the vessel wall after balloon injury[109] and at the endothelial wound edge following catheter injury.[88] Egr-1-dependent gene expression, at least in the context of PDGF-A and PDGF-B, involves the displacement of pre-bound Sp1, which mediates basal gene expression, from overlapping binding sites.[88, 110] Anti-sense strategies targeting Egr-1 inhibit the regrowth of endothelial cells[111] and SMCs[112] after mechanical injury *in vitro*. SMC proliferation is also dependent upon the activation of Egr-1.[112] Recent strategies targeting Egr-1 using catalytic DNA molecules demonstrate that Egr-1 activation is required for neointima formation in rat carotid arteries after balloon injury.[113] These findings are consistent with the rapid activation by arterial injury of extracellular signal-regulated kinase (ERK) and c-Jun N-terminal kinase (JNK).[114] Which phosphorylate nuclear factors required for Egr-1 transcription.[111] Egr-1 has recently been localized to smooth muscle cells and endothelial cells in human atherosclerotic lesions.[115] Strategies targeting Egr-1, and/or other transcription factors regulating the expression of growth factor, and indeed cytokine genes, may serve as useful therapeutic tools in human vascular disease.

12.9 Conclusion

Atherogenesis is a complex process involving numerous cell types and regulatory molecules. Initial events involve the local expression of cell-surface adhesion molecules and the secretion of chemoattractants which facilitate the recruitment of inflammatory cells to the lesion. Atheroma development is controlled by the interplay of local cytokine and growth factor networks mediating cell migration, proliferation and matrix deposition, and further gene expression. An improved understanding of the mechanisms involved in the pathogenesis of atherosclerosis would have important implications to the development of novel therapeutic strategies for the management of the disease.

Acknowledgements

The authors regret that for reasons of space only a limited number of references could be cited. The authors' research was supported by grants from the National Health and Medical Research Council of Australia, National Heart Foundation and Australian Research Council.

References

1. Ross, R. (1993). The pathogenesis of atherosclerosis: a perspective for the 1990s. *Nature*, **362**, (6423), 801–9.
2. Elneihoum, A.M., Falke, P., Hedblad, B., Lindgarde, F., and Ohlsson, K. (1997). Leukocyte activation in atherosclerosis: correlation with risk factors. *Atherosclerosis*, **131**, (1), 79–84.
3. Libby, P. (1996). Vascular cells produce and respond to cytokines. In *Immune functions of the vessel wall* (ed. G.K. Hansson and P. Libby), pp. 45–64. Overseas Publishers Association, Amsterdam. (Vadas, M.A. and Haralan, J. (ed.), *Advances in vascular biology*, Vol. 2.)
4. Geczy, C.L. (1994). Cellular mechanisms for the activation of blood coagulation. *International Review of Cytology*, **152**, 49–108.
5. DeGraba, T.J. (1997). Expression of inflammatory mediators and adhesion molecules in human atherosclerotic plaque. *Neurology*, **49**, (5 Suppl. 4), S15–9.
6. Clausell, N., Molossi, S., Sett, S., and Rabinovitch, M. (1994). *In vivo* blockade of tumor necrosis factor-alpha in cholesterol-fed rabbits after cardiac transplant inhibits acute coronary artery neointimal formation. *Circulation*, **89**, (6), 2768–79.
7. Elhage, R., Maret, A., Pieraggi, M.T., Thiers, J.C., Arnal, J.F., and Bayard, F. (1998). Differential effects of interleukin-1 receptor antagonist and tumor necrosis factor binding protein on fatty-streak formation in apolipoprotein E-deficient mice. *Circulation*, **97**, (3), 242–4.
8. Thornberry, N.A., Bull, H.G., Calaycay, J.R. *et al.* (1992). A novel heterodimeric cysteine protease is required for interleukin-1 beta processing in monocytes. *Nature*, **356**, (6372), 768–74.
9. Schonbeck, U., Mach, F., Bonnefoy, J.Y., Loppnow, H., Flad, H.D., and Libby, P. (1997). Ligation of CD40 activates interleukin 1beta-converting enzyme (caspase-1) activity in vascular smooth muscle and endothelial cells and promotes elaboration of active interleukin 1beta. *Journal of Biological Chemistry*, **72**, (31), 19569–74.
10. Loppnow, H. and Libby, P. (1990). Proliferating or interleukin 1-activated human vascular smooth muscle cells secrete copious interleukin 6. *Journal of Clinical Investigation*, **85**, (3), 731–8.
11. Funakoshi, Y., Ichiki, T., Ito, K., and Takeshita, A. (1999). Induction of interleukin-6 expression by angiotensin II in rat vascular smooth muscle cells. *Hypertension*, **34**, (1), 118–25.

12. Zhou, X. and Hansson, G.K. (1999). Detection of B cells and proinflammatory cytokines in atherosclerotic plaques of hypercholesterolaemic apolipoprotein E knockout mice. *Scandinavian Journal of Immunology*, **50**, (1), 25–30.

13. Kinlay, S., Selwyn, A.P., Libby, P., and Ganz, P. (1998). Inflammation, the endothelium, and the acute coronary syndromes. *Journal of Cardiovascular Pharmacology*, **32**, (Suppl. 3), S62–6.

14. Libby, P. and Hansson, G.K. (1991). Involvement of the immune system in human atherogenesis: current knowledge ands unanswered questions. *Laboratory Investigation*, **64**, (1), 5–15.

15. Chia, M.C. (1998). The role of adhesion molecules in atherosclerosis. *Critical Reviews in Clinical Laboratory Sciences*, **35**, (6), 573–602.

16. Barks, J.L., McQuillan, J.J., and Iademarco, M.F. (1997). TNF-alpha and IL-4 synergistically increase vascular cell adhesion molecule-1 expression in cultured vascular smooth muscle cells. *Journal of Immunology*, **159**, (9), 4532–8.

17. Nageh, M.F., Sandberg, E.T., Marotti, K.R. *et al.* (1997). Deficiency of inflammatory cell adhesion molecules protects against atherosclerosis in mice. *Arteriosclerosis, Thrombosis & Vascular Biology*, **17**, (8), 1517–20.

18. Mantovani, A., Allavena, P., Colotta, F., Sozzani, A. (1996). Chemokines in vascular pathophysiology. In *Immune functions of the vessel wall* (ed. G.K. Hansson and P. Libby), pp. 65–76. Overseas Publishers Association, Amsterdam. (Vadas, M.A. and Haran, J. (ed.), *Advances in vascular biology*, Vol. 2.)

19. Boring, L., Gosling, J., Cleary, M., and Charo, I.F. (1998). Decreased lesion formation in CCR2 −/− mice reveals a role for chemokines in the initiation of atherosclerosis. *Nature*, **394**, (6696), 894–7.

20. Takeya, M., Yoshimura, T., Leonard, E.J., and Takahashi, K. (1993). Detection of monocyte chemoattractant protein-1 in human atherosclerotic lesions by an anti-monocyte chemoattractant protein-1 monoclonal antibody. *Human Pathology*, **24**, (5), 534–9.

21. Cushing, S.D., Berliner, J.A., Valente, A.J. *et al.* (1990). Minimally modified low density lipoprotein induces monocyte chemotactic protein 1 in human endothelial cells and smooth muscle cells. *Proceedings of the National Academy of Sciences of the United States of America*, **87**, (13), 5134–8.

22. Colotta, F., Sciacca, F.L., Sironi, M., Luini, W., Rabiet, M.J., and Mantovani, A. (1994). Expression of monocyte chemotactic protein-1 by monocytes and endothelial cells exposed to thrombin. *American Journal of Pathology*, **144**, (5), 975–85.

23. Kovanen, P.T., Kaartinen, M., and Paavonen, T. (1995). Infiltrates of activated mast cells at the site of coronary atheromatous erosion or rupture in myocardial infarction (see comments). *Circulation*, **92**, (5), 1084–8.

24. Bromley, M., Fisher, W.D., and Woolley, D.E. (1984). Mast cells at sites of cartilage erosion in the rheumatoid joint. *Annals of the Rheumatic Diseases*, **43**, (1), 76–9.

25. Ceponis, A., Konttinen, Y.T., Takagi, M. *et al.* (1998). Expression of stem cell factor (SCF) and SCF receptor (c-kit) in synovial membrane in arthritis: correlation with synovial mast cell hyperplasia and inflammation. *Journal of Rheumatology*, **25**, (12), 2304–14.

26. Galli, S.J. (1993). New concepts about the mast cell. *New England Journal of Medicine*, **328**, (4), 257–65. Issn: 0028-4793.

27. de Mora, F., Williams, C.M., Frenette, P.S., Wagner, D.D., Hynes, R.O., and Galli, S.J. (1998). P- and E-selectins are required for the leukocyte recruitment, but not

the tissue swelling, associated with IgE- and mast cell-dependent inflammation in mouse skin. *Laboratory Investigation*, **78**, (4), 497–505.

28. Kubes, P. and Granger, D.N. (1996). Leukocyte–endothelial cell interactions evoked by mast cells. *Cardiovascular Research*, **32**, (4), 699–708.

29. Mekori, Y.A. and Metcalfe, D.D. (1999). Mast cell–T cell interactions. *Journal of Allergy and Clinical Immunology*, **104**, (3 Pt 1), 517–23.

30. Galli, S.J., Maurer, M., and Lantz, C.S. (1999). Mast cells as sentinels of innate immunity. *Current Opinion in Immunology*, **11**, (1), 53–9.

31. Nilsson, G., Mikovits, J.A., Metcalfe, D.D., and Taub, D.D. (1999). Mast cell migratory response to interleukin-8 is mediated through interaction with chemokine receptor CXCR2/interleukin-8RB. *Blood*, **93**, (9), 2791–7.

32. Gruber, B.L., Marchese, M.J., and Kew, R.R. (1994). Transforming growth factor-beta 1 mediates mast cell chemotaxis. *Journal of Immunology*, **152**, (12), 5860–7.

33. Nilsson, G., Butterfield, J.H., Nilsson, K., and Siegbahn, A. (1994). Stem cell factor is a chemotactic factor for human mast cells. *Journal of Immunology*, **153**, (8), 3717–23.

34. Meininger, C.J., Yano, H., Rottapel, R., Bernstein, A., Zsebo, K.M., and Zetter, B.R. (1992). The c-kit receptor ligand functions as a mast cell chemoattractant. *Blood*, **79**, (4), 958–63.

35. Frangogiannis, N.G., Youker, K.A., Rossen, R.D. *et al.* (1998). Cytokines and the microcirculation in ischemia and reperfusion. *Journal of Molecular & Cellular Cardiology*, **30**, (12), 2567–76.

36. Witt, D.P. and Lander, A.D. (1994). Differential binding of chemokines to glycosaminoglycan subpopulations. *Current Biology*, **4**, (5), 394–400.

37. Kuschert, G.S., Hoogewerf, A.J., Proudfoot, A.E. *et al.* (1998). Identification of a glycosaminoglycan binding surface on human interleukin-8. *Biochemistry*, **37**, (32), 11193–201.

38. Kato, M., Wang, H., Bernfield, M., Gallagher, J.T., and Turnbull, J.E. (1994). Cell surface syndecan-1 on distinct cell types differs in fine structure and ligand binding of its heparan sulfate chains. *Journal of Biological Chemistry*, **269**, (29), 18881–90.

39. Kakakios, A.M., Ryan, J., and Geczy, C.L. (1990). Effect of locally administered heparins on delayed-type hypersensitivity reactions. *International Archives of Allergy & Applied Immunology*, **93**, (4), 300–7.

40. He, S. and Walls, A.F. (1998). Human mast cell chymase induces the accumulation of neutrophils, eosinophils and other inflammatory cells *in vivo*. *British Journal of Pharmacology*, **125**, (7), 1491–500.

41. Tetlow, L.C., Harper, N., Dunningham, T., Morris, M.A., Bertfield, H., and Woolley, D.E. (1998). Effects of induced mast cell activation on prostaglandin E and metalloproteinase production by rheumatoid synovial tissue *in vitro*. *Annals of the Rheumatic Diseases*, **57**, (1), 25–32.

42. Bidri, M., Vouldoukis, I., Mossalayi, M.D. *et al.* (1997). Evidence for direct interaction between mast cells and Leishmania parasites. *Parasite Immunology*, **19**, (10), 475–83.

43. Metcalfe, D.D., Baram, D., and Mekori, Y.A. (1997). Mast cells. *Physiological Reviews*, **77**, (4), 1033–79.

44. Jordan, N.J., Watson, M.L., Williams, R.J., Roach, A.G., Yoshimura, T., and Westwick, J. (1997). Chemokine production by human vascular smooth

muscle cells: modulation by IL-13. *British Journal of Pharmacology*, **122**, (4), 749–57.

45. Arock, M., Zuany-Amorim, C., Singer, M., Benhamou, M., and Pretolani, M. (1996). Interleukin-10 inhibits cytokine generation from mast cells. *European Journal of Immunology*, **26**, (1), 166–70.

46. Mallat, Z., Heymes, C., Ohan, J., Faggin, E., Leseche, G., and Tedgui, A. (1999). Expression of interleukin-10 in advanced human atherosclerotic plaques: relation to inducible nitric oxide synthase expression and cell death. *Arteriosclerosis, Thrombosis & Vascular Biology*, **19**, (3), 611–16.

47. Vadas, M.A., Gamble, J.R., Rye, K., and Barter, P. (1997). Regulation of leucocyte–endothelial interactions of special relevance to atherogenesis. *Clinical & Experimental Pharmacology & Physiology*, **24**, (5), A33–5.

48. Di Febbo, C., Baccante, G., Reale, M. *et al.* (1998). Transforming growth factor beta1 induces IL-1 receptor antagonist production and gene expression in rat vascular smooth muscle cells. *Atherosclerosis*, **136**, (2), 377–82.

49. Donato, R. (1999). Functional roles of S100 proteins, calcium-binding proteins of the EF-hand type. *Biochimica et Biophysica Acta*, **1450**, (3), 191–231.

50. Newton, R.A. and Hogg, N. (1998). The human S100 protein MRP-14 is a novel activator of the beta 2 integrin Mac-1 on neutrophils. *Journal of Immunology*, **160**, (3), 1427–35.

51. Kassam, G., Le, B.H., Choi, K.S. *et al.* (1998). The p11 subunit of the annexin II tetramer plays a key role in the stimulation of t-PA-dependent plasminogen activation. *Biochemistry*, **37**, (48), 16958–66.

52. Hessian, P.A., Edgeworth, J., and Hogg, N. (1993). MRP-8 and MRP-14, two abundant Ca(2+)-binding proteins of neutrophils and monocytes. *Journal of Leukocyte Biology*, **53**, (2), 197–204.

53. Sohnle, P.G. (1997). Antimicrobial defence through competition for zinc. *Reviews in Medical Microbiology*, **8**, 217–24.

54. McNutt, N.S. (1998). The S100 family of multipurpose calcium-binding proteins. *Journal of Cutaneous Pathology*, **25**, (10), 521–9.

55. Lackmann, M., Rajasekariah, P., Iismaa, S.E. *et al.* (1993). Identification of a chemotactic domain of the pro-inflammatory S100 protein CP-10. *Journal of Immunology*, **150**, (7), 2981–91.

56. Devery, J.M., King, N.J., and Geczy, C.L. (1994). Acute inflammatory activity of the S100 protein CP-10. Activation of neutrophils *in vivo* and *in vitro*. *Journal of Immunology*, **152**, (4), 1888–97.

57. Hofmann, M.A., Drury, S., Fu, C. *et al.* (1999). RAGE mediates a novel proinflammatory axis: a central cell surface receptor for S100/calgranulin polypeptides. *Cell*, **97**, (7), 889–901.

58. Geczy, C.L. (1996). Regulation and proinflammatory properties of the chemotactic protein, CP-10. *Biochimica et Biophysica Acta*, **1313**, 246–52.

59. Harrison, C.A., Raftery, M.J., Alewood, P., and Geczy, C.L. (1999). Structure/function studies of S100A8/A9. *Letters in Peptide Science*, **5**, 1–11.

60. Passey, R.J., Xu, K., Hume, D.A., Geczy, C.L. (1990). S100A8: emerging functions and regulation. *Journal of Leukocyte Biology*, **66**, 549–56.

61. Kocher, M., Kenny, P.A., Farram, E., Abdul Majid, K.B., Finlay-Jones, J.J., and Geczy, C.L. (1996). Functional chemotactic factor CP-10 and MRP 14 are abundant in murine abscesses. *Infection and Immunity*, **64**, 1342–50.

62. Kumar, R.K., Harrison, C.A., Cornish, C.J., Kocher, M., and Geczy C.L. (1998). Immunodetection of the murine chemotactic protein CP-10 in bleomycin-induced pulmonary injury. *Pathology*, **30**, 51–6.

63. Lackmann, M., Cornish, C.J., Simpson, R.J., Moritz, R.L., and Geczy, C.L. (1992). Purification and structural analysis of a murine chemotactic cytokine (CP-10) with sequence homology to S100 proteins. *Journal of Biological Chemistry*, **267**, (11), 7499–504.

64. Lau, W., Devery, J.M., and Geczy, C.L. (1995). A chemotactic S100 peptide enhances scavenger receptor and Mac-1 expression and cholesteryl ester accumulation in murine peritoneal macrophages *in vivo. Journal of Clinical Investigation*, **95**, (5), 1957–65.

65. Passey, R.J., Williams, E., Lichanska, A.M. *et al.* (1990). A null mutation in the inflammation-associated S100 protein S100A8 causes early resorption of the mouse embryo. *Journal of Immunology*, **163**, 2209–16.

66. Geng, Y., Kodama, T., and Hansson, G. (1994). Differential expression of scavenger receptors on monocytes-macrophages differentiation and foam cell formation. *Arteriosclerosis, Thrombosis & Vascular Biology*, **14**, 798–806.

67. Ishibashi, S., Inaba, T., Shimano, H. *et al.* (1990). Monocyte colony-stimulating factor enhances uptake and degradation of acetylated low density lipoproteins and cholesterol esterification in human monocyte-derived macrophages. *Journal of Biological Chemistry*, **265**, 14109–17.

68. Barrett, T.B. and Benditt, E.P. (1998). Platelet-derived growth factor gene expression in human atherosclerotic plaques and normal artery wall. *Proceedings of the National Academy of Sciences of the United States of America*, **85**, (8), 2810–14.

69. Wilcox, J.N., Smith, K.M., Williams, L.T., Schwartz, S.M., and Gordon, D. (1988). Platelet-derived growth factor mRNA detection in human atherosclerotic plaques by *in situ* hybridization. *Journal of Clinical Investigation*, **82**, (3), 1134–43.

70. Ross, R., Masuda, J., and Raines, E.W. *et al.* (1990). Localization of PDGF-B protein in macrophages in all phases of atherogenesis. *Science*, **248**, (4958), 1009–12.

71. Betsholtz, C. and Raines, E.W. (1997). Platelet-derived growth factor: a key regulator of connective tissue cells in embryogenesis and pathogenesis. *Kidney International*, **51**, (5), 1361–9.

72. Evanko, S.P., Raines, E.W., Ross, R., Gold, L.I., and Wight, T.N. (1998). Proteoglycan distribution in lesions of atherosclerosis depends on lesion severity, structural characteristics, and the proximity of platelet-derived growth factor and transforming growth factor-beta. *American Journal of Pathology*, **152**, (2), 533–46.

73. Bahadori, L., Milder, J., Gold, L., and Botney, M. (1995). Active macrophage-associated TGF-beta co-localizes with type I procollagen gene expression in atherosclerotic human pulmonary arteries. *American Journal of Pathology*, **146**, (5), 1140–9.

74. Bobik, A., Agrotis, A., Kanellakis, P. *et al.* (1999). Distinct patterns of transforming growth factor-beta isoform and receptor expression in human atherosclerotic lesions. Colocalization implicates TGF-beta in fibrofatty lesion development. *Circulation*, **99**, (22), 2883–91.

75. Ueda, M., Becker, A.E., Kasayuki, N., Kojima, A., Morita, Y., and Tanaka, S. (1996). *In situ* detection of platelet-derived growth factor-A and -B chain mRNA in human coronary arteries after percutaneous transluminal coronary angioplasty. *American Journal of Pathology*, **149**, (3), 831–43.

76. Hajjar, K.A., Hajjar, D.P., Silverstein, R.L., and Nachman, R.L. (1987). Tumor necrosis factor-mediated release of platelet-derived growth factor from cultured endothelial cells. *Journal of Experimental Medicine*, **166**, (1), 235–45.

77. Gay, C.G. and Winkles, J.A. (1990). Heparin-binding growth factor-1 stimulation of human endothelial cells induces platelet-derived growth factor A-chain gene expression. *Journal of Biological Chemistry*, **265**, (6), 3284–92.

78. Hughes, S.E., Crossman, D., and Hall, P.A. (1993). Expression of basic and acidic fibroblast growth factors and their receptor in normal and atherosclerotic human arteries. *Cardiovascular Research*, **27**, (7), 1214–19.

79. Brogi, E., Winkles, J.A., Underwood, R., Clinton, S.K., Alberts, G.F., and Libby, P. (1993). Distinct patterns of expression of fibroblast growth factors and their receptors in human atheroma and nonatherosclerotic arteries. Association of acidic FGF with plaque microvessels and macrophages. *Journal of Clinical Investigation*, **92**, (5), 2408–18.

80. Inoue, M., Itoh, H., Ueda, M. *et al.* (1998). Vascular endothelial growth factor (VEGF) expression in human coronary atherosclerotic lesions: possible pathophysiological significance of VEGF in progression of atherosclerosis. *Circulation*, **98**, (20), 2108–16.

81. Plenz, G., Koenig, C., Severs, N.J., and Robenek, H. (1997). Smooth muscle cells express granulocyte-macrophage colony-stimulating factor in the undiseased and atherosclerotic human coronary artery. *Arteriosclerosis, Thrombosis & Vascular Biology*, **17**, (11), 2489–99.

82. Donnelly, L.H., Bree, M.P., Hunter, S.E., Keith, J.C., Jr, and Schaub, R.G. (1997). Immunoreactive macrophage colony-stimulating factor is increased in atherosclerotic lesions of Watanabe heritable hyperlipidemic rabbits after recombinant human macrophage colony-stimulating factor therapy. *Molecular Reproduction & Development*, **46**, (1), 92–5.

83. Grant, M.B., Wargovich, T.J., Ellis, E.A. *et al.* (1996). Expression of IGF-I, IGF-I receptor and IGF binding proteins-1, -2, -3, -4 and -5 in human atherectomy specimens. *Regulatory Peptides*, **67**, (3), 137–44.

84. Clowes, A.W., Reidy, M.A., and Clowes, M.M. (1990). Kinetics of cellular proliferation after arterial injury. I. Smooth muscle growth in the absence of endothelium. *Laboratory Investigation*, **49**, (3), 327–33.

85. Majesky, M.W., Reidy, M.A., Bowen-Pope, D.F., Hart, C.E., Wilcox, J.N., and Schwartz, S.M. (1990). PDGF ligand and receptor gene expression during repair of arterial injury. *Journal of Cell Biology*, **111**, (5 Pt 1), 2149–58.

86. Lindner, V., Giachelli, C.M., Schwartz, S.M., and Reidy, M.A. (1995). A subpopulation of smooth muscle cells in injured rat arteries expresses platelet-derived growth factor-B chain mRNA. *Circulation Research*, **76**, (6), 951–7.

87. Lindner, V. and Reidy, M.A. (1995). Platelet-derived growth factor ligand and receptor expression by large vessel endothelium *in vivo*. *American Journal of Pathology*, **146**, (6), 1488–97.

88. Khachigian, L.M., Lindner, V., Williams, A.J., and Collins, T. (1996). Egr-1-induced endothelial gene expression: a common theme in vascular injury. *Science*, **271**, (5254), 1427–31.

89. Silverman, E.S., Khachigian, L.M., Lindner, V., Williams, A.J., and Collins, T. (1997). Inducible PDGF A-chain transcription in smooth muscle cells is mediated by Egr-1 displacement of Sp1 and Sp3. *American Journal of Physiology*, **273**, (3 Pt 2), H1415–26.

90. Ferns, G.A., Raines, E.W., Sprugel, K.H., Motani, A.S., Reidy, M.A., and Ross, R. (1991). Inhibition of neointimal smooth muscle accumulation after angioplasty by an antibody to PDGF. *Science*, **253**, (5024), 1129–32.

91. Jawien, A., Bowen-Pope, D.F., Lindner, V., Schwartz, S.M., and Clowes, A.W. (1992). Platelet-derived growth factor promotes smooth muscle migration and intimal thickening in a rat model of balloon angioplasty. *Journal of Clinical Investigation*, **89**, (2), 507–11.

92. Sirois, M.G., Simons, M., and Edelman, E.R. (1997). Antisense oligonucleotide inhibition of PDGFR-beta receptor subunit expression directs suppression of intimal thickening (see comments). *Circulation*, **95**, (3), 669–76.

93. Nabel, E.G., Yang, Z., Liptay, S. *et al.* (1993). Recombinant platelet-derived growth factor B gene expression in porcine arteries induce intimal hyperplasia *in vivo*. *Journal of Clinical Investigation*, **91**, (4), 1822–9.

94. Pompili, V.J., Gordon, D., San, H. *et al.* (1995). Expression and function of a recombinant PDGF B gene in porcine arteries. *Arteriosclerosis, Thrombosis & Vascular Biology*, **15**, (12), 2254–64.

95. Hart, C.E., Kraiss, L.W., Vergel, S. *et al.* (1999). PDGFbeta receptor blockade inhibits intimal hyperplasia in the baboon. *Circulation*, **99**, (4), 564–9.

96. Majesky, M.W., Lindner, V., Twardzik, D.R., Schwartz, S.M., and Reidy, M.A. (1991). Production of transforming growth factor beta 1 during repair of arterial injury. *Journal of Clinical Investigation*, **88**, (3), 904–10.

97. Ward, M.R., Agrotis, A., Kanellakis, P., Dilley, R., Jennings, G., and Bobik, A. (1997). Inhibition of protein tyrosine kinases attenuates increases in expression of transforming growth factor-beta isoforms and their receptors following arterial injury. *Arteriosclerosis, Thrombosis & Vascular Biology*, **17**, (11), 2461–70.

98. Kanzaki, T., Tamura, K., Takahashi, K. *et al.* (1995). *In vivo* effect of TGF-beta 1. Enhanced intimal thickening by administration of TGF-beta 1 in rabbit arteries injured with a balloon catheter. *Arteriosclerosis, Thrombosis & Vascular Biology*, **15**, (11), 1951–7.

99. Lindner, V., Lappi, D.A., Baird, A., Majack, R.A., and Reidy, M.A. (1991). Role of basic fibroblast growth factor in vascular lesion formation. *Circulation Research*, **68**, (1), 106–13.

100. Reidy, M.A., Irvin, C., and Lindner, V. (1996). Factors controlling the development of arterial lesions after injury. *Circulation*, (Suppl III), III43.

101. Lindner, V., Reidy, M.A., and Fingerle, J. (1989). Regrowth of arterial endothelium. Denudation with minimal trauma leads to complete endothelial cell regrowth. *Laboratory Investigation*, **61**, (5), 556–63.

102. Ruef, J., Hu, Z.Y., Yin, L.Y. *et al.* (1997). Induction of vascular endothelial growth factor in balloon-injured baboon arteries. A novel role for reactive oxygen species in atherosclerosis. *Circulation Research*, **81**, (1), 24–33.

103. Hausner, E.A., Orsini, J.A., Foster, L.L. *et al.* (1999). Vascular endothelial growth factor in porcine coronary arteries following balloon angioplasty. *Journal of Investigative Surgery*, **12**, (1), 15–23.

104. Couper, L.L., Bryant, S.R., Eldrup-Jorgensen, J., Bredenberg, C.E., and Lindner, V. (1997). Vascular endothelial growth factor increases the mitogenic response to fibroblast growth factor-2 in vascular smooth muscle cells *in vivo* via expression of fms-like tyrosine kinase-1. *Circulation Research*, **81**, (6), 932–9.

105. Tsurumi, Y., Murohara, T., Krasinski, K. *et al.* (1997). Reciprocal relation between VEGF and NO in the regulation of endothelial integrity. *Nature Medicine*, **3**, (8), 879–86.

106. Asahara, T., Bauters, C., Pastore, C. *et al.* (1995). Local delivery of vascular endothelial growth factor accelerates reendothelialization and attenuates intimal hyperplasia in balloon-injured rat carotid artery (see comments). *Circulation*, **91**, (11), 2793–801.

107. Asahara, T., Chen, D., Tsurumi, Y. *et al.* (1996). Accelerated restitution of endothelial integrity and endothelium-dependent function after phVEGF165 gene transfer. *Circulation*, **94**, (12), 3291–302.

108. Gashler, A. and Sukhatme, V.P. (1995). Early growth response protein 1 (Egr-1): prototype of a zinc-finger family of transcription factors. *Progress in Nucleic Acid Research & Molecular Biology*, **50**, 191–224.

109. Kim, S., Kawamura, M., Wanibuchi, H. *et al.* (1995). Angiotensin II type 1 receptor blockade inhibits the expression of immediate-early genes and fibronectin in rat injured artery. *Circulation*, **92**, (1), 88–95.

110. Khachigian, L.M., Williams, A.J., and Collins, T. (1995). Interplay of Sp1 and Egr-1 in the proximal platelet-derived growth factor A-chain promoter in cultured vascular endothelial cells. *Journal of Biological Chemistry*, **270**, (46), 27679–86.

111. Santiago, F.S., Lowe, H.C., Day, F.L., Chesterman, C.N., and Khachigian, L.M. (1999). Egr-1 induction by injury is triggered by release and paracrine activation by fibroblast growth factor-2. *American Journal of Pathology*, **154**, 937–44.

112. Santiago, F.S., Atkins, D.G., and Khachigian, L.M. (1999). Vascular smooth muscle cell proliferation and regrowth after injury *in vitro* is dependent upon NGFI-A/Egr-1. *American Journal of Pathology*, **155**, 897–905.

113. Santiago, F.S., Lowe, H.C., Kavurma, M.M. *et al.* (1999). Novel DNA enzyme targeting Egr-1 mRNA inhibits vascular smooth muscle proliferation and regrowth factor injury. *Nature Medicine*, **11**, 1264–69.

114. Hu, Y., Cheng, L., Hochleitner, B.W., and Xu, Q. (1997). Activation of mitogen-activated protein kinases (ERK/JNK) and AP-1 transcription factor in rat carotid arteries after balloon injury. *Arteriosclerosis, Thrombosis & Vascular Biology*, **17**, (11), 2808–16.

115. McCaffrey, T.A., Fu, C., Du, B. *et al.* (2000). High level expression of Egr-1 and Egr-1-inducible genes in mouse and human atherosclerosis. *Journal of Clinical Investigation*, **105**, (5), 653–62.

13 Lipoprotein influx and efflux in the atherosclerotic intima

Leonard Kritharides
Clinical Research Group, The Heart Research Institute, Camperdown, Sydney,
NSW 2050, Australia
Department of Cardiology, Concord Hospital, The University of Sydney, Sydney,
NSW 2139, Australia

Trevor G. Redgrave
Department of Physiology, The University of Western Australia, Nedlands,
WA 6907, Australia

13.1 Introduction

Influx and efflux of lipids in the arterial wall are life-long phenomena. Progressive
focal accumulations of lipid could even perhaps be regarded as physiological, since
aortic fatty streaks are observed in infancy and are present in all individuals by 10
years of age.[1] Coronary lesions are detected by the age of 15.[2] These early lesions
consist mostly of cholesteryl esters, and undoubtedly originate mainly by influx
from plasma lipoproteins.[3] Influx is probably continual for low-density lipoprotein
(LDL), but may be episodic for post-prandial remnant particles. Net accumulation
in the arterial wall reflects the balance between influx and efflux.

13.2 Lipoprotein influx is essential for atherogenesis

In the early fatty streak, the trapping of LDL in the sub-endothelial space leads to
a chronic inflammatory reaction and activation of NFκB-like transcription factors
that cause the expression of products mediating monocyte binding, chemotaxis and
conversion into macrophages destined to become foam cells.[4] Calculated aortic
retention of LDL averages $<0.01\%$ of the plasma pool per gram aortic wet weight
per day, and the rate of LDL entry into the arterial wall vastly exceeds the LDL

accumulation rate as estimated using the tyramine cellobiose labelling technique.[5] This led Williams and Tabas to postulate that atherosclerosis is fundamentally determined by factors affecting the retention of lipoproteins in the sub-intima, the so-called 'response to retention'.[6, 7]

13.2.1 apoB-containing particles in human and animal atherosclerosis

There is uniform agreement on the contribution of plasma apoB-containing lipoproteins, and especially LDL, to the accumulation of lipids in arterial lesions.[8–11] The accumulation of extracellular, cholesterol-rich liposomes containing apoB precedes fatty streak formation and appears essential for monocyte recruitment and fatty streak formation.[12, 13] Human aorta contains LDL demonstrated by immuno-electrophoresis, and is concentrated in regions of intimal thickening.[14] By immunohistochemical methods, apo(a) from lipoprotein(a) was also detected in human aortas.[15] The unequivocal role of LDL in pathogenesis is confirmed by epidemiological, animal and pathological studies as well as the atherosclerosis seen in patients with Familial Hypercholesterolemia.[16] Importantly, other apoB-containing apolipoproteins are probably also pathogenic.

Remnants of triglyceride-rich lipoproteins labelled with a fluorescent marker rapidly penetrate arterial tissue and accumulate in lesions *in vivo*[17, 18] and in isolated macrophages *in vitro*.[19, 20] Remnants are not only taken up by fibroblasts and arterial smooth muscle cells with greater affinity than LDL but are also cytotoxic.[21–24] This evidence is consistent with a pathological role of remnant lipoproteins in atherosclerosis in man.[25–29]

Mice deficient in apoE demonstrate abnormal remnant metabolism and accelerated atherosclerosis,[30] confirming a pathogenic role for remnants. That most of the apoB in these mice is apoB48,[7] supports the notion that chylomicron remnant-derived, apoB-containing lipoproteins may be atherogenic. Regression of atherosclerotic lesions in apoE-deficient mice follows adenovirus-mediated hepatic expression of apoE,[31] and also by bone marrow transplantation of macrophages expressing apoE,[32–34] with greater effect in younger animals with fatty lesions than in older mice.[35] This suggests that diminished influx of atherogenic lipoproteins may permit spontaneous regression of atherosclerotic lesions *in vivo*.

13.3 Kinetic studies

Animals with hypercholesterolemia demonstrate pre-lesional transport of excess plasma lipoproteins across the endothelium, followed by intimal accumulation of modified and reassembled lipoproteins.[12, 13] Immunostaining identifies apoB-containing lipoproteins as being important in pre-lesional changes. However, all lipoprotein classes have been shown to enter the artery wall.[36] Endothelial denudation is not a prerequisite for such accumulation to occur, but alterations in endothelial permeability have been related to increases in accumulation.

13.3.1 Regulation of lipoprotein influx

Modelling studies have shown interaction between lipoprotein influx and increases in endothelial permeability, elevated transmural pressure, the serum concentration of the particle (e.g. lipoproteins), fenestration of the internal elastic lamina, and pre-existing intimal thickening.[37, 38] Synergy between transmural pressure and lipoprotein particle concentration in promoting lipoprotein influx may mechanistically explain the pro-atherogenic interactions of risk factors such as hypercholesterolemia and hypertension.[39] Open intercellular junctions and cellular transcytotic, plasmalemmal vesicles both contribute to transport across the endothelium.[40]

13.3.2 Endothelial permeability

The permeability of the endothelium can be measured using Evans blue albumin conjugates.[41] LDL transport is at least in part dependent on transport between endothelial cell (EC), as can be assessed by the effect of agents which increase endothelial permeability such as calcium ionophores, and the coupling of LDL flux to transvascular water flow.[42] Experimentally observed minor degrees of endothelial detachment and increased macromolecular permeability across intercellular clefts would increase overall permeability by 50–100%.[43] EC turnover contributes to permeability between EC, with almost all EC in the mitotic phase showing increased permeability, and dividing cells appear to account for 30–45% of increased permeability of EC layer to LDL in some states.[41]

13.3.3 Lipoprotein size, influx and efflux (stochastic analyses)

Stender and Nordestgaard have provided direct *in vivo* measurements of lipoprotein flux using radiolabelled HDL and apoB-containing lipoproteins injected at varying intervals before surgery (in humans) or sacrifice (in animals). There is a close positive association between the apparent permeability to LDL of a given aortic segment and the cholesterol accumulation in that same aortic segment after cholesterol feeding.[44] The intimal uptake of high density, low density, and very low density lipoproteins decreases linearly with the logarithm of the macromolecular diameter. This indicates that the arterial influx of three plasma lipoprotein fractions and of plasma proteins proceeds by similar, size-dependent mechanisms, and supports general permeability as being an important determinant of lipoprotein uptake.[45, 46]

Lipoprotein size determines both influx and efflux from the arterial wall.[47, 48] While small dense LDL may penetrate the arterial intima more readily than large buoyant LDL, or the remnants of triglyceride-rich lipoproteins, the larger particles deliver more cholesterol to the sub-endothelium. Furthermore, retention of particles is relatively small compared with total flux, so it appears likely that factors mediating accumulation of lipoproteins in a lesion are more pertinent than minor differences in permeability of the arterial wall.

13.3.4 Lipoprotein accumulation is incompletely explained by rates of influx

The net accumulation of lipoproteins over short intervals is greater than that anticipated given the actual accumulation of lipoprotein lipids in advanced atherosclerotic lesions. This implies that removal or egress of lipoprotein lipids is an ongoing process, and that the balance of uptake, degradation and egress may be influential in regulating net lipoprotein accumulation. The difficulty of extracting LDL from plaque (see below) raises uncertainty as to whether the increased accumulation of lipoprotein lipids is cause or consequence of lesion formation. If retention of LDL is greater in plaque than in normal tissue, identical permeability would generate very different net accumulation in diseased and normal segments. Thus, decreased lipoprotein egress, as a consequence of greater binding in plaque, may subsequently mediate the apparently greater nett uptake in lesion-prone areas.

13.3.5 Internal elastic lamina and lipoprotein retention

Simple sieving would be expected to increase arterial exposure to smaller molecules relative to larger molecules. However, net retention of lipoproteins differs markedly according to lipoprotein size, with LDL being far more avidly retained than HDL or albumin.[36] LDL concentration in intimal fluid is twice that in plasma, and this appears to relate in part to the function of the internal elastic lamina (IEL) preventing movement of sub-endothelial LDL to the outer media, and preventing the egress of sub-endothelial LDL back across the EC. In capillaries, which do not have an IEL, net lipoprotein transport is simply inversely related to size.[36] The integrity of the IEL also contributes to the overall permeability of arteries and particular arterial segments (such as branch points).[9, 36]

The localized nature of lesion development near branch points and near points of high shear has not been adequately explained, but probably reflects cell responses to the local physical environment, such as is now established for expression of nitric oxide synthase. Theoretical modelling of the effects of various physical and fluid mechanical factors such as wall shear rate, diffusivity of LDL and filtration velocity of water at the vessel wall predicted that, under physiological conditions, flow-dependent concentration or dilution of LDL occurs at a blood/endothelium surface. Compared with bulk flow in the lumen of the vessel, LDL could be more concentrated at the artery wall by up to several-fold, depending on local flow factors.[49]

13.3.6 Focal accumulation of LDL follows lesion formation

In the pigeon, the rate of LDL influx appears to increase after atherosclerosis has developed; before atherosclerotic lesion formation, uptake is as great or greater in non-atherosclerosis-prone regions.[50] Net influx of lipoproteins increases at branch points, but not in other susceptible areas in which cholesterol accumulates preferentially.[51, 52] This may imply that saturation of normal clearance or degradation mechanisms may determine net cholesterol accumulation in the arterial wall, and post-lesional changes in these parameters may be important. Net influx of lipoproteins (VLDL, IDL and LDL) in arterial tissues is many fold greater for atherosclerotic aorta than for normal arterial tissue.[53] Calculated LDL residence times in normolipidemic and hyperlipidemic rabbits are similar in lesion-prone and -resistant sites before lesion formation, but with development of lesions increase to 10–25-fold that of normolipidemic tissue.[54] This is consistent with LDL retention and apparently greater uptake being a consequence rather than a cause of lesion formation.

13.3.7 Cellular metabolism contributes to net arterial accumulation of lipoprotein lipids

Injections of LDL in animal atherosclerosis models found that arterial accumulation of injected LDL principally occurred within macrophages and the extracellular matrix, with greater accumulation of injected LDL in atherosclerotic lesions than in normal tissue.[55] The fact that macrophages rapidly accumulate LDL indicates that cellular accumulation and degradation of LDL may contribute to the overall arterial degradation and net accumulation of lipids in atherosclerotic lesions.

Cell metabolism may also contribute to uptake of remnant particles. Macrophage foam cells and smooth muscle cell foam cells express alpha 2-macroglobulin receptor/LDL receptor-related protein (alpha 2 MR/LRP) mRNA and protein.[56, 57] Macrophages in lesions also express scavenger receptor mRNA and protein,[56, 58] but not LDL receptors. Expression of another member of the LDL receptor family, LR11, was found to be increased in aortic lesions of cholesterol-fed rabbits.[59]

The factors regulating accumulation and degradation of lipoproteins in cells or matrix of the artery wall may not be mutually independent. Carew postulated that HDL may be anti-atherogenic as a result of interfering with SMC-mediated uptake of LDL in the artery wall.[60] Such an arterial wall process would add to the cellular interactions of HDL. These include the ability of HDL to remove cholesterol from foam cells within lesions (so-called cholesterol efflux[61]), and the ability to inhibit endothelial cell expression of cellular adhesion molecules.[62]

13.4 Retention and non-oxidative modification of LDL in atherosclerotic plaque

13.4.1 Extraction of lipoproteins from atherosclerotic tissue

Numerous studies have isolated lipoproteins from human atherosclerotic lesions, providing additional evidence that lipoproteins other than LDL contribute to the accumulation of apoB-containing lipoproteins in plaque. Triglyceride-rich lipoproteins containing apoB and apoE have been identified and extracted from human atherosclerotic plaque.[63] Similarly, Lp(a), containing LDL and apo(a), has been recovered from human normal and atherosclerotic tissue.[64]

Although fibrofatty atherosclerotic lesions contain more immunohistochemically detectable apolipoprotein B (apoB) than normal aortae, the efficiency of recovery of apoB during gentle saline extraction of plaque is less from atheromatous tissues than normal aortae.[65] The recovery of apoB-containing lipoproteins from plaque is increased by detergent extraction, confirming both their presence and avid binding to components of atherosclerotic tissue.[66] More vigorous extraction procedures modify extracted lipoprotein particles. Conditions of most efficient recovery alter the nature of lipoproteins in lesions, and gentle extractions with incomplete recovery may potentially leave in plaque the most modified and pathogenic lipoprotein derivatives,[67] thereby complicating evaluation of the relative importance of different atherogenic modifications. Some differences reported between plasma and vascular tissue LDL (extracted by relatively vigorous homogenization) include lower cholesteryl ester (CE) content, mild lipid peroxidation, increased sphingomyelin content and decreased phosphatidylcholine content than plasma LDL.[68]

13.4.2 Remodelling of lipoproteins in the interstitial space

Egress of lipoproteins from the arterial wall could involve their movement back into the arterial lumen across the endothelial layer, or flux via interstitial fluid in the arterial wall into lymphatics and drainage into lymph before re-entering the circulation. apoB- and apoA-I-containing lipoproteins in the lymphatics demonstrate substantial remodelling, with both larger and smaller particles appearing. Interstitial fluid is generally believed to contain 10% of the protein content of plasma; however, the relative concentrations of different plasma proteins vary substantially.[36] HDLs of greater density are well recognized in interstitial fluid,[36] and these have been more recently characterized as preβ-HDL and apoA-I-only particles.[69] These findings support the notion that the interaction of small HDL-derived cholesterol acceptors with cells[70] is likely to be of major importance in the removal of cholesterol from cells in the artery wall.[71]

13.4.3 Factors contributing to retention of lipoproteins in the artery wall

Understanding the factors that promote LDL retention and modification within plaque may point to prospects for reversal or prevention of disease. As discussed above, retention within the artery wall may be a more important consideration than overall permeability or flux of lipoproteins through the sub-intimal space.

13.4.3.1 Lipoprotein aggregation

Aggregation is a consistently identified modification of LDL in human plaque, present even in studies showing minimal evidence of LDL oxidation.[72] Aggregation conceivably involves interaction between various lipoprotein particles which could be similar (e.g. LDL with LDL) or different (e.g. LDL with remnants). A number of routes of LDL aggregation have been suggested (discussed below). By increasing the effective diameter of lipoprotein particles, and by binding of lipoproteins, aggregates encourage lipoprotein retention in the artery wall.

13.4.3.2 Enzymes and proteoglycans can promote binding and retention of LDL

Aggregation of LDL can be enhanced by many factors, which include mild oxidation, modification of phospholipids and by binding to enzymes and proteoglycans. Within 2 h of injection of rabbits, focal clustering of LDL with extracellular matrix filaments in the artery wall has been demonstrated by freeze etching studies.[73]

Binding of LDL to other molecules which stimulate aggregation, such as proteoglycans[74] and enzymes such as lipoprotein lipase (LPL) and sphingomyelinase (SPMase),[75, 76] directly promote LDL retention, and can promote macrophage uptake of lipoproteins.[77] Mouse models which over-express secretory SPMase demonstrated increased LDL retention compared with controls.[78] In addition, apoE KO mice with variable copies of the gene for secretory SPMase demonstrated a dose-dependent protective effect of SPMase deficiency.[78] These recent studies suggest that future strategies to modulate LDL retention and modification may have dramatic effects on atherogenesis.

In addition to enhancing lipoprotein retention, extracellular enzymes such as LPL can increase endothelial permeability as a consequence of the products released during lipolysis of lipoproteins.[76] Non-pancreatic PLA2 hydrolyses phospholipids at the sn-2 position to release free fatty acids and lysophospholipids, both of which may act as inflammatory agents, and has been isolated from atherosclerotic plaque.[79] Binding of LDL on specific matrix components facilitates the interaction of PLA2 with the lipoproteins.[80]

LDL forms complexes with arterial wall proteoglycans (PGs) which aggregate, and promote the degradation of LDL by macrophages and the cellular accumulation of lipoprotein lipid.[81] PGs derived from atherosclerotic plaque tissue appear to bind LDL with greater affinity than PGs derived from normal arteries.[82] The yield of apoB-containing apolipoproteins extracted from plaque is markedly increased by treating plaque with collagenase and elastase, emphasizing the importance of LDL–proteoglycan in the retention of LDL within arterial lesions.[83] Hyaluronan-rich and versican-rich proteoglycans appear to be favourable for smooth muscle migration,[84] and various cell wall factors, such as free fatty acid concentration, appear to modulate cell expression of proteoglycan core proteins.[85] Consequently, intramural enzymic lipolysis may further contribute to lipoprotein retention by encouraging matrix deposition.

Mice transgenic for a human apoB-100 variant with a single point mutation in one basic cluster produce LDL with normal LDL receptor binding and no affinity for a range of arterial proteoglycans.[86] Importantly, mice transgenic for unmutated apoB-100 develop severe atherosclerosis with dietary hypercholesterolemia, whereas mice transgenic for mutated apoB-100 develop almost no atherosclerosis despite similar hypercholesterolemia (Boren *et al.* published in abstract form and cited in Ref. (7)). This supports intra-arterial binding of LDL by proteoglycans as being important in LDL retention and atherogenesis.

13.4.4 The necrotic core and lipoprotein cholesterol

Intralesional lipoproteins may undergo many non-oxidative modifications, including aggregation and vesicle formation,[87] and may directly generate the formation of cholesterol crystals.[88] The ultrastructural similarity of LDL aggregated *in vitro* and the deposits of extracellular lipid found in atherosclerotic lesions is consistent with their direct extracellular formation from lipoproteins.[89]

There is a high concentration of tissue factor (TF) in the necrotic core, and in the atherosclerotic gruel of the necrotic core there is a correlation of TF with macrophage number.[90] This supports a contribution of foam cell necrosis to the genesis of the necrotic core. This is also supported by *in vitro* studies which demonstrate that cholesterol accumulated in lysosomal and cytosolic compartments can crystallize and cause cell necrosis.[91, 92]

Early crystal deposition can be found in lesions resembling fatty streaks before apparent cell necrosis.[93] This suggests that lipoprotein aggregation and fusion, in combination with other processes, may contribute to the generation of the crystal-containing core. This cell-free model of core formation poses difficulties, even to its proponents, as plaque core deposits contain 60% free cholesterol whereas lipoproteins predominantly contain esterified cholesterol.

There is currently no completely satisfactory mechanism for spontaneous extracellular hydrolysis of lipoprotein CE. There is the potential for cellular release of CE hydrolases with activity at neutral pH, which may convert lipoprotein CE into FC, and would not require prior intracellular accumulation and hydrolysis. One such enzyme may be the bile salt-stimulated cholesteryl esterase secreted by human monocyte-derived macrophages in response to exposure to modified LDL.[94] Other potential mechanisms for extracellular stimulation of lipid droplet formation include the proteolytic exposure of LDL to chymotrypsin from mast cells,[95] hydrolysis of LDL during non-endocytic association of aggregated LDL with cells,[96] or secretion of free cholesterol–protein aggregates by cells. Non-endocytic processes may thus contribute to cell-mediated lipoprotein modification in the artery wall, but at this time are incompletely resolved.

13.5 Relative significance of cellular cholesterol efflux and arterial lipoprotein efflux

13.5.1 The requirement for cholesterol acceptors in peripheral cells

Reverse cholesterol transport (RCT) is the process by which cholesterol is removed from peripheral non-steroidogenic cells and cleared by the liver after transfer of cholesterol to other lipoproteins. Because peripheral cells cannot catabolize cholesterol once accumulated, as occurs after lipoprotein endocytosis, clearance by the cell is required. Relatively slow basal constitutive release of cholesterol can occur from cells such as macrophages, particularly during the secretion of apoE,[97] and after hydroxylation of cholesterol to 27-hydroxycholesterol.[98] It is generally believed that lipid-poor or lipid-free apoA-I-containing particles are major initial acceptors of cholesterol in the artery wall.[70] Cholesterol efflux is regulated by physico-chemical and receptor-mediated processes, and these are discussed in some detail in Chapter 9 (macrophages and atherosclerosis).

13.5.2 Relative importance of cellular cholesterol efflux and lipoprotein egress

Potential strategies for the treatment of atherosclerosis include the enhanced removal of accumulated lipoprotein cholesterol from the artery wall. If cellular cholesterol accumulation is quantitatively important (e.g. a substantial proportion of cholesterol in the artery wall being accumulated within cells), or qualitatively important (e.g. by increasing cell secretion of inflammatory factors), then optimal cholesterol efflux and RCT are desirable. If lipoprotein influx, egress and accumulation are more important, then blocking influx or enhancing egress are desirable. At this time, it is difficult to rank the relative importance of the two processes. Most likely, combinations of both will be required to achieve early stabilization of inflammatory lipid-rich macrophages (e.g. by cholesterol efflux from cells), and to prevent the long-term residence and modification of plasma-derived lipoproteins (e.g. by reducing endothelial permeability and lipoprotein retention). The interactions of cellular and extracellular factors in the artery wall ensures that any therapeutic strategies will have implications for both cellular and extracellular lipid accumulation.

13.6 Concluding comments

Knowledge of fluxes of lipoproteins and lipids in the genesis of atherosclerotic lesions is currently benefiting from a growing understanding of the underlying cellular mechanisms. With new therapies directed at regression of existing lesions, measurements in living humans are needed to follow the responses of individual patients. New methods such as near-infrared diffuse reflection spectroscopy in fibre-optic systems appear to be capable of measuring the cholesterol content of plaques of the aortic wall and of coronary vessels with acceptable precision.[99] Plaque measurement *in vivo* would greatly assist in assessment of new therapies, and could be the next advance in understanding the role of lipoprotein metabolism in lesion development.

References

1. Day, A.J. and Wahlqvist, M.L. (1970). Cholesterol ester and phospholipid composition of normal aortas and of atherosclerotic lesions in children. *Exp. Mol. Pathol.*, **13**, 199.
2. Stary, H.C. (1994). Changes in components and structure of atherosclerotic lesions developing from childhood to middle age in coronary arteries. *Basic Res. Cardiol.*, **89**, 17.
3. Dayton, S. and Hashimoto, S. (1970). Recent advances in molecular pathology: a review. Cholesterol flux and metabolism in arterial tissue and in atheromata. *Exp. Mol. Pathol.*, **13**, 253.
4. Navab, M., Fogelman, A.M., Berliner, J.A., Territo, M.C., Demer, L.L., Frank, J.S., Watson, A.D., Edwards, P.A., and Lusis, A.J., (1995). Pathogenesis of atherosclerosis. *Am. J. Cardiol.*, **76**, 18C.
5. Carew, T.E., Pittman, R.C., Marchand, E.R., and Steinberg, D. (1984). Measurement *in vivo* of irreversible degradation of low density lipoprotein in the rabbit aorta. Predominance of intimal degradation. *Arteriosclerosis*, **4**, 214.
6. Williams, K.J. and Tabas, I. (1995). The response-to-retention hypothesis of early atherogenesis. *Arterioscler. Thromb. Vasc. Biol.*, **15**, 551.
7. Williams, K.J. and Tabas, I. (1998). The response-to-retention hypothesis of atherogenesis reinforced. *Curr. Opin. Lipidol.*, **9**, 471.
8. Hoff, H., Gaubatz, J.W., and Gotto, A.M. (1978). Apo B concentration in the normal human aorta. *Biochem. Biophys. Res. Commun.*, **85**, 1424.
9. Smith, E.B. and Ashall, C. (1983). Low-density lipoprotein concentration in interstitial fluid from human atherosclerotic lesions. Relation to theories of endothelial damage and lipoprotein binding. *Biochim. Biophys. Acta*, **754**, 249.

10. Mora, R., Lupu, F., and Simionescu, N. (1987). Prelesional events in atherogenesis-colocalisation of apolipoprotein B, unesterified cholesterol and extracellular phospholipid liposomes in the aorta of hyperlipidemic rabbit. *Atherosclerosis*, **67**, 143.

11. Yla-Herttuala, S., Jaakola, O., Enholm, C., Tikkanen, M.J., Solakivi, T., Sarkioja, T., and Nikkari, T. (1988). Characterization of two lipoproteins containing apolipoproteins B and E from lesion-free human aortic intima. *J. Lipid Res.*, **29**, 563.

12. Simionescu, M. and Simionescu, N. (1993). Proatherosclerotic events: pathobiochemical changes occurring in the arterial wall before monocyte migration. *FASEB J.*, **7**, 1359.

13. Simionescu, N., Vasile, E., Lupu, F., Popescu, G., and Simionescu, M. (1986). Prelesional events in atherogenesis-accumulation of cholesterol-rich liposomes in the arterial intima and cardiac valves of the hyperlipidemic rabbit. *Am. J. Path.*, **123**, 109.

14. Spring, P.M. and Hoff, H.F. (1989). LDL accumulation in the grossly normal human iliac bifurcation and common iliac arteries. *Exp. Mol. Pathol.*, **51**, 179.

15. Jurgens, G., Chen, Q., Esterbauer, H., Mair, S., Ledinski, G., and Dinges, H.P. (1993). Immunostaining of human autopsy aortas with antibodies to modified apolipoprotein B and apoprotein(a). *Arterioscler. Thromb.*, **13**, 1689.

16. Goldstein, J.L. and Brown, M.S. (1987). Regulation of low-density lipoprotein receptors: implications for pathogenesis and therapy of hypercholesterolemia and atherosclerosis. *Circulation*, **76**, 504.

17. Proctor, S.D. and Mamo, J.C. (1996). Arterial fatty lesions have increased uptake of chylomicron remnants but not low-density lipoproteins. *Coron. Artery Dis.*, **7**, 239.

18. Proctor, S.D. and Mamo, J.C. (1998). Retention of fluorescent-labelled chylomicron remnants within the intima of the arterial wall – evidence that plaque cholesterol may be derived from post-prandial lipoproteins. *Eur. J. Clin. Invest.*, **28**, 497.

19. Yu, K.C. and Mamo, J.C. (1997). Regulation of cholesterol synthesis and esterification in primary cultures of macrophages following uptake of chylomicron remnants. *Biochem. Mol. Biol. Int.*, **41**, 33.

20. Mamo, J.C., Elsegood, C.L., Gennat, H.C., and Yu, K. (1996). Degradation of chylomicron remnants by macrophages occurs via phagocytosis. *Biochemistry*, **35**, 10210.

21. Redgrave, T.G., Fidge, N.H., and Yin, J. (1982). Specific, saturable binding and uptake of rat chylomicron remnants by rat skin fibroblasts. *J. Lipid Res.*, **23**, 638.

22. Yu, K.C. and Mamo, J.C. (1997). Binding and uptake of chylomicron remnants by cultured arterial smooth muscle cells from normal and Watanabe-heritable-hyperlipidemic rabbits. *Biochim. Biophys. Acta*, **1346**, 212.

23. Yu, K.C. and Mamo, J.C. (1996). Killing of arterial smooth muscle cells by chylomicron remnants. *Biochem. Biophys. Res. Commun.*, **220**, 68.

24. Yu, K.C., Smith, D., Yamamoto, A., Kawaguchi, A., Harada-Shiba, M., Yamamura, T., and Mamo, J.C. (1997). Phagocytic degradation of chylomicron remnants by fibroblasts from subjects with homozygous familial hypercholesterolaemia. *Clin. Sci. (Colch)*, **92**, 197.

25. Masuoka, H., Ishikura, K., Kamei, S., Obe, T., Seko, T., Okuda, K., Koyabu, S., Tsuneoka, K., Tamai, T., Sugawa, M. *et al.* (1998). Predictive value of remnant-like particles cholesterol/high-density lipoprotein cholesterol ratio as a new indicator of coronary artery disease. *Am. Heart J.*, **136**, 226.

26. Groot, P.H., van Stiphout, W.A., Krauss, X.H., Jansen, H., van Tol, A., van Ramshorst, E., Chin-On, S., Hofman, A., Cresswell, S.R., and Havekes, L. (1991). Postprandial lipoprotein metabolism in normolipidemic men with and without coronary artery disease. *Arterioscler. Thromb.*, **11**, 653.

27. Gronholdt, M.L., Nordestgaard, B.G., Nielsen, T.G., and Sillesen, H. (1996). Echolucent carotid artery plaques are associated with elevated levels of fasting and postprandial triglyceride-rich lipoproteins. *Stroke*, **27**, 2166.

28. Devaraj, S., Vega, G., Lange, R., Grundy, S.M., and Jialal, I. (1998). Remnant-like particle cholesterol levels in patients with dysbetalipoproteinemia or coronary artery disease. *Am. J. Med.*, **104**, 445.

29. Meyer, E., Westerveld, H.T., de Ruyter-Meijstek, F.C., van Greevenbroek, M.M., Rienks, R., van Rijn, H.J., Erkelens, D.W., and de Bruin, T.W. (1996). Abnormal postprandial apolipoprotein B-48 and triglyceride responses in normolipidemic women with greater than 70% stenotic coronary artery disease: a case–control study. *Atherosclerosis*, **124**, 221.

30. Plump, A.S., Smith, J.D., Hayek, T., Aalto-Setala, K., Walsh, A., Verstuyft, J.G., Rubin, E., and Breslow, J.L. (1992). Severe hypercholesterolemia and atherosclerosis in apolipoprotein E-deficient mice created by homologous recombination in ES cells. *Cell*, **71**, 343.

31. Tsukamoto, K., Tangirala, R., Chun, S.H., Pure, E., and Rader, D.J. (1999). Rapid regression of atherosclerosis induced by liver-directed gene transfer of ApoE in ApoE-deficient mice. *Arterioscler. Thromb. Vasc. Biol.*, **19**, 2162.

32. Linton, M.F., Atkinson, J.B., and Fazio, S. (1995). Prevention of atherosclerosis in apolipoprotein E-deficient mice by bone marrow transplantation. *Science*, **267**, 1034.

33. Van Eck, M., Herijgers, N., Yates, J., Pearce, N.J., Hoogerbrugge, P.M., Groot, P.H., and Van Berkel, T.J. (1997). Bone marrow transplantation in apolipoprotein E-deficient mice. Effect of ApoE gene dosage on serum lipid concentrations, (beta)VLDL catabolism, and atherosclerosis. *Arterioscler. Thromb. Vasc. Biol.*, **17**, 3117.

34. Boisvert, W.A., Spangenberg, J., and Curtiss, L.K. (1995). Treatment of severe hypercholesterolaemia in apolipoprotein E-deficient mice by bone marrow transplantation. *J. Clin. Invest.*, **96**, 1118.

35. Hasty, A.H., Linton, M.F., Brandt, S.J., Babev, V.R., Gleaves, L.A., and Fazio, S. (1999). Retroviral gene therapy in apo E-deficient mice-apo E expression in the artery wall reduces early foam cell lesion formation. *Circulation*, **99**, 2571.

36. Smith, E.B. (1990). Transport, interactions and retention of plasma proteins in the intima: the barrier function of the internal elastic lamina. *Eur. Heart J.*, **11**, E72.

37. Fry, D.L. (1987). Mass transport, atherogenesis, and risk. *Arteriosclerosis*, **7**, 88.

38. Friedman, M.H. and Fry, D.L. (1993). Arterial permeability dynamics and vascular disease. *Atherosclerosis*, **104**, 189.

39. Chobanian, A.V., Prescott, M.F., and Haudenschild, C.C. (1984). Recent advances in molecular pathology. The effects of hypertension on the arterial wall. *Exp. Mol. Pathol.*, **41**, 153.

40. Kao, C.H., Chen, J.K., Kuo, J.S., and Yang, V.C. (1995). Visualization of the transport pathways of low density lipoproteins across the endothelial cells in the branched regions of rat arteries. *Atherosclerosis*, **116**, 27.

41. Lin, S.J., Jan, K.M., Weinbaum, S., and Chien, S. (1989). Transendothelial transport of low density lipoprotein in association with cell mitosis in rat aorta. *Arteriosclerosis*, **9**, 230.

42. Rutledge, J.C., Curry, F.-R.E., Lenz, J.F., and Davis, P.A. (1990). Low density lipoprotein transport across a microvascular endothelial barrier after permeability is increased. *Circulation Res.*, **66**, 486.

43. Weinbaum, S., Tzeghai, G., Ganatos, P., Pfeffer, R., and Chien, S. (1985). Effect of cell turnover and leaky junctions on arteriolar macromolecular transport. *Am. J. Physiol.*, **248**, H945.

44. Nielsen, L.B., Nordestgaard, B.G., Stender, S., and Kjeldsen, K. (1992). Aortic permeability to LDL as a predictor of aortic cholesterol accumulation in cholesterol-fed rabbits. *Arterioscler. Thromb.*, **12**, 1402.

45. Stender, S. and Zilversmit, D.B. (1981). Transfer of plasma lipoprotein components and of plasma proteins into aortas of cholesterol-fed rabbits. Molecular size as a determinant of plasma lipoprotein influx. *Arteriosclerosis*, **1**, 38.

46. Stender, S. and Zilversmit, D.B. (1981). *In vivo* influx, tissue esterification and hydrolysis of free and esterified plasma cholesterol in the cholesterol-fed rabbit. *Biochim. Biophys. Acta*, **663**, 674.

47. Nordestgaard, B.G., Tybjaerg-Hansen, A., and Lewis, B. (1992). Influx *in vivo* of low density, intermediate density, and very low density lipoproteins into aortic intimas of genetically hyperlipidemic. *Arterioscler. Thromb.*, **12**, 6.

48. Nordestgaard, B.G., Wootton, R., and Lewis, B. (1995). Selective retention of VLDL, IDL, and LDL in the arterial intima of genetically hyperlipidemic rabbits *in vivo*. Molecular size as a determinant of fractional loss from the intima-inner media. *Arterioscler. Thromb. Vasc. Biol.*, **15**, 534.

49. Wada, S. and Karino, T. (1999). Theoretical study on flow-dependent concentration polarization of low density lipoproteins at the luminal surface of a straight artery. *Biorheology*, **36**, 207.

50. Schwenke, D.C. and St Clair, R.W. (1993). Influx, efflux, and accumulation of LDL in normal arterial areas and atherosclerotic lesions of white Carneau pigeons with naturally occurring and cholesterol-aggravated aortic atherosclerosis. *Arterioscler. Thromb.*, **13**, 1368.

51. Schwenke, D.C. and Carew, T.E. (1989). Initiation of atherosclerotic lesions in cholesterol-fed rabbits. II. Selective retention of LDL vs selective increases in LDL permeability in susceptible sites of arteries. *Arteriosclerosis*, **9**, 908.

52. Schwenke, D.C. and Carew, T.E. (1989). Initiation of atherosclerotic lesions in cholesterol-fed rabbits. Focal increases in arterial LDL concentration precede development of fatty streak lesions. *Arteriosclerosis*, **9**, 895.

53. Nicoll, A., Duffield, R., and Lewis, B. (1981). Flux of plasma lipoproteins into human arterial intima. Comparison between grossly normal and atheromatous intima. *Atherosclerosis*, **39**, 229.

54. Tozer, E.C. and Carew, T.E. (1997). Residence time of low-density lipoprotein in the normal and atherosclerotic rabbit aorta. *Circ. Res.*, **80**, 208.

55. Rosenfeld, M.E., Carew, T.E., von Hodenberg, E., Pittman, R.C., Ross, R., and Steinberg, D. (1992). Autoradiographic analysis of the distribution of 125I-tyramine–cellobiose–LDL in atherosclerotic lesions of the WHHL rabbit. *Arterioscler. Thromb.*, **12**, 985.

56. Luoma, J., Hiltunen, T., Sarkioja, T., Moestrup, S.K., Gliemann, J., Kodama, T., Nikkari, T., and Yla-Herttuala, S. (1994). Expression of alpha 2-macroglobulin receptor/low density lipoprotein receptor-related protein and scavenger receptor in human atherosclerotic lesions. *J. Clin. Invest.*, **93**, 2014.

57. Leppanen, P., Luoma, J.S., Hofker, M.H., Havekes, L.M., and Yla-Herttuala, S. (1998). Characterization of atherosclerotic lesions in apo E3-Leiden transgenic mice. *Atherosclerosis*, **136**, 147.

58. de Villiers, W.J. and Smart, E.J. (1999). Macrophage scavenger receptors and foam cell formation. *J. Leukoc. Biol.*, **66**, 740.

59. Kanaki, T., Bujo, H., Hirayama, S., Ishii, I., Morisaki, N., Schneider, W.J., and Saito, Y. (1999). Expression of LR11, a mosaic LDL receptor family member, is markedly increased in atherosclerotic lesions. *Arterioscler. Thromb. Vasc. Biol.*, **19**, 2687.

60. Carew, T.E., Koschinsky, T., Hayes, S.B., and Steinberg, D. (1976). A mechanism by which high-density lipoproteins may slow the atherogenic process. *Lancet*, **1**, 1315.

61. Rothblat, G.H., de la Llera-Moya, M., Atger, V., Kellner-Weibel, G., Williams, D.L., and Phillips, M.C. (1999). Cell cholesterol efflux: integration of old and new observations provides new insights. *J. Lipid Res.*, **40**, 781.

62. Cockerill, G.W., Rye, K.A., Gamble, J.R., Vadas, M.A., and Barter, P.J. (1995). High-density lipoproteins inhibit cytokine-induced expression of endothelial cell adhesion molecules. *Arterioscler. Thromb. Vasc. Biol.*, **15**, 1987.

63. Rapp, J.H., Lespine, A., Hamilton, R.L., Colyvas, N., Chaumeton, A.H., Tweedie-Hardman, J., Kotite, L., Kunitake, S.T., Havel, R.J., and Kane, J.P. (1994). Triglyceride-rich lipoproteins isolated by selected-affinity anti-apolipoprotein B immunosorption from human atherosclerotic plaque. *Arterioscler. Thromb.*, **14**, 1767.

64. Reblin, T., Meyer, N., Labeur, C., Henne-Bruns, D., and Beisiegel, U. (1995). Extraction of lipoprotein(a), apo B, and apo E from fresh human arterial wall and atherosclerotic plaques. *Atherosclerosis*, **113**, 179.

65. Hoff, H.F., Heideman, C.L., Gotto, A.M., Jr, and Gaubatz, J.W. (1977). Apolipoprotein B retention in the grossly normal and atherosclerotic human aorta. *Circ. Res.*, **41**, 684.

66. Hoff, H.F., Heideman, C.L., Gaubatz, J.W., Scott, D.W., and Gotto, A.M., Jr (1978). Detergent extraction of tightly-bound apoB from extracts of normal aortic intima and plaques. *Exp. Mol. Pathol.*, **28**, 290.

67. Yla-Herttuala, S., Palinski, W., Rosenfeld, M.E., Steinberg, D., and Witztum, J.L. (1990). Lipoproteins in normal and atherosclerotic aorta. *Eur. Heart J.*, **11** (Suppl. E), 88.

68. Daugherty, A., Zweifel, B.S., Sobel, B.E., and Schonfeld, G. (1988). Isolation of low density lipoprotein from atherosclerotic vascular tissue of Watanabe heritable hyperlipidaemic rabbits. *Arteriosclerosis*, **8**, 768.

69. Asztalos, B.F. and Roheim, P.S. (1995). Presence and formation of 'free apolipoprotein-A-I-like' particles in human plasma. *Arterioscler. Thromb. Vasc. Biol.*, **15**, 1419.

70. Fielding, C.J. and Fielding, P.E. (1995). Molecular physiology of reverse cholesterol transport. *J. Lipid Res.*, **36**, 211.

71. Reichl, D. (1994). Extravascular circulation of lipoproteins: their role in reverse transport of cholesterol. *Atherosclerosis*, **105**, 117.

72. Steinbrecher, U.P. and Lougheed, M. (1992). Scavenger receptor-independent stimulation of cholesterol esterification in macrophages by low density lipoprotein extracted from human aortic intima. *Arterioscler. Thromb.*, **12**, 608.

73. Nievelstein, P.F.E.M., Fogelman, A.M., Mottino, G., and Frank, J.S. (1991). Lipid accumulation in rabbit aortic intima 2 hours after bolus infusion of low density lipoprotein. *Arterioscler. Thromb.*, **11**, 1795.

74. Mawhinney, T.P., Augustyn, J.M., and Fritz, K.E. (1978). Glycosaminoglycan–lipoprotein complexes from aortas of hypercholesterolemic rabbits. Part 1. Isolation and characterization. *Atherosclerosis*, **31**, 155.

75. Tabas, I., Li, Y., Brocia, R.W., Xu, S.W., Swenson, T.L., and Williams, K.J. (1993). Lipoprotein lipase and sphingomyelinase synergistically enhance the association of atherogenic lipoproteins with smooth muscle cells and extracellular matrix. A possible mechanism for low density lipoprotein and lipoprotein(a) retention and macrophage foam cell formation. *J. Biol. Chem.*, **268**, 20419.

76. Rutledge, J.C., Woo, M.M., Rezai, A.A., Curtiss, L.K., and Goldberg, I.J. (1997). Lipoprotein lipase increases lipoprotein binding to the artery wall and increases endothelial layer permeability by formation of lipolysis products. *Circ. Res.*, **80**, 819.

77. Hurt, E. and Camejo, G. (1987). Effect of arterial proteoglycans on the interaction of LDL with human monocyte-derived macrophages. *Atherosclerosis*, **67**, 115.

78. Wong, M.-L., Xie, B., Beatii, N., Phu, P., Marathe, S., Johns, A., Hirsch, E., Williams, K.J., and Tabas, I. (1999). Sphingomyelinase transgenic and knockout mice: direct evidence that sphingomyelinase is atherogenic *in vivo*. *Circulation*, **100**, I.

79. Hurt-Camejo, E., Andersen, S., Standal, R., Rosengren, B., Sartipy, P., Stadberg, E., and Johansen, B. (1997). Localization of nonpancreatic secretory phospholipase A2 in normal and atherosclerotic arteries. Activity of the isolated enzyme on low-density lipoproteins. *Arterioscler. Thromb. Vasc. Biol.*, **17**, 300.

80. Sartipy, P., Bondjers, G., and Hurt-Camejo, E. (1998). Phospholipase A2 type II binds to extracellular matrix biglycan: modulation of its activity on LDL by colocalization in glycosaminoglycan matrixes. *Arterioscler. Thromb. Vasc. Biol.*, **18**, 1934.

81. Vijayagopal, P., Srinivasan, S.R., Jones, K.M., Radhakrishnamurthy, B., and Berenson, G.S. (1985). Complexes of low-density lipoproteins and arterial proteoglycan aggregates promote cholesteryl ester accumulation in mouse macrophages. *Biochim. Biophys. Acta*, **837**, 251.

82. Vijayagopal, P., Figueroa, J.E., Fontenot, J.D., and Glancy, D.L. (1996). Isolation and characterization of a proteoglycan variant from human aorta exhibiting a marked affinity for low density lipoprotein and demonstration of its enhanced expression in atherosclerotic plaques. *Atherosclerosis*, **127**, 195.

83. Srinivasan, S.R., Radhakrishnamurthy, B., Dalferes, E.R., Jr, and Berenson, G.S. (1979). Collagenase-solubilized lipoprotein–glycosaminoglycan complexes of human aortic fibrous plaque lesions. *Atherosclerosis*, **34**, 105.

84. Evanko, S.P., Angello, J.C., and Wight, T.N. (1999). Formation of hyaluronan-and versican-rich pericellular matrix is required for proliferation and migration of vascular smooth muscle cells. *Arterioscler. Thromb. Vasc. Biol.*, **19**, 1004.

85. Olsson, U., Bondjers, G., and Camejo, G. (1999). Fatty acids modulate the composition of extracellular matrix in cultured human arterial smooth muscle cells by altering the expression of genes for proteoglycan core proteins. *Diabetes*, **48**, 616.

86. Boren, J., Olin, K., Lee, I., Chait, A., Wight, T.N., and Innerarity, T.L. (1998). Identification of the principal proteoglycan-binding site in LDL. A single-point mutation in apo-B100 severely affects proteoglycan interaction without affecting LDL receptor binding. *J. Clin. Invest.*, **101**, 2658.

87. Guyton, J.R. and Klemp, K.F. (1988). Ultrastructural discrimination of lipid droplets and vesicles in atherosclerosis: value of osmium– thiocarbohydrazide–osmium and tannic acid–paraphenylenediamine techniques. *J. Histochem. Cytochem.*, **36**, 1319.

88. Guyton, J.R. and Klemp, K.F. (1989). The lipid-rich core region of human atherosclerotic fibrous plaques. Prevalence of small lipid droplets and vesicles by electron microscopy. *Am. J. Pathol.*, **134**, 705.

89. Guyton, J.R., Klemp, K.F., and Mims, M.P. (1991). Altered ultrastructural morphology of self-aggregated low density lipoproteins: coalescence of lipid domains forming droplets and vesicles. *J. Lipid Res.*, **32**, 953.

90. Moreno, P.R., Bernardi, V.H., Lopez-Cuellar, J., Murcia, A.M., Palacios, I.F., Gold, H.K., Mehran, R., Sharma, S.K., Nemerson, Y., Fuster, V. *et al.* (1996). Macrophages, smooth muscle cells, and tissue factor in unstableangina. Implications for cell-mediated thrombogenicity in acute coronary syndromes. *Circulation*, **94**, 3090.

91. Tangirala, R.K., Jerome, W.G., Jones, N.L., Small, D.M., Johnson, W.J., Glick, J.M., Mahlberg, F.H., and Rothblat, G.H. (1994). Formation of cholesterol monohydrate crystals in macrophage-derived foam cells. *J. Lipid Res.*, **35**, 93.

92. Kellner-Weibel, G., Yancey, P.G., Jerome, W.G., Walser, T., Mason, R.P., Phillips, M.C., and Rothblat, G.H. (1999). Crystallization of free cholesterol in model macrophage foam cells. *Arterioscler. Thromb. Vasc. Biol.*, **19**, 1891.

93. Guyton, J.R. and Klemp, K.F. (1993). Transitional features in human atherosclerosis. Intimal thickening, cholesterol clefts, and cell loss in human aortic fatty streaks. *Am. J. Pathol.*, **143**, 1444.

94. Li, F. and Hui, D.Y. (1997). Modified low density lipoprotein enhances the secretion of bile-salt-stimulated cholesterol esterase by human monocyte-macrophages. *J. Biol. Chem.*, **272**, 28666.

95. Piha, M., Lindstedt, L., and Kovanen, P.T. (1995). Fusion of proteolyzed low-density lipoprotein in the fluid phase: a novel mechanism generating atherogenic lipoprotein particles. *Biochemistry*, **34**, 10120.

96. Buton, X., Mamdouh, Z., Ghosh, R., Du, H., Kuriakose, G., Beatini, N., Grabowski, G.A., Maxfield, F.R., and Tabas, I. (1999). Unique cellular events occurring during the initial interaction of macrophages with matrix-retained or methylated aggregated low density lipoprotein (LDL). Prolonged cell-surface contact during which ldl-cholesteryl ester hydrolysis exceeds ldl protein degradation. *J. Biol. Chem.*, **274**, 32112.

97. Kruth, H.S., Skarlatos, S.I., Gaynor, P.M., and Gamble, W. (1994). Production of cholesterol-enriched nascent high density lipoproteins by human monocyte-derived macrophages is a mechanism that contributes to macrophage cholesterol efflux. *J. Biol. Chem.*, **269**, 24511.

98. Bjorkhem, I., Andersson, O., Diczfalusy, U., Sevastik, B., and Xiu, R.-J. (1994). Atherosclerosis and sterol 27-hydroxylase: evidence for a role of this enzyme in elimination of cholesterol from human macrophages. *Proc. Natl. Acad. Sci. USA*, **91**, 8592.

99. Jaross, W., Neumeister, V., Lattke, P., and Schuh, D. (1999). Determination of cholesterol in atherosclerotic plaques using near infrared diffuse reflection spectroscopy. *Atherosclerosis*, **147**, 327.

14 Inhibition of the cytokine-induced expression of endothelial cell adhesion molecules by high-density lipoproteins

Paul Baker, Moira Clay, Kerry-Anne Rye, and Philip Barter
Department of Medicine, University of Adelaide and the Cardiovascular Investigation
Unit, Royal Adelaide Hospital, Adelaide, Australia

14.1 Introduction

High-density lipoproteins (HDLs) protect against the development of atherosclerosis.[1] In human populations the risk of coronary heart disease (CHD) correlates inversely with the concentration of HDL cholesterol[2–6] and, in intervention studies in humans, treatment that increases the HDL level is accompanied by a reduction in future coronary events.[7] In studies of atherosclerosis-prone animals, the increase in HDL cholesterol level that follows the introduction and over-expression of the human apolipoprotein (apo) A-I gene results in a marked reduction in the extent of the atherosclerosis.[8–10]

The mechanism by which HDLs protect against atherosclerosis is uncertain, although it almost certainly relates to one or more of the known functions of these lipoproteins (Table 14. 1). For example, the well-known action of HDL in promoting the efflux of cholesterol from cells[11] has an obvious potential to inhibit atherosclerosis. However, HDLs have actions in addition to their involvement in cholesterol transport which may also contribute to their anti-atherogenic properties.

This paper focuses on the ability of HDL to inhibit the adhesion of blood monocytes to endothelial cells. Specifically, it describes how HDLs inhibit the cytokine-induced expression of endothelial cell adhesion molecules and discusses how this action of HDL contributes to the anti-atherogenic properties of these lipoproteins.

Table 14. 1 Known functions of HDLs

Function	Reference
Reverse cholesterol transport	(11)
Inhibit oxidative modification of LDL	(21, 22)
Mitogenic activity	(12)
Bind lipopclysaccharide (LPS), including LPS that has associated with macrophages	(13, 14)
Protect erythrocytes against pro-coagulant activity	(15)
Maintain normal vasoconstriction in early stages of atherosclerosis	(20)
Reduce epidermal growth factor-induced DNA synthesis in smooth muscle cells	(19)
Stimulate endothelial prostacyclin synthesis	(17)
Bind prostacyclin, enhancing its biological half-life	(18)
Enhance the anti-coagulant activities of plasma protein S and activated protein C	(16)
Inhibit oxidized LDL-induced monocyte transmigration	(23)
Inhibit cytokine-induced endothelial adhesion molecule expression	(24–28)
Inhibit monocyte adhesion to encothelial cells	(29)

14.2 Role of endothelial cell adhesion molecules in atherogenesis

Atherosclerosis is a chronic inflammatory disorder characterized by the accumulation of macrophages and T-lymphocytes in the arterial intima.[30] The macrophages are derived from blood monocytes that adhere to endothelial cells and then, in response to chemotactic factors, migrate into the sub-endothelial space. Within the sub-endothelial space, the monocytes are transformed to macrophages which take up (oxidized) LDL, initiating the process of atherosclerosis (Fig. 14.1). A class of proteins called adhesion molecules[31] mediates the first step in this process, the adhesion of blood monocytes to endothelial cells. Three of these – vascular cell adhesion molecule-1 (VCAM-1), intercellular adhesion molecule-1 (ICAM-1) and E-selectin – are expressed in endothelial cells which have been activated by cytokines such as tumour necrosis factor-α (TNF-α) and interleukin-1 (IL-1).[32] E-selectin retards ·monocytes that are in blood and causes them to roll along the endothelial cell surface, while ICAM-1 and VCAM-1 bind the monocytes more tightly[33] making them accessible to the action of chemoattractants that promote their migration into the arterial intima (Fig. 14.2). Increased plasma levels of these

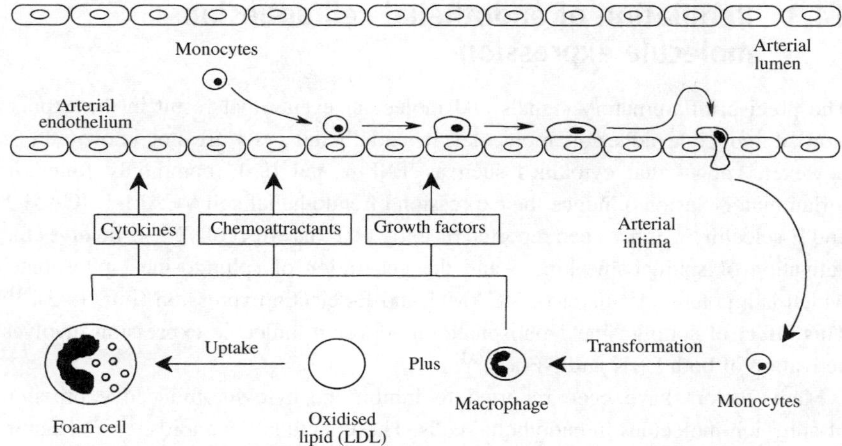

Fig. 14.1 Migration of monocytes into the arterial intima during the early stages of atherogenesis.

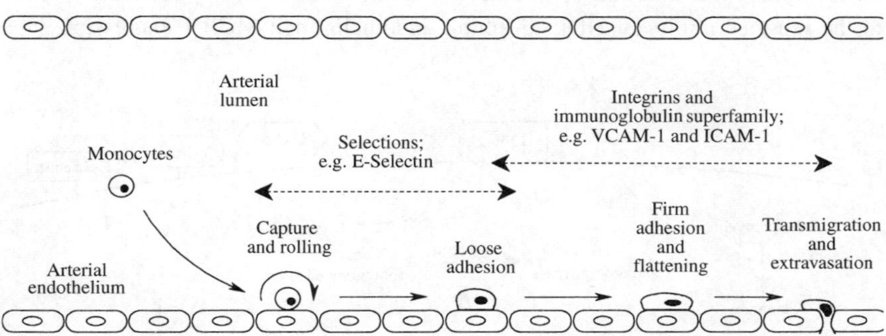

Fig. 14.2 Adhesion molecules implicated in the recruitment of monocytes into the arterial intima during the early stages of atherogenesis.

adhesion molecules have been reported to be predictive of CHD in human subjects.[34–37] Furthermore, they have been shown in animal studies to be expressed in endothelial cells *in vivo* at sites of developing atherosclerosis.[38-40]

The observation that human HDLs inhibit the cytokine-induced expression of endothelial cell VCAM-1, ICAM-1 and E-selectin[24–28] is thus of potentially great importance.

14.3 Regulation of endothelial cell adhesion molecule expression

The precise inflammatory signals and molecular events that result in the expression of inducible adhesion molecules in endothelial cells are not known. It is, however, known that cytokines such as TNF-α and IL-1 (commonly found in inflammatory lesions) induce the expression of endothelial cell VCAM-1, ICAM-1 and E-selectin.[32] It has been reported recently that this effect of TNF-α involves the activation of sphingosine kinase and the generation of sphingosine-1-phosphate, which is a potent stimulator of VCAM-1 and E-selectin expression (Fig. 14.3).[41] This effect of sphingosine-1-phosphate on adhesion molecule expression involves activation of both ERK and NF-κB.[41]

Many factors have been reported to inhibit the cytokine-induced expression of adhesion molecules in endothelial cells. These include flavonoids,[42, 43] vitamin E,[44] non-esterified fatty acids[45, 46] and HDL.[24–28]

14.4 Inhibition of endothelial cell adhesion molecule expression by human plasma HDL

HDLs inhibit the adhesion of monocytes to endothelial cells.[29] This may relate to the reported ability of HDL to inhibit endothelial cell MCP-1 expression,[23] a

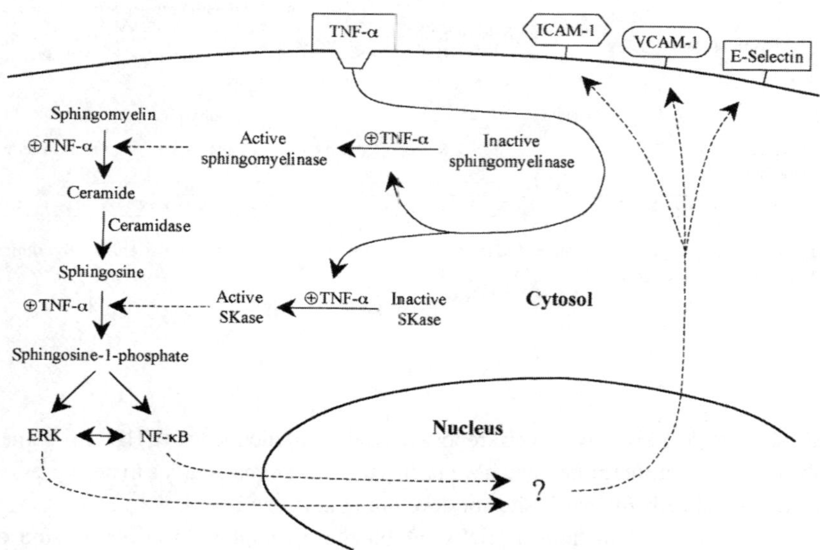

Fig. 14.3 Postulated mechanism by which TNF-α induces the expression of adhesion molecules in endothelial cells.

protein that is known to attract monocytes into the sub-endothelial space,[47, 48] or to the HDL-mediated inhibition of endothelial cell adhesion molecules.[24–28]

When human umbilical vein endothelial cells (HUVECs) are activated by TNF-α or IL-1, the expression of VCAM-1, ICAM-1 and E-selectin is inhibited by human plasma HDL in a concentration-dependent manner. The HDLs do not interfere with the binding of TNF-α to its receptor.[24] The inhibition is apparent whether the HDLs are added to the cells several hours prior to their activation by the cytokine or whether they are added at the same time as the cytokine. Furthermore, when the HDLs are present in a pre-incubation with the endothelial cells but are removed prior to the addition of the cytokine, the inhibition remains apparent[24, 25] (Fig. 14.4), implying that the HDLs have in some way modified the cells so as to render them resistant to cytokine-induced adhesion molecule expression.

The extent of the inhibition of endothelial cell VCAM-1 is greater with the HDL_3 sub-fraction than with HDL_2.[25] In contrast, the apolipoprotein composition of human HDLs appears to have no demonstrable effect on their ability to inhibit the cytokine-induced expression of VCAM-1 in endothelial cells.[25] Replacement of all the HDL apoA-I with either apoA-II[25] or even serum amyloid A protein (SAA) (Ashby and Barter, unpublished) has no effect on the inhibitory activity of the HDL.

Preparations of HDL isolated from different human subjects vary widely in their ability to inhibit endothelial VCAM-1 expression.[25] The explanation for this variation is not known. Human HDLs are heterogeneous, comprising several sub-populations of particles that vary in shape (discoidal or spherical), size, surface charge and in the composition of both their lipids and apolipoproteins (Fig. 14.5). The inhibitory activity of a given preparation of HDLs may be a function of the composition and sub-population distribution of the particles. However, this is difficult to investigate because it is not possible to isolate human HDL sub-populations that are

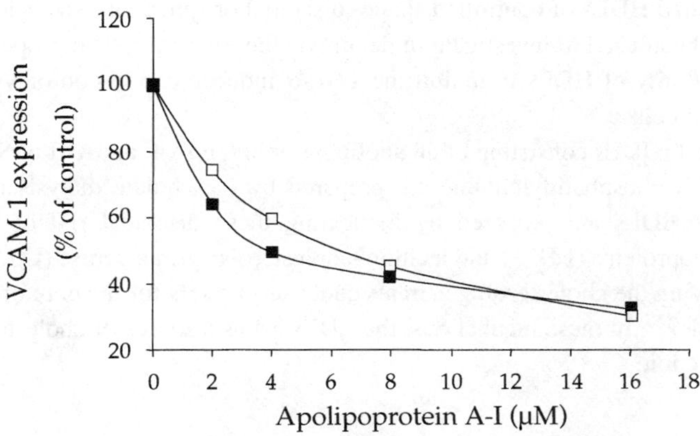

Fig. 14.4 Effect of discoidal (A-I)rHDL being present or absent during stimulation of cells with TNF-α. HUVECs were pre-incubated with discoidal (A-I)rHDL for 16 h before stimulating the cells with TNF-α. The rHDLs were either removed before the addition of TNF-α (□) or remained in the culture media during stimulation with TNF-α (■). VCAM-1 expression was determined by flow cytometry 5 h after TNF-α addition.

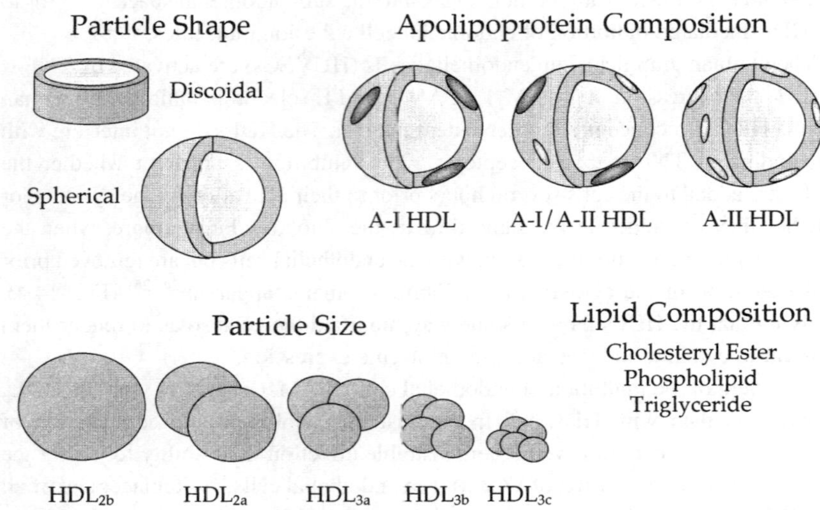

Fig. 14.5 Heterogeneity of native HDL.

homogeneous in shape, size and composition. One way around this problem is to use reconstituted HDLs (rHDLs) that are indistinguishable from native HDL in structure and function but which can be prepared as homogeneous populations of particles with a shape, size and composition that can be tightly regulated.

14.5 Inhibition of endothelial cell adhesion molecule expression by reconstituted HDL

Reconstituted HDLs of controlled shape (discoidal or spherical), size and composition have been used to investigate in detail whether these physical properties influence the ability of HDLs to inhibit the TNF-α-induced expression of VCAM-1 in endothelial cells.[26]

Discoidal rHDLs consisting of an apolipoprotein (apoA-I, apoA-II or SAA) complexed with phosphatidylcholine are prepared by the cholate dialysis method.[49] Spherical rHDLs are prepared by incubating these discoidal rHDLs with low-density lipoproteins (LDLs) and lecithin–cholesterol acyltransferase (LCAT) which, by esterifying the cholesterol, generates cholesteryl esters for the core of the spherical particle.[50] In these incubations, the LDLs act as a source of cholesterol for the LCAT reaction.

14.5.1 Effect of rHDL shape

Inhibition of cytokine-induced VCAM-1 expression in HUVECs has been observed with both discoidal[24, 26] and spherical[26] rHDLs. In one study, the inhibition at a given apolipoprotein concentration was greater with spherical than with discoidal rHDL.[26] Spherical and discoidal particles, however, differ in more than shape, with the spherical rHDL containing a core of cholesteryl esters that is absent in discoidal rHDL. Spherical rHDLs also contain sphingomyelin and possibly other phospholipid species.[51]

14.5.2 Effect of varying the size of spherical rHDL

When spherical rHDLs containing three molecules of apoA-I per particle are incubated with cholesteryl ester transfer protein (CETP), the particles undergo a process of fusion and rearrangement of constituents. This rearrangement generates a larger number of smaller particles that contain two rather than three molecules of apoA-I per particle.[52] In terms of their ability to inhibit VCAM-1 expression in HUVECs, the larger and smaller rHDLs are identical when equated for rHDL particle concentration.[26] This suggests that the differing inhibitory activities of HDL_2 and HDL_3[25] may relate to factors other than the well-known differences in particle size of the two HDLs sub-fractions.

14.5.3 Effect of varying the apolipoprotein composition of rHDL

Studies with both discoidal and spherical rHDLs have investigated the effects of varying the apolipoprotein composition on the inhibitory activity, and concluded that particles in which the sole apolipoprotein is apoA-I, apoA-II and SAA are equivalent in terms of their abilities to inhibit VCAM-1 expression in HUVECs.[25, 26]

14.5.4 Effect of varying the core lipid composition of rHDL

Spherical rHDLs are typically prepared with cholesteryl esters as the sole core lipid. Replacement of most of the cholesteryl esters by triglyceride in a process mediated by CETP has no effect on the ability of the resultant particles to inhibit endothelial cell VCAM-1 expression.[26]

14.5.5 Effect of the phospholipid composition of rHDL

In contrast to the lack of effect of changes in size, shape, apolipoprotein composition and the core lipid composition, the inhibitory activity of rHDL is greatly altered by changing the phospholipid composition of the particles (Baker, Rye, and Barter, unpublished).

14.6 Phospholipid composition of human HDL

Phospholipids account for about 30% by mass of human HDLs. About 80% of these phospholipids are phosphatidylcholines (PCs); another 15% are sphingomyelin with the remainder being made up of small amounts of phosphatidylethanolamine, phosphatidylinositol and phosphatidylglycerol.[53] There is considerable variation in the acyl chain composition of the PCs in human HDL. In general, they have a saturated fatty acid (mainly palmitic or stearic acid) in the sn-1 position, with an unsaturated fatty acid in the sn-2 position.[54] The three most abundant HDLs PCs are palmitoyl-linoleoyl phosphatidylcholine (PLPC), palmitoyl-oleoyl phosphatidylcholine (POPC) and palmitoyl-arachidonyl phosphatidylcholine (PAPC) which collectively account for more than half the PCs in human HDL.[54] Another PC species of interest is palmitoyl-docosahexaenoyl phosphatidylcholine (PDPC) which contains the omega-3 polyunsaturated fatty acid docosahexaenoic acid (DHA) in the sn-2 position. PDPC represents 3–8% of the HDL PCs.

14.6.1 Effects of diet on HDL phospholipid composition

The relative proportions of PLPC, POPC and PDPC in plasma lipoproteins from a given subject can be altered by changing the proportions of linoleic acid (an omega-6 polyunsaturated fatty acid), oleic acid (a monounsaturated fatty acid) and docosahexaenoic acid (an omega-3 polyunsaturated fatty acid) in the diet.[55–57] This has been shown in the case of consumption of high levels of fish oils, which result in enrichment of HDL with PDPC.[57]

14.6.2 Effects of HDLs phospholipid composition on adhesion molecule inhibition

We have conducted studies with rHDLs which contain a series of different PC species. We have used both discoidal rHDLs (which contain only apoA-I and PC) and spherical rHDLs (which contain apoA-I, PC, unesterified cholesterol and cholesteryl esters). By making these rHDLs with a series of different PC species, it has been possible to document precisely how phospholipid composition influences the ability of the rHDLs to inhibit endothelial cell adhesion molecule expression. In studies with both discoidal and spherical rHDLs, particles containing PCs in which the fatty acid in the sn-2 position is polyunsaturated (PLPC or PAPC) are highly effective as inhibitors of the TNF-α-induced expression of VCAM-1 in HUVECs. In contrast, PCs such as POPC (in which the monounsaturated oleic acid occupies the sn-2 position) or DPPC (in which saturated palmitic acid occupies the sn-2 position) are much less effective (Fig. 14.6).

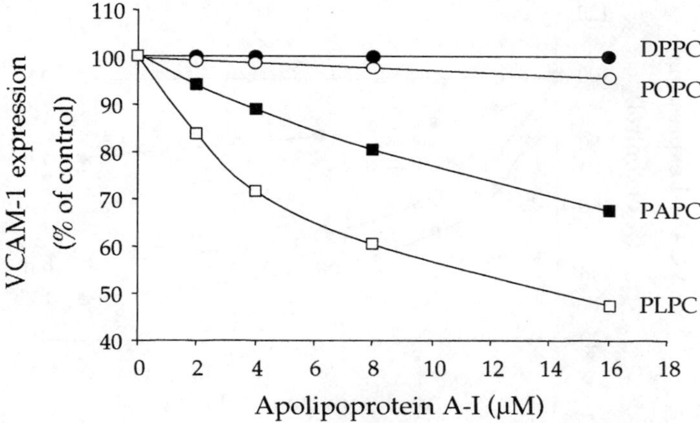

Fig. 14.6 Effect of the sn-2 fatty acyl moiety of phospholipid on the inhibitory activity of discoidal (A-I)rHDL. HUVECs were pre-incubated with discoidal (A-I)rHDL for 16 h before removing the HDL and stimulating the cells with TNF-α. The rHDLs were prepared with either DPPC (●), POPC (O), PAPC (■) or PLPC (□). VCAM-1 expression was determined by flow cytometry 5 h after TNF-α addition.

14.6.3   Effects of phosphatidylcholine vesicles

Small unilamellar vesicles of PLPC (but not POPC) are also able to inhibit the cytokine-induced expression of VCAM-1 in HUVECs. However, this occurs only when the vesicles are protected against oxidation by an anti-oxidant, e.g. butylated hydroxytoluene (Baker *et al.*, unpublished observation). When the anti-oxidant is not present, the PLPC becomes oxidized during vesicle preparation or during the subsequent incubation and is cytotoxic to the cells. This observation suggests that phospholipids are a major active component of HDLs responsible for their inhibitory activity. As phospholipid vesicles are not known to exist *in vivo*, this observation has little physiological significance. The apoA-I in discoidal rHDL containing PLPC may protect the PLPC from oxidation during culture, enabling it to retain its anti-inflammatory properties. Such an anti-oxidant role of apoA-I has been reported for both native HDL and spherical rHDL.[21, 22, 58]

14.6.4   Implications of the effects of different phospholipids

The observation that phospholipid composition impacts on the ability of HDL to inhibit endothelial cell adhesion molecule expression provides a possible link between the type of dietary fat consumed and the development of atherosclerosis. For example, diets rich in linoleic acid that enrich HDL with PLPC would be predicted to enhance the inhibition of endothelial cell adhesion molecule expression to a greater extent and thus provide greater protection against atherosclerosis than when the HDLs are enriched with POPC by eating a diet rich in oleic acid.

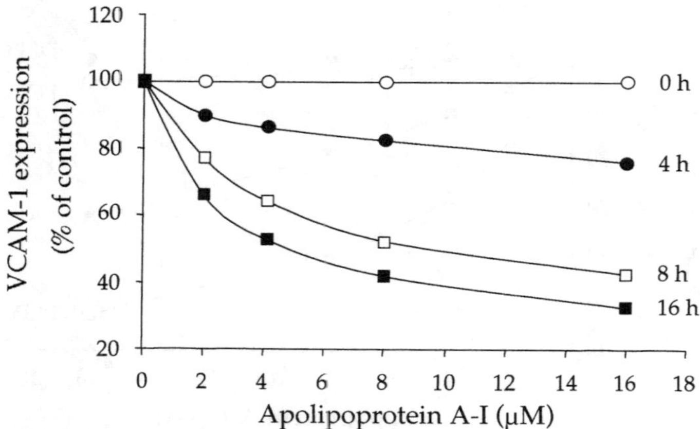

Fig. 14.7 Pre-incubation of HUVECs with discoidal (A-I)rHDL is necessary to obtain maximal inhibition of VCAM-1 expression. HUVECs were pre-incubated with discoidal (A-I)rHDL for either 5 min (O), 4 h(●), 8 h (□) or 16 h (■) before removing the HDL and stimulating the cells with TNF-α. VCAM-1 expression was determined by flow cytometry five hours after TNF-α addition.

14.7 Time sequence of the HDL-mediated inhibition of endothelial cell adhesion molecule expression

Studies have been conducted to determine (i) whether the inhibition of cytokine-induced expression of endothelial cell adhesion molecules mediated by HDL requires prior exposure of the cells to the HDL and (ii) if so, to discover whether the magnitude of the inhibition is influenced by the duration of the pre-incubation of the cells with HDL. In studies conducted with discoidal rHDL, the effect of the rHDL on expression of adhesion molecules in activated HUVECs was found not to be instantaneous but to require a period of pre-incubation of the cells with the rHDL. It was also found that the magnitude of the inhibition increased with increasing duration (up to 16 h) of the pre-incubation (Fig. 14.7). However, having undergone a period of pre-incubation, it was found that the rHDL can be removed from the cells before adding the TNF-α without any apparent loss of the inhibition of adhesion molecule expression. Furthermore, once the inhibitory effect of rHDL on HUVECs has been achieved by pre-incubation, it persists for several hours after the rHDLs have been removed. These findings imply that rHDLs modify the cells so as to make them resistant to cytokine-induced expression of VCAM-1 (and E-selectin) in a time-dependent process.

14.8 Mechanism of the HDL-mediated inhibition of endothelial cell adhesion molecule expression

Having established that activation of sphingosine kinase and the consequent generation of sphingosine-1-phosphate are key elements in the process by which TNF-α induces the expression of VCAM-1 and E-selectin in endothelial cells,[41] it is of great interest to discover that HDLs inhibit both the activity of sphingosine kinase and the generation of sphingosine-1-phosphate in endothelial cells.[27] Furthermore, the inhibitory activity of HDL is mimicked by dimethyl sphingosine, a known competitive inhibitor of sphingosine kinase.[27] The HDL-mediated inhibition of endothelial sphingosine kinase is accompanied by a reduction in the TNF-α medi-ated activation of ERK and NF-κB, both of which are activated by sphingosine-1-phosphate.[27] A schematic diagram of the proposed mechanism by which HDLs inhibit endothelial cell adhesion molecule expression is shown in Fig. 14.8.

The mechanism by which HDL phospholipids interact with endothelial cells to inhibit their sphingosine kinase activity is not known. The fact that the HDL-mediated inhibition of VCAM-1 expression occurs much sooner than is observed with non-esterified fatty acids (NEFA)[45, 46] suggests that hydrolysis of the phospholipids with subsequent release of NEFA is not involved. Rather, the time course studies are more consistent with the effect of NEFA being secondary to their incorporation into membrane phospholipids.

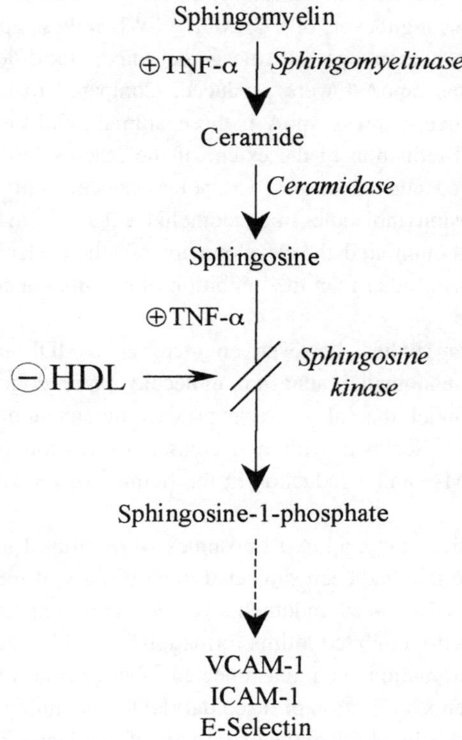

Fig. 14.8 Postulated mechanism by which HDL inhibits the TNF-α-induced expression of endothelial cell VCAM-1, ICAM-1 and E-selectin.

14.9 Relevance of the inhibition of adhesion molecule expression to atherosclerosis

The relationship between HDL and CHD risk relates to the concentration of HDL in the circulating plasma. How this equates with the concentration of HDL in the artery wall is uncertain. This issue is of potential importance when considering that both the effect of HDL on the efflux of cholesterol from foam cells and the inhibition of the oxidative modification of LDL are dependent on the presence of HDL in the arterial intima. In contrast, endothelial cells are exposed to whatever is the level of HDL present in the circulating plasma, implying that what is observed *in vitro* may translate directly into what occurs *in vivo*. It should be emphasized that observations made *in vitro* can be extrapolated to what may be occurring *in vivo* with only the greatest of caution.

To date, there have been four reported studies investigating the possible *in vivo* relevance of HDL-mediated inhibition of endothelial cell adhesion molecule expression (Table 14.2). One of these studies failed to demonstrate an effect of HDL on endothelial cell adhesion molecule expression *in vivo*, while three did show an effect. In the negative study, apoE knockout mice were used as a model of spontaneous atherosclerosis. The endothelial cells at the site of the atherosclerotic lesions in these animals express high levels of VCAM-1.[59] When these apoE knockout mice were crossed with human apoA-I transgenic mice, apoE-deficient mice that over-expressed human apoA-I were produced. Compared to the apoE knockout mice that did not over-express apoA-I, these animals had an elevated level of HDL and a marked reduction in the extent of the atherosclerosis. Unexpectedly, however, the reduced atherosclerosis was not associated with a decrease in the expression of adhesion molecules in endothelial cells.[59] On this basis, it was concluded that HDL inhibited the development of atherosclerosis in this mouse model by a mechanism other than the inhibition of endothelial cell adhesion molecule expression.

In the three other studies, however, an increase in HDL levels *in vivo* was found to decrease endothelial adhesion molecule expression. In another apoE knockout mouse model that also over-expressed human apoA-I, the increased level of HDL was associated with a decreased expression of endothelial cell VCAM-1 and ICAM-1 and a reduction in the homing of macrophages to sites of atherosclerosis.[60]

In a second positive study, apoA-I liposomes were infused into cholesterol-fed apoE knockout mice that had been subjected to cuff injury of their carotid arteries. Infusion of the apoA-I reduced endothelial cell VCAM-1 expression at the site of the lesion and markedly inhibited intima formation.[61] A third study used a porcine model of acute inflammation in animals injected subcutaneously with interleukin-1. In this study, intravenous injection of discoidal rHDL containing apoA-1 and phosphatidylcholine markedly inhibited the expression of E-selectin in the subcutaneous endothelial cells of these animals.[62]

Table 14.2 Studies investigating the effects of HDL on endothelial adhesion molecule expression *in vivo*

Model	Finding	Reference
Endothelial adhesion molecule expression in apoE knockout and apoE knockout/ apoA-I transgenic mice.	Diminished lesion formation in the apoE knockout/apoA-I transgenic mice with no evidence of reduced endothelial VCAM-1 expression.	59
Endothelial adhesion molecule expression and monocyte homing to the endothelium in apoE knockout and apoE knockout/apo A-I transgenic mice.	Reduced endothelial VCAM-1and ICAM-1 expression and endothelial monocyte homing in the apoE knockout/ apoA-I transgenic mice	60
Endothelial adhesion molecule expression in cuff-injured carotid arteries of apoE knockout mice following apoA-I liposome infusion.	Reduced endothelial VCAM-1 expression, monocyte infiltration and lesion formation in the apoE knockout infused with apoA-I liposome.	61
Subcutaneous IL-1 induced endothelial adhesion molecule expression following subcutaneous injection of exogenous rHDL.	Reduced subcutaneous endothelial E-selectin expression following subcutaneous rHDL injection.	62

Thus, the balance of evidence suggests that the HDL-mediated inhibition of endothelial cell adhesion molecule expression observed *in vitro* also operates *in vivo*.

14.10 Conclusion

The ability of HDL to inhibit the cytokine-induced expression of adhesion molecules in endothelial cells has been demonstrated both *in vitro* and *in vivo*. The precise mechanism is not known but appears to involve the ability of HDL to inhibit endothelial sphingosine kinase. The magnitude of the inhibition varies with changes to the phospholipid composition of the HDL, but appears to be unaffected by the composition of other HDL lipids or apolipoproteins. These studies have a number of potential therapeutic implications. For example, strategies designed to increase the concentration of the HDL_3 sub-fraction of HDL or to increase the PLPC content of HDL may be highly anti-atherogenic. The studies also identify sphingosine kinase as a possible pro-atherogenic enzyme that may be a potential therapeutic target for inhibition not only by HDL but also by other agents. The recent *in vivo* results clearly indicate that this is an area that warrants much further investigation.

References

1. Gordon, D.J., Probstfield, J.L., Garrison, R.J., Neaton, J.D., Castelli, W.P., Knoke, J.D., Jacobs, D.R., Bangdiwala, S., and Tyroler, H.A. (1989). High-density lipoprotein cholesterol and cardiovascular disease. Four prospective American studies. *Circulation*, **79**, 8.
2. Gordon, T., Kannel, W.B., Castelli, W.P., and Dawber, T.R. (1981). Lipoproteins, cardiovascular disease, and death. The Framingham Study. *Arch. Intern. Med.*, **141**, 1128.
3. Miller, M., Seidler, A., Kwiterovich, P.O., and Pearson, T.A. (1992). Long-term predictors of subsequent cardiovascular events with coronary artery disease and "desirable" levels of plasma total cholesterol. *Circulation*, **86**, 1165.
4. Wilson, P.W.F., Abbot, R.D., and Castelli, W.P. (1988). High density lipoprotein cholesterol and mortality. The Framingham Heart Study. *Arteriosclerosis*, **8**, 737.
5. Pekkanen, J., Linn, S., Heiss, P.H.G., Suchindran, C.M., Leon, A., and Rifkind, B.M. (1990). Ten-year mortality from cardiovascular disease in relation to cholesterol level among men with and without pre-existing cardiovascular disease. *N. Engl. J. Med.*, **322**, 1700.
6. Assmann, G. and Schulte, H. (1993). Results and conclusions of the Prospective Cardiovascular Münster (PROCAM) study. In *Lipid metabolism disorders and coronary heart disease* (ed. G. Assman), p. 19. MMV Medizin Verlag, Munich.
7. Manninen, V., Elo, O., Frick, M.H., Happa, K., Heinonen, O.P., Heinsalmi, P., Helo, P., Huttunen, J.K., Kaitaniemi, P., Koskinen, P., Maenpaa, H., Malkonen, M., Manttari, M., Norola, S., Pasternack, A., Pikkarainen, J., Romo, M., Sjoblom, T., and Nikkila, E.A. (1988). Lipid alterations and decline in the incidence of coronary heart disease in the Helsinki Heart Study. *JAMA*, **260**, 641.
8. Duverger, N., Kruth, H., Emmanuel, F., Caillaud, J.-M., Viglietta, C., Castro, G., Taileux, A., Fievert, C., Fruchart, J.-C., Houdebine, L.M., and Denefle, P. (1996). Inhibition of atherosclerosis development in cholesterol-fed human apolipoprotein A-I-transgenic rabbits. *Circulation*, **94**, 713.
9. Pastzy, C., Maeda, N., Verstuyft, J., and Rubin, E.M. (1994). Apolipoprotein AI transgene corrects apolipoprotein E deficiency induced atherosclerosis in mice. *J. Clin. Invest.*, **94**, 899.

10. Rubin, E.M., Krauss, R.M., Spangler, E.A., Verstuyft, J.G., and Clift, S.M. (1991). Inhibition of early atherogenesis in transgenic mice by human apolipoprotein AI. *Nature*, **53**, 265.

11. Fielding, C.J. and Fielding, P.E. (1995). Molecular physiology of reverse cholesterol transport. *J. Lipid Res.*, **36**, 211.

12. Darbon, J.M., Tournier, J.F., Tauber, J.P., and Bayard, F. (1986). Possible role of protein phosphorylation in the mitogenic effect of high density lipoproteins on cultured vascular endothelial cells. *J. Biol. Chem.*, **261**, 8002.

13. Baumberger, C., Ulevitch, R.J., and Dayer, J.M. (1991). Modulation of endotoxic activity of lipopolysaccharide by high-density lipoprotein. *Pathobiology*, **59**, 378.

14. Tahri-Jouti, M.A. and Chaby, R. (1991). Binding of endotoxin to macrophages: distinct effects of serum constituents. *Immunol. Invest.*, **20**, 377.

15. Rosenson, R.S. and Lowe, G.D. (1998). Effects of lipids and lipoproteins on thrombosis and rheology. *Atherosclerosis*, **140**, 271.

16. Griffin, J.H., Kojima, K., Banka, C.L., Curtiss, L.K., and Fernandez, J.A. (1999). High-density lipoprotein enhancement of anticoagulant activities of plasma protein S and activated protein C. *J. Clin. Invest.*, **103**, 219.

17. Fleisher, L.N., Tall, A.R., Witte, L.D., Miller, R.W., and Cannon, P.J. (1982). Stimulation of arterial endothelial cell prostacyclin synthesis by high density lipoproteins. *J. Biol. Chem.*, **257**, 6653.

18. Yui, Y., Aoyama, T., Morishita, H., Takahashi, M., Takatsu., Y., and Kawai, C. (1988). Serum prostacyclin stabilising factor is identical to apolipoprotein A-I (Apo A-I). A novel function of apo A-I. *J. Clin. Invest.*, **82**, 803.

19. Ko, Y., Haring, R., Stiebler, H., Wieczorek, A.J., Vetter, H., and Sachindis, A. (1993). High-density lipoprotein reduces epidermal growth factor-induced DNA synthesis in vascular smooth muscle cells. *Atherosclerosis*, **99**, 253.

20. Zeiher, A.M., Schachlinger, V., Hohnloser, S.H., Saurbier, B., and Just, H. (1994). Coronary atherosclerosis wall thickening and vascular reactivity in humans. Elevated high-density lipoprotein levels ameliorate abnormal vasoconstriction in early atherosclerosis. *Circulation*, **89**, 2525.

21. Banka, C.L. (1996). High density lipoprotein and lipoprotein oxidation. *Curr. Opin. Lipidol.*, **7**, 139.

22. Parthasarathy, S., Barnett, J., and Fong, L.G. (1990). High-density lipoprotein inhibits the oxidative modification of low-density lipoprotein. *Biochim. Biophys. Acta*, **1044**, 275.

23. Navab, M., Imes, S.S., Hama, S.Y., Hough, G.P., Ross, L.A., Bork, R.W., Valente, A.J., Berliner, J.A., Drinkwater, D.C., Laks, H., and Fogelman, A.M. (1991). Monocyte transmigration induced by modification of low density lipoprotein in cocultures of human aortic wall cells is due to induction of monocyte chemotactic protein 1 synthesis and is abolished by high density lipoprotein. *J. Clin. Invest.*, **88**, 2039.

24. Cockerill, G.W., Rye, K.-A., Gamble, J.R., Vadas, M.A., and Barter, P.J. (1995). High-density lipoproteins inhibit cytokine-induced expression of endothelial cell adhesion molecules. *Arterioscler. Thromb. Vasc. Biol.*, **15**, 1987.

25. Ashby, D.T., Rye, K.-A., Clay, M.A., Vadas, M.A., Gamble, J.R., and Barter, P.J. (1998). Factors influencing the ability of high density lipoproteins to inhibit the expression of vascular cell adhesion molecule-1 in endothelial cells. *Arterioscler. Thromb. Vasc. Biol.*, **18**, 1450.

26. Baker, P.W., Rye, K.-A., Gamble, J.R., Vadas, M.A., and Barter, P.J. (1999). Ability of reconstituted high density lipoproteins to inhibit cytokine-induced expression of vascular cell adhesion molecule-1 in human umbilical vein endothelial cells. *J. Lipid Res.*, **40**, 1.

27. Xia, P., Vadas, M.A., Rye, K.-A., Barter, P.J., and Gamble, J.R. (1999). High density lipoproteins (HDL) interrupt the sphingosine kinase signalling pathway. *J. Biol. Chem.*, **274**, 33143.

28. Cockerill, G.W., Saklatvala, J., Ridley, S.H., Yarwood, H., Miller, N.E., Oral, B., Nithyanathan, S., Taylor, G., and Haskard, D.O. (1999). High-density lipoproteins differentially modulate cytokine-induced expression of E-selectin and cyclooxygenase-2. *Arterioscler. Thromb. Vasc. Biol.*, **19**, 910.

29. Maier, J.A.M., Barenghi, L., Pagani, F., Bradmante, S., Comi, P., and Ragnotti, G. (1994). The protective role of high-density lipoprotein on oxidized-low-density-lipoprotein-induced U937/endothelial cell interactions. *Eur. J. Biochem.*, **221**, 35.

30. O'Brien, K.D. (1994). The biology of the artery wall in atherogenesis. *Med. Clin. North Am.*, **78**, 41.

31. Faruqi, R.M. and DiCorleto, P.E. (1993). Mechanisms of monocyte recruitment and accumulation. *Am. Heart J.*, **69**, S19.

32. Klein, C., Kohler, H., Bittinger, F., Wagner, M., Hermanns, I., Grant, K., Lewis, J.C., and Kirkpatrick, C.J. (1994). Comparative studies on vascular endothelium *in vitro*. I. Cytokine effects on the expression of adhesion molecules by human umbilical vein, saphenous vein and femoral artery endothelial cells. *Pathobiology*, **62**, 199.

33. Faull, R.J. (1995). Adhesion molecules in health and disease. *Aust. N.Z. J. Med.*, **25**, 720.

34. Peter, K., Nawroth, P., Condradt, C., Nordt, T., Weiss, T., Boehme, M., Wunsch, A., Allenberg, J., Kubler, W., and Bode, C. (1997). Circulating vascular cell adhesion molecule-1 correlates with the extent of human atherosclerosis in contrast to circulating intracellular adhesion molecule-1, E-selectin, P-selectin, and thrombomodulin. *Arterioscler. Thromb. Vasc. Biol.*, **17**, 505.

35. Rohde, L.E., Lee, R.T., Rivero, J., Jamacochian, M., Arroyo, L.H., Briggs, W., Rifai, N., Libby, P., Creager, M.A., and Ridker, P.M. (1998). Circulation cell adhesion molecules are correlated with ultrasound-based assessment of carotid atherosclerosis. *Arterioscler. Thromb. Vasc. Biol.*, **18**, 1765.

36. Morisaki, N., Saito, I., Tamura, K., Tashiro, J., Masuda, M., Kanzaki, T., Watanabe, S., Masuda, Y., and Saito, Y. (1997). New indices of ischemic heart disease and aging: studies on the serum levels of soluble intracellular adhesion molecule-1 (ICAM-1) and soluble vascular cell adhesion molecule-1 (VCAM-1) in patients with hypercholesterolemia and ischemic heart disease. *Atherosclerosis*, **131**, 43.

37. Ridker, P.M., Hennekens, C.H., Roltman-Johnson, B., Stampfer, M.J., and Allen, J. (1998). Plasma concentration of soluble intracellular adhesion molecule 1 and risks of future myocardial infarction in apparently healthy men. *Lancet*, **351**, 88.

38. Li, H., Cybulsky, M.I., Gimbrone, M.A., and Libby, P. (1993). An atherogenic diet rapidly induces VCAM-1, a cytokine-regulable mononuclear leukocyte adhesion molecule, in rabbit aortic endothelium. *Arterioscler. Thromb.*, **13**, 197.

39. Sakai, A., Kume, N., Nishi, E., Tanoue, K., Miyasaka, M., and Kita, T. (1997). P-selectin and vascular cell adhesion molecule-1 are focally expressed in aortas of

hypercholesterolemic rabbits before intimal accumulation of macrophages and T lymphocytes. *Arterioscler. Thromb. Vasc. Biol.*, **17**, 310.

40. Richardson, M., Hadcock, S.J., DeReske, M., and Cybulsky, M.I. (1994). Increased expression *in vivo* of VCAM-1 and E-selectin by the aortic endothelium of normolipemic and hyperlipemic diabetic rabbits. *Arterioscler. Thromb.*, **14**, 760.

41. Xia, P., Gamble, J.R., Rye, K.-A., Wang, L., Hii, C.S.T., Cockerill, P., Khew-Goodall, Y., Bert, A.G., Barter, P.J., and Vadas, M.A. (1998). Tumor necrosis factor-α induces adhesion molecule expression through the sphingosine kinase pathway. *Proc. Natl. Acad. Sci. USA*, **95**, 14196.

42. Gerritsen, M.E., Carley, W.W., Ranges, G.E., Shen, C.-P., Phan, S.A., Ligon, G.F., and Perry, C.A. (1995). Flavonoids inhibit cytokine induced endothelial cell adhesion protein gene expression. *Am. J. Pathol.*, **147**, 278.

43. Wölle, J., Hill, R.R., Ferguson, E., Devall, L.J., Trivedi, B.K., Newton, R.S., and Saxena, U. (1996). Selective inhibition of tumor necrosis factor-induced vascular cell adhesion molecule-1 gene expression by a novel flavonoid. Lack of effect on transcription factor NF-κB. *Arterioscler. Thromb. Vasc. Biol.*, **16**, 1501.

44. Martin, A., Foxall, T., Blumberg, J.B., and Meydani, M. (1997). Vitamin E inhibits low-density lipoprotein-induced adhesion of monocytes to human aortic endothelial cells *in vitro*. *Arterioscler. Thromb. Vasc. Biol.*, **17**, 429.

45. De Caterina, R. and Libby, P. (1996). Control of endothelial leukocyte adhesion molecules by fatty acids. *Lipids*, **31**, S57.

46. De Caterina, R., Bernini, W., Carluccio, M.A., Liao, J.K., and Libby, P. (1998). Structural requirements for inhibition of cytokine-induced endothelial activation by unsaturated fatty acids. *J. Lipid Res.*, **39**, 1062.

47. Charo, I.F. (1992). Monocyte-endothelial expression of a mononuclear leukocyte adhesion molecule during atherogenesis. *Curr. Opin. Lipidol.*, **3**, 335.

48. Rollins, B.J. (1991). JE/MCP-1: an early-response gene encodes a monocyte-specific cytokine. *Cancer Cells*, **3**, 517.

49. Matz, C.E. and Jonas, A. (1982). Micellar complexes of human apolipoprotein A-I with phosphatidylcholines and cholesterol prepared from cholate-lipid dispersions. *J. Biol. Chem.*, **257**, 4535.

50. Rye, K.-A., Garrety, K.H., and Barter, P.J. (1993). Preparation and characterization of spheroidal, reconstituted high-density lipoproteins with apolipoprotein A-I only or with apolipoprotein A-I and A-II. *Biochim. Biophys. Acta*, **1167**, 316.

51. Rye, K.-A., Hime, N.J., and Barter, P.J. (1996). The influence of sphingomyelin on the structure and function of reconstituted high density lipoproteins. *J. Biol. Chem.*, **217**, 4243.

52. Rye, K.-A., Hime, N.J., and Barter, P.J. (1997). Evidence that cholesterol ester transfer protein-mediated reductions in reconstituted high density lipoprotein size involve particle fusion. *J. Biol. Chem.*, **272**, 3953.

53. Bagdade, J.D., Buchanan, W.F., Pollare, T., and Lithell, H. (1995). Abnormal lipoprotein phospholipid composition in patients with essential hypertension. *Atherosclerosis*, **117**, 209.

54. Subbaiah, P.V. and Haydn Pritchard, P. (1989). Molecular species of phosphatidylcholine in familial lecithin–cholesterol acyltransferase deficiency: effect of enzyme supplementation. *Biochim. Biophys. Acta.*, **1003**, 145.

55. James, M.J., Gibson, R.A., D'Angelo, M., Neumann, M.A., and Cleland, L.G. (1993). Simple relationships exist between dietary linoleate and the n-6 fatty acids of human neutrophils and plasma. *Am. J. Clin. Nutr.*, **58**, 497.

56. Cleland, L.F., James, M.J., Neumann, M.A., D'Angelo, M., and Gibson, R.A. (1992). Linoleate inhibits EPA incorporation from dietary fish oil supplements in human subjects. *Am. J. Clin. Nutr.*, **55**, 395.

57. Hodge, J., Sanders, K., and Sinclair, A.J. (1993). Differential utilization of eicosapentaenoic acid and docosahexaenoic acid in human plasma. *Lipids,* **28**, 525.

58. Garner, B., Waldeck, A.R., Witting, P.K., Rye, K.-A., and Stocker, R. (1998). Oxidation of high density lipoproteins. II. Evidence for direct reduction of lipid hydroperoxides by methionine residues of apolipoproteins A-I and A-II. *J. Biol. Chem.*, **273**, 6088.

59. Dansky, H.M., Charlton, S.A., Barlow, C.B., Tamminen, M., Smith, J.D., Frank, J.S., and Breslow, J.L. (1999). ApoA-I inhibits foam cell formation in apoE-deficient mice after monocyte adherence to endothelium. *J. Clin. Invest.*, **104**, 31.

60. Theilmeier, G., Van Veldhoven, P.P., Michiels, C., Collen, D., Himpens, B., and Holvoet, P. (1999). Human-like HDL – *in vivo* – reduces oxidative stress, Ca^{2+} transients, adhesion molecule expression and macrophage homing in apoE deficient mice. *Circulation*, **100** (Suppl.), I-189.

61. Dimayuga, P., Zhu, J., Oguchi, S., Chyu, K.-Y., Xu, X.H., Yano, J., Shah, P.K., Nilsson, J., and Cercek, B. (1999). Reconstituted HDL containing human apolipoprotein A-I reduces VCAM-1 expression and neointima formation following periadventitial cuff-induced carotid injury in apoE null mice. *Biochem. Biophys. Res. Commun.*, **264**, 465.

62. Cockerill, G.W., Huehns, T.Y., Miller, N.E., and Haskard, D.O. (1999). Exogenous reconstituted high-density lipoprotein (HDL) discoidal particles ablate the IL-1-induced expression of E-selectin in microvascular endothelium in a porcine model of acute inflammation *in vivo. Circulation*, **100** (Suppl.), I-813.

15 Protein oxidation in atherogenesis

Shanlin Fu and Roger T. Dean
Cell Biology Group, The Heart Research Institute, Camperdown,
Sydney, NSW 2050, Australia

Michael J. Davies
EPR Group, The Heart Research Institute, Camperdown,
Sydney, NSW 2050, Australia

Jay W. Heinecke
Division of Atherosclerosis, Nutrition and Lipid Research, Washington University
School of Medicine, St Louis, Missouri, 63110, USA

15.1 Introduction: a brief synopsis of current knowledge on protein oxidation

Protein oxidation by free radicals and other oxidants has been studied since the beginning of the twentieth century (Ref. (1), reviewed in Ref. (2)). Because of constraints on space and numbers of references, we largely reference substantial review articles and the sole monograph available on protein oxidation, and concentrate on primary references pertaining to the central area of protein oxidation in atherosclerosis.

The basic molecular consequences of protein oxidation are now quite well understood, and include amino acid modification, polypeptide fragmentation and cross-linking (reviewed in Refs (2–5)). This complex of reactions leads to protein unfolding, alteration or loss of protein activity, and altered susceptibility to proteolytic enzymes, depending upon the degree of oxidation. It is now also well documented that oxidized moieties on proteins are not always inert, but often reactive and able to initiate secondary damage to proteins and other biomolecules.[2, 5–7] Radical-mediated protein oxidation has been associated with many physiological and pathological events such as aging, cancer, cataract, diabetes, as well as atherosclerosis, although its exact roles in such processes are still largely unknown.[2, 5, 8–13]

15.1.1 Chemistry

The most studied biologically relevant oxidants include free radical species (one-electron oxidants), notably hydroxyl radical, and non-radical species (two-electron oxidants) such as hypochlorite and peroxynitrite.

15.1.1.1 One-electron oxidation

The hydroxyl radical is the most reactive and damaging among the numerous radicals which may be of biological significance. A major source for its formation is from the transition metal ion-catalysed Fenton–Haber–Weiss reactions involving superoxide radicals, the primary free radicals generated in most oxygenated biological systems as a result of electron leakages from the electron transport chains of mitochondria, chloroplasts and the endoplasmic reticulum. The hydroxyl radical exerts a diverse action on proteins, ranging from modification of amino acid side chains (e.g. hydroxylation and peroxidation) and backbone cleavage, to both intra- and inter-molecular cross-linking.

Aromatic amino acids have long been regarded as a major 'sink' for hydroxyl radicals due to their high reaction rates towards them.[14] One characteristic reaction of such radical attack is hydroxylation of the aromatic ring. For instance, reaction of hydroxyl radical with phenylalanine gives rise to p-, m- and o-tyrosines (Fig. 15.1).[15] While p-tyrosine is the biologically 'native' tyrosine, both m- and o-tyrosines have been widely used as biomarkers for assessing *in vivo* protein oxidation (reviewed in Ref. (16)). In the case of tyrosine, 3,4-dihydroxyphenylalanine (DOPA) is the predominant hydroxylated product and can be generated both in the presence and absence of oxygen (Fig. 15.2). Reaction of Tyr with hydroxyl, and a wide variety of other radicals, also generates tyrosyl radicals which can undergo radical–radical coupling reactions to form o,o'-dityrosine (assuming they are vicinal to each other), especially if there are no reductants (e.g. thiols, vitamin E) present to repair the tyrosyl radicals (reviewed in Refs (2, 5)). Free L-tyrosine is converted, in the presence of H_2O_2, to o,o'-dityrosine by myeloperoxidase secreted from activated human phagocytes, via the formation of tyrosine phenoxyl radical.[17–19] Plasma concentrations of Cl^- have been shown to inhibit the formation

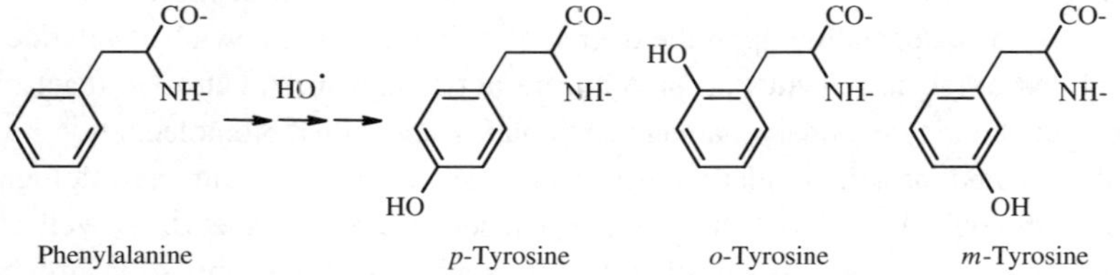

Phenylalanine   p-Tyrosine   o-Tyrosine   m-Tyrosine

Fig. 15.1 Formation of tyrosine analogues from radical-mediated oxidation of phenylalanine residues.

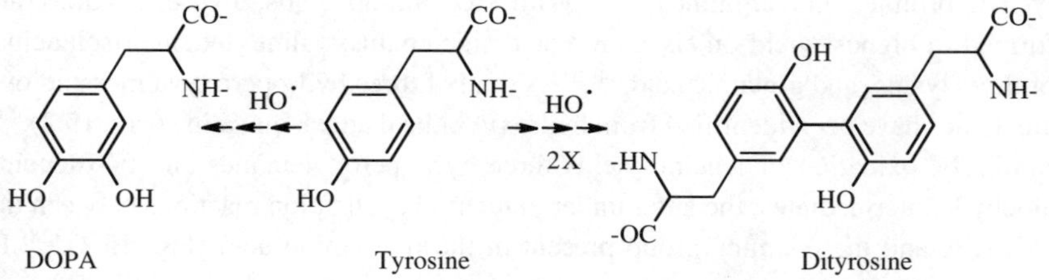

CO-
NH-
HO·
HO OH
DOPA
CO-
NH-
HO·
HO
Tyrosine
2X   -HN
-OC
OH
CO-
NH-
HO
Dityrosine

Fig. 15.2 Oxidation products of tyrosyl residues in proteins from the action of hydroxyl radicals.

of o,o'-dityrosine modestly, and when amino acids, Cl^- and L-tyrosine were all present in the reaction mixture at physiologically relevant concentrations, 30% of the H_2O_2 was converted to o,o'-dityrosine.[17] Thus, free L-tyrosine in human blood plasma may be an important substrate for myeloperoxidase.

o,o'-Dityrosine cross-links have been shown to be abundant in the fertilization envelope of the nascent sea urchin embryo[20] and the spore wall coat of the yeast *Saccharomyces cerevisiae*,[21] suggesting that myeloperoxidase may induce o,o'-dityrosine cross-links in proteins. Our results indicate that the myeloperoxidase–H_2O_2 system of activated neutrophils can use L-tyrosine to form o,o'-dityrosine adducts in proteins.[18] The requirement for L-tyrosine suggests that the tyrosyl radical serves as a diffusible catalyst for the reaction. o,o'-Dityrosine is perhaps the best characterized protein cross-link formed as a result of oxidative damage. Hydroxyl radicals are also known to attack tryptophan and histidine residues to form a complex mixture of oxidation products including hydroxylated compounds, carbonyl functions, and ring-opened products (reviewed in Refs (2, 16)).

Sulphur-containing amino acids are also important sites for radical attack.[22] Reaction of cysteine with free radicals such as hydroxyl radicals occurs extremely rapidly and preferentially at the sulphur centre to give a thiyl radical ($RS^•$) by hydrogen abstraction. The thiyl radical can subsequently react further to give species such as $RSOO^•$, $RSOOH$, disulphides, and oxyacids.[23] In the absence of oxygen, thiyl radicals dimerize to form S–S bridges, leading to protein cross-linking. The thioether side chain of methionine is also readily damaged by a wide variety of oxidants, yielding products such as the sulphoxide and sulphone.[24]

Reaction of radicals with side chains of aliphatic amino acid residues in the presence of oxygen gives rise to a multitude of products, including hydroperoxides, alcohols, and carbonyl compounds (reviewed in Refs (2, 16)). The initial reaction involves the formation of a carbon-centred radical species (via hydrogen atom abstraction), which subsequently reacts rapidly with oxygen to give a peroxyl radical. The chemistry of such peroxyl radicals is relatively well understood and gives rise to oxygenated products via hydrogen abstraction, fragmentation, and dimerization reactions (reviewed in Refs (2, 3, 15)). In recent studies, the yields and nature of some of these materials have been studied in some detail for valine, leucine, lysine, proline, and arginine.[25–29] With free amino acids, hydroperoxides are formed in highest yield on six amino acid side chains: valine, leucine, isoleucine, proline, lysine, and glutamic acid.[30, 31] A total of three hydroperoxyvalines and two aldehydes have been identified from hydroxyl radical attack on valine (Fig. 15.3).[26] Similarly, oxidation of leucine yields three hydroperoxyleucines and two leucine aldehyde intermediates; the latter undergo internal cyclization reactions between the aldehyde and the α-amine group present in the free amino acid (Fig. 15.4).[28] In

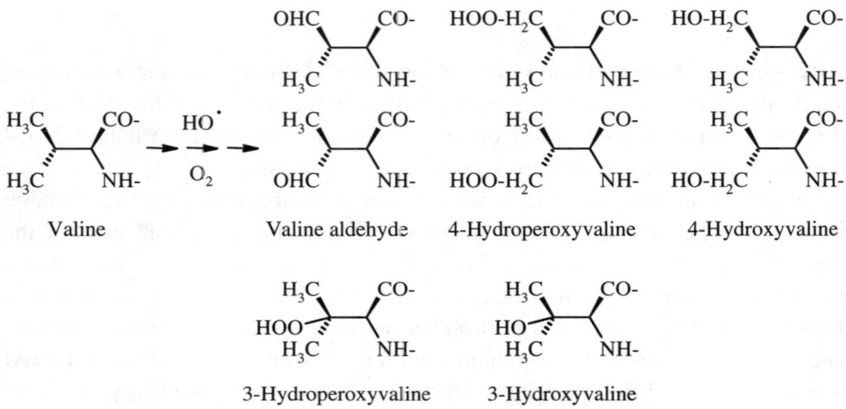

Fig. 15.3 Oxidation products of valine residues in proteins from the action of hydroxyl radicals.

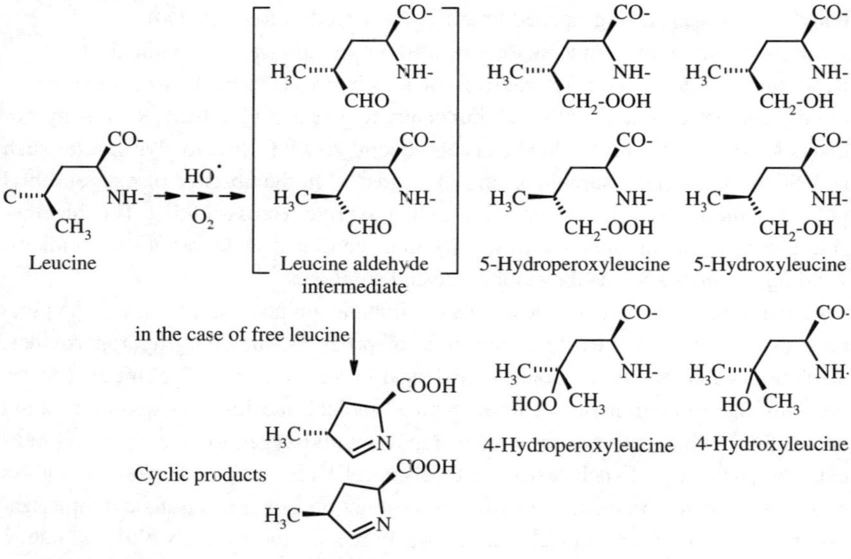

Fig. 15.4 Oxidation products of leucine in proteins from the action of hydroxyl radicals.

lysine, hydroxyl radical attack at positions 3, 4, and 5 yields the corresponding hydroxylated compounds,[29] though only the first of these can be used as a quantitative marker of damage as both 4- and 5-hydroxylysines are natural products.[16] Another novel hydroxylated product, 5-hydroxy-2-aminovaleric acid, has also been identified in radical-damaged proteins.[27, 32] This substance is thought to arise from the oxidation of proline and arginine residues, involving ring opening for the former, and loss of the guanidinium group for the latter.

Hydroxyl and other oxyl radicals are also known to attack the α-carbon site along the protein backbone, giving rise to backbone peroxyl radicals in the presence of oxygen.[33] Subsequent reactions of these peroxyl radicals lead to the formation of backbone hydroperoxides,[34] fragmentation of the protein backbone, and generation of new moieties such as dicarbonyl groups and amide groups (Fig. 15.5; reviewed in Refs (2, 3, 34, 35)).

15.1.1.2 Two-electron oxidation

Two important classes of non-radical oxidants that can induce protein oxidation via the two-electron oxidation process are hypochlorite and peroxynitrite. Hypochlorite can be generated *in vivo* by the action of myeloperoxidase on hydrogen peroxide in the presence of chloride ion; this species is in equilibrium with hypochlorous acid at physiological pH values.[36] Hypochlorite is known to react rapidly with a number of amino acid side chains, including Cys (to give disulphides and oxygenated products), Met (to give the sulphoxide), and Lys (to give chloramine species).[37, 38] The chloramines are not stable products, and decompose to give nitrogen-centred radicals.[39, 40] Hypochlorite also reacts with tyrosine, yielding 3-chlorotyrosine and 3,5-dichlorotyrosine (Fig. 15.6) in addition to *o,o'*-dityrosine.[17, 41–43] Similar chlorination reactions would be expected with other aromatic side chains in proteins, though this has not been extensively studied as yet. When free tyrosine is exposed to hypochlorite, 4-hydroxyphenylacetaldehyde becomes the predominant

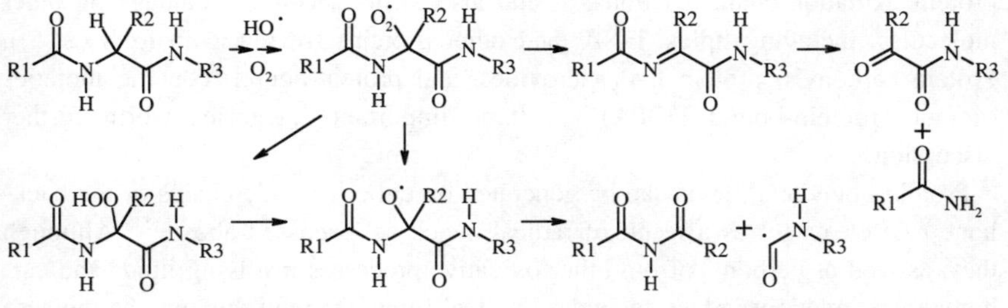

Fig. 15.5 Proposed mechanisms for radical-induced protein strand fragmentation involving peroxyl and alkoxyl radical species.

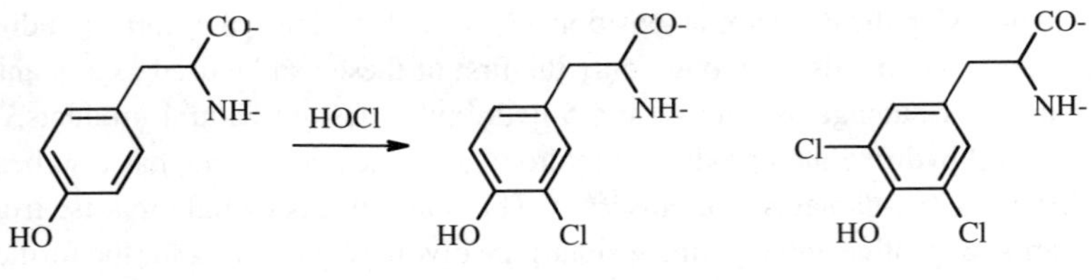

CO-   CO-   CO-

NH-   NH-   NH-

HOCl   Cl—

HO   HO   Cl   HO   Cl

Tyrosine   3-Chlorotyrosine   3,5-Dichlorotyrosine

Fig. 15.6 Oxidation reactions of tyrosyl residues in proteins with hypochlorite.

product.[43, 44] This material is not generated from tyrosyl residues in proteins, perhaps because this reaction requires chloramine formation at the free amino group of the amino acid.[45]

Peroxynitrite is formed from the fast reaction between superoxide and nitric oxide radicals. Nitric oxide is an important biological signalling molecule and is a relatively unreactive radical. In contrast, peroxynitrite reacts with many protein moieties.[46] One well-characterized product is 3-nitrotyrosine,[47–49] a nitration product of tyrosine, though it is not clear whether this is a consequence of radical or non-radical processes. *o,o'*-Dityrosine and 3,5-dinitrotyrosine can also be produced (reviewed in Ref. (16)). It should be borne in mind that nitration of tyrosine is not unique to peroxynitrite, since other reactive nitrogen species can also be effective (reviewed in Ref. (16)).

15.1.2 Secondary reactions of initial products of protein oxidation

The products of protein oxidation are not all inert and non-reactive. Rather, some are capable of further reactions, which may lead to adduct formation or generate secondary radical fluxes. These reactive intermediates may therefore be part of protein oxidation chain reactions,[50] and also inflict secondary damage on other molecules, including lipids, DNA, and other proteins (reviewed in Refs (2, 5)). Protein carbonyls, protein hydroperoxides, and protein-bound reducing moieties (notably protein-bound DOPA) are three important categories worth further discussion.

Protein-bound carbonyls can be generated on either the side chains or the backbone (via cleavage) as a result of radical attack as discussed above.[51] Although they, as well as carbonyls from other oxidative processes involving lipids and carbohydrates, may not induce secondary radical fluxes (though this may be the case in certain scenarios[52]), they do form Schiff bases with vicinal amino groups, and can give rise to certain other cross-links (reviewed in Refs (2, 5, 12)). One example of such a reaction is the formation of ring structures during the oxidation of leucine (Fig. 15.4).[28, 53, 54] The relevance of such reactions in intact proteins is unknown.

Protein hydroperoxides are capable of oxidizing, and hence depleting, important cellular reductants such as glutathione and ascorbate, with their concomitant conversion to the corresponding protein hydroxides.[55] Transition metals, glutathione reductase, and cellular systems can also reduce amino acid and protein hydroperoxides to hydroxides.[55] Secondary radical fluxes (including carbon-centred, alkoxyl, and peroxyl species) can result from the reaction of protein and amino acid hydroperoxides with transition metal ions; these have been detected by EPR spin trapping.[34, 56] Figure 15.7 summarizes the key radicals detected, and the reactions that are believed to account for the observed products. These secondary reactions occur with both side chains and α-carbon hydroperoxides. In the latter case, protein main chain fragmentation is one of the consequences, with the α-carbon alkoxyl radical as a key intermediate.[34] There is experimental evidence to demonstrate that the secondary radical fluxes generated during protein hydroperoxide degradation can damage other molecules, including DNA.[57]

Protein-bound reducing moieties are able to reduce transition metal ions and metalloproteins such as cytochrome c.[30, 58] A large proportion of this reductive activity is due to protein-bound DOPA, formed by hydroxylation of tyrosine (Fig. 15.2).[58] DOPA, being a catechol, is prone to further oxidation and this can result in the generation of both quinone and cyclized products, and the release of up to six electrons (Fig. 15.8).[58] This susceptibility to further oxidation, which is particularly marked in the presence of redox active metal ions, can result in further radical formation and damage of other biomolecules such as DNA.[59] DOPA quinone can also undergo Michael addition reaction with nucleophiles such as thiols, and this can result in the formation of adduct species such as 5-cysteinyl-DOPA.[60] It is not known whether such reactions contribute significantly to protein cross-linking.

$$ROOH + e^- \longrightarrow RO^{\bullet} + HO^-$$

$$RO^{\bullet} + R_1H \longrightarrow ROH + R_1^{\bullet}$$

$$R_1^{\bullet} + O_2 \longrightarrow R_1OO^{\bullet}$$

$$R_1OO^{\bullet} + R_2H \longrightarrow R_1OOH + R_2^{\bullet}$$

$$R(R_1)(R_2)CO^{\bullet} \longrightarrow R(R_1)CO + R_2^{\bullet}$$

$$R(R_1)HCO^{\bullet} \longrightarrow R(R_1)(HO)C^{\bullet}$$

$$R(R_1)(HO)C^{\bullet} + O_2 \longrightarrow R(R_1)(HO)COO^{\bullet}$$

$$R(R_1)(HO)COO^{\bullet} \longrightarrow R(R_1)CO + HOO^{\bullet}$$

$$R(R_1)(HO)C^{\bullet} + R_2H \longrightarrow R(R_1)(HO)CH + R_2^{\bullet}$$

Fig. 15.7 Postulated mechanism for the generation of secondary radical fluxes during the decomposition of protein hydroperoxides (adapted from Ref. (2)).

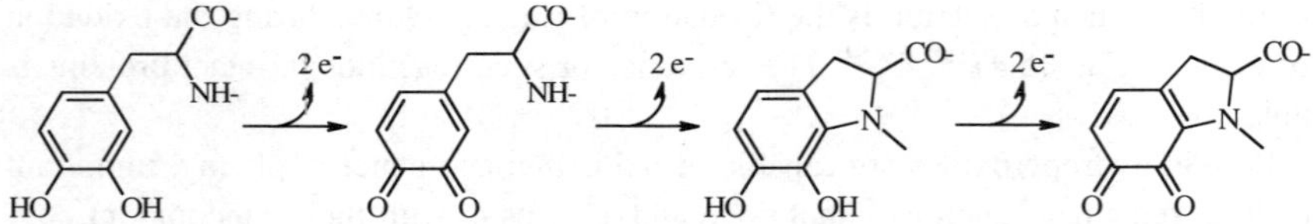

Fig. 15.8 Further oxidation of protein-bound DOPA may be a ready source for secondary radical generation involving redox cycling of transition metals.

In vitro, tyrosyl radicals generated by myeloperoxidase can initiate the peroxidation of low-density lipoprotein lipid by a pathway that is independent of free metal ions.[18] The reaction bears similarities to the enzymatic initiation of peroxidation of polyunsaturated fatty acids by cyclooxygenase, a process that involves tyrosyl radicals.[61, 62] Both *o,o'*-dityrosine formation and initiation of low-density lipoprotein oxidation by myeloperoxidase are greatly enhanced by plasma concentrations of tyrosine.[18, 19] Such reactions may take place in protected extracellular microenvironments where phagocytes are abundant and local antioxidants that scavenge tyrosyl radical may be depleted.

15.1.3 The biochemistry of protein oxidation

15.1.3.1 Limited protein oxidation

It has long been appreciated that functional inactivation is a common consequence of protein oxidation initiated by radical attack. Early anaerobic radiation studies showed that the hydroxyl radical was the most effective inactivator for lysozyme, ribonuclease, chymotrypsin, and other enzymes (e.g. Ref. (63), reviewed in Ref. (15)). Since 1981, Stadtman and colleagues have studied the inactivation of proteins in cell-free systems by metal ion-catalysed oxidations, usually involving metal ion-dependent autoxidations of ascorbate and/or hydrogen peroxide.[4, 64–66] Although metal ion-catalysed oxidation systems are mechanistically diverse, they all involve hydrogen peroxide, and produce other reactive intermediates akin to Fenton systems. They inactivate a wide range of proteins such as glutamine synthetase and cytochrome c. Despite the fact that many metal ion-catalysed oxidation systems generate a complex mixture of radical species, some systems can cause selective damage in cell-free systems, such as to the histidine residues in glutamate synthetase (reviewed in Ref. (67)). However, this modification is accompanied by other alterations, as indicated in this case by both protein fragmentation and by complex changes in hydrophobicity. Limited oxidation increases hydrophilicity of the enzyme, while further oxidation increases hydrophobicity.[68] In a careful study by Chevalier *et al.*[69] on the oxidative inactivation of enzymes in living *Klebsiella pneumoniae*, the authors noted that several enzymes, including glycerol dehydrogenase, were inactivated when the living organism was switched from anaerobic to aerobic conditions (or exposed to hydrogen peroxide). The glycerol dehydrogenase isolated from cells exposed to peroxide (90% inactivated) was compared with the native enzyme. There were no detectable differences in either subunit molecular size or amino acid composition (including sulphydryl groups). The lack of gross changes suggests that very limited modifications are sufficient for enzyme inactivation.[69] Non-enzymic proteins such as α_1-proteinase inhibitor and serpin (a neutrophil cytosolic serine-proteinase inhibitor) are also inactivated by metal ion-catalysed oxidation, possibly via selective oxidation of methionine residues.[70, 71] In contrast, Fenton systems inactivate α_1-proteinase inhibitor by generating a wide range of amino acid derivatives.[72]

A number of enzymes such as ribonucleotide reductase contain an intrinsic radical species while in their active form (reviewed in Refs (73, 74)). These radicals can be associated with tyrosine, tryptophan, glycine, or thiols. It has been reported that some radical versions of ribonucleotide reductase undergo reaction with oxygen which can cause enzyme inactivation by backbone cleavage rather than side chain modifications (reviewed in Ref. (74)). The inactivation can sometimes be reversed by loss of oxygen and regeneration of the original protein radical, as in the case of pyruvate formate lyase which contains an intrinsic glycyl α-carbon radical.[75, 76] Many oxidative enzymes, such as cyclooxygenase, lipoxygenases and the cytochrome P-450 family, can also generate radical species during their catalytic interactions with substrates, resulting in their 'suicide' or self-inactivation.[61, 77] Protein inactivation can also be triggered by radicals or carbonyls arising from other biological processes. For example, radicals produced during lipid autoxidation can inactivate α_1-proteinase inhibitor by oxidation of methionine-358.[78] Carbonyls derived from autoxidation of lipids[79] and sugars,[80] or from oxidation of amino acid residues[81] can inactivate proteins probably via Schiff base formation.

Limited protein oxidation is sometimes transiently reversible, and such reversal may cause protein reactivation. One of the earliest events during radical-mediated protein oxidation was the conversion of sulphydryl groups into disulphides and other oxidized species such as oxyacids, resulting in conformational changes. For instance, treatment of aldose reductase with oxidized glutathione (GSSG), an oxidant with limited oxidation reactions, leads to mixed disulphide formation, conformational alterations, and enzyme inactivation; this damage can be reversed by GSH.[82] Reversible S-thiolation has also been observed on proteins in cells exposed to radicals.[83] Evidence suggests that biological thiols such as GSH and cysteine can reactivate proteins either directly by reacting with radical fluxes, or indirectly by forming reversible bonds with normally free thiols on proteins.

The oxidation of methionine residues as in the proteinase inhibitors discussed above can be reversible in the early oxidation stages by two possible pathways. Firstly, the initial radical species derived from methionine by one-electron oxidation may be reduced by suitable electron donors (e.g. Trolox C, a water-soluble tocopherol analogue[84]). Secondly, the first stable oxidation product of methionine, methionine sulphoxide, can be reduced by methionine sulphoxide reductase, a specialized reductase present in many mammalian cells.[85] This process may be another biological control mechanism, in addition to the reversible thiolation phenomenon.

Schiff bases formed from the reaction of an amine with a carbonyl are unstable, and thus may play a role in reversible reactivation. However, the formed Schiff bases often undergo rapid Amadori rearrangement reactions, so that the importance of this reversal route is usually modest.

15.1.3.2 Extensive protein oxidation

Extensive oxidation of proteins is usually associated with increased unfolding, fragmentation, and polymerization. These changes are unlikely to be reversible, and usually lead to enhanced susceptibility to degradation by proteinases (reviewed in Refs (5, 86)). It is also known that more extensively oxidized proteins may be denatured and aggregated to such an extent that proteolytic access is restricted rather than enhanced, although the distinction between limited and extensive oxidation is an arbitrary division of a continuum.[86] It is important to distinguish the difference between altered susceptibility of oxidized proteins to proteolysis, and altered rates of intracellular proteolysis of these molecules, which is a function of the susceptibility plus the disposition within the cell. For instance, those proteinases shown to be active in cell-free systems are not necessarily the ones responsible for degradation within living cells. Issues of subcellular location, and of substrate and proteinase concentration are also important.[86] The supply of oxidized protein substrates via endocytosis is one of few approaches in which one can control the quantities of substrates, and investigate their catabolic rates and locations within the cell. Substantially oxidized BSA, or apoB of substantially oxidized LDL are poorly catabolized by cells after endocytosis and accumulate to an elevated level compared with their unoxidized native counterparts.[87]

The catabolism of intracellularly generated oxidized proteins is a more challenging issue to investigate (reviewed in Ref. (86)). Such degradation may require intracellular transport of degradable molecules, even from organelle to organelle. To define the absolute rates of catabolism and half-lives of oxidized proteins, knowledge of the pool sizes of such oxidized proteins is required.[86] Although microinjection techniques have been employed to deliver known quantities of oxidized proteins into cells, artificial conditions (such as microinjection itself, and inappropriate pool sizes) make them difficult to relate to normal conditions. In a recent effort to generate controlled amounts of endogenous 'oxidized' proteins without oxidative affront to the cell, purified oxidized amino acids have been used as substrates for intracellular protein synthesis (Rodgers *et al.*, unpublished). We have demonstrated the incorporation of oxidized amino acids including 3-hydroxyvaline, 5-hydroxyleucine, DOPA, *m*-tyrosine, and *o*-tyrosine into mouse macrophages in culture via protein synthesis, even in the presence of the parent amino acids. The levels achieved can be controlled. We are exploiting this as a promising approach to study the kinetics and mechanisms of degradation of oxidized proteins. Experiments to date utilizing radiolabelled oxidized amino acids have shown that the rates of protein degradation are enhanced for the oxidized species compared to the native counterparts, as was observed previously when amino acid analogues were incorporated into proteins.[88]

15.1.3.3 Control of protein oxidation

Cells have several protective mechanisms for limiting the extent of protein oxidation. The first line of defence is the restriction of radical generation by sequestration of transition metal ions. For example, the specialized proteins transferrin, ferritin, and ceruloplasmin control the availability of iron and copper, which are critical for radical generation and transformation. Non-specialized proteins such as albumin, the most abundant protein in blood plasma, can also bind metal ions by means of histidine, cysteine, and other residues. Radical generation and radical-mediated damage can thus either be restricted or localized so that other molecules may be protected.

A second preventive mechanism relies upon S-thiolation of proteins, as discussed above, by glutathione[83] and thiol-specific antioxidant proteins.[89] One such protein has been isolated, which may be an analogue of superoxide dismutase.[89]

The remaining primary defences are the enzymatic and the sacrificial antioxidants. The enzymatic antioxidants, such as superoxide dismutase, catalase, glutathione peroxidase and phospholipid peroxidase, act together to remove peroxides and intermediate radicals. The sacrificial antioxidants, such as aqueous ascorbate, bilirubin and urate, and lipophilic tocopherols and quinols, can limit peroxidation chains by scavenging and removing chain-carrying radical species. Because of the significant differences in the chemistry of lipid and protein oxidation (in which peroxyl and alkoxyl radicals, respectively, have key roles), it cannot be assumed that lipid antioxidants are effective against protein oxidation.[90] Indeed it seems possible that novel protein antioxidation mechanisms may exist.

Cells also have mechanisms for the detoxification of the reactive moieties on oxidized proteins, such as protein hydroperoxides, so that deleterious secondary radical reactions can be restricted. For example, protein hydroperoxides can be reduced efficiently by J774 mouse macrophage cells in culture to the unreactive hydroxides.[55]

15.2 Occurrence of protein oxidation in diseases other than atherogenesis

Protein oxidation has been reported to occur in many diseases, such as cataract, Alzheimer's, diabetes and inflammatory diseases (reviewed in Refs (2, 5)), in addition to atherosclerosis which will be discussed below. Specific protein oxidation products, such as hydroxyvaline, hydroxyleucine, DOPA, *o,o'*-dityrosine, *o*-tyrosine and *m*-tyrosine, accumulate in human cataractous lenses,[91] and are accompanied by loss of cysteine and methionine residues.[92] The levels of these oxidized products increase with the development of the disease by up to 15-fold in the case of DOPA. The product profile is similar to that induced by hydroxyl radical damage to proteins and different from that initiated by other oxidants such as hypochlorite and peroxynitrite.[91] More interestingly, such extensive protein oxidation including cross-linking could explain many features of the disease, such as protein aggregation, and lens colouration and opacification.

Alzheimer's disease is another example in which protein oxidation is regarded as central since the discovery of the accumulation of cytotoxic amyloid β-protein in diseased brains.[93] The accumulation of the amyloid proteins as neurofibrillary tangles can occur through oxidation,[94] and protein carbonyls and glycoxidation products have been shown to accumulate in Alzheimer's brains.[95] It has been generally accepted that radical metabolism and protein oxidation may contribute specifically to the progression of Alzheimer's disease, but it remains to be established whether protein oxidation is a key initiating factor.

15.3 Occurrence of protein oxidation in atherogenesis

Many lines of evidence support the hypothesis that oxidation of low-density lipoprotein (LDL), the major carrier of blood cholesterol, plays a role in the pathogenesis of atherosclerosis, though much of the evidence is indirect. One potential pathogenic mechanism involves the unregulated uptake of oxidized LDL by tissue macrophages, leading to the formation of foam cells.[96] If this occurs within the vessel walls, wall thickening and ultimately plaque formation may be consequent. Most studies on LDL oxidation have concentrated on oxidation of lipid. However, it is known that apolipoprotein B-100 (apoB), the sole protein component of most LDL particles, can also be oxidized.[97] It was not until recently that the two separate oxidation processes, i.e. direct lipid oxidation and direct protein oxidation, were distinguished. There is now ample evidence to indicate that direct protein oxidation takes place in LDL (reviewed in Refs (5, 9, 98)), in addition to the indirect protein modifications induced by secondary reactions of oxidized lipids or other molecules within the LDL particles,[99] but the kinetic relationships between protein and lipid oxidation still require close scrutiny.

In vitro experiments have shown that apoB in LDL can be readily oxidized by a variety of radicals, giving rise to hydroxy and hydroperoxy species such as DOPA, *o*-tyrosine, *m*-tyrosine,[100] and hydro(pero)xy-valines and leucines,[101] and that it can be fragmented by radiolysis[102] or metal ion-catalysed oxidation.[97] LDL exposed to tyrosyl radical generated by myeloperoxidase exhibits a selective increase in protein-bound *o,o'*-dityrosine, with no change in its content of *o*-tyrosine or *m*-tyrosine, two markers of damage induced by other oxidants.[100] It can also be oxidized by non-radical oxidants such as hypochlorite and peroxynitrite, yielding characteristic products, notably 3-chlorotyrosine and 3-nitrotyrosine, respectively.[42, 43, 103–105] *o,o'*-Dityrosine is a common oxidation product for each oxidation system investigated (reviewed in Ref. (16)). Native plasma LDL apoB also contains detectable amounts of oxidized amino acids, with a spectrum of relative abundance paralleling that which can be caused by the attack of hydroxyl radicals and similar oxidant systems.[101]

Chemical analysis of these oxidized species on proteins isolated from the atherosclerotic vessel walls may provide valuable information on the mechanisms of protein oxidation in atherogenesis. Significant advances in this regard have been made recently by the authors' laboratories. The levels of the oxidized amino acids in atherosclerotic tissues have been determined and carefully compared with those in control groups. The results obtained from these studies are summarized in Tables 15.1–15.3 (reviewed in Refs (16, 106)).

In one study, normal iliac arteries from liver transplant donors, and advanced human atherosclerotic plaques from patients undergoing carotid endarterectomy were examined. The levels of six oxidized amino acids (i.e. DOPA, *o,o'*-dityrosine, *o*-tyrosine, *m*-tyrosine, 3-hydroxyvaline, and 5-hydroxyleucine) in the intimal proteins are significantly elevated in plaques in comparison to the normal controls as determined by HPLC (Table 15.1).[107] The relative abundance of these oxidized products are in the order of DOPA > *o-/m*-tyrosine, *o,o'*-dityrosine > hydroxyvaline/leucine, similar to that obtained from *in vitro* oxidation of LDL or BSA by hydroxyl radicals (DOPA > *o-/m*-tyrosine > hydroxyvaline/leucine, *o,o'*-dityrosine).[107] The levels of 3-chlorotyrosine and 3-nitrotyrosine are below the detection limit of the method used (400 µmol per mol of tyrosine). The data suggest a role for oxy radical-mediated protein oxidation in atherogenesis and hence possibly a role for metal ion-catalysed chemistry.[107] The relatively higher level of *o,o'*-dityrosine in plaque than that observed after *in vitro* exposure of proteins to

Table 15.1 Protein oxidation products in intimal proteins of normal human arteries and advanced human atherosclerotic plaques*

Protein oxidation product	Normal artery	Carotid plaque
DOPA	464	1034
o,o'-Dityrosine	31	297
m-Tyrosine	22	121
o-Tyrosine	176	467
3-Hydroxyvaline	20	47
5-Hydroxyleucine	10	85

*Data are expressed as µmol of the oxidized amino acid per mol of the parent amino acid.[107]

hydroxyl radicals suggests the additional involvement of (at least one) other oxidant, such as tyrosyl radical, nitrogen dioxide radical, or hypochlorite.

In related studies, the levels of oxidized amino acids were measured in aortic arterial tissues from either surgical tissue or post-mortem donors (Table 15.2) and in LDL isolated from human atherosclerotic lesions (Table 15.3) by GS/MS methods.[42, 100, 105] Four types of arterial tissues (normal tissue, fatty streak, intermediate lesion, and advanced lesion) were obtained from each post-mortem donor. The results show significant elevations for o,o'-dityrosine, 3-chlorotyrosine, and 3-nitrotyrosine in lesions and lesion LDL when compared to the normal tissues and normal circulating LDL.[42, 104] A significant increase in the levels of o-tyrosine and m-tyrosine was only noted in the advanced lesions, and not in the fatty streaks

Table 15.2 Protein oxidation products in aortic arterial tissue at various stages of atherogenesis*

Protein oxidation product	Normal	Fatty streak	Intermediate lesion	Advanced lesion	Lesion
o-Tyrosine	280	310	400	760	
m-Tyrosine	120	120	140	250	
o,o'-Dityrosine	30	330	130	200	
3-Chlorotyrosine	80				420

*Data (from Refs (42,100)) are expressed as μmol of the oxidized amino acid per mol of the parent amino acid.

Table 15.3 Protein oxidation products in LDL of normal plasma and advanced human lesions*

Protein oxidation product	Normal plasma LDL	Lesion LDL
DOPA	640[†]	NA[§]
o-Tyrosine	270[‡], 220[†]	300[‡]
m-Tyrosine	130[‡], 130[†]	120[‡]
o,o'-Dityrosine	2[‡], 2[†]	250[‡]
3-Nitrotyrosine	10[‡]	840[‡]
3-Chlorotyrosine	10[‡]	300[‡]
3-Hydroxyvaline	30[†]	NA[§]
5-Hydroxyleucine	10[†]	NA[§]

*Data (from Refs (42, 100, 101,104)) are expressed as μmol of the oxidized amino acid per mol of the parent amino acid.
[†] Data from HPLC methods.
[‡] Data from GC-MS methods.
[§] Not available.

and intermediate lesions, when compared to vicinal sections of 'normal' tissues.[100] These observations suggest that metal ions are not the most likely candidate to promote LDL oxidation early in atherogenesis, at least in aortic tissue. The detection of an elevated level of *o,o'*-dityrosine suggests that other oxidative mechanisms also promote protein oxidation.

Atherosclerotic lesions removed at surgery show a marked increase in 3-chlorotyrosine in both lesion LDL and total atherosclerotic tissue.[42] This suggests that myeloperoxidase can mediate oxidation occurring during atherogenesis, consistent with immunohistochemical studies in which hypochlorite-oxidized proteins and apoB peptides were detected in plaques.[108] 3-Nitrotyrosine levels in LDL isolated from atherosclerotic lesions were 80-fold higher than in circulating LDL, an observation consistent with a role for reactive nitrogen species in modifying LDL *in vivo*.[104]

The data obtained from such studies using different analytical methods show a high degree of consistency and are complementary. It is likely that protein oxidation in atherogenesis is a complex process, featuring both radical oxidation (possibly metal ion-catalysed) and non-radical oxidation (e.g. via hypochlorite or peroxynitrite). However, hypochlorite cannot be the sole player in protein oxidation in atherogenesis, as it does not generate hydroxylated amino acids, such as hydroxyvaline and hydroxyleucine, which seem to be the best fingerprints for oxy radical/Fenton systems.[107] It is not yet known whether these hydroxylated aliphatic amino acids are elevated in early lesions, and therefore it is not yet clear which oxidation system(s) dominates in the early stages of atherogenesis. On the other hand, collective analysis of the currently available data (reviewed in Ref. (16)) seems to suggest that oxy radicals play a substantial role in protein oxidation in advanced plaques. Reactive nitrogen species are also involved in protein oxidation, as indicated by the occurrence of 3-nitrotyrosine in atherosclerotic lesions.[103, 104] Peroxynitrite has been previously regarded as responsible for nitrating tyrosines in protein, though it should be noted that myeloperoxidase (and other peroxidases) can also use nitrite, from the decomposition of nitric oxide or other sources, to generate potent nitrating intermediates.[109, 110]

It is worth emphasizing that all these determinations of tissue levels of oxidized protein moieties simply reflect levels at particular time points. These levels might be transient and do not necessarily indicate the relative rates of production and removal. The pool sizes of oxidized proteins in tissues are governed by many factors, such as proteolysis, re-utilization, and tissue distribution. It is easier to understand the tissue accumulation of oxidized proteins in cataractous lenses, since the lens proteins undergo little protein turnover, and thus damage accumulates. In the case of atherosclerosis and other diseases, understanding of cellular handling of oxidized proteins has become very important and is currently lacking.

Nevertheless, several lines of evidence can be gathered to argue that the observed elevations of protein oxidation products identified here are most likely consequent on atherogenesis rather than on aging. Firstly, even in long-lived proteins such as those of the lens, phenylalanine oxidation products do not show substantial or even often statistically significant age-related accumulation.[111] Secondly, the analysis of age relationships within our limited range of materials is not indicative of any significant correlation with age.[91] Nor are the observed enhanced accumulations of oxidized protein products in diseased tissues associated with variation of dietary intake, though one cannot eliminate diet as a contributor for the basal level of oxidized protein components observed in normal physiological tissues (possibly via incorporation during protein synthesis as discussed above). Issues of dietary input require further investigation.

15.4　On the possible significance of protein oxidation in atherogenesis

As demonstrated, protein oxidation occurs in many diseases, including atherosclerosis. A fundamental biological question needs to be asked: what is the possible physiological or pathological significance of protein oxidation in atherogenesis? There are some examples in which protein oxidation in atherogenesis may be beneficial: for example, HDL oxidized by peroxidase-generated tyrosyl radicals *in vitro* may enhance cholesterol efflux from cultured fibroblasts and macrophage foam cells compared with its native counterpart.[112] One interesting observation is that protein oxidation products are found in plaques even when both aqueous and lipophilic antioxidants such as ascorbate and tocopherols are present at normal concentrations.[113] This may be consequent on the limited capacity of these known antioxidants to restrict protein oxidation.[114, 115] Proteolytic elimination of such oxidized proteins may well be more economic than cellular handling of secondary damage. Equally plausibly, the oxidation events may take place separated in time and space from the antioxidant supply.

There is evidence to argue that the oxidized proteins can be damaging and may be pro-atherogenic. As discussed earlier, many oxidized components of proteins such as protein carbonyls, protein-bound DOPA, and protein hydroperoxides are reactive, and can initiate secondary damage to other target molecules (reviewed in Refs (2, 5, 7)). It is therefore plausible that the reactive species present in the oxidized proteins may inflict secondary damage to other biological molecules and contribute to the progression of the disease. The stored and enlarged oxidized protein pools may also perturb normal tissue metabolism. Based on the limited evidence, we postulate that oxidized proteins in arterial vessel walls play active roles in atherogenesis. However, the exact nature of the pathological impact can only be assessed by experimental data from direct perturbation studies, in which a stimulation or inhibition of protein oxidation, independent of other oxidative processes, is achieved, and its effect on atherogenesis determined.

In this context, antioxidant intervention studies from The Heart Research Institute are of particular relevance. These have established that in some animal models of atherogenesis, antioxidants, concomitant with inhibiting lipid peroxidation (as judged by levels of primary and early products of cholesterol ester fatty acid oxidation), can inhibit the development of disease.[116] However, as we have noted already, the inhibition of lipid peroxidation cannot be expected necessarily to also cause the inhibition of protein oxidation. Consistent with this point, and with the possibility that agents which inhibit lipid peroxidation may also act on other oxidant-sensitive regulatory processes, these intervention studies have shown that in some models the inhibition of lipid peroxidation is not accompanied by alterations in atherogenesis.[117] The questions of whether either lipid peroxidation or protein oxidation are sufficient for atherogenesis remain to be assessed. The pressing need for both selective inhibitors of protein oxidation, and for techniques for the selective experimental initiation of protein and lipid oxidation in arterial sites, are re-emphasized by these studies, and are prime targets for current research.

15.5 Implications for diagnosis, prevention and reversal of atherogenesis

As discussed earlier, oxidized proteins in most cases are not repairable, and the main biological defence mechanism against such events relies on their proteolytic removal. Extrapolating from this, the degradation products of the oxidized proteins may be excreted in urine; the limited data available to date from both human and animal studies is consistent with the rapid excretion of oxidized amino acids via this route. Thus the presence of oxidized amino acids in the urine of exercised or aging rats is believed to correlate with the tissue levels of protein-bound oxidized amino acids,[115, 118] and it has been suggested that measurement of the oxidized and modified amino acids in urine may provide a non-invasive method of measuring oxidative stress *in vivo*.[118, 119] Working on the same principle, novel and unique oxidation products might be generated from the oxidation of apoB or other proteins in the site of atherogenesis and subsequently be released into the bloodstream or urine. Interestingly, *in vitro* oxidation of apoB of LDL by transition metals has been reported to give rise to oxidized amino acids,[97, 120] in addition to small peptide fragments which also contain some oxidized entities.[101] Large-scale screening of blood and urine samples from atherosclerotic patients for unique protein oxidation products may provide useful leads for developing potential diagnostics for atherogenesis.

A huge effort has been made worldwide in studying atherogenesis, with an ultimate goal of finding a means of prevention, and clinical reversal, of the disease. Until now almost all studies concerning the role of antioxidants in the prevention of atherosclerosis have focused on anti-lipid oxidation. It is now clear that protein and lipid oxidation coexist in atherosclerotic tissues, and the molecular aspects of protein oxidation are different from those of the lipid one. In general, with the sole exception of cysteine, amino acids are less reactive than lipids containing bis-allylic hydrogens, so that reactive oxidants, such as hydroxyl radicals and alkoxyl radicals, are of greater relevance to protein than lipid oxidation. A good anti-lipid oxidation antioxidant may provide no protection against protein oxidation.[90] The chain-carrying species in protein oxidation are also different from those in lipid oxidation. For example, in lipoproteins, α-tocopherol radicals are the likely chain carriers as long as vitamin E is present, whereas, after tocopherol depletion, lipid peroxyl radicals become the chain-carrying species (reviewed in Ref. (121)). Thus an antioxidant for lipids can focus on the elimination of a single reactant, independent of the nature of the peroxidation-initiating radical. In contrast, there are several chain-carrying species (such as peroxyl and alkoxyl intermediates) involved during radical-induced protein oxidation (reviewed in Refs (2, 5)), such that individual anti-oxidants may only attenuate particular reactions, and not the overall process. On the other hand, the apparently more substantial role of alkoxyl radicals in protein than in lipid peroxidation may offer at least a selective target for inhibition of protein rather than lipid oxidation.

The ability to quantify distinct amino acid oxidation products in proteins and tissues is invaluable for exploring the roles of various oxidation pathways in the pathogenesis of atherosclerosis. This approach cannot establish the relevance of oxidation chemistry to the pathogenesis of disease, however. The role of non-enzymic free radical processes can only be revealed by their selective inhibition by means of future antioxidant interventions. Animal models that over-express or lack enzymes that generate specific oxidants or oxidation products can be used to point to the role of specific enzymatic steps in oxidation. In concert with the use of HPLC and mass spectrometry to detect specific chemical markers *in vivo*, such approaches should provide further insights into the contributions of oxidative damage to the onset and progression of atherosclerosis.

References

1. Dakin, H.D. (1905–6). The oxidation of amino acids with the production of substances of biological importance. *J. Biol. Chem.*, **1**, 171–6.
2. Davies, M.J. and Dean, R.T. (1997). In *Radical-mediated protein oxidation: from chemistry to medicine*, pp. 1–443. Oxford University Press, Oxford.
3. Garrison, W.M. (1987). Reaction mechanisms in the radiolysis of peptides, polypeptides, and proteins. *Chem. Rev.*, **87**, 381–98.
4. Stadtman, E.R. (1990). Metal ion-catalyzed oxidation of proteins: biochemical mechanism and biological consequences. *Free Rad. Biol. and Med.*, **9**, 315–25.
5. Dean, R.T., Fu, S., Stocker, R., and Davies, M.J. (1997). The biochemistry and pathology of radical-mediated protein oxidation. *Biochem. J.*, **324**, 1–18.
6. Dean, R.T., Gebicki, J., Gieseg, S., Grant, A.J., and Simpson, J.A. (1992). Hypothesis: a damaging role in aging for reactive protein oxidation products? *Mutat. Res.*, **275**, (3–6), 387–93.

7. Dean, R.T., Gieseg, S., and Davies, M.J. (1993). Reactive species and their accumulation on radical-damaged proteins. *Trends Biochem. Sci.*, **18**, (11), 437–41.
8. Wells-Knecht, M.C., Thorpe, S.R., and Baynes, J.W. (1995). Pathways of formation of glycoxidation products during glycation of collagen. *Biochemistry*, **34**, (46), 15134–41.
9. Berliner, J.A. and Heinecke, J.W. (1996). The role of oxidized lipoproteins in atherogenesis. *Free Rad. Biol. Med.*, **20**, 707–27.
10. Heinecke, J.W. (1987). Free radical modification of low density lipoprotein: mechanisms and biological consequences. *Free Rad. Biol. Med.*, **3**, 65–73.
11. Stadtman, E.R., Oliver, C.N., Levine, R.L., Fucci, L., and Rivett, A.J. (1988). Implication of protein oxidation in protein turnover, aging, and oxygen toxicity. *Basic Life Science*, **49**, (331), 331–9.
12. Onorato, J.M., Thorpe, S.R., and Baynes, J.W. (1998). Immunohistochemical and ELISA assays for biomarkers of oxidative stress in aging and disease. *Ann. NY Acad. Sci.*, **854**, 277–90.
13. Stadtman, E.R. and Berlett, B.S. (1998). Reactive oxygen-mediated protein oxidation in aging and disease. *Drug Metab. Rev.*, **30**, (2), 225–43.
14. Buxton, G.V., Greeenstock, C.L., Helman, W.P., and Ross, A.B. (1988). Critical review of rate constants for reactions of hydrated electrons, hydrogen atoms and hydroxyl radicals ($^{\bullet}OH/^{\bullet}O^{-}$) in aqueous solution. *Journal of Physical Chemistry Reference Data*, **17**, 513–886.
15. von Sonntag, C. (1987). *The chemical basis of radiation biology*, Taylor and Francis, p. 515. London.
16. Davies, M.J., Fu, S., Wang, H., and Dean, R.T. (1999). Stable markers of oxidant damage to proteins and their application in the study of human disease. *Free Rad. Biol. Med.*, **27**, 1151–62.
17. Heinecke, J.W., Li, W., Daehnke, H.L., and Goldstein, J.A. (1993). *o,o'*-Dityrosine, a specific marker of oxidation, is synthesized by the myeloperoxidase-hydrogen peroxide system of human neutrophils and macrophages. *J. Biol. Chem.*, **268**, (6), 4069–77.
18. Savenkova, M.L., Mueller, D.M., and Heinecke, J.W. (1994). Tyrosyl radical generated by myeloperoxidase is a physiological catalyst for the initiation of lipid peroxidation in low density lipoprotein. *J. Biol. Chem.*, **269**, (32), 20394–400.
19. McCormick, M.L., Gaut, J.P., Lin, T.S., Britigan, B.E., Buettner, G.R., and Heinecke, J.W. (1998). Electron paramagnetic resonance detection of free tyrosyl radical generated by myeloperoxidase, lactoperoxidase, and horseradish peroxidase. *J. Biol. Chem.*, **273**, (48), 32030–7.
20. Foerder, C.A. and Shapiro, B.M. (1977). Release of ovoperoxidase from sea urchin eggs hardens the fertilization membrane with tyrosine crosslinks. *Proc. Natl. Acad. Sci. USA*, **74**, (10), 4214–18.
21. Briza, P., Winkler, G., Kalchhauser, H., and Breitenbach, M. (1986). *o,o'*-Dityrosine is a prominent component of the yeast ascospore wall. A proof of its structure. *J. Biol. Chem.*, **261**, (9), 4288–94.
22. von Sonntag, C. (1990). Free-radical reactions involving thiols and disulphides. In *Sulfur-centered reactive intermediates in chemistry and biology* (ed. C. Chatgilialoglu and K.-D. Asmus), pp. 359–66. Plenum Press, New York.
23. Wardman, P. and von Sonntag, C. (1995). Kinetic factors that control the fate of thiyl radicals in cells. *Meth. Enzymol.*, **251**, 31–45.

24. Schoneich, C. (1999). Reactive oxygen species and biological aging: a mechanistic approach. *Exp. Gerontol.*, **34**, (1), 19–34.

25. Trelstad, R.L., Lawley, K.R., and Holmes, L.B. (1981). Nonenzymatic hydroxylations of proline and lysine by reduced oxygen derivatives. *Nature*, **289**, (5795), 310–22.

26. Fu, S., Hick, L.A., Sheil, M.M., and Dean, R.T. (1995). Structural identification of valine hydroperoxides and hydroxides on radical-damaged amino acid, peptide, and protein molecules. *Free Rad. Biol. Med.*, **19**, (3), 281–92.

27. Ayala, A. and Cutler, R.G. (1996). The utilization of 5-hydroxyl-2-amino valeric acid as a specific marker of oxidized arginine and proline residues in proteins. *Free Rad. Biol. Med.*, **21**, 65–80.

28. Fu, S.L. and Dean, R.T. (1997). Structural characterization of the products of hydroxyl-radical damage to leucine and their detection on proteins. *Biochem. J.*, **324**, (Pt 1), 41–8.

29. Morin, B., Bubb, W.A., Davies, M.J., Dean, R.T., and Fu, S. (1998). 3-Hydroxylysine, a potential marker for studying radical-induced protein oxidation. *Chem. Res. Toxicol.*, **11**, 1265–73.

30. Simpson, J.A., Narita, S., Gieseg, S., Gebicki, S., Gebicki, J.M., and Dean, R.T. (1992). Long-lived reactive species on free-radical-damaged proteins. *Biochem. J.*, **282**, 621–4.

31. Gebicki, S. and Gebicki, J.M. (1993). Formation of peroxides in amino acids and proteins exposed to oxygen free radicals. *Biochem. J.*, **289**, 743–9.

32. Ayala, A. and Cutler, R.G. (1996). Comparison of 5-hydroxy-2-amino valeric acid with carbonyl group content as a marker of oxidized protein in human and mouse liver tissues. *Free Rad. Biol. Med.*, **21**, 551–8.

33. Garrison, W.M., Kland-English, M., Sokol, H.A. and Jayko, M.E. (1970). Radiolytic degradation of the peptide main chain in dilute aqueous solution containing oxygen. *J. Phys. Chem.*, **74**, (26), 4506–9.

34. Davies, M.J. (1996). Protein and peptide alkoxyl radicals can give rise to C-terminal decarboxylation and backbone cleavage. *Arch. Biochem. Biophys.*, **336**, 163–72.

35. Davies, K.J. and Delsignore, M.E. (1987). Protein damage and degradation by oxygen radicals. III. Modification of secondary and tertiary structure. *J. Biol. Chem.*, **262**, (20), 9908–13.

36. Kettle, A.J. and Winterbourn, C.C. (1997). Myeloperoxidase: a key regulator of neutrophil oxidant production. *Redox Report*, **3**, 3–15.

37. Thomas, E.L. (1979). Myeloperoxidase, hydrogen peroxide, chloride antimicrobial system: nitrogen-chlorine derivatives of bacterial components in bactericidal action against *Escherichia coli*. *Infect. Immun.*, **23**, (2), 522–31.

38. Winterbourn, C.C. (1985). Comparative reactivities of various biological compounds with myeloperoxidase–hydrogen peroxide–chloride, and similarity of the oxidant to hypochlorite. *Biochim. Biophys. Acta*, **840**, (2), 204–10.

39. Hawkins, C.L. and Davies, M.J. (1998). Hypochlorite-induced damage to proteins: formation of nitrogen-centred radicals from lysine residues and their role in protein fragmentation. *Biochem. J.*, **332**, 617–25.

40. Hawkins, C.L. and Davies, M.J. (1998). Reaction of HOCl with amino acids and peptides: EPR evidence for rapid rearrangement and fragmentation reactions of nitrogen-centered radicals. *J. Chem. Soc. Perkin Trans.*, **2**, 1937–45.

41. Kettle, A.J. (1996). Neutrophils convert tyrosyl residues in albumin to chlorotyrosine. *Federation of the European Biochemist Society Letters*, **379**, (1), 103–6.

42. Hazen, S.L. and Heinecke, J.W. (1997). 3-Chlorotyrosine, a specific marker of myeloperoxidase-catalysed oxidation, is markedly elevated in low density lipoprotein isolated from human atherosclerotic intima. *J. Clin. Invest.*, **99**, 2075–81.

43. Fu, S., Wang, H., Davies, M.J., and Dean, R.T. (2000). Reaction of hypochlorous acid with tyrosine and peptidyl-tyrosyl residues gives dichlorinated and aldehydic products in addition to 3-chlorotyrosine. *J. Biol. Chem.*, in press.

44. Hazen, S.L., Gaut, J.P., Hsu, F.F., Crowley, J.R., d'Avignon, A., and Heinecke, J.W. (1997). p-Hydroxyphenylacetaldehyde, the major product of L-tyrosine oxidation by the myeloperoxidase–H_2O_2–chloride system of phagocytes, covalently modifies epsilon-amino groups of protein lysine residues. *J. Biol. Chem.*, **272**, 16990–8.

45. Hazen, S.L., d'Avignon, A., Anderson, M.M., Hsu, F.F., and Heinecke, J.W. (1998). Human neutrophils employ the myeloperoxidase-hydrogen peroxide-chloride system to oxidize alpha-amino acids to a family of reactive aldehydes. Mechanistic studies identifying labile intermediates along the reaction pathway. *J. Biol. Chem.*, **273**, (9), 4997–5005.

46. Tien, M., Berlett, B.S., Levine, R.L., Chock, P.B., and Stadtman, E.R. (1999). Peroxynitrite-mediated modification of proteins at physiological carbondioxide concentration: pH dependence of carbonyl formation, tyrosine nitration, and methionine oxidation. *Proc. Natl. Acad. Sci. USA*, **96**, (14), 7809–14.

47. Ischiropoulos, H., Zhu, L., Chen, J., *et al.* (1992). Peroxynitrite-mediated tyrosine nitration catalyzed by superoxide dismutase. *Arch. Biochem. Biophys.*, **298**, (2), 431–7.

48. van der Vliet, A., Eiserich, J.P., O'Neill, C.A., Halliwell, B., and Cross, C.E. (1995). Tyrosine modification by reactive nitrogen species: a closer look. *Arch. Biochem. Biophys.*, **319**, (2), 341–9.

49. Ischiropoulos, H. (1998). Biological tyrosine nitration: a pathophysiological function of nitric oxide and reactive oxygen species. *Arch. Biochem. Biophys.*, **356**, 1–11.

50. Neuzil, J., Gebicki, J.M., and Stocker, R. (1993). Radical-induced chain oxidation of proteins and its inhibition by chain-breaking antioxidants. *Biochem. J.*, **293**, 601–6.

51. Levine, R.L., Williams, J.A., Stadtman, E.R., and Shacter, E. (1994). Carbonyl assays for determination of oxidatively modified proteins. *Meth. Enzymol.*, **233**, (346), 346–57.

52. Fu, M.-X., Wells-Knecht, K.J., Blackledge, J.A., Lyons, T.J., Thorpe, S.R., and Baynes, J.W. (1994). Glycation, glycoxidation, and cross-linking of collagen by glucose. Kinetics, mechanisms, and inhibition of late stages of the Maillard reaction. *Diabetes*, **43**, 676–83.

53. Dakin, H.D. (1944). Hydroxyleucines. *J. Biol. Chem.*, **154**, 549–55.

54. Dakin, H.D. (1946). γ-Methyl-proline. *J. Biol. Chem.*, **164**, 615–20.

55. Fu, S., Gebicki, S., Jessup, W., Gebicki, J.M., and Dean, R.T. (1995). Biological fate of amino acid, peptide and protein hydroperoxides. *Biochem. J.*, **311**, 821–7.

56. Davies, M.J., Fu, S., and Dean, R.T. (1995). Protein hydroperoxides can give rise to reactive free radicals. *Biochem. J.*, **305**, 643–9.

57. Luxford, C., Morin, B., Dean, R.T., and Davies, M.J. (1999). Histone H1- and other protein- and amino acid-hydroperoxides can give rise to free radicals which oxidise DNA. *Biochem. J.*, **344**, 125–34.

58. Gieseg, S.P., Simpson, J.A., Charlton, T.S., Duncan, M.W., and Dean, R.T. (1993). Protein-bound 3,4-dihydroxyphenylalanine is a major reductant formed during hydroxyl radical damage to proteins. *Biochemistry*, **32**, (18), 4780–6.

59. Morin, B., Davies, M.J., and Dean, R.T. (1998). The protein oxidation product 3,4-dihydroxyphenylalanine (DOPA) mediates oxidative DNA damage. *Biochem. J.*, **330**, 1059–67.

60. Ito, S., Kato, T., and Fujita, K. (1988). Covalent binding of catechols to protein through the sulphydryl group. *Biochem. Pharmacol.*, **37**, 1707–10.

61. Smith, W.L. and Marnett, L.J. (1991). Prostaglandin endoperoxide synthase: structure and catalysis. *Biochim. Biophys. Acta*, **1083**, (1), 1–17.

62. Dietz, R., Nastainczyk, W., and Ruf, H.H. (1988). Higher oxidation states of prostaglandin H synthase. Rapid electronic spectroscopy detected two spectral intermediates during the peroxidase reaction with prostaglandin G2. *Eur. J. Biochem.*, **171**, (1–2), 321–8.

63. Adams, G.E. (1972). Radiation mechanisms in radiation biology. *Adv. Radiation Chem.*, **3**, 125–208.

64. Levine, R.L., Oliver, C.N., Fulks, R.M., and Stadtman, E.R. (1981). Turnover of bacterial glutamine synthetase: oxidative inactivation precedes proteolysis. *Proc. Natl. Acad. Sci. USA*, **78**, (4), 2120–4.

65. Stadtman, E.R. (1991). Ascorbic acid and oxidative inactivation of proteins. *Am. J. Clin. Nutr.*, **54**, (Suppl. 6), 1125S–8S.

66. Stadtman, E.R. and Oliver, C.N. (1991). Metal-catalyzed oxidation of proteins. Physiological consequences. *J. Biol. Chem.*, **266**, (4), 2005–8.

67. Rivett, A.J. and Levine, R.L. (1990). Metal-catalyzed oxidation of *Escherichia coli* glutamine synthetase: correlation of structural and functional changes. *Arch. Biochem. Biophys.*, **278**, (1), 26–34.

68. Cervera, J. and Levine, R.L. (1988). Modulation of the hydrophobicity of glutamine synthetase by mixed-function oxidation. *FASEB J.*, **2**, (10), 2591–5.

69. Chevalier, M., Lin, E.C., and Levine, R.L. (1990). Hydrogen peroxide mediates the oxidative inactivation of enzymes following the switch from anaerobic to aerobic metabolism in *Klebsiella pneumoniae*. *J. Biol. Chem.*, **265**, (1), 40–6.

70. Thomas, R.M., Nauseef, W.M., Iyer, S.S., Peterson, M.W., Stone, P.J., and Clark, R.A. (1991). A cytosolic inhibitor of human neutrophil elastase and cathepsin G. *J. Leukoc. Biol.*, **50**, (6), 568–79.

71. Moreno, J.J. and Pryor, W.A. (1992). Inactivation of alpha 1 proteinase inhibitor by peroxynitrite. *Chem. Res. Toxicol.*, **5**, (3), 425–31.

72. Kwon, N.S., Chan, P.C., and Kesner, L. (1990). Inactivation of alpha 1-proteinase inhibitor by Cu(II) and hydrogen peroxide. *Agents Actions*, **29**, (3–4), 388–93.

73. Pedersen, J.Z. and Finazzi Agro, A. (1993). Protein-radical enzymes. *FEBS Lett.*, **325**, (1–2), 53–8.

74. Stubbe, J. and Riggs-Gelasco, P. (1998). Harnessing free radicals: formation and function of the tyrosyl radical in ribonucleotide reductase. *Trends Biochem. Sci.*, **23**, (11), 438–43.

75. Brush, E.J., Lipsett, K.A., and Kozarich, J.W. (1988). Inactivation of *Escherichia coli* pyruvate formate-lyase by hypophosphite: evidence for a rate-limiting phosphorus-hydrogen bond cleavage. *Biochemistry*, **27**, (6), 2217–22.

76. Wagner, A.F., Frey, M., Neugebauer, F.A., Schafer, W., and Knappe, J. (1992). The free radical in pyruvate formate-lyase is located on glycine-734. *Proc. Natl. Acad. Sci. USA*, **89**, (3), 996–1000.

77. Hartel, B., Ludwig, P., Schewe, T., and Rapoport, S.M. (1982). Self-inactivation by 13-hydroperoxylinoleic acid and lipohydroperoxidase activity of the reticulocyte lipoxygenase. *Eur. J. Biochem.*, **126**, (2), 353–7.

78. Mohsenin, V. and Gee, J.L. (1989). Oxidation of alpha 1-protease inhibitor: role of lipid peroxidation products. *J. Appl. Physiol.*, **66**, (5), 2211–15.

79. Kautiainen, A. (1992). Determination of hemoglobin adducts from aldehydes formed during lipid peroxidation *in vitro*. *Chemico-Biological Interactions*, **83**, (1), 55–63.

80. Baynes, J.W. (1996). Reactive oxygen in the aetiology and complications of diabetes. In *Drugs, diet and disease*, Vol. 2, *Mechanistic approaches to diabetes* (ed. C. Ioannides), pp. 201–40. Pergamon, London.

81. Hazell, L.J., van den Berg, J.J., and Stocker, R. (1994). Oxidation of low-density lipoprotein by hypochlorite causes aggregation that is mediated by modification of lysine residues rather than lipid oxidation. *Biochem. J.*, **302**, 421–8.

82. Cappiello, M., Voltarelli, M., Giannessi, M., *et al.* (1994). Glutathione dependent modification of bovine lens aldose reductase. *Exp. Eye Res.*, **58**, (4), 491–501.

83. Rokutan, K., Thomas, J.A., and Sies, H. (1989). Specific S-thiolation of a 30-kDa cytosolic protein from rat liver under oxidative stress. *Eur. J. Biochem.*, **179**, (1), 233–9.

84. Bisby, R.H., Ahmed, S., and Cundall, R.B. (1984). Repair of amino acid radicals by a vitamin E analogue. *Biochem. Biophys. Res. Commun.*, **119**, (1), 245–51.

85. Moskovitz, J., Berlett, B.S., Poston, J.M., and Stadtman, E.R. (1999). Methionine sulfoxide reductase in antioxidant defense. *Meth. Enzymol.*, **300**, 239–44.

86. Dean, R.T., Armstrong, S.G., Fu, S., and Jessup, W. (1994). Oxidised proteins and their enzymatic proteolysis in eucaryotic cells: a critical appraisal. In *Free radicals in the environment, medicine and toxicology* (ed. H. Nohl, H. Esterbauer, and C. Rice- Evans), pp. 47–79. Richelieu Press, London.

87. Grant, A.J., Jessup, W., and Dean, R.T. (1993). Inefficient degradation of oxidized regions of protein molecules. *Free Rad. Res. Commun.*, **18**, (5), 259–67.

88. Dean, R.T. and Riley, P.A. (1978). The degradation of normal and analogue-containing proteins in MRC-5 fibroblasts. *Biochim. Biophys. Acta*, **539**, (2), 230–7.

89. Yim, M.B., Chae, H.Z., Rhee, S.G., Chock, P.B., and Stadtman, E.R. (1994). On the protective mechanism of the thiol-specific antioxidant enzyme against the oxidative damage of biomacromolecules. *J. Biol. Chem.*, **269**, (3), 1621–6.

90. Hazell, L.J. and Stocker, R. (1997). α-Tocopherol does not inhibit hypochlorite-induced oxidation of apolipoprotein B-100 of low-density lipoprotein. *FEBS Lett.*, **414**, 541–4.

91. Fu, S., Dean, R., Southan, M., and Truscott, R. (1998). The hydroxyl radical in lens nuclear cataractogenesis. *J. Biol. Chem.*, **273**, 28603–9.

92. Truscott, R.J. and Augusteyn, R.C. (1977). Oxidative changes in human lens proteins during senile nuclear cataract formation. *Biochim. Biophys. Acta*, **492**, 43–52.

93. Smith, C.D., Carney, J.M., Tatsumo, T., Stadtman, E.R., Floyd, R.A., and Markesbery, W.R. (1992). Protein oxidation in aging brain. *Ann. NY Acad. Sci.*, **663**, (110), 110–19.

94. Dyrks, T., Dyrks, E., Masters, C.L., and Beyreuther, K. (1993). Amyloidogenicity of rodent and human beta A4 sequences. *Federation of the European Biochemist Society Letters*, **324**, (2), 231–6.

95. Smith, C.D., Carney, J.M., Starke, R.P. *et al.* (1991). Excess brain protein oxidation and enzyme dysfunction in normal aging and in Alzheimer disease. *Proc. Natl. Acad. Sci. USA*, **88**, (23), 10540–3.

96. Steinberg, D., Parthasarathy, S., Carew, T.E., Khoo, J.C., and Witztum, J.L. (1989). Beyond cholesterol: modifications of low-density lipoprotein that increase its atherogenicity. *N. Engl. J. Med.*, **320**, 915–24.

97. Fong, L.G., Parthasarathy, S., Witztum, J.L., and Steinberg, D. (1987). Nonenzymatic oxidative cleavage of peptide bonds in apoprotein B-100. *J. Lipid Res.*, **28**, (12), 1466–77.

98. Heinecke, J.W. (1998). Oxidants and antioxidants in the pathogenesis of atherosclerosis: implications for the oxidized low density lipoprotein hypothesis. *Atherosclerosis*, **141**, (1), 1–15.

99. Yla-Herttuala, S., Luoma, J., Viita, H., Hiltunen, T., Sisto, T., and Nikkari, T. (1995). Transfer of 15-lipoxygenase gene into rabbit iliac arteries results in the appearance of oxidation-specific lipid–protein adducts characteristic of oxidized low density lipoprotein. *J. Clin. Invest.*, **95**, (6), 2692–8.

100. Leeuwenburgh, C., Rasmussen, J.E., Hsu, F.F., Mueller, D.M., Pennathur, S., and Heinecke, J.W. (1997). Mass spectrometric quantification of markers for protein oxidation by tyrosyl radical, copper, and hydroxyl radical in low density lipoprotein isolated from human atherosclerotic plaques. *J. Biol. Chem.*, **272**, (6), 3520–6.

101. Bruce, D., Fu, S., Armstrong, S., and Dean, R.T. (1999). Human Apo-lipoprotein B from normal human plasma contains oxidised peptides. *Int. J. Biochem. Cell Biol.*, **31**, 1409–20.

102. Bedwell, S., Dean, R.T., and Jessup, W. (1989). The action of defined oxygen-centred free radicals on human low-density lipoprotein. *Biochem. J.*, **262**, (3), 707–12.

103. Beckman, J.S., Ye, Y.Z., Anderson, P.G. *et al.* (1994). Extensive nitration of protein tyrosines in human atherosclerosis detected by immunohistochemistry. *Biol. Chem. Hoppe-Seyler*, **375**, (2), 81–8.

104. Leeuwenburgh, C., Hardy, M.M., Hazen, S.L. *et al.* (1997). Reactive nitrogen intermediates promote low density lipoprotein oxidation in human atherosclerotic intima. *J. Biol. Chem.*, **272**, 1433–6.

105. Heinecke, J.W. (1999). Mass spectrometric quantification of amino acid oxidation products in proteins: insights into pathways that promote LDL oxidation in the human artery wall. *FASEB J.*, **13**, (10), 1113–20.

106. Heinecke, J.W., Hsu, F.F., Crowley, J.R. *et al.* (1999). Detecting oxidative modification of biomolecules with isotope dilution mass spectrometry: sensitive and quantitative assays for oxidized amino acids in proteins and tissues. *Meth. Enzymol.*, **300**, 124–44.

107. Fu, S., Davies, M.J., Stocker, R., and Dean, R.T. (1998). Evidence for roles of radicals in protein oxidation in advanced human atherosclerotic plaque. *Biochem. J.*, **333**, 519–25.

108. Hazell, L.J., Arnold, L., Flowers, D., Waeg, G., Malle, E., and Stocker, R. (1996). Presence of hypochlorite-modified proteins in human atherosclerotic lesions. *J. Clin. Invest.*, **97**, (6), 1535–44.

109. Kettle, A.J., van Dalen, C.J., and Winterbourn, C.C. (1997). Peroxynitrite and myeloperoxidase leave the same footprint in protein nitration. *Redox Report*, **3**, 257–8.

110. Sampson, J.B., Ye, Y.-Z., Rosen, H., and Beckman, J.S. (1998). Myeloperoxidase and horseradish peroxidase catalyze tyrosine nitration in proteins from nitrite and hydrogen peroxide. *Arch. Biochem. Biophys.*, **356**, 207–13.

111. Wells-Knecht, M.C., Huggins, T.G., Dyer, D.G., Thorpe, S.R., and Baynes, J.W. (1993). Oxidized amino acids in lens protein with age. Measurement of *o*-tyrosine and *o,o'*-dityrosine in the aging human lens. *J. Biol. Chem.*, **268**, (17), 12348–52.

112. Francis, G.A., Mendez, A.J., Bierman, E.L., and Heinecke, J.W. (1993). Oxidative tyrosylation of high density lipoprotein by peroxidase enhances cholesterol removal from cultured fibroblasts and macrophage foam cells. *Proc. Natl. Acad. Sci. USA*, **90**, (14), 6631–5.

113. Suarna, C., Dean, R.T., May, J., and Stocker, R. (1995). Human atherosclerotic plaque contains both oxidized lipids and relatively large amounts of alpha-tocopherol and ascorbate. *Arterioscler. Thromb. Vasc. Biol.*, **15**, (10), 1616–24.

114. Dean, R.T., Hunt, J.V., Grant, A.J., Yamamoto, Y., and Niki, E. (1991). Free radical damage to proteins: the influence of the relative localization of radical generation, antioxidants, and target proteins. *Free Rad. Biol. Med.*, **11**, (2), 161–8.

115. Leeuwenburgh, C., Hansen, P.A., Holloszy, J.O., and Heinecke, J.W. (1999). Oxidized amino acids in the urine of aging rats: potential markers for assessing oxidative stress *in vivo*. *Am. J. Physiol.*, **276**, (1 Pt 2), R128–35.

116. Witting, P.K., Pettersson, K., Ostlund-Lindqvist, A.M., Westerlund, C., Eriksson, A.W., and Stocker, R. (1999). Inhibition by a coantioxidant of aortic lipoprotein lipid peroxidation and atherosclerosis in apolipoprotein E and low density lipoprotein receptor gene double knockout mice. *FASEB J.*, **13**, (6), 667–75.

117. Witting, P., Pettersson, K., Ostlund-Lindqvist, A.M., Westerlund, C., Wagberg, M., and Stocker, R. (1999). Dissociation of atherogenesis from aortic accumulation of lipid hydro(pero)xides in Watanabe heritable hyperlipidemic rabbits. *J. Clin. Invest.*, **104**, (2), 213–20.

118. Leeuwenburgh, C., Hansen, P.A., Holloszy, J.O., and Heinecke, J.W. (1999). Hydroxyl radical generation during exercise increases mitochondrial protein oxidation and levels of urinary *o,o'*-dityrosine. *Free Rad. Biol. Med.*, **27**, (1–2), 186–92.

119. Mahmoodi, H., Hadley, M., Chang, Y.X., and Draper, H.H. (1995). Increased formation and degradation of malondialdehyde-modified proteins under conditions of peroxidative stress. *Lipids*, **30**, (10), 963–6.

120. Steinbrecher, U.P., Witztum, J.L., Parthasarathy, S., and Steinberg, D. (1987). Decrease in reactive amino groups during oxidation or endothelial cell modification of LDL. Correlation with changes in receptor-mediated catabolism. *Arteriosclerosis*, **7**, (2), 135–43.

121. Stocker, R., Upston, J.M., Niu, X. *et al.* (1998). Lipoprotein oxidation in human atherosclerosis. *Atherosclerosis*, **XI**, 475–82.

16 Antioxidants and co-antioxidation in lipoproteins and the intima

Roland Stocker and Joanne M. Upston
Biochemistry Group, The Heart Research Institute,
Camperdown, Sydney, NSW 2050, Australia

16.1 Introduction

Oxidative damage to lipoproteins within the arterial intima, by an (as yet) undefined mechanism, is widely purported to contribute to and/or initiate atherosclerosis.[1, 2] This damage is most likely to occur when the balance between antioxidants and pro-oxidants shifts in favour of the latter. Many substances may prevent, or significantly delay, the oxidation of other substrates. However, an antioxidant is characterized as being effective against oxidative damage when present in much smaller quantity than the substance that is protected. In atherosclerosis, vascular antioxidants need to protect against a range of oxidative mechanisms, as studies suggest that both radical (one-electron) and two-electron oxidants are produced (Table 16.1). Vascular antioxidants also need to be protective of damage to lipoproteins, particularly low-density lipoproteins (LDLs), as the biological activities of 'oxidized LDL' are pro-atherogenic and the cellular accumulation of oxidized LDL is a hallmark of atherosclerosis.[1, 2] We describe herein the events that precipitate oxidative damage to intimal lipoproteins and discuss the antioxidants in the arterial intima and lipoproteins, especially LDL, to counteract such damage and hence, potentially, atherosclerosis.

Despite the fact that inflammatory cells laden with intracellular LDL lipid are found early in atherosclerosis,[3] the potential pro-atherogenic activities of oxidized LDL are associated with extracellular lipoproteins.[1, 2] This, together with the notion that extracellular antioxidants are inferior to intracellular ones (see below), suggests that oxidative modification of LDL occurs outside cells and, further, that extracellular antioxidants present in LDL, and in the surrounding vascular milieu, may represent the primary defence against oxidants. The properties of these

Table 16.1 Potential one-electron oxidants that induce TMP of lipoprotein lipid in the vascular wall

TMP oxidants (radical or one-electron)	Non-TMP oxidants* (two-electron oxidants)
Transition metal ions	
Hydroxyl $+/-$ superoxide anion radical	Singlet oxygen
Reactive nitrogen species	Peroxynitrite
15-Lipoxygenase	
Myeloperoxidase/H_2O_2/Cl^-/tyrosine	Hypochlorous acid
Aqueous or lipophilic peroxyl radicals	
Metal-containing media and inflammatory cells	

*These oxidants largely (though not exclusively) oxidize apolipoprotein B-100. Lipids also become oxidized and this can involve secondary, radical-mediated reactions, part of which proceed via TMP.

proteinaceous and small molecular weight antioxidants will be discussed in detail below; however, it is first necessary to describe the molecular mechanisms underlying the early stages of oxidative modification of LDL in atherosclerosis.

The most important antioxidant for prevention of lipid peroxidation in biological tissues is α-tocopherol (α-TOH), the major antioxidant-active component of vitamin E.[4] Lipid peroxidation in homogeneous phase occurs via a radical chain mechanism in which lipid peroxyl radicals (LOO$^•$) are the chain-propagating species. α-TOH may intercept initiating radicals before they attack lipid or act as a powerful, lipid peroxidation chain-breaking antioxidant by donating its phenolic hydrogen to form lipid hydroperoxides (LOOH) and a less reactive α-tocopheroxyl radical (α-TO$^•$). In homogeneous phase, there are no physical barriers to restrict the involvement of α-TO$^•$ in bimolecular reactions. Thus, LOO$^•$ may also be eliminated by radical termination reaction with α-TO$^•$. Each α-TOH molecule can scavenge two initiating radicals, resulting in $\leqslant 1$ LOOH molecule formed per α-TOH consumed. Under these conditions, the antioxidant activity exerted by α-TOH results in a well-defined 'lag period' before lipid peroxidation proceeds at a substantial rate, and any increase in α-TOH content will prolong this period of inhibition.

In light of the many and varied studies of LDL oxidation, it has become apparent that lipoprotein-containing fluid is an emulsion of lipid droplets, the radical-induced oxidation of which proceeds via a mechanism that resembles emulsion polymerization rather than lipid peroxidation in homogeneous phase.[5, 6] This important distinction allows α-TOH, a powerful antioxidant in homogeneous phase, to exhibit pro-oxidant activity for lipids and proteins in lipoproteins such as LDL exposed to radical oxidants via formation of the α-TO$^•$. Such pro-oxidant activity of α-TOH is described by tocopherol-mediated peroxidation (TMP), a general model for the α-TOH-containing period of radical-induced lipoprotein lipid peroxidation (Fig. 16.1).[5]

16.2 TMP of lipoprotein lipid

TMP is a mechanism of controlled lipid peroxidation (Fig. 16.1).[5] In this model, the fate of the α-TO$^\bullet$, once formed in the lipid droplet, determines whether significant lipid peroxidation occurs.[5–8] TMP was described initially for LDL, but was subsequently shown to be relevant for lipid peroxidation in all α-TOH-containing lipoproteins.[9–11]

Exposure of vitamin E-containing lipoproteins to radical oxidants (Table 16.1) results in the formation of α-TO$^\bullet$ independent of the oxidant involved (Fig. 16.1), as α-TO$^\bullet$ is thermodynamically the most stable radical that can be formed.[4] Once formed, this radical is physically segregated from other radicals in the aqueous phase and in other oxidizing lipid particles (unlike in homogeneous phase), and can initiate lipid peroxidation via abstraction of bis-allylic hydrogen atom from LDL core or surface lipid (Fig. 16.1). Unless α-TO$^\bullet$ is eliminated, a substantial proportion of lipoprotein lipid may become oxidized via the TMP cycle without loss

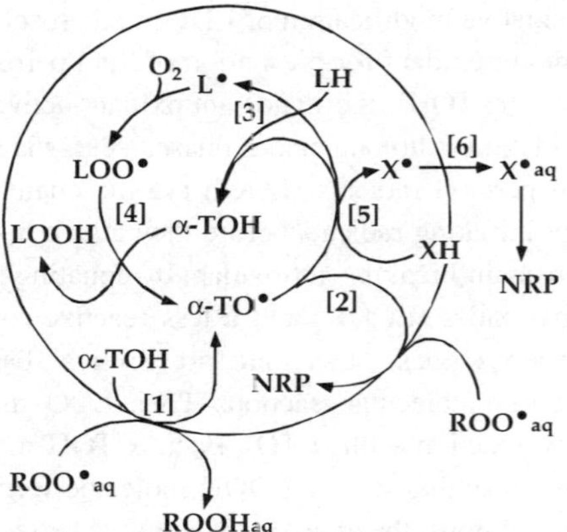

Fig. 16.1 Tocopherol-mediated peroxidation of lipoprotein lipid. Lipid peroxidation in LDL and other lipoproteins proceeds via TMP that may be initiated when a radical oxidant, such as ROO$^\bullet$aq, is scavenged by α-TOH (reaction 1), quantitatively the most abundant antioxidant in LDL. This reaction is favoured over reaction of ROO$^\bullet$aq with bis-allylic hydrogens (LH) of LDL core and surface lipid. Under conditions of high fluxes of initiating radicals, a second ROO$^\bullet$aq will frequently enter LDL particles undergoing oxidation, resulting in termination of lipid peroxidation via radical–radical termination (reaction 2). Under conditions of mild radical flux, the relatively stable α-TO$^\bullet$ reacts with LH to produce LOO$^\bullet$ (reaction 3). Reaction of LOO$^\bullet$ with another molecule of α-TOH gives rise to LOOH and another α-TO$^\bullet$ (reaction 4). The cyclic nature of TMP (reactions 3–4) allows many molecules of LOOH to be formed with relatively little consumption of α-TOH. Co-antioxidants (XH) inhibit TMP by reacting with α-TO$^\bullet$ (to regenerate α-TOH, reaction 5) followed by export of radical (present as the co-antioxidant-derived radical, X$^\bullet$) from the lipoprotein particle into the aqueous space (reaction 6) to yield non-radical products (NRP). Elimination of the radical from LDL is crucial.

of the vitamin (Fig. 16.1). However, the rate of hydrogen abstraction by α-TO$^\bullet$ is approximately three orders of magnitude lower than that catalysed by LOO$^\bullet$.[6] Thus, TMP represents controlled or retarded lipid peroxidation when compared with uninhibited peroxidation that occurs in the absence of α-TOH.

The ability of α-TO$^\bullet$ to act as a lipid peroxidation chain-carrying moiety (Fig. 16.1) refers to the chain transfer activity of α-TOH. This activity is reflected by the comparable fractional peroxidation rates of core and surface lipid in LDL.[5, 12, 13] α-TOH also exerts a phase transfer activity in LDL (Fig. 16.1), i.e. α-TOH transfers exogenous radicals into the lipoprotein. Such activity is best demonstrated by exchange of α-TOH's phenolic hydrogen with deuterium, resulting in a decrease in the pro-oxidant activity of the vitamin in LDL.[14] This contrasts the situation in homogeneous phase, where replacing the hydrogen with deuterium decreases the antioxidant activity of α-TOH.[15] Where studied, the oxidizability of LDL lipid decreases proportionally with α-TOH content and complete removal of the vitamin can circumvent oxidative damage induced by some oxidants.[12] These last two features are also at odds with the classical antioxidant activity of α-TOH. Thus the presence of α-TOH in lipoproteins affects the overall oxidizability of the particle.

The reaction of α-TO$^\bullet$ with LDL lipid is the rate-limiting reaction of TMP. This has important consequences as the two processes that inhibit TMP, radical–radical termination and co-antioxidation, target this reaction (Fig. 16.1). Radical termination of α-TO$^\bullet$ predominates under conditions of frequent radical attack, as this increases the probability of another radical entering LDL to combine with α-TO$^\bullet$, which subsequently results in cessation of chain-propagated lipid peroxidation (Fig. 16.1). Concordant with suppression of TMP, consumption of α-TOH also occurs under high radical flux conditions.

Co-antioxidation is defined as that process which allows the removal of the radical character from α-TO$^\bullet$ (via reduction) in an oxidizing lipoprotein particle to the aqueous phase where it is converted into non-radical product(s) (Fig. 16.1).[8] By reducing α-TO$^\bullet$, co-antioxidants prevent the consumption of α-TOH and prevent chain-propagated lipid peroxidation. Thus, the most efficient protection of LDL lipid against oxidative attack is produced in the presence of multiple antioxidants, i.e. the lipoproteins' endogenous α-TOH and suitable endogenous or exogenous co-antioxidants for α-TOH. As co-antioxidation may be an important and amenable anti-atherogenic strategy, its concept and two effective natural co-antioxidants are discussed in detail below.

16.3 Co-antioxidation of lipoprotein lipid

As discussed above, the α-TOH content influences the extent of oxidative damage to LDL. Co-antioxidants inhibit TMP and lipoprotein lipid peroxidation. The more vitamin E the lipoprotein contains, the greater the susceptibility to oxidative attack and the greater the need for co-antioxidants to prevent lipid peroxidation. Hence, the balance of available co-antioxidants *and* α-TOH will determine whether TMP of lipoproteins occurs in biological systems.

Several natural co-antioxidants exist, which may be found in the vascular wall. These include ascorbate, α-tocopheryl hydroquinone, ubiquinol-10 ($CoQ_{10}H_2$), 3-hydroxyanthranilic acid, and albumin-bound bilirubin.[16, 17] Some co-antioxidants, for example $CoQ_{10}H_2$, may be endogenous to plasma lipoproteins (see below). A number of synthetic co-antioxidants have also been described[16] and these include butylated hydroxytoluene, hydroquinones and Trolox. The co-antioxidant efficacy of the dietary compounds, ascorbate and $CoQ_{10}H_2$ may be of importance in the vascular wall and will be discussed further.

16.3.1 Ascorbate

The recycling of α-TOH from α-TO$^{\bullet}$ by ascorbate (the reduced, antioxidant active form of vitamin C), with the subsequent generation of the relatively non-reactive aqueous ascorbyl radical, occurs in micellar, lipid bilayer and lipoprotein systems.[18, 19] This reaction is thought to be an important physiological recycling mechanism, and several enzymatic and cellular processes exist that reduce oxidized vitamin C to ascorbate.[20]

The co-antioxidant action of ascorbate for α-TOH in LDL undergoing oxidation is perhaps best demonstrated by the example of horse radish peroxidase / hydrogen peroxide (HRP/H_2O_2)-induced LDL lipid peroxidation.[21] α-TO$^{\bullet}$ formed under these conditions can be measured directly by electron paramagnetic resonance spectroscopy. Addition of catalase (to remove H_2O_2) and either urate or ascorbate (to eliminate the initiating radical oxidant formed by HRP/H_2O_2) results in cessation of α-TOH consumption. Addition of ascorbate also eliminates α-TO$^{\bullet}$ (with concomitant formation of ascorbyl radical) and halts LDL lipid peroxidation. In contrast, addition of urate fails to eliminate α-TO$^{\bullet}$ so that lipid peroxidation proceeds until α-TO$^{\bullet}$ has decayed.[21]

16.3.2 $CoQ_{10}H_2$

Although $CoQ_{10}H_2$ (the reduced form of coenzyme Q_{10}, CoQ_{10}) in isolation is a lipophilic antioxidant,[22] in LDL it appears to primarily act as a co-antioxidant for α-TOH. Dietary supplementation with CoQ_{10} increases the content of $CoQ_{10}H_2$ in human LDL from <1 to 2–2.4 molecules per particle[23, 24] and the resistance of LDL to peroxyl radical- or transition metal ion-induced lipid peroxidation.[24] An increased $CoQ_{10}H_2$ content also negates the pro-oxidant effect observed with vitamin E supplementation.[24] The exact co-antioxidant mechanism exerted by $CoQ_{10}H_2$ is unknown. It is postulated that $CoQ_{10}H_2$ removes the radical character from α-TO$^{\bullet}$ via semiquinone radical formation and subsequent generation of superoxide anion radical that may diffuse from the lipoprotein.[6] Similarly to ascorbate, there is some evidence to suggest that cellular systems exist to recycle oxidized CoQ_{10} and maintain the antioxidant active hydroquinone form.[25, 26]

Each molecule of $CoQ_{10}H_2$ can potentially terminate up to two lipid peroxidation chain reactions, each producing 20–40 lipid hydroperoxide molecules per α-TO$^•$ generated.[5, 27] Thus, the presence of $CoQ_{10}H_2$ can potentially decrease the extent of LDL lipid peroxidation 40–80 times. In addition, the degree of inhibition decreases only as the square root of the co-antioxidant concentration decreases,[5] explaining why only small amounts of co-antioxidants are required to substantially inhibit LDL lipid peroxidation under conditions of low initiating radical flux.

16.4 Lipoprotein changes during atherosclerosis and consequences for oxidation and antioxidation

Atherosclerosis is associated with complex changes to the content, structure, chemical composition and location of lipid/lipoprotein particles within the vascular wall (reviewed in Chapter 1 and Ref. 28). Subsequently, the mechanism(s) controlling lipid oxidation, and the antioxidants required to defend against this in the vascular wall, may fluctuate as the disease progresses. The remainder of this chapter will discuss the changes to vascular wall lipid and antioxidants during atherogenesis.

The earliest morphological change to the artery wall, the fatty streak, is characterized by the accumulation of cholesterol and cholesteryl oleate. Prior to extracellular deposition, the lipid accumulates as intracellular droplets[29, 30] in inflammatory cells that have migrated into the sub-endothelial space. At this early stage it is thought that oxidative modification to lipoproteins allows the unregulated cellular uptake of lipid.[1] Cholesteryl oleate, normally a minor component of LDL, is likely generated by intracellular re-esterification of cholesterol following hydrolysis of the internalized lipoprotein cholesteryl esters.[31] As the fatty streaks progress into fibro-fatty and complex plaques, cholesteryl esters in the lipid deposits become more heterogeneous until cholesteryl linoleate dominates. At these late stages, the vascular wall concentration of apolipoprotein B-100-containing lipoproteins increases severalfold over that found in normal blood vessels.[29] The increase in lipoprotein concentration is relatively specific for LDL, and retention of HDL, for example, is much less pronounced.[30]

Extracellular accumulation of lipid is increasingly prominent in the intermediate stages of atherosclerosis, and two different types of lipid particles occur.[29, 32] The first is LDL-like, $1.01\,\mathrm{g\,l^{-1}}$ dense, and composed of a core containing cholesteryl esters and surrounded by a monolayer of phospholipids and cholesterol. Other particles typify unesterified cholesterol-rich vesicles comprising single or multiple lamellas.[32] The particles vary in size up to nine times that of plasma LDL. Oleate and palmitate are found in the phospholipid fraction; however, cholesteryl linoleate is the single, major oxidizable lipid isolated from human atherosclerotic lesions.[33] The esterified lipid may arise from LDL retained in the extracellular matrix of the vascular wall by aggregation or fusion, and/or by cell death and subsequent release of internalized lipoprotein lipid. Thus, extracellular lipid in the vascular wall undergoes significant alteration during disease progression, including changes in type (from saturated and monounsaturated to polyunsaturated), total content, and particle structure. The development of atherosclerosis signals an inability to contend with these changes in the light of other factors, e.g. oxidative events.

Oxidized lipid accumulates as atherosclerotic lesions develop[34] (Upston and Stocker, unpublished), and antioxidants in the vascular wall may therefore be required at all developmental stages. Just as the precise origin and role of lipid particles in the diseased artery wall are unknown, the mechanism(s) of oxidation and antioxidation of these particles is not defined. Whether TMP is relevant to intimal lipoprotein lipid oxidation and atherosclerosis is also not known. However, in the earliest stages of the disease, the accumulating LDL-derived lipoproteins are likely to contain several molecules of α-TOH per particle and, hence, oxidize via TMP.

Studies show that LDL aggregates are formed in the developing lesion. Injection of human LDL into rabbits generates lipid-containing particles trapped in matrix filaments that fuse to be 7–8 times that of plasma LDL.[35] Each LDL that enters the sub-endothelial space carries 6–10 molecules of α-TOH with it and, as intimal particle size is generally larger than LDL, it is likely that intimal LDL-derived particles contain several α-TOH molecules. Mechanistic studies with LDL aggregates (produced by physical means or by complexation with proteoglycans) indicate that lipids in aggregated LDL lipid peroxidize with typical TMP kinetics (K. Morris and R. Stocker, unpublished). This, together with the fact that gross consumption of α-TOH does not appear to occur (see below), makes TMP a probable mechanism governing lipid peroxidation in vascular wall lipid particles. Hence, antioxidants, and co-antioxidation to prevent TMP may both be crucial to prevent oxidative damage to LDL during atherogenesis, as discussed below.

16.5 Lipoprotein and vascular wall antioxidants

Antioxidants present in the extracellular space in the vascular wall have not been characterized systematically as this material is difficult to obtain. However, it is known that lipoproteins, including LDL, with their full complement of antioxidants, normally transfer to this compartment from plasma. Thus, it is reasonable to assume that, qualitatively, the vascular wall comprises the same antioxidants as those found in plasma[36] and in suction blister fluid, a model of interstitial fluid.[37] The extracellular proteinaceous and small molecular weight antioxidants present in the vascular wall are summarized below with reference to lipoprotein oxidation and atherosclerosis.

16.5.1 Non-proteinaceous antioxidants

Of the non-proteinaceous, aqueous antioxidants, ascorbate is regarded as a powerful free radical scavenger and inhibitor of lipoprotein oxidation, as it prevents oxidative damage to plasma lipids,[36] represses transition metal ion-induced LDL oxidation,[38] and inhibits TMP (see above). Ascorbate may also prevent peroxyl radical-induced lipoprotein lipid peroxidation via direct interception of the oxidant.[12] Thus, ascorbate exhibits versatility and, depending on the oxidizing conditions, may prevent biological lipid oxidation by co-antioxidation or direct radical scavenging.

Urate, a significant component of human plasma, is an antioxidant that can directly scavenge aqueous radicals[39] and certain oxidants produced by enzymes.[21] Urate also binds transition metals[40] and this may be relevant for LDL lipid peroxidation in the vascular wall.[41] However, urate cannot eliminate α-TO$^\bullet$ and thus does not prevent TMP of lipoprotein lipids (see above).

3-Hydroxyanthranilic, a tryptophan-derived aminophenol produced and released by activated human monocytes and macrophages, is a small molecular weight, extracellular antioxidant.[42] This cell-derived metabolite inhibits LDL lipid peroxidation induced by interferon-γ-activated human macrophages;[43] is a highly efficient scavenger of radicals and peroxynitrite;[44] and also reduces α-TO$^\bullet$ thereby acting as a co-antioxidant.[45]

Bilirubin is a strong reducing agent and a potential physiological antioxidant.[46] In extracellular fluids, the pigment is predominantly bound to albumin and this complex retains antioxidant activity. Thus, free and albumin-bound bilirubin are efficient co-antioxidants for α-TOH and inhibit plasma and LDL lipid peroxidation.[47]

In addition to these aqueous antioxidants, lipid-soluble antioxidants also play an important role in preventing oxidative damage in biological tissues. $CoQ_{10}H_2$ and the tocopherols that make up vitamin E are the major lipid-soluble antioxidants in lipoproteins, including LDL present in extracellular fluids. Indeed, lipoproteins are the primary vehicles for the transfer of vitamin E from the liver and plasma to peripheral tissues.

α-TOH is quantitatively and qualitatively the major antioxidant in LDL and central to the control of radical-induced lipid peroxidation. The mechanisms whereby α-TOH acts as a classical antioxidant or exerts a pro-oxidant effect (Fig. 16.1) in lipoprotein lipid oxidation have been described above. In addition to α-TOH, γ-TOH is present in lipoproteins, but is less antioxidant active than α-TOH and, hence, less able to control radical-induced LDL lipid peroxidation.[14] However, γ-TOH may be important in the detoxification of nitrogen dioxide,[48] although the physiological significance of this remains unknown.

$CoQ_{10}H_2$ is a potent antioxidant and scavenges peroxyl radicals with slightly higher efficacy than α-TOH. Although $CoQ_{10}H_2$ is present in LDL in smaller quantity than α-TOH, it strongly inhibits lipoprotein lipid peroxidation initiated by a variety of radical oxidants and is consumed before α-TOH (see above). $CoQ_{10}H_2$ reduces α-TO$^\bullet$, and its ability to interact with α-TOH[6] probably accounts for its strong inhibition of lipid peroxidation. As detailed above, $CoQ_{10}H_2$ is an endogenous co-antioxidant for α-TOH in lipoproteins.

All the above low molecular weight, non-proteinaceous antioxidants may be present in arterial walls as plasma constituents are present in normal and atherosclerotic vessels.[29, 33] The transfer of aqueous antioxidants, such as ascorbate and urate, to the intima likely occurs via simple diffusion. Thus, the concentration of these antioxidants in the extracellular space of the vascular wall may approximate that of the lumen. Indeed, the concentration of ascorbate and urate in interstitial fluid[37] and lymph[11] is similar to that in plasma. In contrast, the concentrations of the lipophilic, lipoprotein-associated antioxidants α-TOH and $CoQ_{10}H_2$ are considerably lower (8–20% of respective plasma levels) and resemble the concentration of lipoprotein lipid present in interstitial fluid (13–23%).[37]

The content of non-proteinaceous antioxidants in the extracellular fluid of the vascular wall has only been studied in tissue homogenates prepared from normal or diseased tissue.[29, 33] Normal arteries contain approximately one-third the concentration of ascorbate and significantly less urate than that found in normal human plasma[33] (Table 16.2). It is not clear whether ascorbate originates from extra- or intracellular compartments. However, given that interstitial fluid contains ascorbate and urate and that plasma proteins are found in the vascular wall even in healthy arteries, it is reasonable to expect a portion of the ascorbate detected to be extracellular. The majority of α-TOH and CoQ_{10} detected (Table 16.2) are probably located within cells in the healthy vascular wall where the concentration of lipoproteins is relatively (to albumin) low.[29]

Table 16.2 Non-enzymatic antioxidants in human blood plasma and homogenates of normal human arteries and atherosclerotic plaques

	Plasma	Normal artery*	Plaque†
Water-soluble (nmol/mg protein)			
Ascorbate	0.33–1.88	0.12 ± 0.07	1.30 ± 0.87
Urate	2.0–5.6	0.11 ± 0.09	3.32 ± 2.26
Lipid-soluble			
α-TOH (mmol/mol C)	8.9–23.7	4.2 ± 1.7	6.3 ± 4.8
α-TOH (mmol/mol C18:2)	8.9–16.3	–	28.6 ± 21.8
CoQ_{10} (mmol/mol C)	0.3–0.8	1.5 ± 1.5	1.5 ± 0.15

Data (range or mean $\pm$ SD) are taken from Ref. (33). For comparison, lipid-soluble antioxidants are expressed per unesterified cholesterol (C). As cholesteryl esters become hydrolysed and free fatty acids, particularly linoleic acid, are metabolized in a developing lesion, the content of vitamin E is also expressed per readily oxidizable, bis-allylic hydrogen-containing lipid, i.e. α-TOH per cholesteryl linoleate (C18:2).
*Normal iliac arteries ($n=6$).
†Advanced fibro-fatty lesions obtained from patients undergoing surgery of the femoral artery ($n=7$) or carotid endarterectomy ($n=4$).

16.5.2 Enzymatic and proteinaceous antioxidants

The classical antioxidant enzymes are largely cell-associated proteins whose functions are to maintain a reducing tone intracellularly. They include Cu, Zn- and Mn-superoxide dismutase (SOD), glutathione peroxidase, glutathione reductase, glutathione transferase, phospholipid glutathione-dependent peroxidase, and catalase. Many of these enzymatic antioxidants have been isolated from normal arteries (Table 16.3), although they are likely present within vascular wall cells as extracellular fluid is largely devoid of enzymatic antioxidants (Table 16.3) (reviewed in Ref. (49)).

The exception to the above is extracellular SOD (EC-SOD), which is present in significant amounts in the normal arterial wall.[50] This enzyme is localized to the connective tissue matrix and is produced by smooth muscle cells in human aorta.[51] Vascular SODs are thought to protect nitric oxide against superoxide anion radical as these enzymes have been shown to maintain endothelial-dependent vasodilation.[52] In the presence of superoxide anion radicals, nitric oxide may be converted to peroxynitrite, a strong oxidant. Hence, the major antioxidant role of EC-SOD may be the prevention of peroxynitrite-induced oxidation of lipoproteins.

Despite the overall lack of the major antioxidant enzymes in extracellular fluid, hydroperoxides of phospholipids and cholesteryl esters may be reduced to the

Table 16.3 Non-proteinaceous antioxidants in human plasma and homogenates of normal human arteries and atherosclerotic plaques

	Plasma	Normal artery	Plaque
Glutathione-related[†](mU/mg protein)			
Selenium-dependent glutathione peroxidase	5	11.9 ± 3.83	$55 \pm 2.1^*$
Glutathione reductase	0.38	3.2 ± 1.2	$1.07 \pm 0.4^*$
Selenium-independent glutathione peroxidase	0	ND	$1.37 \pm 0.35^*$
Total glutathione transferase	0.06	22.5^6	20.9^5.3
Superoxide dismutases[‡] (U/mg protein)			
EC-SOD	0.06 – 0.25	104 ± 49	77 ± 55
Cu,Zn-SOD	ND	100 ± 50	90 ± 42
Mn-SOD	ND	2.5 ± 1.4	3.6 ± 1.9

Results shown (mean $\pm$ SD) are taken from Refs (49) (plasma), (88) (glutathione-related enzymes), and (89) (SODs). ND: not detectable.

$^*P < 0.05$ versus normal artery

[†] Data obtained from 13 internal mammary arteries (Normal artery) and 13 endarterectomy-derived carotid plaques (Plaque).

[‡] Data obtained from post-mortem aortic samples classified by macroscopic examination into Normal artery ($n = 8$) and Plaque (lesion types IV, V, and Vc) ($n = 8$). For comparative purposes, the results shown here are adjusted to protein and presented without statistical analysis as this appears to have been carried out on non-adjusted values (see Ref. (89)).

corresponding alcohols. This conversion represents an antioxidant defence as the potential to form pro-oxidants from hydroperoxide degradation is negated. Phospholipid glutathione-dependent peroxidase, the only enzyme known to reduce complex lipid hydroperoxides in lipoproteins, is a membrane-bound protein,[53] is not found in extracellular fluid, and glutathione peroxidase, which may be found extracellularly, acts only on non-esterified fatty acid hydroperoxides.

The reduction of lipoprotein and complex lipid hydroperoxides in plasma appears to be mediated via a non-enzymatic reaction[54] that nonetheless requires apolipoproteins. Recent evidence suggests that methionine residues in apolipoproteins A-I, A-II[55] and, to a lesser extent, in apolipoprotein B-100[56] reduce lipoprotein lipid hydroperoxides. In exchange for hydroperoxide reduction, the methionine residues are oxidized to methionine sulphoxide.[55] The physiological relevance of apolipoprotein methionine residue-mediated hydroperoxide reduction has not been demonstrated. Further, it is not known whether the formation of methionine sulphoxide (representative of oxidative damage to the apolipoproteins) also results in some loss in apolipoprotein function which may invalidate the antioxidant activity. *In vitro*, the above reaction is slow. However, it is accelerated by unknown factor(s) in the liver.[57] A slow reduction could still be important however as the average residence time of lipoproteins in the intima is long.

Thioredoxin and glutaredoxin, protein disulphide oxidoreductases, also reduce lipid hydroperoxides.[58] These proteins are produced and secreted by various cells and are present in extracellular fluid.[59] It is unknown whether thioredoxin or glutaredoxin can mediate lipid hydroperoxide reduction in lipoproteins. However, there is evidence to suggest that these proteins are released into extracellular fluid in response to oxidative stress.[60]

Several studies have shown that transition metal ions induce oxidative damage to lipoproteins and may be relevant to atherosclerosis. Indeed, copper is commonly used as the 'gold standard' *in vitro* oxidant for LDL.[61] Thus, binding of adventitious transition metal ions to inactive chelates in the vascular wall may be an antioxidant defence. Several proteins present in extracellular fluid specifically bind biological iron and copper complexes, such as ceruloplasmin, transferrin, haptoglobin, and hemopexin. Many more proteins are also found that bind metals non-specifically. For example, albumin has several metal binding sites and is able to inhibit *in vitro* lipoprotein oxidation induced by transition metals[62] and ceruloplasmin[63] at a concentration similar to that in the vascular wall.[29] Similarly, effective antioxidation of lipoproteins in the presence of Cu^{2+} ions is mediated by the high molecular weight fraction of human suction blister fluid.[37]

The above discussion of antioxidants present in the vascular wall focused on defence against radical oxidants. This is because relatively little is known of antioxidants effective against two-electron oxidants such as hypochlorous acid (HOCl) which appears to be produced during atherogenesis.[64, 65] Protein thiols provide the primary defence against HOCl,[66] and albumin in the vascular wall may be effective against this oxidant. However, in the early stages of atherosclerosis, where the albumin concentration is low,[29] other protein thiols may be targets for hypochlorous acid. Thus, this type of antioxidation could become limited or negated by loss of function to essential proteins, e.g. those found in extracellular matrix and cell plasma membranes.

16.6 Changes in antioxidant status of the vascular wall during atherosclerosis

Compared to our knowledge of lipid changes in atherosclerosis, little is known of the accompanying changes (and if they occur) to vascular wall antioxidants. This is true even in studies of animal models of atherosclerosis. Despite this, the notion that oxidative damage to LDL is an early and causative effect of atherosclerosis prevails and vitamin E supplementation to subvert oxidation has proved a popular anti-atherogenic strategy, albeit with limited success.[67] The data currently available is primarily restricted to studies of advanced vascular disease and components released from homogenates of large pieces of aortic tissue. Thus, systematic studies on antioxidant changes at various disease stages, and in focal areas of the intima, are not readily available. With these limitations, the following discussion serves as an indication of the gross changes in vascular wall antioxidants that may occur during disease progression. Recent studies addressing the stage-dependent changes in antioxidant content and quality in diseased vessels will also be highlighted.

16.6.1 Changes in vascular wall non-enzymatic antioxidants

Changes to enzymatic antioxidants that may occur during atherosclerosis most likely take place within vascular wall cells. While an altered intracellular redox state may be atherogenic,[68] stronger evidence suggests that changes in extracellular and non-enzymatic (see the above discussion) antioxidants play a role in disease initiation and/or progression.[1] In particular, antioxidants that are associated with LDL and are required to inhibit lipoprotein lipid peroxidation are thought to be deficient and/or ineffective.

A comparison of the levels of non-enzymatic antioxidants in human plasma, normal arteries, and advanced atherosclerotic lesions is shown in Table 16.2. An early study found ascorbate levels in diseased human aorta to be comparable to those found in plasma.[69] This finding was validated recently by a study of atherosclerotic plaque[33] in which urate concentrations were also determined to be comparable to those in plasma (Table 16.2). Further, only small amounts of vitamin C were present as dehydroascorbic acid (the two-electron oxidation product of vitamin C) in atherosclerotic plaque. This data, together with the finding that urate and ascorbate were present at elevated levels compared to normal arteries[33] argue against a deficiency in aqueous antioxidants in human plaque.

Concordant with the above data for aqueous antioxidants, the levels of α-TOH in homogenates of advanced (Table 16.2) and early to intermediate human atherosclerotic lesions (J. Upston and R. Stocker, unpublished) are comparable to plasma levels. This is true whether α-TOH is expressed per free cholesterol molecule or per cholesteryl linoleate, the major readily oxidizable lipid in lesions (Table 16.2), and argues against significant vitamin E depletion even at advanced stages of atherosclerosis, as previously proposed.[70] Thus, these studies undermine the argument used to suggest that oxidative damage to LDL must occur in the intima rather than plasma, i.e. that intima has a relatively (to plasma) low antioxidant capacity.

Although not shown to date, it is possible that 3-hydroxyanthranilic acid is also present in the extracellular fluid of diseased arteries. Thus, interferon-γ, required to stimulate production of 3-hydroanthranilic acid by cells, is present in human atherosclerotic lesions.[71] Also, preliminary studies indicate the expression in human lesions of indoleamine 2,3-dioxygenase, the rate-limiting enzyme in oxidative tryptophan metabolism that leads to the formation of 3-hydroxyanthranilic acid (S.R. Thomas and R. Stocker, unpublished).

Despite the presence of apparently adequate levels of non-enzymatic antioxidants and co-antioxidants, a significant proportion of cholesteryl linoleate in advanced human lesions is nonetheless oxidized.[33] As discussed above, so long as ascorbate and α-TOH are present, TMP of lipoprotein lipid is effectively prevented *in vitro*. Thus, an explanation for the co-existence of large amounts of oxidized lipid and antioxidants in advanced atherosclerosis seems elusive. However, the above does not reflect oxidized lipid and antioxidant levels at focal areas of lesions as whole tissue homogenates were used for analysis. Hence it may be that oxidative damage occurs at sites remote from available antioxidants, for example, in the extracellular matrix where LDL may become trapped and fuses to large vesicles.[35] This rationale does not likely extend to the co-antioxidant, $CoQ_{10}H_2$, which associates with lipoproteins. However, there is no data available on the amount of $CoQ_{10}H_2$ in lesions[33] (Table 16.2).

To address the above conundrum, that is, antioxidants and oxidized lipids apparently co-exist in lesions, and the fact that the levels of reduced, lipophilic co-antioxidants for α-TOH in atherosclerosis are presently unknown, animal intervention studies employing co-antioxidants have been carried out, or are in progress, in our laboratory. As indicated above, the physiological levels of natural co-antioxidants, in particular ascorbate and $CoQ_{10}H_2$, can be manipulated by dietary supplementation. Thus co-antioxidation is amenable as a potential anti-atherogenic strategy. Preliminary results with synthetic compounds suggest that aortic lipoprotein lipid peroxidation can be effectively inhibited by lipophilic co-antioxidants (see Chapter 3 of this book) and, if confirmed, could have important implications for future intervention trials.

16.6.2 Changes in vascular wall enzymatic antioxidants

The levels of some important proteinaceous antioxidants may also be altered in the atherosclerotic vascular wall. Table 16.3 depicts alterations in glutathione-dependent enzymes and SOD (reviewed in Ref. (72)). Compared to normal arteries, selenium-dependent glutathione peroxidase and glutathione reductase activities are decreased in advanced human plaque, whereas selenium-independent glutathione peroxidase activity is increased in plaque. In lesions of cholesterol-fed rabbits, the activity of the selenium-dependent glutathione peroxidase is increased, whereas that of glutathione reductase and glutathione transferase is decreased. When Japanese quails were fed a cholesterol-enriched diet, comparable levels of glutathione-related enzymes were found between control and atherosclerotic arteries and between arteries derived from cholesterol-fed animals whether or not they developed macroscopic lesions. Thus, the changes in glutathione-related antioxidant enzymes during atherosclerosis are clearly inconsistent between species and between animal models of atherosclerosis. The latter study suggests that atherogenesis proceeds in the absence of gross changes to glutathione-dependent enzymatic antioxidants; however the relevance to human atherosclerosis remains to be elucidated.

SOD activity does not appear to be radically altered in advanced human lesions compared to normal arteries, with perhaps the exception of a slight decrease in EC-SOD (reviewed in Ref. (72)). Decreased SOD activity could lead to an increased extent of reaction of superoxide anion radical with nitric oxide and, hence, dysfunctional endothelium-dependent vasodilation (see above) and increased production of peroxynitrite. Evidence for the latter includes the presence of nitrated proteins in plaque[73] and LDL isolated from such lesions.[74] Chronic inhibition of Cu,Zn-SOD in rats decreases the efficacy of endothelium-derived nitric oxide *ex vivo* and increases non-enzymatic lipid peroxidation,[52] indicating a protective role for SOD in atherosclerosis. In contrast, over-expression of Cu,Zn-SOD in fat-fed C57BL/6 mice increases lesion formation.[75] Hence, the role of SOD in intimal LDL oxidation and atherogenesis remains unknown.

Increased aortic activity of inducible nitric oxide synthase in lesions (reviewed in Ref. (72)) may alter endothelium-dependent vasorelaxation and/or increase peroxynitrite formation. However, it should also be considered that increased nitric oxide production represents an antioxidant defence as nitric oxide inhibits cell-induced LDL lipid peroxidation.[76–78] This antioxidant activity is prevented by inhibition of nitric oxide formation, yet is unaffected by the simultaneous production of superoxide by these cells.[78] Thus, in contrast to the situation where these two radicals are produced simultaneously from the sydnonimine SIN-1,[76] cellular production of superoxide and nitric oxide may not necessarily lead to peroxynitrite formation and LDL oxidation.

Developing lesions have decreased ratios of albumin and apolipoprotein A-1 to LDL.[29, 30] Relative decreases in albumin may promote the availability of transition metal ions via decreased metal binding or lower the concentration of sacrificial thiols to scavenge two-electron oxidants (see above). Decreased apolipoprotein A-1 infers a relative decrease in HDL in the vascular wall. This could also favour oxidative events as a number of antioxidant activities have been assigned to HDL.[79]

The molecular basis for an antioxidant action of paraoxonase (an aryl esterase) is obscure, yet in HDL it exerts antioxidant activity including inhibition of lipid peroxidation.[79, 80] Paraoxonase is present in interstitial fluid associated with HDL[81] and gene knockout studies demonstrate increased (although small) lesion formation with decreased paraoxonase.[82] Nonetheless, direct interaction of HDL paraoxonase with LDL lipid oxidative events in developing lesions remains to be shown. Of equal, or perhaps greater, importance, a decrease in the relative concentration of intimal HDL would decrease methionine residue availability for lipid hydroperoxide reduction and removal of potential pro-oxidants (see above).

Heme oxygenase-1 (HO-1) is induced in oxidative stress[83] and is expressed in cells of human atherosclerotic lesions.[84] Thus, increased HO-1 could represent a local antioxidant response,[85] particularly if biliverdin reductase activity was present together with ferritin synthesis.[83, 86] This scenario would result in removal of heme and, hence, the removal of a potential pro-oxidant, the generation of bilirubin, a co-antioxidant for α-TOH, and the sequestration of iron.[86] Interestingly, *in vitro* oxidized LDL induces HO-1 in macrophages[84] and plasma bilirubin is inversely associated with the risk of cardiovascular disease.[87] For HO-1 and all of the above enzymatic antioxidants, the use of gene knockout techniques will prove useful in determining whether lack of these enzymes results in lesion formation at least in animal models of atherosclerosis.

16.7   Conclusions

Despite extensive literature on LDL lipid oxidation and intense interest in the 'oxidative theory' of atherosclerosis, we still do not fully appreciate the changes to the nature and location of the antioxidants in the vascular wall during atherogenesis. Indeed, where shown, the relevance of changes in enzymatic antioxidants during atherosclerotic lesion formation and for intimal lipoprotein oxidation is unclear. TMP of lipoprotein lipid is a general model that accounts for the pro-oxidant activity of α-TOH in oxidizing lipoproteins that is not explained by the vitamin's classical antioxidant activity. While the presence of α-TOH and oxidized lipid in atherosclerotic lesions may signal the possibility of TMP occurring *in vivo*, the additional co-existence of co-antioxidants that prevent TMP in lesions is confounding. What is clear is that substantial lipid oxidation occurs in the absence of gross vitamin E consumption. Clearly, further systematic and stage-dependent studies of enzymatic and non-enzymatic and non-proteinaceous antioxidants in human lesions are required to resolve these issues and adequately address the validity of the 'oxidation theory' of atherosclerosis.

Acknowledgements

Past and present support from The National Health & Medical Research Council and the National Heart Foundation of Australia is acknowledged.

References

1. Steinberg, D., Parthasarathy, S., Carew, T.E., Khoo, J.C., and Witztum, J.L. (1989). Beyond cholesterol: modifications of low-density lipoprotein that increase its atherogenicity. *N. Engl. J. Med.*, **320**, 915.
2. Berliner, J.A. and Heinecke, J.W. (1996). The role of oxidized lipoproteins in atherogenesis. *Free Radic. Biol. Med.*, **20**, 707.
3. Guyton, J.R. and Klemp, K.F. (1993). Transitional features in human atherosclerosis. Intimal thickening, cholesterol clefts, and cell loss in human aortic fatty streaks. *Am. J. Pathol.*, **143**, 1444.
4. Burton, G.W. and Ingold, K.U. (1986). Vitamin E: application of the principles of physical organic chemistry to the exploration of its structure and function. *Acc. Chem. Res.*, **19**, 194.
5. Bowry, V.W. and Stocker, R. (1993). Tocopherol-mediated peroxidation. The pro-oxidant effect of vitamin E on the radical-initiated oxidation of human low-density lipoprotein. *J. Am. Chem. Soc.*, **115**, 6029.
6. Ingold, K.U., Bowry, V.W., Stocker, R., and Walling, C. (1993). Autoxidation of lipids and antioxidation by α-tocopherol and ubiquinol in homogeneous solution and in aqueous dispersions of lipids. The unrecognized consequences of lipid particle size as exemplified by the oxidation of human low density lipoprotein. *Proc. Natl. Acad. Sci. USA*, **90**, 45.
7. Bowry, V.W., Ingold, K.U., and Stocker, R. (1992). Vitamin E in human low-density lipoprotein. When and how this antioxidant becomes a pro-oxidant. *Biochem. J.*, **288**, 341.
8. Bowry, V.W., Mohr, D., Cleary, J., and Stocker, R. (1995). Prevention of tocopherol-mediated peroxidation of ubiquinol-10-free human low density lipoprotein. *J. Biol. Chem.*, **270**, 5756.
9. Bowry, V.W., Stanley, K.K., and Stocker, R. (1992). High density lipoprotein is the major carrier of lipid hydroperoxides in fasted human plasma. *Proc. Natl. Acad. Sci. USA*, **89**, 10316.
10. Mohr, D. and Stocker, R. (1994). Radical-mediated oxidation of isolated human very low density lipoprotein. *Arterioscl. Thromb.*, **14**, 1186.
11. Mohr, D., Umeda, Y., Redgrave, T.G., and Stocker, R. (1999). Antioxidant defenses in rat intestine and mesenteric lymph. *Redox Report*, **4**, 79.
12. Neuzil, J., Thomas, S.R., and Stocker, R. (1997). Requirement for, promotion, or inhibition by α-tocopherol of radical-induced initiation of plasma lipoprotein lipid peroxidation. *Free Radic. Biol. Med.*, **22**, 57.
13. Upston, J.M., Neuzil, J., Witting, P.K., Alleva, R., and Stocker, R. (1997). 15-Lipoxygenase-induced enzymic oxidation of low density lipoprotein associated free fatty acids stimulates nonenzymic, α-tocopherol-mediated peroxidation of cholesteryl esters. *J. Biol. Chem.*, **272**, 30067.
14. Witting, P.K., Bowry, V.W., and Stocker, R. (1995). Inverse deuterium kinetic isotope effect for peroxidation in human low-density lipoprotein (LDL): a simple test for tocopherol-mediated peroxidation of LDL lipids. *FEBS Lett.*, **375**, 45.

15. Ingold, K.U. and Howard, J.A. (1962). Reaction of phenols with peroxy radicals. *Nature*, **195**, 280.

16. Witting, P.K., Westerlund, C., and Stocker, R. (1996). A rapid and simple screening test for potential inhibitors of tocopherol-mediated peroxidation of LDL lipids. *J. Lipid Res.*, **37**, 853.

17. Neuzil, J., Witting, P.K., and Stocker, R. (1997). α-Tocopheryl hydroquinone is an efficient multifunctional inhibitor of radical-initiated oxidation of low-density lipoprotein lipids. *Proc. Natl. Acad. Sci. USA*, **94**, 7885.

18. Doba, T., Burton, G.W., and Ingold, K.U. (1985). Antioxidant and co-antioxidant activity of vitamin C. The effect of vitamin C, either alone or in the presence of vitamin E or a water-soluble vitamin E analogue, upon the peroxidation of aqueous multilamellar phospholipid liposomes. *Biochim. Biophys. Acta*, **835**, 298.

19. Kagan, V.E., Serbinova, E.A., Forte, T., Scita, G., and Packer, L. (1992). Recycling of vitamin E in human low density lipoproteins. *J. Lipid Res.*, **33**, 385.

20. Upston, J.M., Karjalainen, A., Bygrave, F.L., and Stocker, R. (1999). Efflux of hepatic ascorbate. A potential contributor to the maintenance of plasma vitamin C. *Biochem. J.*, **342**, 49.

21. Witting, P.K., Upston, J.M., and Stocker, R. (1997). The role of α-tocopheroxyl radical in the initiation of lipid peroxidation in human low density lipoprotein exposed to horse radish peroxidase. *Biochemistry*, **36**, 1251.

22. Frei, B., Kim, M., and Ames, B.N. (1990). Ubiquinol-10 is an effective lipid-soluble antioxidant at physiological concentrations. *Proc. Natl. Acad. Sci. USA*, **87**, 4879.

23. Mohr, D., Bowry, V.W., and Stocker, R. (1992). Dietary supplementation with coenzyme Q10 results in increased levels of ubiquinol-10 within circulating lipoproteins and increased resistance of human low density lipoprotein to the initiation of lipid peroxidation. *Biochim. Biophys. Acta*, **1126**, 247.

24. Thomas, S.R., Neuzil, J., and Stocker, R. (1996). Co-supplementation with coenzyme Q prevents the pro-oxidant effect of α-tocopherol and increases the resistance of low-density lipoprotein towards transition metal-dependent oxidation initiation. *Arterioscl. Thromb. Vasc. Biol.*, **16**, 687.

25. Stocker, R. and Suarna, C. (1993). Extracellular reduction of ubiquinone-1 and -10 by human Hep G2 and blood cells. *Biochim. Biophys. Acta*, **1158**, 15.

26. Kozlov, A.V., Gille, L., Staniek, K., and Nohl, H. (1999). Dihydrolipoic acid maintains ubiquinone in the antioxidant active form by two-electron reduction of ubiquinone and one-electron reduction of ubisemiquinone. *Arch. Biochem. Biophys.*, **363**, 148.

27. Stocker, R., Bowry, V.W., and Frei, B. (1991). Ubiquinol-10 protects human low density lipoprotein more efficiently against lipid peroxidation than does α-tocopherol. *Proc. Natl. Acad. Sci. USA*, **88**, 1646.

28. Guyton, J.R. and Klemp, K.F. (1994). Development of the atherosclerotic core region. Chemical and ultrastructural analysis of microdissected atherosclerotic lesions from human aorta. *Arterioscler. Thromb.*, **14**, 1305.

29. Smith, E.B. (1974). The relationship between plasma and tissue lipids in human atherosclerosis. *Adv. Lipid Res.*, **12**, 1.

30. Ylä-Herttuala, S. (1991). Biochemistry of the arterial wall in developing atherosclerosis. *Ann. NY Acad. Sci.*, **623**, 40.

31. Cignarella, A., Brennhausen, B., von Eckardstein, A., Assmann, G., and Cullen, P. (1998). Differential effects of lovastatin on the trafficking of endogenous and lipoprotein-derived cholesterol in human monocyte-derived macrophages. *Arterioscler. Thromb. Vasc. Biol.*, **18**, 1322.

32. Chao, F.-F., Blanchette-Mackie, E.J., Chen, Y.-J., Dickens, B.F., Berlin, E., Amende, L.M., Skarlatos, S.I., Gamble, W., Resau, J.H., Mergner, W.T., and Kruth, H.S. (1990). Characterization of two unique cholesterol-rich lipid particles isolated from human atherosclerotic lesions. *Am. J. Pathol.*, **136**, 169.

33. Suarna, C., Dean, R.T., May, J., and Stocker, R. (1995). Human atherosclerotic plaque contains both oxidized lipids and relatively large amounts of α-tocopherol and ascorbate. *Arterioscler. Thromb. Vasc. Biol.*, **15**, 1616.

34. Letters, J.M., Witting, P.K., Christison, J.K., Westin Eriksson, A., Pettersson, K., and Stocker, R. (1999). Changes to lipids and antioxidants in plasma and aortae of apoE-deficient mice. *J. Lipid Res.*, **40**, 1104.

35. Frank, J.S. and Fogelman, A.M. (1989). Ultrastructure of the intima in WHHL and cholesterol-fed rabbit aortas prepared by ultra-rapid freezing and freeze-etching. *J. Lipid Res.*, **30**, 967.

36. Frei, B., Stocker, R., and Ames, B.N. (1988). Antioxidant defenses and lipid peroxidation in human blood plasma. *Proc. Natl. Acad. Sci. USA*, **85**, 9748.

37. Dabbagh, A.J. and Frei, B. (1995). Human suction blister interstitial fluid prevents metal ion-dependent oxidation of low density lipoprotein by macrophages and in cell-free systems. *J. Clin. Invest.*, **96**, 1958.

38. Retsky, K.L., Freeman, M.W., and Frei, B. (1993). Ascorbic acid oxidation product(s) protect human low density lipoprotein against atherogenic modification. Anti- rather than prooxidant activity of vitamin C in the presence of transition metal ions. *J. Biol. Chem.*, **268**, 1304.

39. Ames, B.N., Cathcart, R., Schwiers, E., and Hochstein, P. (1981). Uric acid provides an antioxidant defence in humans against oxidant- and radical-caused aging and cancer: a hypothesis. *Proc. Natl. Acad. Sci. USA*, **78**, 6858.

40. Davies, K.J.A., Sevanian, A., Muakkassah-Kelly, S.F., and Hochstein, P. (1986). Uric acid-iron ion complexes. A new aspect of the antioxidant functions of uric acid. *Biochem. J.*, **235**, 747.

41. Leake, D.S. (1997). Does an acidic pH explain why low density lipoprotein is oxidised in atherosclerotic lesions? *Atherosclerosis*, **129**, 149.

42. Christen, S., Peterhans, E., and Stocker, R. (1990). Antioxidant activities of some tryptophan metabolites: possible implication for inflammatory diseases. *Proc. Natl. Acad. Sci. USA*, **87**, 2506.

43. Christen, S., Thomas, S.R., Garner, B., and Stocker, R. (1994). Inhibition by interferon-γ of human mononuclear cell-mediated low density lipoprotein oxidation. Participation of tryptophan metabolism along the kynurenine pathway. *J. Clin. Invest.*, **93**, 2149.

44. Thomas, S.R., Davies, M.J., and Stocker, R. (1998). Oxidation and antioxidation of human low-density lipoprotein and plasma exposed

to 3-morpholinosydnonimine and reagent peroxynitrite.
Chem. Res. Toxicol., **11**, 484.

45. Thomas, S.R., Witting, P.K., and Stocker, R. (1996). 3-Hydroxyanthranilic acid is an efficient, cell-derived co-antioxidant for α-tocopherol, inhibiting human low density lipoprotein and plasma lipid peroxidation. *J. Biol. Chem.*, **271**, 32714.

46. Stocker, R., Yamamoto, Y., McDonagh, A.F., Glazer, A.N., and Ames, B.N. (1987). Bilirubin is an antioxidant of possible physiological importance. *Science*, **235**, 1043.

47. Neuzil, J. and Stocker, R. (1994). Free and albumin-bound bilirubin is an efficient co-antioxidant for α-tocopherol, inhibiting plasma and low density lipoprotein lipid peroxidation. *J. Biol. Chem.*, **269**, 16712.

48. Cooney, R.V., Franke, A.A., Harwood, P.J., Hatch-Pigott, V., Custer, L.J., and Mordan, L.J. (1993). γ-Tocopherol detoxification of nitrogen dioxide: superiority to α-tocopherol. *Proc. Natl. Acad. Sci. USA*, **90**, 1771.

49. Stocker, R. and Frei, B. (1991). Endogenous antioxidant defenses in human blood plasma. In *Oxidative stress: oxidants and antioxidants* (ed. H. Sies), p. 213. Academic Press, London.

50. Strålin, P., Karlsson, K., Johansson, B.O., and Marklund, S.L. (1995). The interstitium of the human arterial wall contains very large amounts of extracellular superoxide dismutase. *Arterioscl. Thromb. Vasc. Biol.*, **15**, 2032.

51. Oury, T.D., Day, B.J., and Crapo, J.D. (1996). Extracellular superoxide dismutase in vessels and airways of humans and baboons. *Free Radic. Biol. Med.*, **20**, 957.

52. Lynch, S.M., Frei, B., Morrow, J.D., Roberts, L.J., 2nd, Xu, A., Jackson, T., Reyna, R., Klevay, L.M., Vita, J.A., and Keaney, J.F., Jr (1997). Vascular superoxide dismutase deficiency impairs endothelial vasodilator function through direct inactivation of nitric oxide and increased lipid peroxidation. *Arterioscler. Thromb. Vasc. Biol.*, **17**, 2975.

53. Thomas, J.P., Maiorino, M., Ursini, F., and Girotti, A.W. (1990). Protective action of phospholipid hydroperoxide glutathione peroxidase against membrane-damaging lipid peroxidation. *In situ* reduction of phospholipid and cholesterol hydroperoxides. *J. Biol. Chem.*, **265**, 454.

54. Sattler, W., Maiorino, M., and Stocker, R. (1994). Reduction of HDL- and LDL-associated cholesterylester- and phospholipid hydroperoxides by phospholipid hydroperoxide glutathione peroxidase and Ebselen (PZ 51). *Arch. Biochem. Biophys.*, **309**, 214.

55. Garner, B., Waldeck, A.R., Witting, P.K., Rye, K.-A., and Stocker, R. (1998). Oxidation of high density lipoproteins. II. Evidence for direct reduction of HDL lipid hydroperoxides by methionine residues of apolipoproteins AI and AII. *J. Biol. Chem.*, **273**, 6088.

56. Sattler, W., Christison, J.K., and Stocker, R. (1995). Cholesterylester hydroperoxide reducing activity associated with isolated high- and low-density lipoproteins. *Free Radic. Biol. Med.*, **18**, 421.

57. Christison, J.K., Karjalainen, A., Brauman, J., Bygrave, F., and Stocker, R. (1996). Rapid reduction and removal of HDL- but not LDL-associated

cholesterylester hydroperoxides by *in situ* perfused rat liver. *Biochem. J.*, **314**, 739.

58. Björnstedt, M., Hamberg, M., Kumar, S., Xue, J., and Holmgren, A. (1995). Human thioredoxin reductase directly reduces lipid hydroperoxides by NADPH and selenocystine strongly stimulates the reaction via catalytically generated selenols. *J. Biol. Chem.*, **270**, 11761.

59. Rubartelli, A., Bajetto, A., Allavena, G., Wollman, E., and Sitia, R. (1992). Secretion of thioredoxin by normal and neoplastic cells through a leaderless secretory pathway. *J. Biol. Chem.*, **267**, 24161.

60. Nakamura, H., Vaage, J., Valne, G., Padilla, C.A., Björnstedt, M., and Holmgren, A. (1998). Measurement of plasma glutaredoxin and thioredoxin in healthy volunteers and during open-heart surgery. *Free Radic. Biol. Med.*, **24**, 1176.

61. Esterbauer, H., Gebicki, J., Puhl, H., and Jürgens, G. (1992). The role of lipid peroxidation and antioxidants in oxidative modification of LDL. *Free Radic. Biol. Med.*, **13**, 341.

62. van Hinsburg, V.W.M., Scheffer, M., Havekes, L., and Kempen, H.J.M. (1986). Role of endothelial cells and their products in the modification of low-density lipoproteins. *Biochim. Biophys. Acta*, **878**, 49.

63. Ehrenwald, E., Chisolm, G.M., and Fox, P.L. (1994). Intact human ceruloplasmin oxidatively modifies low density lipoprotein. *J. Clin. Invest.*, **93**, 1493.

64. Daugherty, A., Dunn, J.L., Rateri, D.L., and Heinecke, J.W. (1994). Myeloperoxidase, a catalyst for lipoprotein oxidation, is expressed in human atherosclerotic lesions. *J. Clin. Invest.*, **94**, 437.

65. Hazell, L.J., Arnold, L., Flowers, D., Waeg, G., Malle, E., and Stocker, R. (1996). Presence of hypochlorite-modified proteins in human atherosclerotic lesions. *J. Clin. Invest.*, **97**, 1535.

66. Hu, M.L., Louie, S., Cross, C.E., Motchnik, P., and Halliwell, B. (1993). Antioxidant protection against hypochlorous acid in human plasma. *J. Lab. Clin. Med.*, **121**, 257.

67. Stocker, R. (1999). The ambivalence of vitamin E in atherogenesis. *TiBS*, **24**, 219.

68. Offermann, M.K. and Medford, R.M. (1994). Antioxidants and atherosclerosis: a molecular perspective. *Heart Disease and Stroke*, **3**, 52.

69. Willis, G.C. and Fishman, S. (1955). Ascorbic acid content of human arterial tissue. *Canad. Med. Assoc. J.*, **72**, 500.

70. Carpenter, K.L., Cheeseman, K.H., van der Veen, C., Taylor, S.E., Walker, M.K., and Mitchinson, M.J. (1995). Depletion of alpha-tocopherol in human atherosclerotic lesions. *Free Radic. Res.*, **23**, 549.

71. Hansson, G.K., Holm, J., and Jonasson, L. (1989). Detection of activated T lymphocytes in the human atherosclerotic plaque. *Am. J. Pathol.*, **135**, 169.

72. Stocker, R. (1999). Antioxidant defenses in the vascular wall. In *Oxidative stress and vascular disease* (ed. J.F., Jr Keaney), p. 27. Kluwer Academic Publishers, Boston.

73. Beckman, J.S., Ye, Y.Z., Anderson, P.G., Chen, J., Accavitti, M.A., Tarpey, M.M., and White, C.R. (1994). Extensive nitration of protein tyrosine in human

atherosclerosis detected by immunohistochemistry. *Biol. Chem. Hoppe Seyler*, **375**, 81.

74. Leeuwenburgh, C., Hardy, M.M., Hazen, S.L., Wagner, P., Oh-ishi, S., Steinbrecher, U.P., and Heinecke, J.W. (1997). Reactive nitrogen intermediates promote low density lipoprotein oxidation in human atherosclerotic intima. *J. Biol. Chem.*, **272**, 1433.

75. Tribble, D.L., Gong, E.L., Leeuwenburgh, C., Heinecke, J.W., Carlson, E.L., Verstuyft, J.G., and Epstein, C.J. (1997). Fatty streak formation in fat-fed mice expressing human copper-zinc superoxide dismutase. *Arterioscler. Thromb. Vasc. Biol.*, **17**, 1734.

76. Jessup, W., Mohr, D., Gieseg, S.P., Dean, R.T., and Stocker, R. (1992). The participation of nitric oxide in cell free- and its restriction of macrophage-mediated oxidation of low-density lipoprotein. *Biochim. Biophys. Acta*, **1180**, 73.

77. Yates, M.T., Lambert, L.E., Whitten, J.P., McDonald, I., Mano, M., Ku, G., and Mao, S.J. (1992). A protective role for nitric oxide in the oxidative modification of low density lipoproteins by mouse macrophages. *FEBS Lett.*, **309**, 135.

78. Jessup, W. and Dean, R.T. (1993). Autoinhibition of murine macrophage-mediated oxidation of low-density lipoprotein by nitric oxide synthesis. *Atherosclerosis*, **101**, 145.

79. Mackness, M.I., Arrol, S., Abbott, C., and Durrington, P.N. (1993). Protection of low-density lipoprotein against oxidative modification by high-density lipoprotein associated paraoxonase. *Atherosclerosis*, **104**, 129.

80. Watson, A.D., Berliner, J.A., Hama, S.Y., La Du, B.N., Faull, K.F., Fogelman, A.M., and Navab, M. (1995). Protective effect of high density lipoprotein associated paraoxonase. Inhibition of the biological activity of minimally oxidized low density lipoprotein. *J. Clin. Invest.*, **96**, 2882.

81. Mackness, M.I., Mackness, B., Arrol, S., Wood, G., Bhatnagar, D., and Durrington, P.N. (1997). Presence of paraoxonase in human interstitial fluid. *FEBS Lett.*, **416**, 377.

82. Shih, D.M., Gu, L., Xia, Y.-R., Navab, M., Li, W.-F., Hama, S., Castellani, L.W., Furlong, C.E., Costa, L.G., Fogelman, A.M., and Lusis, A.J. (1998). Mice lacking serum paraoxonase are susceptible to organophosphate toxicity and atherosclerosis. *Nature*, **394**, 284.

83. Vile, G.F., Basu-Modak, S., Waltner, C., and Tyrrell, R.M. (1994). Heme oxygenase 1 mediates an adaptive response to oxidative stress in human skin fibroblasts. *Proc. Natl. Acad. Sci. USA*, **91**, 2607.

84. Wang, L.J., Lee, T.S., Lee, F.Y., Pai, R.C., and Chau, L.Y. (1998). Expression of heme oxygenase-1 in atherosclerotic lesions. *Am. J. Pathol.*, **152**, 711.

85. Stocker, R. (1990). Induction of haem oxygenase as a defence against oxidative stress. *Free Radic. Res. Commun.*, **9**, 101.

86. Balla, G., Jacob, H.S., Balla, J., Rosenberg, M., Nath, K., Apple, F., Eaton, J.W., and Vercellotti, G.M. (1992). Ferritin: a cytoprotective antioxidant strategem of endothelium. *J. Biol. Chem.*, **267**, 18148.

87. Schwertner, H.A., Jackson, W.G., and Tolan, G. (1994). Association of low serum concentration of bilirubin with increased risk of coronary artery disease. *Clin. Chem.*, **40**, 18.

88. Lapenna, D., de Gioia, S., Ciofani, G., Mezzetti, A., Ucchino, S., Calafiore, A.M., Napolitano, A.M., Di Ilio, C., and Cuccurullo, F. (1998). Glutathione-related antioxidant defenses in human atherosclerotic plaques. *Circulation*, **97**, 1930.
89. Luoma, J.S., Strålin, P., Marklund, S.L., Hiltunen, T.P., Sarkioja, T., and Ylä-Herttuala, S. (1998). Expression of extracellular SOD and iNOS in macrophages and smooth muscle cells in human and rabbit atherosclerotic lesions: colocalization with epitopes characteristic of oxidized LDL and peroxynitrite-modified proteins. *Arterioscler. Thromb. Vasc. Biol.*, **18**, 157.

17 Oxysterols in atherogenesis

Malcolm A. Lyons and Andrew J. Brown
Cell Biology Group, The Heart Research Institute, Camperdown,
Sydney, NSW 2050, Australia

Cholesterol oxidation products, or oxysterols, are derivatives of cholesterol that have been modified by addition of one or more oxygen functional groups (Table 17.1, Fig. 17.1). Ever since they were first discovered in atherosclerotic lesions more than 50 years ago, controversy has surrounded their role in atherosclerosis. In this chapter, we will critically evaluate the evidence for oxysterols playing a role in the initiation and/or development of atherosclerosis. For a more comprehensive and general discussion of oxysterols, the reader is referred elsewhere.[1, 2]

17.1 Oxysterols in human atherosclerotic lesions

Central to the proposition that oxysterols play an important role in atherogenesis, is their presence in human atherosclerotic lesions. The oxysterol to cholesterol ratio in lesions is several orders of magnitude higher than in normal tissues or plasma and is even greater in lipid-laden macrophage-foam cells (Fig. 17.2). In foam cells, approximately 5% of the sterol is oxidized compared with less than 0.01% in plasma (reviewed in Ref. (2)). The enzymically formed 27-hydroxycholesterol is a major oxysterol in lesions and foam cells along with oxysterols believed to be primarily derived non-enzymically (either endogenously or from the diet). These include 7-ketocholesterol, 7β-hydroxycholesterol, and cholesterol α- and β-epoxide. 7α-Hydroxycholesterol could originate either by the action of cholesterol 7α-hydroxylase or by a non-enzymic route. The apparent concentration of non-enzymically produced oxysterols in macrophage-foam cells, the predominant cell-type in early lesions, probably explains why their levels (in particular, 7-ketocholesterol and 7β-hydroxycholesterol) are elevated in fatty streaks (the earliest detectable lesion) compared to advanced lesions.[3] A concentration of oxysterols in macrophage-foam cells, generally regarded as the progenitor cell-type in atherosclerosis, is

Table 17. 1 Nomenclature of some common oxysterols

Common name(s)	Chemical name
Cholestenoic acid	3β-Hydroxy-(25*R*)-cholest-5-en-26-oic acid
Cholesterol α-epoxide	Cholestan-5α,6α-epoxy-3β-ol
Cholesterol β-epoxide	Cholestan-5β,6β-epoxy-3β-ol
Cholestane-triol	Cholestan-3β,5α,6β-triol
7α-Hydroxycholesterol	Cholest-5-en-3β,7α-diol
7β-Hydroxycholesterol	Cholest-5-en-3β,7β-diol
7α-Hydroperoxycholesterol	Cholest-5-en-7α-hydroperoxy-3β-ol
7β-Hydroperoxycholesterol	Cholest-5-en-7α-hydroperoxy-3β-ol
7-Ketocholesterol*	Cholest-5-en-3β-ol-7-one
7-Oxocholesterol	
–	Cholest-3,5-diene-7-one
22*R*-Hydroxycholesterol	(22*R*)-Cholest-5-en-3β,22-diol
24*S*-Hydroxycholesterol	(24*S*)-Cholest-5-en-3β,24-diol
24*S*,25-Epoxycholesterol	(24*S*)-Cholest-5-en-24,25-epoxy-3β-ol
25-Hydroxycholesterol	Cholest-5-en-3β,25-diol
26-Hydroxycholesterol	(25*S*)-Cholest-5-en-3β,26-diol
27-Hydroxycholesterol	(25*R*)-Cholest-5-en-3β,26-diol

* 7-Ketocholesterol and 7-oxocholesterol are synonyms for this oxysterol.

consistent with the idea that oxysterols may play a role in the initial development of this disease.

17.2   Evidence from studies *in vitro*

The notion that oxysterols are more than innocent bystanders in the atherosclerotic disease process is largely based on data from an abundance of *in vitro* studies. The current dominance of the Oxidation Hypothesis of Atherosclerosis, and more specifically the role of oxidatively modified low-density lipoprotein (LDL), has further intensified research into oxysterols since many of the atherogenic properties ascribed to oxidized LDL are shared by oxysterols. The oxysterols formed when LDL is subjected to oxidation *in vitro*[2, 4] are the same as those non-enzymically produced oxysterols observed in lesion foam cells but the levels can be far higher, comprising more than 50% of total sterol (Fig. 17.2). The early intermediates of cholesterol oxidation, 7α- and β-hydroperoxycholesterol, present at only trace levels in human lesions have often been overlooked in preparations of *in vitro*

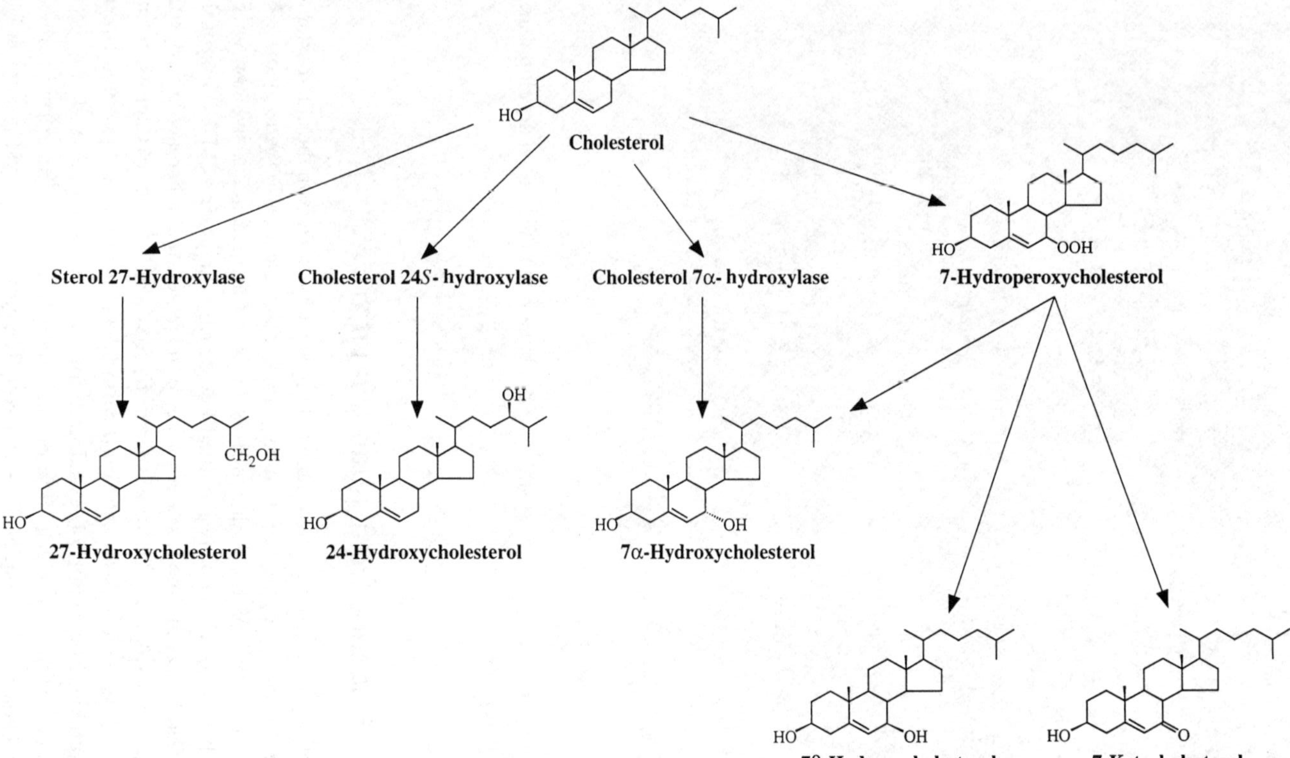

Fig. 17.1 Chemical structures and formation of some important oxysterols. Cholesterol may be acted upon by a number of cytochrome P450 enzymes to generate biologically active oxysterols. Alternatively the epimeric 7-hydroperoxycholesterols may be formed by free radical attack upon cholesterol. 7-Hydroperoxycholesterol can dissociate to form the epimeric 7-hydroxycholesterols or 7-ketocholesterol. 7α-Hydroxycholesterol can be formed by both enzymic and non-enzymic processes.

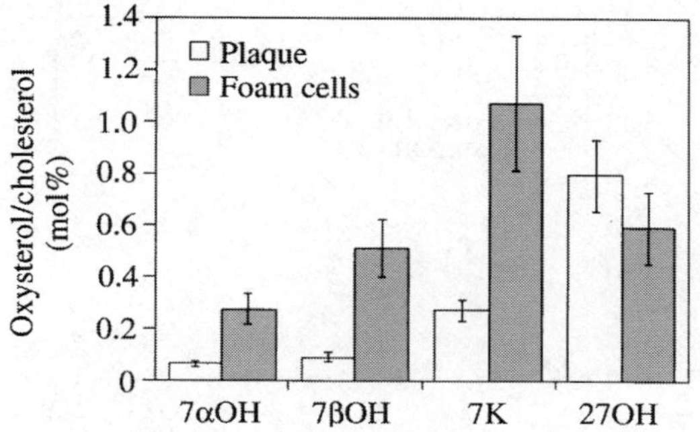

Fig. 17.2 Oxysterols are concentrated in foam cells isolated from human atherosclerotic plaque relative to human carotid atherosclerotic plaque. Foam cells were isolated from aortic aneurysm tissue ($n=3$) or carotid lesions ($n=4$) by Percoll gradient ultracentrifugation ($n=3$) or as CD14 positive macrophages isolated by immunomagnetic cell fractionation ($n=4$). Lipid extracts from all samples were subjected to cold alkaline saponification and analysed by either normal-phase high-performance liquid chromatography (plaque) or as the trimethylsilyl ether derivatives by gas chromatography-mass spectrometry with selected ion monitoring (foam cells). Values are mean $\pm$ SEM, plaque $n=8$, foam cells $n=7$. The molar ratios of oxysterols to cholesterol are expressed as a percentage. Plaque data were recalculated from Brown *et al.*[4] and foam cell data were from Brown and Jessup.[2]

oxidized LDL where they can be significant components.[4] Under various oxidizing conditions including copper-mediated oxidation, the 7-hydroperoxycholesterol species are longer lived than cholesteryl linoleate hydroperoxide or hydroxide (Fig. 17.3)[4] (in that they are slower to peak and slower to decay) and may have far more profound effects than their decomposition products, 7-ketocholesterol and 7-hydroxycholesterol.[5]

Although there are stages of LDL oxidation *in vitro* that reflect the profile of lipid oxidation products in lesions,[4] minimal oxysterol formation occurs *in vitro* in the presence of α-tocopherol[6] while there is an apparent abundance of this antioxidant in advanced lesions.[7] Therefore, if *in situ* formation of oxysterols is a major source of oxysterols in atherosclerotic lesions, then it suggests that it may be preceded by depletion of local antioxidant defences.

From the plethora of *in vitro* studies, many potential pro-atherogenic properties have been attributed to oxysterols, including cytotoxicity, induction of apoptosis, inhibition of endothelial-derived relaxation, possessing pro-coagulant activity, increasing expression of certain cytokines (e.g. interleukin-8), and the ability to perturb sterol metabolism in cells (reviewed in Ref. (2)). A major problem with most of these studies is that the doses of oxysterols used were pharmacological and levels achieved in the cells, in the rare instances where these have been measured, were far higher than would be encountered *in vivo*. For example, studies in human endothelial cells indicate that the lowest concentration of 7-ketocholesterol and 7β-hydroxycholesterol demonstrated to induce apoptosis and inhibit nitric oxide

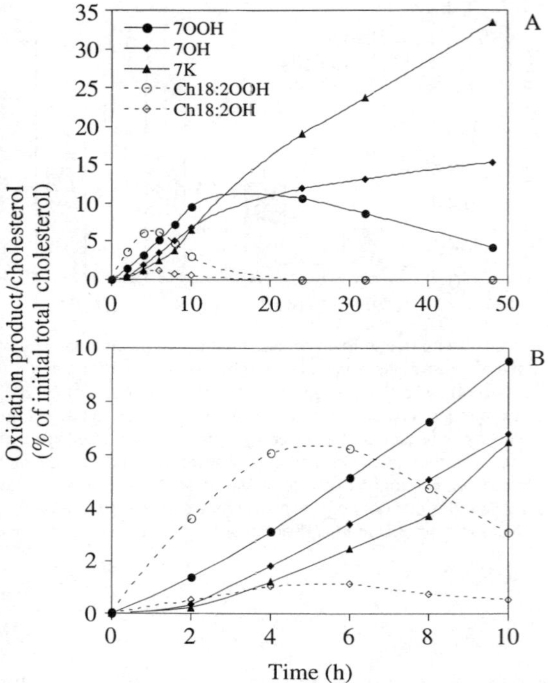

Fig. 17.3 Kinetics of appearance of oxysterols and cholesterol hydro(pero)xides in LDL oxidized *in vitro*. LDL (1.0 mg protein/ml) was incubated with Cu^{2+} (20 μM) at 37 °C for 48 h (panel A). Panel B shows the first 10 h of oxidation Oxysterols were analysed by normal-phase high-performance liquid chromatography. Values are mean of triplicate oxidations from one experiment and standard deviations were < 18% of the mean. The molar ratios of oxidation product to initial total cholesterol level are expressed as a percentage. All α-tocopherol is consumed within the first hour under these conditions.

release was $25 \mu\text{mol}\,1^{-1}$ [8] and $150 \mu\text{mol}\,1^{-1}$,[9] respectively. These effective concentrations are far greater than circulating levels of these two oxysterols (approximately $0.1 \mu\text{mol}\,1^{-1}$).[10]

In the Cell Biology Group at The Heart Research Institute, we have made a point of measuring the levels of oxysterols in our cell culture systems and relating this to what is known about the *in vivo* situation. Hence, we have shown[11] that at high intracellular concentrations, 7-ketocholesterol impairs cholesterol export from mouse peritoneal macrophages to lipid-free apolipoprotein A-I, the principal protein in high-density lipoprotein (HDL). However, much lower levels of 7-ketocholesterol, within the range found in authentic foam cells, are required to inhibit cholesterol export from human monocyte-derived macrophages (see Chapter 9). Thus, there may be some pro-atherogenic properties attributed to oxysterols that may occur at pathophysiologically relevant levels.

17.3 Evidence from studies in animals

17.3.1 Effects of injected oxysterols

Injecting oxysterols into rabbits has proved to be a very reliable way to demonstrate toxic effects to the vasculature *in vivo*.[2] Injection of 2.5 mg kg^{-1} body weight of either cholestan-3β,5α,6β-triol or 25-hydroxycholesterol produced aortic surface blebbing after 24 h[12] while 10 mg kg^{-1} body weight of 7-ketocholesterol was required to produce gross histological changes.[13] However, even a dose of 2.5 mg kg^{-1} body weight equates to 175 mg or ≈35% of daily sterol consumed for a 70 kg human consuming 500 mg sterol per day.[2] More recently, injections of an oxysterol mixture into normocholesterolaemic rabbits (1.4 g over a 70 d period) were shown to increase vascular permeability to macromolecules and to result in cholesterol accumulation in the aorta.[14] Injection of the same oxysterol mixture into cholesterol-fed rabbits increased aortic atherosclerosis compared with vehicle-injected control rabbits.[15] The dose given in these two studies was designed to simulate plasma levels of oxysterols found previously by this group in hypercholesterolaemic rabbits, but still in human terms equates to 39 g for a 70 kg person or ~560 mg d^{-1}. To put these amounts of oxysterols into context with the human diet, a test meal designed to study oxysterol absorption and comprising 150 g each of Parmesan cheese and salami contained only 3.5 mg of total oxysterols.[16]

17.3.2 Dietary oxysterols

17.3.2.1 Absorption of dietary oxysterols

Direct intravascular injection bypasses the additional complications of intestinal absorption. Studies in humans (e.g. Ref. (16)) and animals (e.g. Ref. (17)) have demonstrated that dietary oxysterols can be absorbed and are then transported in chylomicrons. However, estimates of the extent to which oxysterols are absorbed vary greatly; though on balance they suggest that oxysterols may not be absorbed as well as cholesterol.[2]

17.3.2.2 Atherogenicity of dietary oxysterols

Much of the current interest in oxysterols can be traced back to the work of two groups of researchers. Altschul and co-workers conducted a number of studies in which it was suggested that heated cholesterol was more potent at inducing 'experimental arteriosclerosis' than cholesterol alone.[18] The second group of researchers took this idea further and proposed that the atherogenic effect of cholesterol may have been due more to the oxysterols contaminating the cholesterol supplied (United States Pharmacopeia (USP)-grade) than to the cholesterol *per se*.[19] Thus, dietary oxysterols occupy a central position in the evolution of the oxysterol hypothesis. However, since the earliest experiment in which oxysterols were deliberately fed to animals,[20] the results have been equivocal. Rabbits and various avian models have been most commonly studied, with individual oxysterols, or sometimes poorly defined mixtures, supplied at relatively high levels (0.5–5% w/w of the diet). Particular attention has been paid to cholestan-3β,5α,6β-triol and 25-hydroxycholesterol although their levels in the human diet, plasma, and tissues are relatively minor. Of 16 studies identified (Table 17.2), 7 had a pro-atherogenic effect, 8 were anti-atherogenic, and 1 had no clear effect. In general, pro-atherogenic effects were not demonstrated when purified oxysterols were fed in the absence of cholesterol. Fed with cholesterol, oxysterols can often blunt the cholesterol-induced hypercholesterolaemia thereby potentially confounding any effect of oxysterols on atherogenesis. Overall, experiments in which oxysterols have been fed to animals do not support a pro-atherogenic effect of dietary oxysterols.

17.3.2.3 Metabolism of oxysterols

There appears to be a widely held view that oxysterols once formed from cholesterol are immutable. However, there are many examples where clearly oxysterols can be altered (reviewed in Refs (2, 21)). We have recently shown in an acute study in rats[22] that exogenous 7-ketocholesterol (a major dietary oxysterol) is rapidly metabolized by the liver relative to cholesterol, and is excreted into the intestine (Fig. 17.4) mostly as polar metabolites (presumably bile acids). We have preliminary evidence that this also occurs in a human liver cell-line in culture.

If oxysterols found in atherosclerotic lesions do originate exogenously, then oxysterols administered to animal models should accumulate within the artery (aortic) wall. Recent evidence does not support this proposal. In our acute study in rats,[22] aortic levels of injected radiolabelled 7-ketocholesterol had disappeared by 24 h whereas aortic levels of radiolabelled cholesterol injected at the same time were still increasing. In three recent studies investigating the role of oxysterols in atherogenesis in rabbits, an increased cholesterol content was found in the aortae of the animals.[14, 15, 23] However, oxysterols were either not measured,[15] not detected at all (despite being detected in circulating triacylglycerol-rich lipoproteins),[23] or were not elevated above that of the control group.[14] Taken together, these findings raise doubts as to the importance of circulating diet-derived oxysterols as a major source of oxysterols in atherosclerotic lesions and hence to the atherosclerotic disease process. Furthermore, this lack of evidence for dietary oxysterols contributing to oxysterols in the artery wall suggests that *in vivo* generation of these oxysterols close to or within the lesion may be the predominant source.

17.3.3 Association of aortic oxysterol levels and the extent of atherosclerosis

One strategy that can be employed to test the role of oxysterols in atherogenesis is to modulate the extent of atherosclerosis in animal models and to determine if this extent is associated with oxysterol levels. Such an approach has been used

Table 17.2 Summary of studies investigating the atherogenicity of dietary oxysterols

Reference	Oxysterol(s)	Test system	Hypocholesterolaemic effect of oxysterol(s)	Results on atherosclerosis
Pro-atherogenic Altschul[49]	Powdered yolk/ yolk cake	Golden hamster and guinea pigs fed powdered yolk and yolk cake		Intimal lipid deposition. Similar response to chol-fed rabbits
Cook and MacDougall[50]	TRIOL	NZ White rabbits (M) fed -0.1%* in diet (=~30 mg/kg body wt/d) for 27–350 d	No difference chol+TRIOL-fed versus chol-fed group but TRIOL-fed hypochol	Greater severity of aortic atherosclerosis in TRIOL-fed animals than animals fed similar amounts of chol
Shih[51]	USP chol which has been oxidized (60% chol), not treated (95% chol) or purified (98% chol)	Japanese quail fed different cholesterol preparations (0.5% of diet) for 12 weeks		No difference in aortic atherosclerosis score between treatments but aortic cholesterol content was greatest in birds fed oxidized chol, least in purified chol-fed birds, and intermediate in USP-fed birds

Table 17.2 (Cont)

Reference	Oxysterol(s)	Test system	Hypocholesterolaemic effect of oxysterol(s)	Results on atherosclerosis
Jacobson et al.[52]	TRIOL	White Carneau pigeons: administration by gavage 3×/week for 3 months; $3\,mg\,g^{-1}$ chol in a diet containing 0.05% chol	No difference chol-fed versus chol + TRIOL-fed	No difference in aortic accumulation of free and esterified chol but aortic calcium content and coronary artery atherosclerosis increased in α-TRIOL-fed animals compared with those fed purified chol
Mahfouz et al.[53]	USP chol stored > 15 years: 90.4% chol, 4.3% 7K, 2.5% TRIOL, 1.9% 26OH[†], 0.6% 7βOH, 0.2% 7αOH, 0.2% β-EPOX. Purified chol contained 0.7% oxysterols.	NZ White rabbits (M) fed diets which contained 0.5% pure chol or oxidized chol for 11 weeks	No difference chol-fed versus ox-chol-fed	Aorta of animals fed purified chol showed less intimal thickening than in rabbits fed oxidized chol despite similar plasma chol levels

Table 17.2 (Cont)

Staprans et al.[54]	Oxidized chol: 7% 7αOH, 20% 7βOH, 16% β-EPOX, 12% α-EPOX, 42% 7K, 3% 25OH. These amounted to 52% total oxysterols.	NZ White rabbits fed 0.33% chol or 0.33% ox-chol (5% oxidized), i.e. 25 mg ox-chol per day for 12 weeks	No difference chol-fed versus ox-chol-fed	Ox-chol-fed group two-fold increase in lesion area compared with chol-fed group
Vine et al.[23]	Oxidized chol: 42% 7K, 18% 7βOH, 4% α-EPOX, 4% β-EPOX, 7% cholest-4-ene-3-one, 4% 25OH, 21% unknown	Semi-Lop rabbits fed 1% purified chol or 1% ox-chol (6% oxidized) for 14 d	ox-chol-fed group hypochol	Aortic chol increased in ox-chol-fed group
No clear effect Sunde and Lalich[55]	(Oxy)sterol mixture: 31.5% chol, 19% α-EPOX, 13% 25OH, 10% 7αOOH, 7% 7K, 4% 7βOH, 2% 7αOH	Chickens and quails fed mixed oxysterols or purified chol at 5% for 13 or 14 weeks		Intimal aortic plaques only developed in the chol-fed quail. Intimal hypertrophy was only observed in the oxysterol-fed quail. Mortality greater in birds fed oxysterols versus chol

Table 17.2 (Cont)

Reference	Oxysterol(s)	Test system	Hypocholesterolaemic effect of oxysterol(s)	Results on atherosclerosis
Anti-atherogenic Kantiengar et al.[56]	7Kdiene	Light Sussex cockerels fed 0.34% 7Kdiene or 2.5% of commercial chol for 8 weeks	7Kdiene-fed group hypochol	Intimal lipid deposition in aorta of chol-fed birds. 7Kdiene – similar to basal-fed birds.
Schwenk et al.[57]	Oxysterols isolated from commercial chol	Rabbits (Dutch Belted, New New Zealand and Chinchilla) fed purified or commercial chol, commercial chol plus isolated impurities or impurities alone for . 12 weeks	Chol + oxysterol-fed group hyperchol relative to pure chol-fed but oxysterol-fed hypochol	Isolated oxysterols alone did not cause lesions. Commercial chol plus oxysterols caused mild lesions but less severe than commercial chol and purified most severe lesions of all.
Aramaki et al.[58]	TRIOL	White rabbits (M) fed 0.5% chol $\pm$ 0.2 or 0.5% TRIOL for 2.5 months	Chol + TRIOL hypochol	Degree of aortic atherosclerosis was least in the group fed the highest level of TRIOL, intermediate in the group fed 0.2% TRIOL, and most in the chol-fed control group. Plasma cholesterol levels also decreased in the TRIOL-fed groups.

Table 17.2 (Cont)

Imai et al.[19]	Autoxidized chol: 38% chol, 13% 25OH, 13% 7K, 7% 7OH, 5% TRIOL, 4% 5OOH or 7OOH (from Ref. (59))	NZ White rabbits (M) administered by gavage; 1 g of autoxidized chol or purified chol/kg body weight /7weeks or 7.7 g of new or 5-year-old chol/kg/6weeks. Old chol had rancid odour.		Autoxidized chol induced intimal, diffuse fibrous lesions without foam cells or hypercholesterolaemia. No lesions induced in rabbits fed purified, new or old chol.
Imai and Lee[60]	25OH	Hybrid hares (F) fed 2 g (1%) USP chol ± 0.5 mg 25OH/d every second week for up to 18 months	No difference	No effect of additional 25OH on diet-induced aortic lesions
Griminger and Fisher[61]	Heat-dried whole egg powder (undefined oxysterol content)	Single Comb White Leghorn chickens (M) fed 20% whole egg powder or equivalent fresh eggs for 24 or 37 weeks	No difference	Less aortic atherosclerosis observed in powdered egg group versus fresh egg group. Similar plasma cholesterol levels between treatments.
Higley et al.[62]	Oxysterol mixture: 26% 7K, 24% β-EPOX, 18% 7αOOH, 17% β-EPOX, 5% 7βOH,	NZ white rabbits (F) fed diets containing 166 mg sterols/kg body weight/d		Severity of atherosclerosis in various arteries (including aorta) least in rabbits fed oxysterol mixture,

Table 17.2 (*Cont*)

Reference	Oxysterol(s)	Test system	Hypocholesterolaemic effect of oxysterol(s)	Results on atherosclerosis
	4.5% 7αOH, 2%. 25OH, 3% unknown. Oxysterol mix+chol: 35% chol, 25% β-EPOX, 10% 7αOOH, 8% 7K, 2% 25OH, 20% unknowns.			intermediate in group fed oxysterol mix+chol, and most in purified chol-fed group
Tipton *et al.*[63]	Chol hydroperoxides: three hydroperoxide bands representing ~40% of total material. Includes 25-hydroperoxycholesterol.	NZ White rabbits fed diets which contained 1% chol±50 mg d^{-1} chol hydroperoxides (with or without prior reduction) for 56 or 61 d		Addition of unreduced (but not reduced) chol hydroperoxides almost completely prevented aortic lesion formation and inhibited calmodulin activity *in vitro*

Abbreviations: TRIOL, cholestan-3α,5β,6β-triol; chol, cholesterol; 26OH, 26-hydroxycholesterol; 27OH, 27-hydroxycholesterol; 25OH, 25-hydroxycholesterol; 7K, 7-ketocholesterol; 7Kdiene, cholest-3,5-dien-7-one; 7αOH, 7α-hydroxycholesterol; 7βOH, 7β-hydroxycholesterol; β-EPOX, cholesterol β-epoxide; α-EPOX, cholesterol α-epoxide; 7αOOH, 7α-hydroperoxycholesterol; 7βOOH, 7β-hydroperoxycholesterol; 5OOH, 5-hydroperoxycholesterol; hypochol, hypocholesterolaemic; hyperchol, hypercholesterolaemic.

* Sterol in diet is expressed as % w/w.

† Reported as 26-hydroxycholesterol but *R/S* configuration not specified.

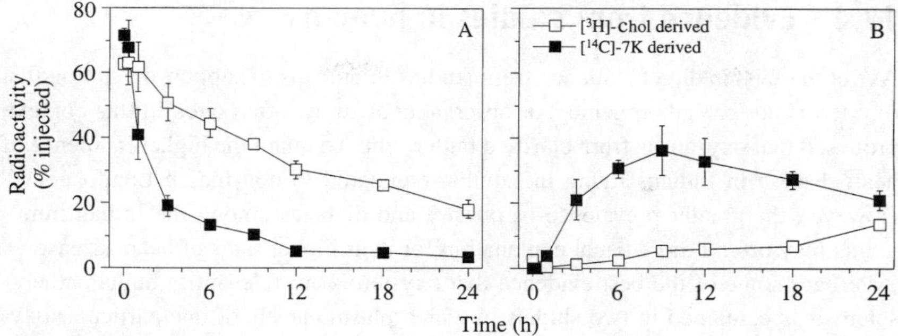

Fig. 17.4 Kinetics of clearance from the liver (panel A) and appearance in the intestine (panel B) of ^{14}C derived from 7-ketocholesterol and ^{3}H derived from cholesterol. Radiolabelled sterols were delivered as the oleate esters incorporated into the core of acetylated LDL, a modified lipoprotein, and administered via a jugular cannula. Total radioactivity was determined by solubilizing the tissue homogenates in a strong organic base followed by addition of scintillant and liquid scintillation counting with a dual-label protocol. Data are derived from Ref. (22) and are expressed as mean $\pm$ SD, $n = 3$ per time point except $t = 1$ h where values are mean $\pm$ range ($n = 2$). Abbreviations: chol, cholesterol; 7K, 7-ketocholesterol.

successfully by Stocker and colleagues to dissociate atherogenesis from aortic accumulation of primary lipid oxidation products, namely lipid hydroperoxides and hydroxides.[24] As is evident from Fig. 17.3, these products can be short-lived and arguably may not reflect cumulative oxidation nor advanced products of oxidation like oxysterols. Modifying the extent of atherosclerosis, almost by definition, means modifying the aortic cholesterol content. Therefore it is necessary that the aortic content of oxysterols be standardized for cholesterol levels. There have been three such studies published to date, each testing the effect of a different antioxidant on atherogenesis in rabbit models. In one of these studies, vitamin C was tested in a hypercholesterolaemic rabbit model where the aorta had been subjected to balloon injury.[25] No differences were found in plasma cholesterol or aortic oxysterol concentrations (cholesterol-standardized) or even in lesion severity. This result does not allow dissociation between the extent of atherosclerosis and oxysterol levels and hence does not refute the proposition that oxysterols are pro-atherogenic. In two studies testing the effects of the lipophilic antioxidants, Probucol[26] and butylated hydroxytoluene (BHT),[27] plasma cholesterol concentrations did not change but the aortic levels of oxysterols (with respect to cholesterol) were lowered in the antioxidant-fed groups and this decrease was associated with a reduction in atherosclerosis (either assessed histologically[27] or on the basis of aortic total cholesterol content.[26]) Although proving nothing more than an association, the studies published to date are consistent with the notion that increased levels of oxysterols in the aorta are pro-atherogenic. On the other hand, if oxysterol levels can be dissociated from the extent of atherosclerosis in further studies as shown for primary products of lipid peroxidation,[24] then the case against oxysterols would be considerably weakened. Further studies examining this association are currently underway at The Heart Research Institute.

17.4   Evidence from studies in humans

As yet there is no direct evidence from studies in humans to support the notion that oxysterols are pro-atherogenic. An observational study, often cited in this context, proposed that oxysterols from clarified butter (ghee) explain the higher incidence of heart disease in Indians living in London compared to non-Indian Londoners.[28] However, the greater prevalence of obesity and diabetes among the Indian immigrants may offer a more banal explanation for their higher rates of heart disease.

Perhaps some of the best evidence that oxysterols are relevant to human atherosclerosis is contained in two studies in which plasma levels of one particular oxysterol, 7β-hydroxycholesterol, were shown to be associated with increased risk of atherosclerosis. In a study of 40 mildly hypercholesterolaemic men from eastern Finland, a high serum concentration of 7β-hydroxycholesterol was the strongest predictor of a three-year increase in carotid wall thickness of more than 30 variables tested in a multivariate analysis.[29] However, levels were not significantly different between those men with the fastest atherosclerotic disease progression compared with those with no progression (78 versus 70 nmol l^{-1}). Another study [30] compared established and putative risk factors between 50 year-old Swedish men and their Lithuanian counterparts: the latter have a four-fold higher risk of heart disease compared to the Swedes. Plasma levels of 7β-hydroxycholesterol were higher in the Lithuanians than in the Swedes (30 versus 22 nmol l^{-1}) despite similar traditional risk factors, including plasma cholesterol levels. It should be noted that levels of 7β-hydroxycholesterol from both these studies were substantially higher than the group average from 31 healthy volunteers (7 nmol l^{-1}) analysed in the same laboratory.[10] There were comparable plasma cholesterol levels in all three studies, the group averages ranging from 4.8 to 5.9 mmol l^{-1}. The isotope dilution mass spectrometric approach employed by this group is arguably the best available method for analysing oxysterols, but artefactual oxidation *ex vivo* (during sample transport, storage, and/or analysis) is always a potential concern when dealing with oxidation products present in samples at levels several orders of magnitude lower than those of the parent compound. Indeed, Kudo *et al.*[31] added highly purified radiolabelled cholesterol to the blood collection tubes for two plasma samples and found that plasma levels of 7β-hydroxycholesterol could be nearly all attributable to *ex vivo* oxidation. Thus, levels of oxysterols in plasma and serum are susceptible to artefact and their use as risk markers should be treated with caution.

17.5   Potentially beneficial aspects of oxysterols

Oxysterols are not universally regarded as being detrimental. They have also been ascribed beneficial properties which may be anti-atherogenic or at least reflect potentially anti-atherogenic processes.

17.5.1 Cholesterol as antioxidant

Conventionally, cholesterol and polyunsaturated fatty acids are viewed as targets of oxidation. But the existence of oxysterols *in vivo* has been put forward as evidence that cholesterol acts as an antioxidant.[1, 32] Given that polyunsaturated fatty acids are far more reactive than cholesterol to the species most likely to initiate lipid oxidation (peroxyl radicals),[33, 34] then by the same token polyunsaturated fatty acids would have to be considered superior antioxidants than cholesterol. The fact that polyunsaturated fatty acids are still viewed as a target of lipid oxidation undermines the proposition that cholesterol rates as an antioxidant. Moreover, an antioxidant is usually considered to be a molecule that protects another at a very low molar ratio.

17.5.2 Oxysterols in the dissolution of cholesterol crystals

There have been many competing hypotheses to explain the emergence of the coronary epidemic in the twentieth century and the cardio-protective effects of the Mediterranean diet. It has been claimed [35] that oxysterols hold the key to both. Krut [35] has proposed that reduced levels of oxysterol in the Western diet (presumed to have fallen since the introduction of refrigeration) may have resulted in the coronary epidemic and linked the consumption of stored meat in Mediterranean countries (presumably enriched with oxysterols) with their lower rates of coronary heart disease. Both observations rely upon the diet being an important source of oxysterols *in vivo*. However, we argue above (Section 17.3.2) that the role of dietary oxysterols may have been overstated. The experimental evidence used to support Krut's hypothesis is that oxysterols, because of their amphipathic nature, can help solubilize crystalline cholesterol in model systems *in vitro* (0.4% oxysterol/cholesterol)[36] and *in vivo*.[37] Oxysterols at sites of atherosclerosis would presumably need to be extracellular and unesterified in order to be effective in this respect. On the contrary, we find that oxysterols are concentrated in macrophage-foam cells[2] and are predominantly esterified in advanced lesions.[4] Moreover, oxysterols have been reported to be concentrated in the core region where cholesterol monohydrate crystals are present, relative to the fibrous cap.[38]

17.5.3 Oxysterols as putative regulators of cholesterol homeostasis

Since oxysterols are formed *in vivo* (both by enzymic and non-enzymic mechanisms), it has been suggested that they may be important physiological regulators of cholesterol homeostasis.[39] Oxysterols exhibit greater potencies than purified cholesterol, perhaps due to the more polar oxysterols gaining entry into cells so as to achieve higher concentrations within the regulatory pool of sterols. There is evidence that the mediator of LDL receptor expression may be a monoxygenated form of cholesterol generated in a cytochrome P450-catalysed reaction.[40] Furthermore, a cytosolic oxysterol binding protein has been identified which displays a relative affinity for oxysterols that correlates with relative potency in suppressing the rate-limiting enzyme in cholesterol biosynthesis, 3-hydroxy-3-methylglutaryl coenzyme A (HMG CoA) reductase.[41] The sterol regulatory element-binding protein (SREBP) pathway by which genes for the LDL receptor and HMG CoA reductase are activated has been elucidated.[42] The SREBP cleavage activating protein (SCAP) allows proteolytic release of the transcription factor, and hence gene activation, and is the target for sterol suppression probably through a sterol-sensing domain. A growing list of proteins share this five-membrane spanning region.[42] Thus with improved molecular understanding of cholesterol homeostasis, a re-evaluation of the possible role for oxysterols is warranted. Which oxysterol(s) may be involved is not known.

25-Hydroxycholesterol has been the most intensively studied oxysterol since it is the most potent oxysterol regulator of gene transcription of the LDL receptor and HMG CoA reductase when assayed *in vitro*. Both human and murine cholesterol 25-hydroxylase cDNA has recently been cloned (interestingly, it is a hydroxylase requiring di-iron co-factors rather than being a cytochrome P450 enzyme). When the murine cDNA was transfected into cells, the cholesterol 25-hydroxylase expression reduced the biosynthesis of cholesterol from acetate and suppressed the cleavage of SREBP-1 and -2.[43] Although 25-hydroxycholesterol is only a minor oxysterol in human tissues (reviewed in Ref. (2)), it could be argued that concentrations of a finely tuned regulator would be best modulated against a low background.

17.5.4 Oxysterols binding to nuclear receptors

25-, 26-, or 27-Hydroxycholesterol selectively enhance transcriptional activity mediated by the nuclear receptor steroidogenic factor 1[44] and it has very recently been shown that oxysterols (including 22*R*-hydroxycholesterol, 24*S*-hydroxycholesterol, and 24*S*,25-epoxycholesterol) are positive effectors of liver X receptors (LXRs) by binding directly to these nuclear receptors at concentrations that occur *in vivo*.[45] Interestingly, LXRα induces a feed-forward pathway of cholesterol catabolism whereby it activates the rate-limiting enzyme of bile acid synthesis, cholesterol 7α-hydroxylase.[46] An attractive hypothesis is that regulation of cholesterol uptake by the LDL receptor and synthesis (both through SREBP) and cholesterol catabolism (through LXRα) is coordinated by a common oxysterol(s).

17.5.5 Oxysterols as intermediates in cholesterol catabolism

The levels of different oxysterols in lesions and foam cells may also reflect very different processes at work. While levels of 7-ketocholesterol and 7β-hydroxycholesterol may reflect non-enzymic oxidation at the site of atherosclerosis, levels of

27-hydroxycholesterol probably reflect the balance between several processes including formation by sterol 27-hydroxylase and subsequent conversion to a more polar product by the same enzyme. Sterol 27-hydroxylase is perhaps best known for its role in catalysing the first step in the alternative or acidic bile acid pathway in the liver, but is particularly highly expressed in macrophages, especially lung macrophages.[47] There is mounting evidence that this enzyme may catalyse an important pathway by which macrophages eliminate excess cholesterol without the need for lipoprotein acceptors (reviewed in Ref. (48)). This is significant considering that lipoprotein acceptors may well be limiting at sites of atherosclerosis. After converting cholesterol to 27-hydroxycholesterol, sterol 27-hydroxylase can subsequently convert 27-hydroxycholesterol to cholestenoic acid. As with free fatty acids, it is believed that this carboxylic acid can bind to albumin for transport back to the liver, where it is further catabolized. The accumulation of 27-hydroxycholesterol, the intermediate in this pathway, suggests that conversion to the carboxylic acid may be impaired in atherosclerosis. It is noteworthy that people who lack this enzyme, a rare condition called Cerebrotendinous Xanthomatosis, tend to develop premature atherosclerosis despite having normal circulating levels of cholesterol.

17.6 Conclusions

Thus far the evidence supporting a role for oxysterols in the initiation and/or development of atherosclerosis is more circumstantial than compelling. Oxysterols are present in human atherosclerotic lesions and appear concentrated in the progenitor cell-type of atherogenesis, the macrophage-foam cell. Oxysterols are abundant in LDL oxidized *in vitro*, but many of the putative pro-atherogenic effects of oxysterols have been achieved at levels of dubious pathophysiological relevance. The case for dietary oxysterols being pro-atherogenic is weak, with as many studies finding anti-atherogenic as pro-atherogenic effects. If oxysterol levels in lesions can be dissociated from the extent of atherosclerosis in animal models, then the case against oxysterols (from whatever source) would be considerably weakened. Beyond the direct question of pro-atherogenicity, oxysterols are still of considerable interest as ligands for nuclear receptors, as potential regulators of cholesterol homeostasis, and intermediates in the disposal of cholesterol. A better understanding of the role of oxysterols in normal physiology is required before we can fully appreciate their potential in pathogenesis.

Acknowledgements

We thank Dr Wendy Jessup for critically reading this manuscript. A.J.B. was supported by a grant from the National Heart Foundation. M.A.L. is the recipient of an Australian Postgraduate Award.

References

1. Smith, L. (1996). Review of progress in sterol oxidations: 1987–1995. *Lipids*, **31**, 453–87.
2. Brown, A.J. and Jessup, W. (1999). Oxysterols and atherosclerosis. *Atherosclerosis*, **142**, 1–28.
3. Carpenter, K.L., Taylor, S.E., van der Veen, C., Williamson, B.K., Ballantine, J.A., and Mitchinson, M.J. (1995). Lipids and oxidised lipids in human atherosclerotic lesions at different stages of development. *Biochim. Biophys. Acta*, **1256**, 141–50.
4. Brown, A.J., Leong, S.-I., Dean, R.T., and Jessup, W. (1997). 7-Hydroperoxycholesterol and its product in oxidized low-density lipoprotein and human atherosclerotic plaque. *J. Lipid Res.*, **38**, 1730–45.
5. Chisolm, G., Ma, G., Irwin, K., Martin, L., Gunderson, K., Linberg, L., Morel, D., and DiCorleto, P. (1994). 7β-Hydroperoxycholest-5-en-3β-ol, a component of human atherosclerotic lesions, is the primary cytotoxin of oxidized human low-density lipoprotein. *Proc. Natl. Acad. Sci. USA*, **91**, 11452–6.
6. Baoutina, A., Dean, R.T., and Jessup, W. (1998). α-Tocopherol supplementation of macrophages does not influence their ability to oxidize LDL. *J. Lipid Res.*, **39**, 114–30.
7. Suarna, C., Dean, R.T., May, J., and Stocker, R. (1995). Human atherosclerotic plaque contains both oxidized lipids and relatively large amounts of alpha-tocopherol and ascorbate. *Arterioscler. Thromb. Vasc. Biol.*, **15**, 1616–24.
8. Lizard, G., Monier, S., Cordelet, C., Gesquière, L., Deckert, V., Gueldry, S., Lagrost, L., and Gambert, P. (1999). Characterization and comparison of the mode of cell death, apoptosis versus necrosis, induced by 7β-hydroxycholesterol and 7-ketocholesterol in the cells of the vascular wall. *Arterioscler. Thromb. Vasc. Biol.*, **19**, 1190–200.
9. Deckert, V., Brunet, A., Lantoine, F., Lizard, G., Brussel, E.M.-v., Monier, S., Lagrost, L., David-Dufilho, M., Gambert, P., and Devynck, M.-A. (1998). Inhibition by cholesterol oxides of NO release from human vascular endothelial cells. *Arterioscler. Thromb. Vasc. Biol.*, **18**, 1054–60.
10. Dzeletovic, S., Breuer, O., Lund, E., and Diczfalusy, U. (1995). Determination of cholesterol oxidation products in human plasma by isotope dilution-mass spectrometry. *Anal. Biochem.*, **225**, 73–80.
11. Gelissen, I. C., Brown, A.J., Mander, E.L., Kritharides, L., Dean, R.T., and Jessup, W. (1996). Sterol efflux is impaired from macrophage foam cells selectively enriched with 7-ketocholesterol. *J. Biol. Chem.*, **271**, 17852–60.
12. Peng, S.K., Taylor, C.B., Hill, J.C., and Morin, R.J. (1985). Cholesterol oxidation derivatives and arterial endothelial damage. *Atherosclerosis*, **54**, 121–33.
13. Imai, H., Werthessen, N.T., Subramanyan, V., LeQuesne, P. W., Soloway, A.H., and Kanisawa, M. (1980). Angiotoxicity of oxygenated sterols and possible precursors. *Science*, **207**, 651–3.
14. Rong, J., Rangaswamy, S., Shen, L., Dave, R., Chang, Y., Peterson, H., Hodis, H., Chisolm, G., and Sevanian, A. (1998). Arterial injury by cholesterol oxidation products causes endothelial dysfunction and arterial wall cholesterol accumulation. *Arterioscler. Thromb. Vasc. Biol.*, **18**, 1885–94.

15. Rong, J.X., Shen, L., Chang, Y.H., Richters, A., Hodis, H.N., and Sevanian, A. (1999). Cholesterol oxidation products induce vascular foam cell lesion formation in hypercholesterolemic New Zealand White rabbits. *Arterioscler. Thromb. Vasc. Biol.*, **19**, 2179–88.

16. Linseisen, J. and Wolfram, G. (1998). Absorption of cholesterol oxidation products from ordinary foodstuff in humans. *Ann. Nutr. Metab.*, **42**, 221–30.

17. Vine, D.F., Croft, K.D., Beilin, L.J., and Mamo, J.C. (1997). Absorption of dietary cholesterol oxidation products and incorporation into rat lymph chylomicrons. *Lipids*, **32**, 887–93.

18. Altschul, R. (1950). *Selected studies on arteriosclerosis.* C.C. Thomas, Springfield, IL.

19. Imai, H., Werthessen, N.T., Taylor, C.B., and Lee, K.T. (1976). Angiotoxicity and arteriosclerosis due to contaminants of USP-grade cholesterol. *Arch. Pathol. Lab. Med.*, **100**, 565–72.

20. Cox, R.H. and Spencer, E.Y. (1949). The effect of 7-ketocholesterol on the rabbit. *Science*, **110**, 11.

21. Lyons, M.A. and Brown, A.J. (1999). Molecules in focus: 7-ketocholesterol. *Int. J. Biochem. Cell Biol.*, **31**, 369–75.

22. Lyons, M.A., Samman, S., Gatto, L., and Brown, A.J. (1999). Rapid hepatic metabolism of 7-ketocholesterol *in vivo*: implications for dietary oxysterols. *J. Lipid Res.*, **40**, 1846–57.

23. Vine, D.F., Mamo, J.C.L., Beilin, L.J., Mori, T.A., and Croft, K.D. (1998). Dietary oxysterols are incorporated in plasma triglyceride-rich lipoproteins, increase their susceptibility to oxidation and increase aortic cholesterol concentration of rabbits. *J. Lipid Res.*, **39**, 1995–2004.

24. Witting, P., Pettersson, K., Ostlund-Lindqvist, A.M., Westerlund, C., Wagberg, M., and Stocker, R. (1999). Dissociation of atherogenesis from aortic accumulation of lipid hydro(pero)xides in Watanabe heritable hyperlipidemic rabbits. *J. Clin. Invest.*, **104**, 213–20.

25. Freyschuss, A., Xiu, R.J., Zhang, J., Ying, X., Diczfalusy, U., Jogestrand, T., Henriksson, P., and Björkhem, I. (1997). Vitamin C reduces cholesterol-induced microcirculatory changes in rabbits. *Arterioscler. Thromb. Vasc. Biol.*, **17**, 1178–84.

26. Hodis, H.N., Chauhan, A., Hashimoto, S., Crawford, D.W., and Sevanian, A. (1992). Probucol reduces plasma and aortic wall oxysterol levels in cholesterol fed rabbits independently of its plasma cholesterol lowering effect. *Atherosclerosis*, **96**, 125–34.

27. Freyschuss, A., Stiko-Rahm, A., Swedenborg, J., Henriksson, P., Björkhem, I., Berglund, L., and Nilsson, J. (1993). Antioxidant treatment inhibits the development of intimal thickening after balloon injury of the aorta of hypercholesterolemic rabbits. *J. Clin. Invest.*, **91**, 1282–8.

28. Jacobson, M.S. (1987). Cholesterol oxides in Indian ghee: possible cause of unexplained high risk of atherosclerosis in Indian immigrant populations. *Lancet*, **ii**, 656–8.

29. Salonen, J.T., Nyyssonen, K., Salonen, R., Porkkala-Sataho, E., Tuomainen, T.-P., Diczfalusy, U., and Björkhem, I. (1997). Lipoprotein oxidation and progression of carotid atherosclerosis. *Circulation*, **95**, 840–5.

30. Zieden, B., Kaminskas, A., Kristenson, M., Kucinskiene, Z., Vessby, B., Olsson, A.G., and Diczfalusy, U. (1999). Increased plasma 7β-hydroxycholesterol concentrations in a population with a high risk for cardiovascular disease. *Arterioscler. Thromb. Vasc. Biol.*, **19**, 967–71.

31. Kudo, K., Emmons, G.T., Casserly, E.W., Via, D.P., Smith, L.C., Pyrek, J.St., and Schroepfer, G.J., Jr (1989). Inhibitors of sterol synthesis. Chromatography of acetate derivatives of oxygenated sterols. *J. Lipid Res.*, **30**, 1097–111.

32. Girao, H., Mota, C., and Pereira, P. (1999). Cholesterol may act as an antioxidant in lens membranes. *Curr. Eye Res.*, **18**, 448–54.

33. Barclay, L.R., Baskin, K.A., Locke, S.J., and Vinqvist, M.R. (1989). Absolute rate constants for lipid peroxidation and inhibition in model biomembranes. *Can. J. Chem.*, **67**, 1366–9.

34. Barclay, L.R., Cameron, R.C., Forrest, B.J., Locke, S.J., Nigam, R., and Vinqvist, M.R. (1990). Cholesterol: free radical peroxidation and transfer into phospholipid membranes. *Biochim. Biophys. Acta*, **1047**, 255–63.

35. Krut, L.H. (1998). Oxidation products of cholesterol: their metabolism, their effect on the solubilization of cholesterol and their postulated role in preventing atherosclerosis. *Recent Res. Devel. Lipids Res.*, **2**, 299–318.

36. Krut, L.H. (1982). Solubility of cholesterol *in vitro* promoted by oxidation products of cholesterol. *Atherosclerosis*, **43**, 95–104.

37. Krut, L.H. (1982). Clearance of subcutaneous implants of cholesterol in the rat promoted by oxidation products of cholesterol. *Atherosclerosis*, **43**, 105–18.

38. Garcia-Cruset, S., Carpenter, K.L.H., Guardiola, F., and Mitchinson, M. J.(1999). Oxysterols in cap and core of human advanced atherosclerotic lesions. *Free Rad. Res.*, **30**, 341–50.

39. Kandutsch, A.A., Chen, H.W., and Heiniger, H.J. (1978). Biological activity of some oxygenated sterols. *Science*, **201**, 498–501.

40. Takagi, K., Alvarez, J.G., Favata, M.F., Trzaskos, J.M., and Strauss, J.F.I. (1989). Control of low density lipoprotein receptor gene promoter activity: ketoconazole inhibits serum lipoprotein but not oxysterol suppression of gene transcription. *J. Biol. Chem.*, **264**, 12352–7.

41. Taylor, F.R., Saucier, S.E., Shown, E.P., Parish, E.J., and Kandutsch, A.A. (1984). Correlation between oxysterol binding to a cytosolic binding protein and potency in the repression of hydroxymethylglutaryl coenzyme A reductase. *J. Biol. Chem.*, **259**, 12382–7.

42. Brown, M.S. and Goldstein, J.L. (1999). A proteolytic pathway that controls the cholesterol content of membranes, cells, and blood. *Proc. Natl. Acad. Sci. USA*, **96**, 11041–8.

43. Lund, E.G., Kerr, T.A., Sakai, J., Li, W.-P., and Russell, D.W. (1998). cDNA cloning of mouse and human cholesterol 25-hydroxylases, polytopic membrane proteins that synthesize a potent oxysterol regulator of lipid metabolism. *J. Biol. Chem.*, **1998**, 34316–27.

44. Lala, D.S., Syka, P.M., Lazarchik, S.B., Mangelsdorf, D.J., Parker, K.L., and Heyman, R.A. (1997). Activation of the orphan nuclear receptor steroidogenic factor 1 by oxysterols. *Proc. Natl. Acad. Sci. USA*, **94**, 4895–900.

45. Janowski, B.A., Grogan, M.J., Jones, S.A., Wisely, G.B., Kliewer, S.A., Corey, E.J., and Mangelsdorf, D.J. (1999). Structural requirements of ligands for the

oxysterol liver X receptors LXRα and LXRβ. *Proc. Natl. Acad. Sci. USA*, **96**, 266–71.

46. Peet, D.J., Turley, S.D., Ma, W., Janowski, B.A., Lobaccaro, J.-M.A., Hammer, R.E., and Mangelsdorf, D.J. (1998). Cholesterol and bile acid metabolism are impaired in mice lacking the nuclear oxysterol receptor LXRα. *Cell*, **93**, 693–704.

47. Babiker, A., Andersson, O., Lindblom, D., van der Linden, J., Wiklund, B., Lütjohann, D., Diczfalusy, U., and Björkhem, I. (1999). Elimination of cholesterol as cholestenoic acid in human lung by sterol 27-hydroxylase: evidence that most of this steroid in the circulation is of pulmonary origin. *J. Lipid Res.*, **40**, 1417–25.

48. Björkhem, I., Diczfalusy, U., and Lütjohann, D. (1999). Removal of cholesterol from extrahepatic sources by oxidative mechanisms. *Curr. Op. Lipidol.*, **10**, 161–5.

49. Altschul, R. (1950). Experimental cholesterol arteriosclerosis. II. Changes produced in Golden hamsters and in Guinea pigs. *Am. Heart J.*, **40**, 401–9.

50. Cook, R.P. and MacDougall, J.D.B. (1968). Experimental atherosclerosis in rabbits after feeding cholestanetriol. *Brit. J. Exp. Pathol.*, **49**, 265–71.

51. Shih, J. C. H. (1980). Increased atherogenicity of oxidized cholesterol. *Fed. Proc.*, **39**, 650a.

52. Jacobson, M.S., Price, M.G., Shamoo, A.E., and Heald, F.P. (1985). Atherogenesis in White Carneau pigeons. Effects of low-level cholestane-triol feeding. *Atherosclerosis*, **57**, 209–17.

53. Mahfouz, M.M., Kawano, H., and Kummerow, F.A. (1997). Effect of cholesterol-rich diets with and without added vitamins E and C on the severity of atherosclerosis in rabbits. *Am. J. Clin. Nutr.*, **66**, 1240–9.

54. Staprans, I., Pan, X.-M., Rapp, J.H., and Feingold, K.R. (1998). Oxidized cholesterol in the diet accelerates the development of aortic atherosclerosis in cholesterol-fed rabbits. *Arterioscler. Thromb. Vasc. Biol.*, **18**, 977–83.

55. Sunde, M.L. and Lalich, J.J. (1985). Renal and blood vessel response to oxidized cholesterol in chicks and quails. *Fed. Proc.*, **44**, 939a.

56. Kantiengar, N.L., Lowe, J.S., Morton, R.A., and Pitt, G.A.J. (1955). The effects of administering cholesterol and cholesta-3:5-dien-7-one to cockerels. *Biochem. J.*, **60**, 34–9.

57. Schwenk, E., Stevens, D.F., and Altschul, R. (1959). Experiments on cholesterol atherosclerosis in rabbits. *Proc. Soc. Exp. Biol. Med.*, **102**, 42–5.

58. Aramaki, Y., Kobayashi, T., Imai, Y., Kikuchi, S., Matsukawa, T., and Kanazawa, K. (1967). Biological studies of cholestane-3β, 5α, 6β-triol and its derivatives. Part 1. Hypocholesterolemic effects in rabbits, chickens and rats on atherogenic diet. *J. Atheroscler. Res.*, **7**, 653–69.

59. Taylor, C.B., Peng, S.-K., Werthessen, N.T., Tham, P., and Lee, K.T. (1979). Spontaneously occurring angiotoxic derivatives of cholesterol. *Am. J. Clin. Nutr.*, **32**, 40–57.

60. Imai, H. and Lee, K.T. (1983). Mosaicism in female hybrid hares heterozygous for glucose-6-phosphate dehydrogenase. IV. Aortic atherosclerosis in hybrid hares fed alternating cholesterol-supplemented and nonsupplemented diets. *Exp. Mol. Pathol.*, **39**, 11–23.

61. Griminger, P. and Fisher, H. (1986). The effect of dried and fresh eggs on plasma cholesterol and atherosclerosis in chickens. *Poultry Sci.*, **65**, 979–82.
62. Higley, N.A., Beery, J.T., Taylor, S.L., Porter, S.L., Dzuiba, J.A., and Lalich, J.J. (1986). Comparative atherogenic effects of cholesterol and cholesterol oxides. *Atherosclerosis*, **62**, 91–104.
63. Tipton, C.L., Leung, P.C., Johnson, J.S., Brooks, R.J., and Beitz, D.C. (1987). Cholesterol hydroperoxides inhibit calmodulin and suppress atherogenesis in rabbits. *Biochem. Biophys. Res. Commun.*, **146**, 1166–72.

18 Intracellular effects of cell surface-localized initiation of blood coagulation

Serena Pringle and Hans Prydz
The Biotechnology Centre of Oslo, University of Oslo, Norway

18.1 Cell biology of tissue factor

The role of tissue factor (TF) in blood coagulation has been known for more than a hundred years. In the early 1970s, TF was shown to be an integral membrane protein of molecular weight around 50 kDa. Cloning revealed the mature protein to consist of 263 amino acids organized into three distinct parts: a 219 amino acid extracellular region of two β-pleated sheets, a hydrophobic transmembrane segment of 23 amino acids, and a short 21 amino acid cytoplasmic tail.[1–4] The extracellular domain contains three possible glycosylation sites, which are not required for TF pro-coagulant function but may have a regulatory function in cell surface expression of the molecule. While several studies have shown the cytoplasmic domain to play little part in the haemostatic function of the TF molecule,[5, 6] recent studies (Section 18.3) suggest that it has an important role in intracellular signalling. A serine residue in the cytoplasmic domain of human TF is phosphorylated by a protein kinase C (PKC) mechanism in response to treatment by phorbol 12-myristate 13-acetate (PMA) supporting this hypothesis.[7] The functional role of this phosphorylation site is unclear; it may be involved in membrane localization and stabilization of the TF molecule at the cell surface, as well as in signalling pathways. The TF molecule contains five conserved cysteines: four in the extracellular domain and one in the cytoplasmic tail that is esterified to inner leaflet membrane lipids, providing an anchor for the TF molecule.[8] The four cysteines in the extracellular domain form two disulphide bridges. Disruption of the disulphide loop proximal to the membrane (Cys186/Cys209) diminishes TF : FVIIa binding. However, disruption of the distal loop (Cys49/Cys57) has no significant effect on TF function.[9]

It is well established that TF acts as the cell surface receptor/cofactor for the circulating plasma protein, FVII, in initiating blood coagulation.[10] Determination of the crystal structure of the TF:FVIIa complex has shown that TF acts as a rigid scaffold at the cell surface, positioning the FVIIa catalytic domain to facilitate substrate recognition and proteolytic activation. Although association of FVIIa with TF is greatly enhanced in the presence of calcium and phospholipid, neither is absolutely essential for binding to occur. However, activation of the TF-bound FVII (by FVIIa, FIX, FX, or thrombin) requires neutral or negatively charged phospholipid, and subsequent activation of TF:FVIIa macromolecular substrates is dependent on membrane anchoring via the interaction of the Gla domain of both FVIIa and FXa with negatively charged phospholipid. For a review, see Martin *et al.*[11]

Direct inhibitors of the TF:FVIIa complex include tissue factor pathway inhibitor-1 (TFPI-1), TFPI-2, annexin V, antithrombin III, sphingosine, and platelet factor 4. TFPI-1, a Kunitz type inhibitor, which is synthesized by endothelial cells, appears to be the principal inhibitor and acts by binding to the TF:FVIIa:FXa complex at the cell surface, simultaneously inactivating the proteolytic sites of both FVIIa and FXa.[12] A homologous Kunitz type inhibitor, TFPI-2, inhibits the TF:FVIIa complex, but the second Kunitz domain does not bind FXa.[13] The precise physiological function of TFPI-2 is not known, but a recent study suggests that it may act as a mitogen for vascular smooth muscle cells (VSMCs).[14]

Cloning of the TF gene in 1987[1–4] revealed structural similarity to the class II cytokine receptor family, which suggested that TF might have a true receptor function and thereby mediate intracellular signalling. Recent studies (Sections 18.3 and 18.4) have shown that TF:FVII may have an important function in processes other than coagulation. It is our intention in this chapter to review such data on the role of the TF:FVII complex in initiating intracellular signalling and consequences thereof for gene regulation. Earlier data in this field are reviewed by Camerer *et al.*[15] Consequences of these novel observations in areas like wound healing, angiogenesis, tumour metastasis and atherogenesis, where TF is known to play a role, are so far subject to speculation only. They do establish a new paradigm for research into the role of initiation of coagulation in these areas.

18.2 Regulation of TF expression

Functional expression of TF at the cell surface is important for maintaining either a pro- or anti-coagulant state and can be regulated at several levels: transcriptional regulation of gene expression; intracellular transport and modification; alterations in the cell membrane of expressing cells; and inactivation of the TF:FVII complex by specific inhibitors (see above) or degradation.

18.2.1 Transcriptional regulation

TF gene regulation has been extensively studied in several cell types exposed to various agonists and data from these studies have been reviewed.[15, 16]

The human TF gene promoter region contains consensus binding sites for the nuclear transcription factors NF-κB (two), AP-1 (two), Sp1 (five), and Egr1 (three) (Fig. 18.1). Basal TF expression in cell types such as fibroblasts, epithelial cells, and smooth muscle cells is controlled by Sp1. Mutational analysis of expression in HeLa cells has shown that the minimal promoter required for TF expression contains three overlapping Sp1/Egr1 sites and the TATA box. The Sp1 sites act synergistically and at least two must be retained to maintain even a basal level of transcription. Negative transcriptional regulatory mechanisms have been proposed to explain the low basal levels of TF expression in quiescent endothelial cells.[17]

When constitutive expressors of TF, such as fibroblasts, are serum starved and then stimulated with serum or mitogens, TF mRNA is rapidly induced (1–2 h) without the requirement for *de novo* protein synthesis. This would classify TF as an 'immediate early gene' or an 'early response gene' like c-*myc* and c-*fos* and suggests that TF may play a role outside coagulation. In quiescent rat VSMCs and in human epithelial cells, both serum growth factors and PMA can induce TF. Both require calcium but only PMA requires PKC, suggesting that serum and PMA induction of TF expression is mediated by different signalling pathways.[18, 19] Both Sp1 and Egr1 mediate PMA- and serum-induced TF expression in epithelial cells, Sp1 being constitutively expressed whereas Egr1 is induced by serum and PMA. In starved fibroblasts, Egr1 expression, upon stimulation, reaches a peak before TF. This may indicate that in this system *de novo* synthesis of Egr1 is required for maximal expression of TF.[16] Egr1 is an immediate early growth response gene induced by cytokines, some growth factors, DNA damaging agents, heat shock, and hypoxia. Camerer *et al.*[20] have shown that Egr1 is rapidly and transiently induced by binding of FVIIa to TF in HaCaT cells, a constitutively TF expressing keratinocyte cell line. This may represent a positive feedback cycle leading to increased levels of TF when coagulation is initiated.

Induction of TF gene transcription in endothelial cells and monocytes occurs in response to a number of stimuli, such as cytokines, lipopolysaccharide (LPS) and PMA, and is mediated by the functional interaction between Fos-Jun and c-Rel-p65 heterodimers.[15] LPS is thought to act via the LPS binding protein/CD14 pathway.[21]

At present little is known about downregulation of TF gene expression. In general, TF synthesis is induced by agents that increase the levels of cytosolic calcium and is suppressed by agents that increase the level of cAMP.[22, 23] An increased level of cAMP is thought to inhibit the transcriptional activity of the NF-κB complex without affecting its nuclear translocation or the proteolytic degradation of IκBα. The immunosuppressive cytokines, IL-4, IL-10 and IL-13, also suppress LPS-induced TF activity in both monocytes and endothelial cells.[24, 25] Their precise mode of action is not known, but available data show that they reduce TF mRNA levels by decreasing the rate of TF gene transcription.[17, 26] Induced TF expression in monocytes and vascular endothelial cells can also be downregulated by all-*trans* retinoic acid.[27] Almus *et al.*[28] showed that human umbilical vein endothelial cells cultured with endothelial cell growth supplement (ECGS) and heparin had an impaired ability to

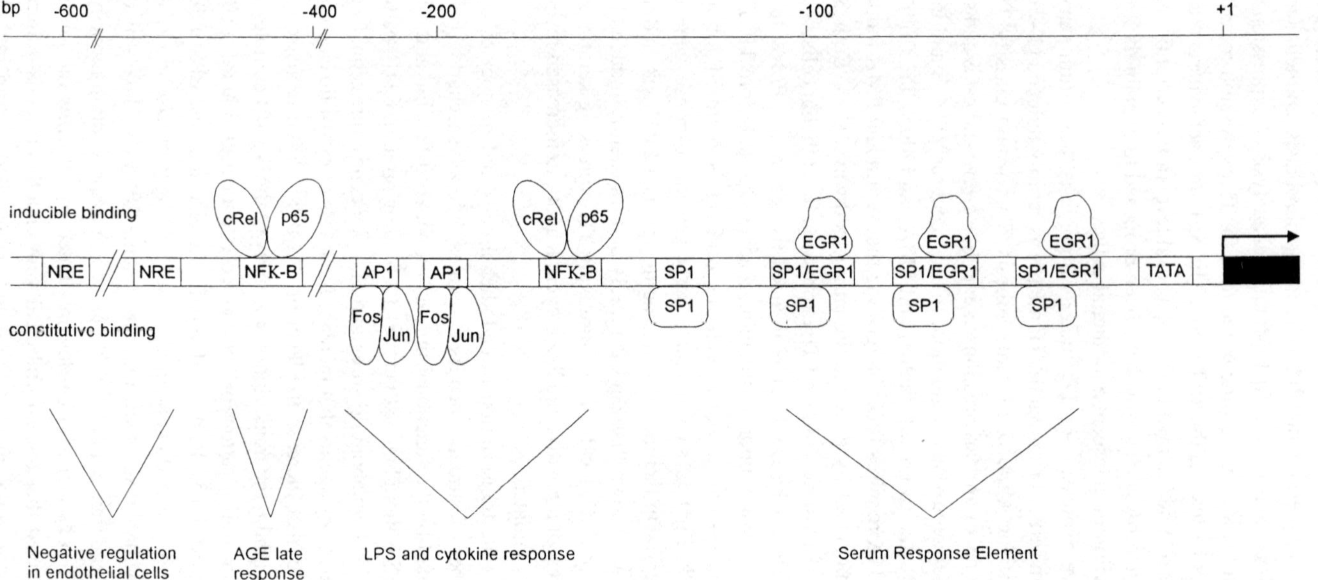

Fig. 18.1 The promoter region of the tissue factor gene. A summary of constitutive and inducible binding of transcription factors to the promoter of the tissue factor gene. The labels indicate the interaction of specific families of transcription factors. The transcription start site is arrowed. AGE, advanced glycosylation endproducts; LPS, lipopolysaccharide. (Reprinted from *Thrombosis Research*, **90**, Martin *et al.*, 1–25, 1998, with permission from Elsevier Science.)

express TF activity and TF mRNA when stimulated with thrombin and other agonists. It was found that both aFGF and bFGF (main constituents of ECGS) suppress PMA-, LPS- and TNF-α-induced TF activity and TF mRNA levels by reducing the rate of gene transcription. This suppressive effect of aFGF and bFGF appears to be endothelial cell specific since growing fibroblasts with aFGF and bFGF did not suppress either the constitutive or the serum-induced TF cell surface and mRNA expression.

The precise mechanism of post-translational control of TF expression is still unclear. After transcription, the first level of control lies in the stabilization/destabilization of the mRNA. The half-life of TF mRNA is very short [$t_{1/2}$ 0.5–1.5 h] and decreases with time after stimulation.[29] Inhibition of protein synthesis with cycloheximide increases the stability of TF mRNA in monocytes, endothelial cells, and fibroblasts. Enhanced TF expression after stimulation of cells with LPS, PMA, and thrombin occurs not only by enhanced transcription of the TF gene but also by mRNA stabilization.[30]

18.2.2 Control of TF expression at the cell membrane

TF is constitutively expressed in many cell types that are not normally in contact with flowing blood, such as the epithelial cells lining the body cavities, and effectively functions as a haemostatic barrier against tissue injury. In monocytes/macrophages and endothelial cells, which are in contact with flowing blood, and under normal conditions have a non-haemostatic surface, TF expression can be induced. In order to serve as a receptor/cofactor for FVIIa, TF must be present on the cell surface in a form that is 'receptive' to FVIIa. *In vitro* studies have shown a discrepancy between the maximal FVIIa binding and maximal FX activation on the surface of unperturbed cells.[31] Even when all the detectable TF antigen is expressed on the cell surface, TF pro-coagulant activity is greatly enhanced by cell lysis.[32, 33] It is worth noting that despite its inability to promote coagulation, this latent or encrypted form of TF is still capable of binding FVII or FVIIa. These observations suggest that an additional factor may be necessary for pro-coagulant function of the TF:FVII complex under certain circumstances. One possible candidate for this role is the anionic phospholipid phosphatidylserine (PS). *In vitro* studies have shown that PS is crucial for TF activity.[34] Under normal conditions, there may be a limited amount of PS in the outer leaflet of the plasma membrane, which could limit the amount of physiologically functional TF:FVIIa complexes that can be formed.[35, 36] Stimulation of endothelial cells may lead to an increase in PS in the membrane outer leaflet.[37] Thus, PS may play an important role in the cell surface activity of the TF:FVIIa complex, but is unlikely to be the only controlling factor. A number of studies have reported the presence of latent TF in caveolae in the cell membrane.[38] These structures are devoid of anionic phospholipids.

In vivo, both epithelial and endothelial cells form tight junction complexes that effectively compartmentalize the cell surface into two distinct domains: apical and basolateral. This is important when considering the function of TF in both induced and constitutively expressing cells. Studies on polarized cells of both induced and basal TF expression have shown discordant expression and activity. In a study by Camerer *et al.*,[39] polarized endothelial cells stimulated with IL-1 showed localization of TF activity and antigen only to the apical (luminal) surface. In the constitutively expressing canine kidney epithelial line (MDCK), it was shown that surface TF was expressed predominantly on the basolateral surface. This basolateral TF was not functionally active, whereas the small amount of TF present on the apical surface of these cells was highly active. Deletion of the TF cytoplasmic domain did not alter the polar distribution of TF in MDCK cells. A recent study by Hansen *et al.*[40] on polarized Caco-2 cells (human intestinal epithelial cells) also demonstrated discordant TF expression and activity; however, in these cells there appeared to be even distribution of apical and basolateral surface TF expression but only the apical TF:FVIIa complexes were active. Polar expression of TFPI-1 did not account for the discrepancy between TF expression and activity in these cells, but should not be discounted in other cell types. The cytoplasmic domain of TF lacks any known active sorting and endocytosis signal sequences and it is possible that TF is subject to surface sorting by default pathways, in which glycosylation may play a role.[41] Hansen *et al.*[40] showed that in the same polarized cells the amount of anionic phospholipid on the basolateral surface was three-fold lower than on the apical surface, which could account for the differences in TF activity. It is worth noting that the size of the apical and basolateral compartments varies between cell types.

There is evidence to suggest that TF forms dimers on the cell surface as well as in reconstituted phospholipid vesicles.[42] In a study by Bach and Moldow[43] of FVIIa binding to TF on ovarian carcinoma cells, two functionally distinct forms of cell surface TF were identified. In these cells, only 10–20% of TF rapidly bound FVIIa and had pro-coagulant activity; the remaining 80–90% bound FVIIa more slowly and had no pro-coagulant activity. Calcium ionophores rapidly increased surface TF activity of fibroblasts without changing the antigen levels. However, in the same study, thrombin stimulation, although increasing cytosolic calcium, did not produce the same effect. A very nice model for TF encryption was proposed[43] in which they suggest that TF exists as dimers on the surface of unperturbed cells. Calcium influx into the cytosol triggers a calmodulin-dependent process, which causes the TF dimers to dissociate. The resulting change in quaternary structure exposes an essential macromolecular substrate binding site on monomeric TF, thus converting inactive TF dimers to pro-coagulant monomers.

Exposure of cultured endothelial cells to haemodynamic forces such as fluid shear stress and cyclic strain, upregulates the expression and biological activity of TF.[44] Elevated blood pressure is a risk factor for vascular diseases, and it has been reported recently that exposure of human endothelial cells to raised ambient pressures results in a transient upregulation (48 h) of TF which is not controlled at the transcriptional or translational levels.[45] In this model the receptor/cofactor activity of TF appears to be modulated by altered lipid–protein interactions due to subtle structural reorganization of the plasma membrane.

A common way of controlling the activity of transmembrane proteins is by internalization and degradation or recycling, and a possible mode of regulation of TF activity is via the surface turnover of the protein or of the TF:FVIIa complex. In endothelial cells, binding of TFPI to the TF:FVIIa:FXa complex leads to a translocation of the complex to caveolae, which may be a mechanism for downregulation of the TF:FVIIa activity.[38] Recently, it has been reported that in monocytes, endocytosis of the TF:FVIIa:FXa complex upon binding of TFPI occurs via interaction with the α-adaptin of clathrin coated pits.[46]

18.3 TF:FVIIa-mediated intracellular signalling

Bazan[47] suggested that TF may have a receptor function due to its structural similarity to the class II cytokine receptor superfamily which includes the interferon (IFN)α/β and γ receptors. The receptor concept requires not only the structural aspect but also the existence of a high affinity and high specificity ligand, which when bound to the receptor will result in an intracellular signal. For TF the only known ligand is FVII/FVIIa. Røttingen et al.[48] were the first to describe such a signal by demonstrating that binding of FVIIa to TF could elicit Ca^{2+} responses in a number of cell types. The signals varied between cell types and between cells within a population. These responses were produced mainly by the release of Ca^{2+} from intracellular stores, but also by the continuous and gradual increase of the basal levels caused by altered membrane Ca^{2+} influx. The Ca^{2+} oscillations in MDCK cells were synchronized in confluent cell monolayers and were also dependent on the concentration of FVIIa. This suggests that binding of FVIIa to TF at the cell surface can generate intracellular signals, thereby modulating cellular functions. This response required the presence of FVIIa in its proteolytically active form but did not require the cytoplasmic tail of TF.[20, 49]

Following this discovery,[48, 49] we demonstrated that addition of FVIIa to TF expressing MDCK cells leads to Ca^{2+} release through a mechanism mediated by phosphatidylinositol-specific phospholipase C (PI-PLC).[49] PI-PLC is expected to be activated either by G-protein-linked or by tyrosine kinase-linked receptors such as the cytokine and growth hormone receptors.[50] Because of its structural similarity with the cytokine receptor superfamily, it would seem likely that the TF:FVIIa Ca^{2+} response would involve tyrosine kinase activation of PLC, but no such involvement has so far been described. Poulsen et al.[51] demonstrated that FVIIa:TF interactions in baby hamster kidney cells (BHK) induced phosphorylation of the extracellular signal regulated kinases 1 and 2 (Erk1/2). The catalytic activity of FVIIa is necessary for this signal transduction pathway but the cytoplasmic domain of TF, including the Cys245 acylation site, is not required.[52]

Camerer et al.[20] demonstrated that in addition to phosphorylation of Erk1/2, FVIIa binding to TF in HaCaT cells also leads to significant phosphorylation of Jun N-terminal kinase and p38^{MAPK} and to upregulation of Egr1. The initial part of the

Ca^{2+} signal is not required for Egr1 induction in these cells. Inhibition of Mek1/2 and thereby of the Erk1/2 kinases completely abolished the upregulation of the Egr1 gene, whereas inhibition of the $p38^{MAPK}$ had no such effect. This shows that TF:FVIIa-mediated upregulation of egr1 in HaCaT cells occurs through a Mek/Erk activation. Recently, several other examples of TF:FVIIa-dependent gene upregulation have been described (Table 18.1 and Ref. (61)).

The proteolytic activity of FVIIa is also an absolute requirement for the signalling event in these cells whereas deletion of the 18 C-terminal amino acids from the cytoplasmic tail of TF did not impair FVIIa-induced Egr1 upregulation. So it seems

Table 18.1 Changes in mRNA levels with time after addition of FVIIa (10 nm) to HaCaT

Time/gene product	0.5	1	1.5	3	6	*n*
egr-1	4.8±0.7*	5.8±0.8*	1.4±0.1	0.5±0.1**	0.5±0.2	3
c-fos	4.3±0.9*	6.1±1.4*	1.7±0.6	2.0±0.4	1.7±0.4	4
ETR101	2.7±0.3**	2.3±0.1**	1.2±0.1	1.0±0.0	1.3±0.2	5
BTEB2	1.1±0.1	1.4±0.1**	2.3±0.2**	1.9±0.2*	1.2±0.2	5
c-myc	0.9±0.1	2.1±0.2**	2.3±0.3**	2.0±0.2**	1.8±0.2*	6
fra-1	1.2±0.1	1.9±0.2*	2.5±0.4*	1.8±0.3	2.0±0.2**	4
TTP	2.4±0.3**	2.8±0.3**	1.9±0.1**	1.4±0.0**	1.3±0.1	5
AR	1.1±0.2	2.6±0.4	3.8±0.4*	2.2±0.0**	1.8±0.1*	3
hbEGF	1.3±0.3	9.9±1.6*	5.0±0.8*	4.8±1.2	3.2±1.1	3
CTGF	3.3±0.1*	7.8±1.0*	7.8±1.3*	1.5±0.3	1.5±0.1	3
FGF-5	0.8±0.1	2.0±0.4	3.5±0.7*	1.8±0.4	1.7±0.4	5
IL-8	2.3±0.3*	18.8±2.2**	14.3±2.4*	4.5±0.9*	2.6±0.3**	4
LIF	3.0±0.6*	4.8±1.6	9.6±1.3**	2.6±0.4*	1.1±0.5	5
MIP2α	3.1±0.9	15.4±4.8*	5.0±1.3*	4.5±1.3*	3.7±1.0*	6
IL-1β	1.1±0.1	2.4±0.1**	2.4±0.1**	1.4±0.1	1.7±0.4	4
RhoE	1.3±0.0*	3.3±0.3**	2.6±0.2**	1.4±0.1*	1.5±0.2	4
uPAR	1.3±0.1	1.5±0.2	2.9±0.5*	2.1±0.2*	2.1±0.3*	4
collagenase 1	0.8±0.1	1.4±0.2*	1.7±0.3*	1.9±0.1**	1.7±0.1**	7
collagenase 3	1.3±0.0**	4.8±1.2	7.0±1.5*	7.0±1.4*	2.9±0.8	4
Jagged1	1.5±0.1*	2.4±0.4*	2.7±0.2**	1.9±0.2*	1.4±0.1	6
PAI-2	1.1±0.1	2.7±0.4*	6.3±1.1*	7.0±0.5**	3.1±0.4*	4
cyclophilin	1.3±0.0**	1.9±0.1**	2.2±0.2**	1.7±0.2*	1.2±0.1	5
GADD45	1.3±0.0**	2.0±0.2*	1.7±0.1**	0.9±0.1	1.0±0.1	4
PGE$_2$R	2.5±0.1**	4.1±0.4*	3.2±0.4*	1.0±0.0	1.7±0.1*	3

n, number of independent mRNA isolations; *, p<0.05; **, p<0.001.

likely that TF-mediated FVIIa signalling can occur by a number of different pathways, which may be cell type dependent.

18.3.1 Signal transduction by the TF:FVIIa complex via a novel cell membrane receptor?

Several studies have shown that TF:FVIIa signalling is dependent on the proteolytic activity of the complex.[20, 49, 55] Since the catalytic activity of TF-bound FVIIa is not directed toward TF itself, it seems likely that another molecule is the target for proteolytic cleavage and hence the signalling event. Control experiments in all of the above-mentioned studies, involving specific inhibitors of FXa and thrombin as well as receptor downregulation,[20] have ruled out the possibility that the signalling events are triggered by downstream protease-receptor activation. This excludes one or more of the thrombin protease-activated receptors (PAR) 1, 3 and 4 as a possible third component in the TF:FVIIa signalling complex, leaving PAR-2[56] as a possible candidate. However, Camerer *et al.*[20] have shown that transfected CHO cells with functional PAR-2 and TF on their surface did not respond to FVIIa with Ca^{2+} signalling. Similar results were obtained with a human kidney cell line (HK-2) which constitutively expresses PAR-2 and TF.[20] Therefore it is unlikely that PAR-2 participates in the TF:FVIIa signalling complex, but this does not rule out the possibility that other, as yet unidentified, PAR receptors are involved, or that PAR-2 may be a necessary but not sufficient component.

18.3.2 Role of the TF cytoplasmic domain in intracellular signalling

The cytoplasmic domain is not required for TF cofactor function in the coagulation process,[57] membrane sorting[39] nor in some signalling pathways.[20, 55] These lead to rapid phosphorylation (of Ser/Thr residues) of a subset of MAP kinases, which are arranged in several parallel signalling pathways.[60] Formation of the TF:FVIIa complex induces phosphorylation of several MAP kinase pathways and leads, through transcription factor activation, to upregulation (2–20-fold) of at least 25 genes,[61] comprising at least 7 transcriptional regulator genes, 4 genes encoding growth factors, 4 inflammatory cytokines, 4 genes involved in matrix degradation as well as TF itself.

There are, however, a number of studies that suggest a role for the cytoplasmic tail in other TF-dependent signalling pathways. Phorbol ester stimulation of TF transfected human kidney cells results in the phosphorylation of a highly conserved serine residue in the cytoplasmic domain.[58] Ligand-induced oligomerization of receptor subunits which juxtapose and engage an intracellular signalling machinery is a common mechanism in signal initiation of cytokine receptors.[59] Cytokine receptors such as TF which lack intrinsic kinase catalytic domains, may couple ligand binding to tyrosine phosphorylation by using non-covalently associated protein kinases, e.g. the Janus kinases, which in turn phosphorylate signal transducers and activators of transcription (STATs) to induce binding to specific DNA elements and activate gene transcription. There is some evidence to suggest that there is cross-talk between the MAPK and the JAK-STAT pathways.[62, 63] The short 21 amino acid residue TF cytoplasmic domain lacks close similarity with cytokine receptor membrane proximal motifs that mediate binding of the non-receptor Janus kinases (JAKS), raising doubts as to whether TF-mediated signalling occurs along the same pathways. Studies mentioned above have shown the involvement of the MAP kinase pathway in TF:FVIIa signalling and also demonstrated that the cytoplasmic domain of TF is not required.

In vivo studies of tumour cell metastasis have provided clear evidence that the TF cytoplasmic domain must actively participate in intracellular signalling. Mutational deletion of the cytoplasmic domain abolishes metastasis in melanoma and TF transfected CHO cells.[64] The binding of catalytically active FVIIa to TF is also an absolute requirement for the metastatic process. In this *in vivo* model, the metastatic function of TF can be initiated by two distinct signalling events: (i) by direct intracellular signalling via the TF cytoplasmic domain, and (ii) by indirect extracellular signalling by thrombin, activated by the proteolytic activity of the TF:FVIIa complex. The latter, not requiring the cytoplasmic tail, thus seems not to be able to promote metastasis by itself.

A recent study of the role of TF in the regulation of vascular endothelial growth factor (VEGF) production in human melanoma cells, suggested that a functional cytoplasmic domain is important for this event but that neither the pro-coagulant function of TF nor the proteolytic activity of the bound FVIIa are required.[65] However, in a similar study of VEGF regulation in human lung fibroblasts, the proteolytic activity of FVIIa was necessary for upregulation of VEGF.[54] The catalytic activity required for signal transduction in these cells was specific for FVIIa, as other serine proteases failed to mimic the response, and the location of the TF:FVIIa complex on the surface of the cell membrane was vital. This data would support the hypothesis that TF:FVIIa might be activating a novel receptor on the cell surface.

Although the evidence for the essential role of the TF cytoplasmic domain in tumour cell metastasis is convincing, a defined intracellular signalling pathway has remained elusive. A recent study, however, identified the actin binding protein 280 (ABP-280) or filamin-1 as a ligand for the TF cytoplasmic domain.[66] This binding occurs upon clustering or cross-linking of TF at the cell surface resulting in phosphorylation of the cytosolic serine residues, which in turn results in the binding of ABP-280 to the TF cytoplasmic tail, leading to reorganization of the actin cytoskeleton which is independent of integrin function. ABP-280 binding to the TF cytoplasmic tail is accompanied by the phosphorylation of the non-receptor focal adhesion kinase (FAK), which is central to cytoskeletal-dependent signalling events. The carboxy terminus of ABP-280 is critical for the activation of the MAP kinase pathway in response to TNF-α stimulation.[67] Both FAK and ABP-280 are critical for cell motility and adhesion, which would suggest that the signalling functions of the TF cytoplasmic domain may be influencing cell migration. In fact, immunofluorescence and EM studies have shown that TF and ABP-280 colocalize at the leading edge of spreading cells.[66, 68]

Data from a recent study show that TF protein lacking the cytoplasmic domain was capable of rescuing TF $-/-$ mouse embryos, providing further evidence that the TF extracellular domain can function independently of the cytoplasmic domain in mediating signal transduction. However, the role of the cytoplasmic domain in other TF-mediated signal transduction pathways and in other cell types seems well established.

18.4 Functions of the TF:FVIIa complex beyond haemostasis

The discovery that the TF:FVIIa complex induces altered gene expression in keratinocytes may have widespread consequences, if we keep in mind that FVII is present in all body fluids and that the complex may form both inside and outside the vessel wall. So far, this is pure speculation, based on the emerging evidence that several cell types respond by altering gene expression. Different processes like angiogenesis, tumour metastasis, wound repair, and atherogenesis may be targeted.

18.4.1 Angiogenesis

The process of angiogenesis and neovascularization starts with proteolysis of the sub-endothelial extracellular matrix followed by migration and proliferation of endothelial cells from pre-existing blood vessels. It is a vital process for many physiological conditions such as embryonic development, wound healing and other repair mechanisms including collateral blood vessel formation in ischaemic heart disease, and for several female reproductive functions.[69] Angiogenesis also occurs in a number of pathological conditions, such as atherosclerosis, diabetic retinopathy, and solid tumour growth.[70] Regulation of angiogenesis is controlled by a family of VEGFs that promote vascular permeability and endothelial cell growth.[71] VEGFs are produced in a variety of cell types including monocytes, macrophages, keratinocytes, fibroblasts, and vascular smooth muscle cells. TF and VEGF expression are linked in a feedback control mechanism, which may play a role in angiogenesis in tumour metastasis. In endothelial cells, VEGF induces Egr1-mediated TF gene expression,[72] whereas in normal human fibroblasts and in tumour cells, VEGF expression is induced by the TF:FVIIa complex.[54] The signalling pathways of this response appear to be cell type dependent.[64, 65, 54]

18.4.2 Embryology

TF plays an important role in embryonic blood vessel development. Deletion of the TF gene in mice results in the death of the embryo between days 8.5 and 10.5. One study suggested that embryonic death occurred as a result of bleeding due to lack of TF-mediated haemostatic function after rupture of the yolk sac vasculature.[73] Another study suggested that TF is important for the development of the yolk sac vessel wall,[74] which argues that, in the embryo, the predominant role of TF may be to modify cellular functions rather than coagulation. However, cellular signalling functions of TF in the embryonic vasculature have not yet been demonstrated. The fact that there are no known human TF mutations, confirms the importance of TF in early embryonic development.

18.4.3 Tumour metastasis

Metastatic disease is a multistep process in which a number of regulated cellular functions (cell adhesion, migration, growth) cooperate to achieve successful tumour growth at the secondary site. Upregulated expression of TF has been demonstrated in a variety of tumours,[75] and an increasing number of experimental findings argue that TF expressed in tumour cells may function predominantly to modify cellular functions rather than mediating haemostasis in the metastatic process.

In pancreatic cancer cells expressing high levels of TF, an upregulation of the gene for urokinase type plasminogen activator receptor (uPAR) was observed after stimulation with FVIIa.[55] uPAR enables invading cells to degrade the extracellular matrix by directing the action of uPA and eventually to tumour cell migration.[76] Over-expression of TF in mouse fibrosarcoma cells results in upregulation of pro-angiogenic VEGF and downregulation of the anti-angiogenic factor, thrombospondin, resulting in improved tumour vascularization.[77] TF:FVIIa interactions are important to the metastatic properties of melanoma cells *in vitro* which are independent of blood coagulation.[64, 65, 78] The recent finding that the cytoplasmic domain of TF interacts with ABP-280, rearranging the cytoskeleton of the cell, thus influencing motility, may provide a molecular pathway by which TF:FVIIa interactions support tumour cell migration in conjunction with TFPI.[66] *In vitro* studies have shown that immobilized TFPI-1 can support tumour cell adhesion via the TF:FVIIa complex at the cell surface, and at the same time allowing intracellular signalling elicited by the TF:FVIIa binding.[79] Confocal microscopy showed that ABP-280 colocalized with TF at the contact site with TFPI-1. The same study demonstrated that phosphorylation of the cell migration promoting FAK is mediated by the binding of TFPI-1 to the TF:FVIIa complex and not by the catalytic activity of FVIIa. This finding could explain the controversy over the role of the proteolytic/catalytic function of TF-bound FVIIa in signalling pathways in tumour cells, the catalytic site of FVIIa being necessary for effective binding of the complex to TFPI-1. From these data, one could speculate on a simple model for the adhesion and migration steps of tumour cell metastasis, where TFPI-1 immobilizes the circulating tumour cell on endothelial or ECM surfaces via the TF:FVIIa complex, initiating a series of coordinated signalling events culminating in the breakdown of the extracellular matrix and the extravasation of the tumour cell from the circulation into the adjacent tissue. The concomitant upregulation of VEGF in the tumour cell may have a dual function by promoting vascular permeability at the site of adhesion facilitating migration while stimulating the neovascularization process necessary for sustained tumour growth. VEGF production by the tumour cell may also stimulate TF expression in endothelial cells activating localized clotting. This simplified view cannot explain the mechanisms of organ specificity or distant site secondary tumour growth.

18.4.4 Cell migration and wound repair

Cell adhesion and migration are central events in angiogenesis, atherosclerosis, and tumour metastasis. Evidence, from the studies discussed above, indicates that the extracellular TF:FVIIa complex functions in cell adhesion while the cytoplasmic tail of TF is vital for cell spreading and motility.[79, 66] Recent data from our laboratory[61] suggest that TF:FVIIa signalling may also play a vital role in wound healing. Using a differential display approach, this study investigated gene transcription in keratinocytes in response to TF:FVIIa signalling. A number of genes which code for proteins related to cellular reorganization and migration were found to be strongly upregulated as well as those for a number of growth factors and pro-inflammatory cytokines, all of which have been shown to be upregulated in various models of epithelial and endothelial wounding. Moreover, the genes induced in the keratinocyte model were classified as immediate early response genes, which would conform to a wound-type response of these cells to FVIIa stimulation. The upregulation of Egr1 in response to TF:FVIIa binding would tie in with this model. Egr1 is an early response gene and is itself responsible for the upregulation of other genes involved in growth and repair.

18.5 Atherosclerosis

A role for TF in cardiovascular disease is well defined by its abundant presence in human atherosclerotic plaques, particularly in their lipid-rich core, and also by its role as a trigger of blood clotting leading to the formation of downstream platelet-active agonists like thrombin and adhesion promoting complexes like TFPI/TF/FVIIa. TF may also be involved on the salutary side in collateral formation through neovascularization. Monocytes, macrophages, and SMCs all contribute to the accumulation of TF at sites of vessel wall injury and plaque formation. In the atheromatous plaque, there is abundant TF antigen and activity, causing the well-known tendency to frequently fatal thrombotic complications to plaque rupture. TF is rapidly induced in the vessel wall as a consequence of acute arterial injury such as that produced by coronary angioplasty, atherectomy and coronary artery stenting,[80] and can be detected in the media as well as in the endothelial cells. Increasing evidence suggests that endothelial cells can be induced to express TF also *in vivo*. In rat aortic and carotid artery injury models, TF mRNA levels were found to be markedly increased 2 h after a single balloon injury, returning to baseline after 24 h.[81] TF antigen and FVIIa binding were undetectable in the endothelium and media during the first 4 h after injury; however, at 24 h after injury TF in the media was abundant and slowly returned to baseline. Accumulation of TF in the developing intimal plaque resulting in high levels detectable 2 weeks after a single injury event, with no detectable TF antigen at the luminal endothelial cell surface, suggest that medial and intimal SMCs are the source of TF. Subsequent injury to the vessel resulted in a rapid and marked exposure of TF antigen, probably due to endothelial denudation and exposure of the underlying SMC.

TF expression on SMC is induced by serum and by a number of mitogenic growth factors such as platelet-derived growth factor (PDGF), epidermal growth factor (EGF), angiotensin II, and α-thrombin. Confocal microscopy studies of human SMC stimulated with PDGF have shown that TF accumulates in the cytoplasm,[82] suggesting that transport of TF to the cell surface may be a critical step in regulating SMC-mediated thrombosis in acute arterial injury. Migration of monocytes to the injured vessel wall in response to monocyte chemoattractant protein (MCP)-1 contributes to the accumulation of TF in early plaque development. MCP-1 is secreted by SMCs and induces TF expression in monocytes and macrophages as well as in the SMCs themselves, acting in an autocrine fashion.[83] As the plaque develops, macrophages and SMCs accumulate lipid and become foam cells, ultimately degenerating into a necrotic core rich in lipid and TF. Recent studies have shown that native and oxidized low-density lipoprotein induced TF gene expression in rat aortic smooth muscle cells by an Egr1/Sp1-mediated mechanism,[84] contributing to the continual accumulation of TF within the plaque. Plaque rupture exposes active TF to circulating FVII, triggering thrombosis. Inhibition of TF-mediated coagulation by TFPI-1 has been shown to be effective in preventing neointimal hyperplasia and stenosis in animal models,[85] possibly by the inhibition of formation of thrombin which is a powerful SMC mitogen. The precise mechanisms of SMC migration from the media to the intima are not known, but a recent study has shown that TFPI-2 is a potent mitogen for VSMCs *in vitro*.[14] This study did not investigate the role of TF or FVIIa in this process.

TF is supposed to have at least three roles in the development of cardiovascular disease. In the case of plaque rupture, TF initiates development of the thrombosis that frequently causes the fatal infarct. Upon induction of TF synthesis in monocytes and endothelial cells, the coagulation process may be activated locally, leading to thrombin formation and activation of thrombocytes and their release reaction, which triggers atherogenic damage in intima and beyond. Through interaction with TFPI-1, the TF:FVIIa complex may be involved in monocyte adhesion to the vessel wall, an early step in atherogenesis.

We suggest that the observed alterations in gene regulation may open a new inroad for studies of the effect of initiation of blood coagulation on the atherogenic process. The upregulation in keratinocytes of numerous genes (so far about 30) encoding matrix degrading enzymes, chemotactic growth factors, inflammatory cytokines and a number of transcription factors[61] allows speculation about possible new mechanisms involving increased cell motility and recruitment, monocyte activation, generation of a cascade(s) of gene activation, and cell proliferation. If similar upregulation takes place in other cell types, this provides a new paradigm for studies on the relation between low-grade continuous coagulation and atherogenesis. Specific non-haemostatic functions for TF in atherogenesis have not yet been proposed, but one could speculate that TF most likely may play a role in the early stages of cellular recruitment and activation in (neo)intimal hyperplasia.

18.6 Concluding remarks

There is now abundant evidence that TF participates in cellular functions quite distinct from blood coagulation. It would appear that the different physiological roles of TF determine whether the signalling mechanism is mediated by its intracellular or extracellular domains. Studies of tumour metastasis show that both the extracellular protease function of TF:FVIIa, and its cytoplasmic domain are equally important for the metastatic process, and may in fact cooperate with each other. The same may be true for TF-mediated cell adhesion and migration in the development of atherosclerotic plaques. The discovery that binding of FVIIa to TF mediates the upregulation of the Egr1 gene in keratinocytes may be significant in wound healing, which would point to a truly bifunctional role for the TF molecule, in which it functions as a cofactor and initiator of haemostasis while simultaneously triggering the processes of repair. Elucidation of the precise signalling mechanisms of the TF molecule and the role of FVIIa in non-haemostatic functions remains an exciting area for future research.

References

1. Morrissey, J.H., Fakhrai, H., and Edgington, T.S. (1987). Molecular cloning of the cDNA for tissue factor, the cellular receptor for the initiation of the coagulation protease cascade. *Cell*, **50**, 129.
2. Fisher, K.L., Gorman, C.M., Vehar, G.A., O'Brien, D.P., and Lawn, R.M. (1987). Cloning and expression of human tissue factor cDNA. *Thromb. Res.*, **48**, 89.
3. Scarpati, E.M., Wen, D., Broze, G.J., Miletich, J.P., Flander-Meyer, R.R., Siegel, N.R., and Sadler, J.E. (1987). Human tissue factor: cDNA sequence and chromosome localisation of the gene. *Biochemistry*, **26**, 5234.
4. Spicer, E.K., Horton, R., Bloem, L., Bach, R., Williams, K.R., Guha, A., Kraus, J., Lin, T.C., Nemerson, Y., and Konigsberg, W.H. (1987). Isolation of cDNA clones coding for human tissue factor: primary structure of the protein and cDNA. *Proc. Natl. Acad. Sci. USA*, **84**, 5148.
5. Neuenschwander, P.F. and Morrissey, J.H. (1992). Deletion of the membrane anchoring region of tissue factor abolishes autoactivation of FVII but not cofactor

function. Analysis of a mutant with a selective deficiency in activity. *J. Biol. Chem.*, **267**, 14477.

6. Paborsky, L.R., Caras, I.W., Fisher, K.L., and Gorman, C.M. (1991). Lipid association, but not the transmembrane domain, is required for tissue factor activity. Substitution of the transmembrane domain with a phosphatidylinositol anchor. *J. Biol. Chem.*, **266**, 21911.

7. Zioncheck, T.F., Roy, S., and Vehar, G.A. (1992). The cytoplasmic domain of tissue factor is phosphorylated by a protein kinase C dependent mechanism. *J. Biol. Chem.*, **276**, 3561.

8. Bach, R., Konigsberg, W.H., and Nemerson, Y. (1988). Human tissue factor contains thioester-linked palmitate and stearate on the cytoplasmic half-cystine. *Biochemistry*, **27**, 4227.

9. Rehemtulla, A., Ruf, W., and Edgington, T.S. (1991). The integrity of the cysteine 186–cysteine 209 bond of the second disulfide loop of tissue factor is required for binding of FVII. *J. Biol. Chem.*, **266**, 10294.

10. Ruf, W. and Edgington, T.S. (1994). Structural biology of tissue factor, the initiator of thrombogenesis *in vivo*. *FASEB J.*, **8**, 385.

11. Martin, D., Wiiger, M.T., and Prydz, H. (1998). Tissue factor and biotechnology (Review). *Thromb. Res.*, **90**, 1.

12. Broze, G.J., Warren, J.R., Novotny, W.F., Higuchi, D.A., Girard, J.J., and Miletich, J.P. (1988). The lipoprotein associated coagulation inhibitor that inhibits the factor VII-tissue factor complex also inhibits factot Xa: insight into its possible mechanism of action. *Blood*, **71**, 335.

13. Sprecher, C.A., Kisiel, W., Mathewes, S., and Foster, D.C. (1994). Molecular cloning, expression and partial characterisation of a second human tissue-factor-pathway inhibitor. *Proc. Natl. Acad. Sci. USA*, **91**, 3353.

14. Shinoda, E., Yui, Y., Hattori, R., Tanaka, M., Inoue, R., Aoyama, T., Takimoto, Y., Mitsui, Y., Miyahari, K., Shizuta, Y., and Sasayama, S. (1999). Tissue factor pathway inhibitor-2 is a novel mitogen for vascular smooth muscle cells. *J. Biol. Chem.*, **274**, 5379.

15. Camerer, E., Kolstø, A.-B., and Prydz, H. (1996). Cell biology of tissue factor, the principal initiator of blood coagulation (Review). *Thromb. Res.*, **81**, 1.

16. Mackman, N. (1995). Regulation of the tissue factor gene (Review). *FASEB J.*, **9**, 883.

17. Ollivier, V., Parry, G.C.N., Cobb, D., and de Prost, D. (1996). Elevated cAMP inhibits NF-κB-mediated transcription in human monocytic cells and endothelial cells. *J. Biol. Chem.*, **271**, 20828.

18. Pettersen, K.S., Wiiger, M.T., Narahara, N., Andoh, K., Gaudernack, G., and Prydz, H. (1992). Induction of tissue factor synthesis in human umbilical vein endothelial cells involves protein kinase C. *Thromb. Haemost.*, **67**, 473.

19. Lyberg, T. (1983). Effect of cyclic AMP and cyclic GMP on thromboplastin (factor III) synthesis in human monocytes *in vitro*. *Thromb. Haemost.*, **50**, 804.

20. Camerer, E., Røttingen, J.A., Gjernes, E., Larsen, K., Skartlien, A.H., Iversen, J.G., and Prydz, H. (1999). Coagulation factors VIIa and Xa induce cell signalling leading to up-regulation of the Egr-1 gene. *J. Biol. Chem.*, **274**, 32225.

21. Steinemann, S., Ulevitch, R.J., and Mackman, N. (1994). Role of the lipopolysaccharide (LPS) binding protein/CD14 pathway in LPS induction of tissue factor expression in monocytic cells. *Arterioscler. Thromb.*, **14**, 1202.

22. Prydz, H. and Lyberg, T. (1980). Effect of some drugs on thromboplastin (factor III) activity of human monocytes *in vitro*. *Biochem. Pharmacol.*, **29**, 9.

23. Galdal, K.S., Lyberg, T., Evensen, S.A., Nilsen, E., and Prydz, H. (1984). Inhibition of the thromboplastin response of endothelial cells *in vitro*. *Biochem. Pharmacol.*, **33**, 2723.

24. Ramani, M., Ollivier, V., Ternisien, C., Vu, T., Elbim, C., Hakim, J., and Deprost, D. (1993). Interleukin-4 prevents the induction of tissue factor messenger RNA in human monocytes in response to LPS or PMA stimulation. *Br. J. Haematol.*, **85**, 462.

25. Ramani, M., Ollivier, V., Khechai, F., Vu, T., Ternisien, C., Bridey, F., and Deprost, D. (1993). Interleukin-10 inhibits endotoxin-induced tissue factor messenger RNA production by human monocytes. *FEBS Lett.*, **334**, 11.

26. Ollivier, V., Houssaye, S., Ternisien, C., Leon, A., Deverneuil, H., Elbim, C., Mackman, N., Edgington, T.S., and Deprost, D. (1993). Endotoxin-induced tissue factor messenger RNA in human monocytes is negatively regulated by a cyclic AMP dependent mechanism. *Blood*, **81**, 973.

27. Barstad, R.M., Hamers, M.J.A.G., Stephens, R.W., and Sakariassen, K. (1995). Retinoic acid reduces induction of monocyte tissue factor and tissue factor VIIa dependent arterial thrombosis formation. *Blood*, **86**, 212.

28. Almus, F.E., Rao, L.V.M., Pendurthi, U.R., Quattrochi, L., and Rapaport, S.I. (1991). Mechanisms for diminished tissue factor expression by endothelial cells cultured with heparin binding growth factor-1 and heparin. *Blood*, **77**, 1256.

29. Ahern, S.M., Miyata, T., and Sadler, J.E. (1993). Regulation of human tissue factor expression by mRNA turnover. *J. Biol. Chem.*, **268**, 2154.

30. Crossman, D.C., Carr, D.P., Tuddenham, E.G., Pearson, J.D., and McVey, J.H. (1990). The regulation of tissue factor mRNA in human endothelial cells in response to endotoxin or phorbol ester. *J. Biol. Chem.*, **265**, 9782.

31. Le, D.T., Rapaport, S.I., and Rao, L.V. (1992). Relations between factor VIIa binding and expression of factor VIIa/tissue factor catalytic activity on cell surfaces. *J. Biol. Chem.*, **267**, 15447.

32. Greeno, E.W., Bach, R.R., and Moldow, C.F. (1996). Apoptosis is associated with increased cell surface tissue factor procoagulant activity. *Lab. Invest.*, **75**, 281.

33. Bombeti, T., Karsan, A., Tait, J.F., and Harlan, J.M. (1997). Apoptotic vascular endothelial cells become procoagulant. *Blood*, **89**, 2429.

34. Bjørklid, E. and Storm, E. (1977). Purification and some properties of the protein component of tissue thromboplastin from human brain. *Biochem. J.*, **165**, 89.

35. Le, D.T., Rapaport, S.I., and Rao, L.V. (1994). Studies on the mechanism for enhanced cell surface factor VIIa tissue factor activation of factor X on fibroblast monolayers after their exposure to *N*-ethylmaleimide. *Thromb. Haemost.*, **72**, 848.

36. Bach, R. and Rifkin, D.B. (1990). Expression of tissue factor procoagulant activity: regulation by cytosolic calcium. *Proc. Natl. Acad. Sci. USA*, **87**, 6995.

37. Ravanat, C., Archipoff, G., Beretz, A., Freund, G., Cazenave, J.P., and Freyssinet, J.M. (1992). Use of annexin-V to demonstrate the role of phosphatidylserine exposure in the maintenance of haemostatic balance by endothelial cells. *Biochem. J.*, **282**, 7.

38. Sevinsky, J.R., Rao, V.M., and Ruf, W. (1996). Ligand induced protease receptor translocation into caveolae: a mechanism for regulating cell surface proteolysis of the tissue factor dependent coagulation pathway. *J. Biol. Chem.*, **133**, 293.

39. Camerer, E., Pringle, S., Wiiger, M.T., Skartlien, A.H., Prydz, K., Kolstø, A.-B., and Prydz, H. (1996). Opposite sorting of tissue factor in human umbilical vein endothelial cells and Madin Darby kidney epithelial cells. *Blood*, **88**, 1339.

40. Hansen, C., van Deurs, B., Petersen, L.C., and Rao, L.V.M. (1999). Discordant expression of tissue factor and its activity in polarized epithelial cells. Asymmetry in anionic phospholipid availability as a possible explanation. *Blood*, **94**, 1657.

41. Keller, O. and Simons, K. (1997). Post-Golgi biosynthetic trafficking. *J. Cell. Sci.*, **110**, 3001.

42. Roy, S., Paborsky, L.R., and Vehar, G.A. (1991). Self association of tissue factor as revealed by chemical cross-linking. *J. Biol. Chem.*, **266**, 4665.

43. Bach, R.R. and Moldow, C.F. (1997). Mechanism of tissue factor activation on HL-60 cells. *Blood*, **89**, 3270.

44. Lin, M.C., Almus-Jacobs, F., Chen, H.H., Parry, G.C., Mackman, N., Shyy, J.Y., and Chien, S. (1997). Shear stress induction of the tissue factor gene. *J. Clin. Invest.*, **9**, 737.

45. Silverman, M.D., Waters, C.R., Hayman, G.T., Wigboldus, J., Samet, M.W., and Lelkers, P.I. (1999). Tissue factor activity is increased in human endothelial cells cultured under elevated static pressure. *Am. J. Physiol.*, **277**, 233.

46. Hamik, A., Setiadi, H., Bu, G., McEver, R.P., and Morrissey, J.H. (1999). Down-regulation of monocyte tissue factor mediated by tissue factor pathway inhibitor and the low density lipoprotein receptor-related protein. *J. Biol. Chem.*, **27**, 4962.

47. Bazan, J.F. (1990). Structural design and molecular evolution of a cytokine receptor superfamily. *Proc. Natl. Acad. Sci. USA*, **87**, 6934.

48. Røttingen, J.-A., Enden, T., Camerer, E., Iversen, J.-G., and Prydz, H. (1995). Binding of human factor VIIa to tissue factor induces cytosolic Ca^{2+} signals in J82 cells, transfected COS-1 cells, Madin-Darby canine kidney cells and in human endothelial cells induced to synthesize tissue factor. *J. Biol. Chem.*, **270**, 4650.

49. Camerer, E., Røttingen, J.-A., Iversen, J.-G., and Prydz, H. (1996). Coagulation factors VII and X induce Ca^{2+} oscillations in Madin-Darby canine kidney cells only when proteolytically active. *J. Biol. Chem.*, **271**, 29034.

50. Berridge, M.J. (1993). Inositol trisphosphate and calcium signalling. *Nature*, **361**, 315.

51. Poulsen, L.K., Jacobsen, N., Sørensen, B.B., Bergenhem, N.C.H., Kelly, J.D., Foster, D.C., Thastrup, O., Ezban, M., and Petersen, L.C. (1998). Signal transduction via the mitogen activated protein kinase pathway induced by binding of coagulation factor VIIa to tissue factor. *J. Biol. Chem.*, **273**, 6228.

52. Sørensen, B.B., Freskgård, P.-O., Nielsen, L.S., Rao, L.V.M., Ezban, M., and Petersen, L.C. (1998). Factor VIIa-induced p44/42 mitogen activated protein kinase activation requires the proteolytic activity of factor VIIa and is independent of the tissue factor cytoplasmic domain. *J. Biol. Chem.*, **274**, 21349.

53. Pendurthi, U.R., Alok, D., and Rao, L.V.M. (1997). Binding of factor VIIa to tissue factor induces alterations in gene expression in human fibroblasts cells: up-regulation of poly(A) polymerase. *Proc. Natl. Acad. Sci. USA*, **94**, 12598.

54. Ollivier, V., Bentilila, S., Chabbat, J., Hakim, J., and de Prost, D. (1998). Tissue factor dependent vascular endothelial growth factor production by human fibroblasts in response to activated factor VII. *Blood*, **91**, 2698.

55. Taniguchi, T., Kakkar, A.K., Tuddenham, E.G.D., Williamson, R.C.N., and Lemoine, N.R. (1998). Enhanced expression of urokinase receptor induced

through the tissue factor – factor VIIa pathway in human pancreatic cancer. *Cancer Res.*, **58**, 4461.

56. Nystedt, S., Emilsson, K., Wahlestedt, C., and Sundelin, J. (1994). Molecular cloning of a potential proteinase activated receptor. *Proc. Natl. Acad. Sci. USA*, **91**, 9208.

57. Paborsky, L.R., Caras, I.W., Fisher, K.L., and Gorman, C.M. (1991). Lipid association, but not the transmembrane domain, is required for tissue factor activity. *J. Biol. Chem.*, **266**, 21911.

58. Zioncheck, T.F., Roy, S., and Vehar, G.A. (1992). The cytoplasmic domain of tissue factor is phosphorylated by a protein kinase C dependent mechanism. *J. Biol. Chem.*, **267**, 3561.

59. Mott, H.R. and Campbell, I.D. (1995). Four-helix bundle growth factors and their receptors: protein–protein interactions. *Curr. Opin. Struct. Biol.*, **5**, 114.

60. Whitmarsh, A.J., Yang, S.H., Su, M.S., Sharrocks, A.D., and Davis, R.J. (1997). Role of p38 and JNK mitogen activated protein kinases in the activation of ternary complex factors. *Mol. Cell Biol.*, **17**, 2360.

61. Camerer, E., Gjernes, E., Wiiger, M., Pringle, S., and Prydz, H. (2000). Binding of factor VIIa to Tissue Factor on keratinocytes induces gene expression. *J. Biol. Chem.*, **275**, in press.

62. Karnitz, L.M. and Abraham, R.T. (1995). Cytokine receptor signaling mechanisms. *Curr. Opin. Immunol.*, **7**, 320.

63. Wen, Z., Zhong, Z., and Darnell, J.E. (1995). Maximal activation of transcription by STAT1 and STAT3 requires both tyrosine and serine phosphorylation. *Cell*, **82**, 241.

64. Mueller, M. and Ruf, W. (1998). Requirement for binding of catalytically active factor VIIa in tissue factor dependent experimental metastasis. *J. Clin. Invest.*, **101**, 1372.

65. Abe, K., Shoji, M., Chen, J., Bierhus, A., Danave, I., Micko, C., Casper, K., Dillehay, D.L., Nawroth, P.P., and Rickles, F.R. (1999). Regulation of vascular endothelial growth factor production and angiogenesis by the cytoplasmic tail of tissue factor. *Proc. Natl. Acad. Sci. USA*, **96**, 8663.

66. Ott, I., Fischer, E.G., Miyagi, Y., Mueller, B.M., and Ruf, W. (1998) A role for tissue factor in cell adhesion and migration mediated by interaction with actin-binding protein 280. *J. Cell Biol.*, **140**, 1241.

67. Marti, A., Luo, Z., Cunningham, C., Ohta, J., Hartwig, J., Stossel, T.P., Kyriakis, J.M., and Avruch, J. (1997). Actin binding protein 280 binds the stress activated protein kinase (SAPK) activator SEK-1 and is required for tumour necrosis factor-α activation of SAPK in melanoma cells. *J. Biol. Chem.*, **272**, 2620.

68. Muller, M., Albrecht, S., Golfert, F., Hofer, A., Funk, R.H., Magdolen, V., Flossel, C., and Luther, T. (1999). Localisation of tissue factor in actin-rich membrane areas of epithelial cells. *Exp. Cell Res.*, **248**, 136.

69. Folkman, J. (1995). Clinical applications of research on angiogenesis. *N. Engl. J. Med.*, **333**, (26), 1757.

70. Hanahan, D. and Folkman, J. (1996). Patterns and emerging mechanisms of the angiogenic switch during tumourigenesis. *Cell*, **86**, 353.

71. Dvorak, H.F., Brown, L.F., Detmar, L.F., and Dvorak, A.M. (1995). Vascular permeability factor/vascular endothelial growth factor, microvascular hyperpermeability and angiogenesis. *Am. J. Pathol.*, **146**, (5), 1029.

72. Mechtcheriakova, D., Wlachos, A., Holzmuller, H., Binder, B.R., and Hofer, E. (1999). Vascular endothelial cell growth factor induced tissue factor expression in endothelial cells is mediated by EGR-1. *Blood*, **93**, 3811.

73. Bugge, T.H., Xiao, Q., Kombrinck, K.W., Flick, M.J., Holmback, K., Danton, H.J.S., Colbert, M.C., Witte, D.P., Fujikawa, K., Davie, E.W., and Degen, J.L. (1996). Fatal embryonic bleeding events in mice lacking tissue factor, the cell associated initiator of blood coagulation. *Proc. Natl. Acad. Sci. USA*, **93**, 6258.

74. Carmeliet, P., Mackman, N., Moons, L., Luther, T., Gressens, P., Van Vlaenderen, I., Demunck, H., Kasper, M., Breier, G., Evrard, P., Muller, M., Risau, W., Edgington, T., and Collen, D. (1996). Role of tissue factor in embryonic blood vessel development. *Nature*, **383**, 73.

75. Callander, N.S., Varki, N., and Rao, L.V.M. (1992). Immunohistochemical identification of tissue factor in solid tumours. *Cancer*, **70**, 1194.

76. Hudson, M.A. and McReynolds, L.M. (1997). Urokinase and the urokinase receptor: association with *in vitro* invasiveness of human bladder cancer cell lines. *J. Natl. Cancer Inst.*, **89**, 709.

77. Zhang, Y., Deng, Y., Luther, T., Muller, M., Ziegler, R., Waldherr, R., and Nawroth, P.P. (1994). Tissue factor controls the balance of angiogenic and anti-angiogenic properties of tumour cells in mice. *J. Clin. Invest.*, **94**, 1320.

78. Bromberg, M.E., Konigsberg, W.H., Madison, J.F., Pawashe, A., and Garen, A. (1995). Tissue factor promotes melanoma metastasis by a pathway independent of blood coagulation. *Proc. Natl. Acad. Sci. USA*, **92**, 8205.

79. Fischer, E.G., Riewald, M., Huang, H.Y., Miyagi, Y., Kubota, Y., Mueller, B.M., and Ruf, W. (1999). Tumour cell adhesion and migration supported by interaction of a receptor protease complex with its inhibitor. *J. Clin. Invest.*, **104**, 1213.

80. Fuster, V., Badimon, L., Badimon, J.J., and Chesebro, J.H. (1992). The pathogenesis of coronary artery disease and the acute coronary syndromes. *N. Engl. J. Med.*, **326**, 310.

81. Marmur, J.D., Rossikhuna, M., Guha, A., Fyfe, B., Friedrich, V., Mendlowitz, M., Nemerson, Y., and Taubman, M.B. (1993). Tissue factor is rapidly induced in arterial smooth muscle after balloon injury. *J. Clin. Invest.*, **91**, 2253.

82. Taubman, M.B., Marmur, J.D., Rosenfield, C.L., Guha, A., Nichtberger, S., and Nemerson, Y. (1993). Agonist mediated tissue factor expression in vascular smooth muscle cells: role of calcium mobilisation and protein kinase C activation. *J. Clin. Invest.*, **91**, 547.

83. Schechter, A.D., Fallon, J.T., Rossikhina, M., Xhang, X., Nemerson, Y., and Taubman, M.B. (1995). Monocyte chemoattractant protein 1 (MCP-1) induces tissue factor in human arterial smooth muscle cells and monocytes. *Blood*, **86**, 286.

84. Cui, M.Z., Penn, M.S., and Chisholm, G.M. (1999). Native and oxidised low density lipoprotein induction of tissue factor gene expression in smooth muscle cells is mediated by both Egr-1 and Sp1. *J. Biol. Chem.*, **274**, 32795.

85. Oltrona, L., Speidal, C.M., Recchia, D., Wickline, S.A., Eisenberg, P.R., and Abendschien, D.R. (1997). Inhibition of tissue factor mediated coagulation markedly attenuates stenosis after balloon induced arterial injury in minipigs. *Circulation*, **96**, 646.

19 Plaque rupture and its consequences

Geoffrey H. Tofler and David T. Kelly
Cardiology Department, Royal North Shore Hospital, St Leonards, NSW 2065, Australia

19.1 Introduction

Plaque disruption and subsequent arterial thrombosis are now recognized as critical to the onset of acute coronary syndromes.[1,2] Factors increasing the likelihood of a thrombotic coronary occlusion include: a plaque that is vulnerable to disruption, acute physiologic changes that induce plaque disruption, and finally a relatively hypercoagulable state and heightened vasomotor tone to promote complete lumen occlusion by thrombus.

19.2 Determinants of plaque rupture

The last 20 years have seen major advances in understanding the mechanism of plaque rupture and thrombus leading to acute myocardial infarction. These processes are now recognized to often occur in lesions previously causing only mild reduction of the diameter of the arterial lumen.[3] For example, Little *et al.* studied the extent of prior stenosis at sites in coronary arteries that subsequently became occluded. In two-thirds of the patients with mild-to-moderate coronary disease, the site of occlusion had less than 50% stenosis on the pre-infarction angiogram.[3] Thus, it is not the severity of stenosis (plaque volume) that determines the outcome but rather the type of stenosis (plaque composition) and its vulnerability to rupture and subsequent thrombosis. In the absence of these complications, coronary atherosclerosis is generally benign, and individuals may remain asymptomatic or have stable angina.

19.2.1 Development of the vulnerable lesion

Although a vulnerable lesion does not develop in a predictable manner, it occurs at the site of an atherosclerotic plaque. An in-depth discussion of atherosclerosis occurs elsewhere in this book; however, in brief, atherosclerotic plaques develop in areas with pre-existing intimal thickening. Increased endothelial permeability and activation leads to monocyte recruitment.[4] Lipoproteins trapped in the intima and endocytosed by the recruited monocyte-derived macrophages give rise to lipid-filled foam cells. Collections of lipid-filled foam cells may be seen macroscopically as yellow streaks. The endothelium is usually intact over early lesions, although it may be activated and dysfunctional. Intimal smooth muscle cell proliferation, migration and matrix synthesis, and extracellular lipid accumulation may give rise to mature and clinically significant atherosclerotic plaques with luminal narrowing progressing over years, frequently associated with calcification and/or ossification.

Ongoing inflammation with recruitment of monocyte-derived macrophages, T lymphocytes and a few mast cells may destroy the adjacent endothelium and lead to focal denuded areas often with platelets adhering to the exposed sub-endothelial tissue.[5] Platelets and microthrombi contribute to plaque growth by stimulating adjacent cells within the plaque, and by mediating proliferative, migratory, synthetic and inflammatory processes. Tissue factor is a major contributor to thrombus formation following plaque injury. Tissue factor is a glycoprotein expressed by circulating monocytes, macrophages, endothelial cells, and smooth muscle cells. Within the plaque itself, almost all of the tissue factor activity results from microparticles of monocyte and lymphocyte origin, which are shed by apoptotic cell death within the atherosclerotic plaque.[6] Circulating monocytes also participate in transient inflammatory responses. Early in the development of atherosclerotic lesions, activated endothelial cells recruit circulating peripheral blood monocytes, which then migrate into the sub-endothelial space. During this invasion into the arterial wall, monocytes interact with components of the extracellular matrix, especially collagen type 1. Once established in the plaque, macrophages accumulate lipid and maintain a continuous state of activation during which they secrete mediators of inflammation, functions that distinguish them from their circulating progenitors. Thus from a short-lived circulating cell, the monocyte differentiates into a resident cell with the potential to divide within the lesion. The presence of modified lipoprotein has long been recognized as a potential stimulus for recruitment and differentiation of circulating monocytes into the foam cell phenotype characteristic for atheroma.

Atherosclerotic plaques have two main components: atheromatous material and sclerotic tissue. The former is lipid-rich and soft, while the latter is collagen-rich and hard. The sclerotic plaque component may stabilize the plaque whereas the atheromatous component is dangerous, because it softens plaques making them prone to rupture. Plaque characteristics – composition, consistency, vulnerability, and thrombogenicity – vary from patient to patient, even from plaque to plaque in the same artery.[7] Serial angiographic studies have revealed that progression of human coronary artery disease is neither linear nor predictable. New high-grade lesions often appear in segments of artery that were normal on previous examination. This unpredictable and episodic progression can be explained by the occurrence of plaque rupture with subsequent haemorrhage into the plaque and/or mural thrombosis leading to stepwise and rapid growth. Disruption of a plaque with thrombosis is the principal mechanism of rapid progression of coronary atherosclerosis and the acute coronary syndromes.[8]

19.2.2 Vulnerability to rupture

A vulnerable plaque typically consists of a pool of soft atheromatous gruel separated from the vascular lumen by a cap of fibrous tissue. Three major determinants of plaque vulnerability have been identified: (i) the size and consistency of the atheromatous plaque; (ii) the thickness and collagen content of the fibrous cap; and (iii) ongoing inflammation superficially in the plaque. The vulnerable plaque is smaller in size, richer in lipids and poorer in smooth muscle cells, and more infiltrated with macrophages than the stable fibromuscular plaque.[9]

19.2.3 Atheromatous core

The mechanical properties of plaques, and the size and consistency of the atheromatous component, are critical in determining the distribution of stresses that may precipitate plaque rupture. Post-mortem studies have shown that plaques undergoing thrombosis have larger lipid pools than intact plaques. For example, Gertz and Roberts[10] reported the composition of plaques from 17 infarct related arteries examined post-mortem and found much larger atheromatous cores in the 39 segments with plaque disruption than in segments with an intact surface (32% versus 12%). Davies[7] determined that intact aortic plaques containing a core occupying more than 40% of the plaque area were considered particularly vulnerable and at high risk of rupture and thrombosis. Lipid in the form of cholesteryl esters softens plaque whereas crystalline cholesterol has the opposite effect.[11] Lipid lowering would deplete plaque lipid leading to a reduction in liquid cholesteryl esters and a relative increase in crystalline cholesterol, theoretically resulting in a stiffer and more stable atheromatous lesion.

19.2.4 Fibrous cap thickness

The atheromatous core is separated from the vascular lumen by a fibrous cap. This cap, which may be very thin and infiltrated with foam cells of macrophage origin, must bear all the biomechanical forces that tend to rupture the plaque. The fibrous cap is usually thinnest and weakest at its junctions with the relatively normal intima, also known as the shoulder regions.[12, 13] Figure 19.1 shows an example of plaque rupture occurring at the shoulder region of a fibrous cap. Collagen is important for tensile strength, and disrupted aortic caps contain less collagen than intact caps.[14] Based on La Place's law, for fibrous caps of the same tensile strength, caps

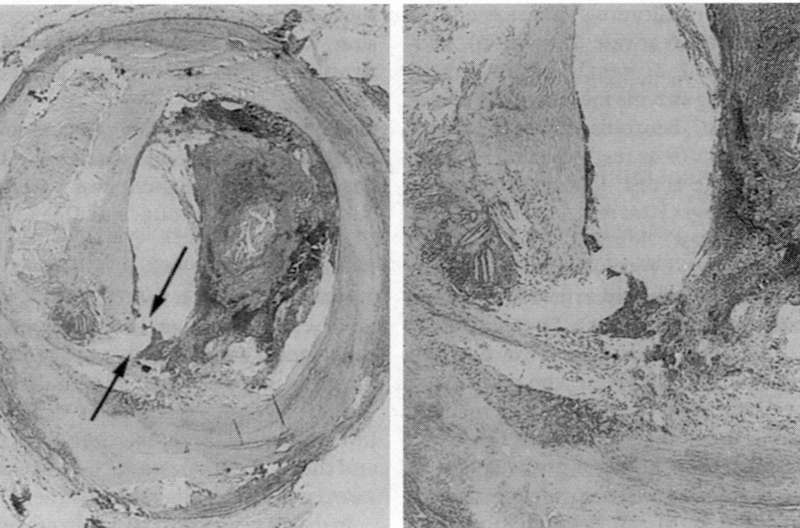

Fig. 19.1 Plaque rupture occurring at the junction of the fibrous cap with the mildly thickened intimal of the relatively normal arterial wall (shoulder region). (*Right*) A higher magnification of the shoulder area showing the rupture site and overlying thrombus. (Reproduced from Burke, A.P., Farb, A., Malcolm, G.T., Liang, Y.-h., Smialek, J.E., and Virmani, R. (1999). Plaque rupture and sudden death related to exertion in men with coronary artery disease. *JAMA*, **281**, 921.)

covering mildly or moderately stenotic plaques are more prone to rupture than caps covering stenotic plaques because the former have to bear a greater circumferential tension.[15]

19.2.5 Ongoing inflammation

Increasing evidence supports the view that acute coronary syndromes reflect exacerbations or activation of a chronic inflammatory response in the fibrous cap with subsequent rupture and exposure of thrombogenic substances to the flowing blood, leading to coronary thrombosis. Systemic markers of inflammation have been identified in patients with unstable angina and acute myocardial infarction (AMI). The acute-phase reactants C-reactive protein and amyloid A are elevated in patients with unstable angina and AMI, and fibrinogen has been shown in several studies to be an independent cardiovascular risk factor.[16, 17] In advanced lesions, macrophages and activated T cells secrete cytokines, such as interferon-γ, interleukin-2, tumour necrosis factor-α, and interleukin-1, all of which can modulate gene expression in vascular endothelial and smooth muscle cells. There is currently great interest in the potential role of matrix metalloproteinases elaborated by macrophages within plaque to degrade the fibrous caps of atherosclerotic lesions.[18] Such degradation of extracellular matrix at macrophage-rich sites could lead to plaque destabilization and rupture. Besides matrix metalloproteinases, macrophages and smooth muscle cells activated by inflammatory cytokines can elaborate other enzymes that weaken the connective tissue structure of the plaque fibrous cap. Secreted lysosomal proteinases and plasminogen activators may also contribute to tissue degradation: the relative importance of the enzymes may vary in individual situations. The presence of activated T cells suggests that specific immune responses may occur in the plaque. Several candidate antigens have been detected in advanced plaques, including oxidized LDL, heat shock proteins, and infective antigens such as *Chlamydia pneumoniae*. Focal expression of HLA-DR inflammatory antigens has been demonstrated at sites of plaque erosion and disruption.[19] It remains uncertain how much of the inflammatory component in an individual plaque is causal, versus how much is a response to the plaque injury.

19.2.6 Mechanical stress

Repeated stretching, compression, bending flexion, shear and pressure fluctuations may fatigue and weaken a fibrous cap predisposing it to rupture. Conversely, lowering the frequency and magnitude of load should reduce risk of rupture. La Place's law relates luminal pressure and radius to wall tension: the higher the blood pressure and the larger the luminal diameter, the more the tension that develops in the wall.[20] Plaque disruption may occur from the lumen into the plaque due to increase in luminal pressure, or the cap may rupture from the plaque into the lumen due to an increase in intraplaque pressure secondary to vasospasm, bleeding from vasa vasorum or plaque oedema. Spasm could theoretically rupture plaques by compressing the atheromatous core, forcing the fibrous cap out into the lumen. Bleeding and oedema into plaques from thin-walled vasa vasorum found at the base of plaques could increase the intraplaque pressure.[21] Vasa vasorum may originate upstream to a coronary stenosis, and, under certain conditions, their pressure could exceed that in the coronary lumen distal to the stenosis, leading to rupture of the plaque into the lumen.[21]

19.2.7 Rupture triggers

A coronary atherosclerotic plaque is exposed to a number of systemic physiologic processes that could, if the plaque were vulnerable, cause disease onset. Many of these processes increase in intensity in the morning. Such increases could, alone or in combination, account for the morning increase in cardiovascular disease onset through a variety of mechanisms.[22] The arterial pressure surge, which is accompanied by a heart rate increase,[23] could cause plaque rupture. The coronary arterial tone increase could worsen the flow reduction produced by a fixed stenosis.[24]

The increase in platelet aggregability, blood viscosity and von Willebrand factor, at a time of reduced fibrinolytic activity, could produce a state of relative hypercoagulability.[22, 25, 26] Such a thrombotic tendency could increase the likelihood that an otherwise harmless mural thrombus overlying a small plaque fissure would propagate and occlude the coronary lumen. The increased serum cortisol levels could increase the sensitivity of the coronary arteries to the vasoconstrictor effects of catecholamines. Many of the physiologic changes that are described in the morning are produced by stressors such as physical activity and anger, particularly in the presence of endothelial dysfunction. Heavy physical exertion has been associated with triggering of acute events. In the Onset study, a 5.9-fold increase in relative risk of myocardial infarction was observed in the hour after exertion.[27] Among sedentary individuals who exercise less than once per week, the relative risk was increased to 107-fold, while in those who exercised 5 or more times per week, the relative risk was reduced to 2-fold (Fig. 19.2). Other triggers that have been documented included a 20-fold increase in relative risk with cocaine usage, and a doubling in risk with episodes of anger[28] or with sexual activity.[29] In addition, an increased frequency of cardiovascular events has been reported in the winter months, possibly related in part to an increase in respiratory infections.[30] Possible triggers of disease onset, also called acute risk factors, have been reported by nearly 50% of patients with myocardial infarction.[31] The pathophysiologic mechanisms responsible for the non-random and often triggered onset of infarction are unknown, but are probably related to plaque disruption, likely caused by surges in sympathetic activity with a sudden increase in blood pressure, pulse rate, heart contraction, and coronary blood

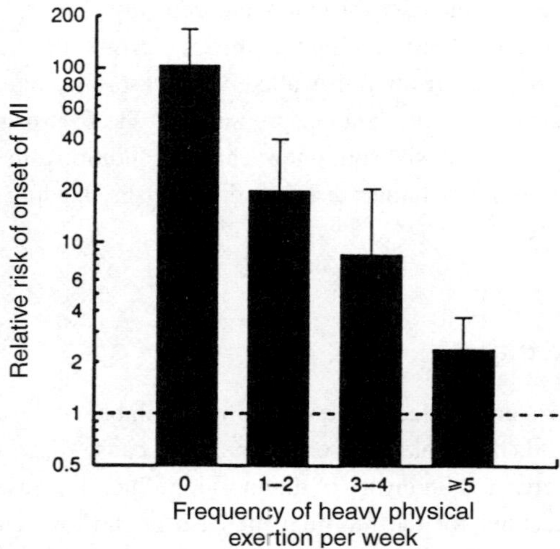

Fig. 19.2 Relative risk of myocardial infarction according to the usual frequency of heavy exertion. Habitually sedentary individuals had a relative risk of 107, whereas those who reported heavy exertion 5 times or more per week had a risk only 2.4 times higher than the baseline risk (adapted from Ref. (28)).

flow. Thrombosis occurs on previously disrupted or intact plaques when the thrombotic tendency is high because of platelet hyperaggregability, hypercoagulability, and/or impaired fibrinolysis and vasoconstriction, occurring locally around a coronary plaque or generalized.

The beneficial effect of β-blockade in the secondary prevention of myocardial infarction provides strong evidence for the theory that plaque disruption may trigger disease onset. β-blocker therapy in particular blunts the morning peak in onset of infarction probably by blunting the sympathetic surge in the morning, indicating that mechanical and haemodynamic forces could be critical in triggering plaque disruption and disease onset.[32–34] Activation of the sympathetic nervous system and increased catecholamines associated with arousal, exercise, emotional stress, and smoking could promote platelet aggregation and vasoconstriction via α-receptors. Sudden thrombus growth on previously disrupted plaques or intact plaques due to changes in platelet function and coagulation is probably an important mechanism responsible for onset of acute coronary syndromes.

Figure 19.3 illustrates the contribution of the vulnerable plaque and external triggers to acute coronary syndromes. The process begins with a non-vulnerable atherosclerotic plaque. This plaque may become vulnerable to disruption, due to increases in macrophage and lipid content and inflammatory cell products, and thinning of the fibrous cap. At this point, haemodynamic or vasoconstrictive forces generated by a physical or psychologic stress might disrupt the plaque. If the plaque disruption produces a major thrombogenic stimulus with extensive exposure of the collagen and atherosclerotic core contents, an occlusive thrombus forms rapidly leading to myocardial infarction or sudden death. If the plaque disruption produces only a minor thrombogenic stimulus, the patient may be asymptomatic, yet platelet aggregates and thrombin generation may contribute to growth of the plaque or cause unstable angina or non-Q wave infarction. At this point, other external triggering activities could cause the artery to constrict and the thrombus to become occlusive, or could lead to increased coagulability that also predisposes to occlusive thrombus formation.

Besides providing a framework for ongoing research, this scheme also leads to some speculations. For example, an inverse relation can be postulated between the degree of plaque vulnerability and intensity of the triggering stimulus required to produce rupture. For example, an elderly individual with a severe fixed stenosis in a coronary artery and an extremely vulnerable plaque may develop thrombosis with only a minor stress, whereas a young individual with only a luminal irregularity may require a major stress (such as heavy lifting or cocaine usage) to trigger plaque rupture and coronary thrombosis.

19.2.8 Identification of vulnerable and disrupted plaques

Although coronary angiography is suitable for the identification of obstructive lesions responsible for myocardial ischaemia and infarction, visualization of the

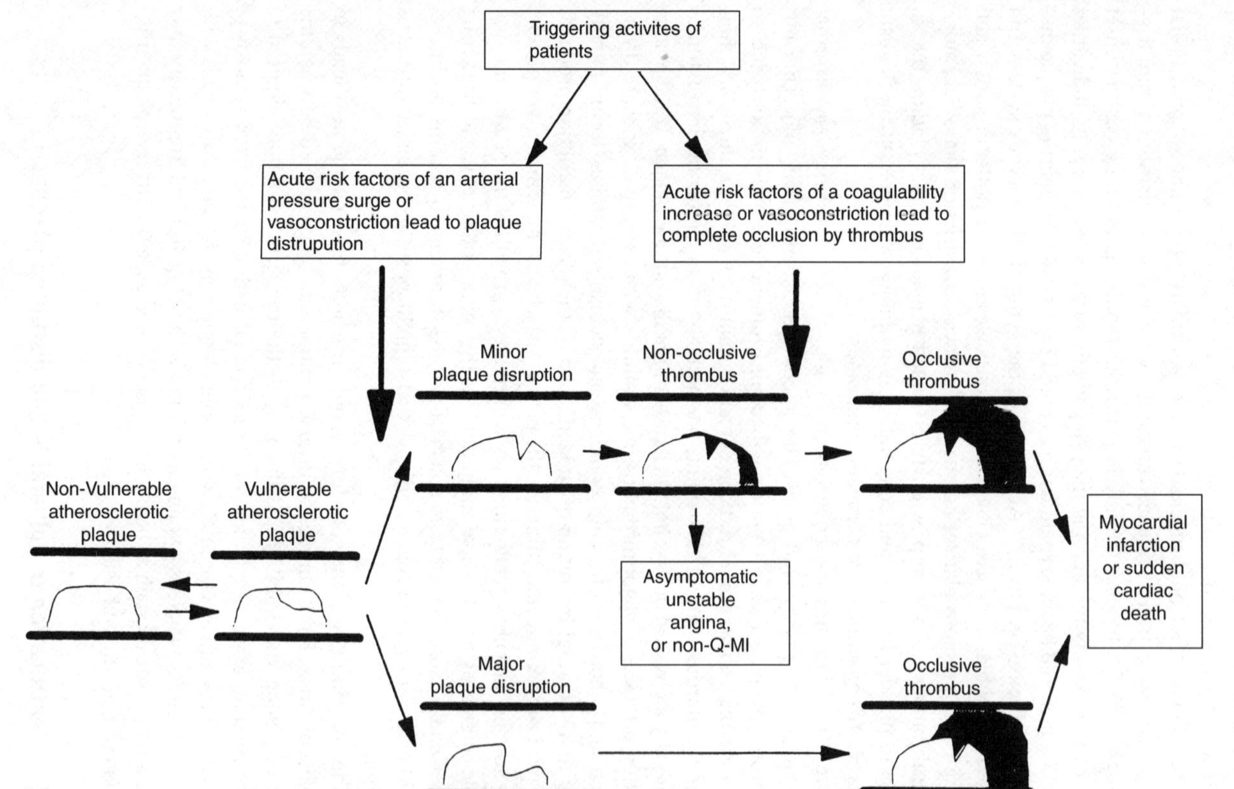

Fig. 19.3 Illustration of how a plaque evolves from non-vulnerable to vulnerable, and how daily activities may trigger plaque rupture and thrombosis. The scheme identifies areas for further research and intervention (adapted from Ref. (27)).

arterial wall rather than the lumen is necessary for the identification of early lesions and non-stenotic vulnerable plaques. Intravascular ultrasound and angioscopy may reveal important plaque and surface features not seen angiographically, and magnetic resonance imaging, spectroscopy, infrared imaging or temperature probes may in the near future improve the *in vivo* identification and characterization of coronary plaques further. A blood sample may reveal signs of inflammation and activation of endothelial cells, leukocytes, coagulation, and fibrinolysis that may prove useful in differentiating between stable and progressing atherosclerosis. Angioscopy, with its ability to detect yellow plaques and red versus white thrombus, may aid in the search for surface characteristics of the vulnerable plaque. Intracoronary ultrasound, with its ability to penetrate beneath the lumen surface and reveal certain tissue characteristics, may have the potential to detect within the arterial wall lipid-rich plaques that are vulnerable to disruption. An animal model of triggering was developed more than 30 years ago by Dr Paris Constantinides in which these ideas can be explored.[35] Atherosclerotic plaques were produced in New Zealand White rabbits by intermittent cholesterol feeding. Triggering of plaque disruption and thrombosis was then produced by intraperitoneal injection of Russell Viper Venom, a procoagulant and endothelial toxin, followed by the intravenous injection of histamine or adrenaline. The aortas of the rabbits were then found to have disrupted atherosclerotic plaques with overlying platelet-rich thrombi. The lesions produced are similar to those observed in patients, with the advantage that a biological intervention is used to trigger plaque disruption and platelet-rich thrombi. A modification by Abela to this model, which increases the yield of disrupted plaques, is to accelerate atherosclerosis by the use of a Fogarty catheter introduced through the femoral artery to produce injury to the aorta several weeks before triggering.[36]

19.2.9 Coronary thrombosis

Coronary thrombosis is the result of a dynamic interplay between the arterial wall and the flowing blood. About 75% of thrombi responsible for acute coronary syndromes are precipitated by plaque rupture whereby thrombogenic material is exposed to the flowing blood. For the remaining 25% of thrombi, superficial intimal erosions are usually found. If a plaque disruption leads to exposure of very thrombogenic material, why does an occlusive thrombus not occur on most occasions? Several factors may influence the final result. Firstly, the degree of initial stenosis and surface irregularities that activate platelets may play an important role. Luminal thrombus formation increases with increasing degree of stenosis, probably due to shear-induced platelet activation. Secondly, the thrombotic–thrombolytic balance at the time of plaque disruption may determine the outcome. Systemic thrombogenic factors such as platelet aggregation, plasma fibrinogen, and impaired fibrinolytic activity are associated with the development of acute infarction – a relative thrombotic imbalance in the morning hours may predispose to thrombosis. Thirdly, the character and extent of the exposed plaque components may vary from lesion to lesion. The atheromatous gruel is not only the most vulnerable plaque component, it is also the most thrombogenic, being up to six-fold more thrombogenic than other plaque components.[37] Also of note, a well-developed collateral circulation at the time of occlusion may lead to protection against ensuing myocardial infarction.

19.3 Consequences of plaque rupture

Following plaque disruption, haemorrhage into the plaque, luminal thrombosis, and/or vasospasm may cause sudden flow obstruction, giving rise to new or changing symptoms. The clinical presentation and the outcome depend on the severity and duration of ischaemia. A non-occlusive or transiently occlusive thrombus most frequently underlies primary unstable angina with pain at rest and non-Q wave MI, whereas Q wave MI will be associated with complete occlusion.

19.3.1 Silent progression

Rupture of the plaque surface occurs frequently during plaque growth, and is probably the most important mechanism of rapid progression of coronary artery disease (CAD). Plaque rupture with non-occlusive progression is clinically silent in the majority of cases. Clinical symptoms will develop only if a major luminal thrombus evolves with subsequent flow obstruction. Autopsy data indicate that 9% of normal healthy persons are walking around with disrupted plaques in their coronary arteries, increasing to 22% in persons with diabetes or hypertension.[38] Many persons dying of ischaemic heart disease have both thrombosed and non-thrombosed disrupted plaques in their coronary arteries. As many as 103 disrupted plaques were identified in 47 people dying of CAD and only 40 were associated with significant luminal thrombosis.[8]

19.4 Protective interventions

The best approach to prevention of acute coronary syndromes is primary prevention of atherosclerosis, and extensive research towards this goal must continue. However this approach must be complementary to pursuit of the mechanisms by which more advanced plaques become vulnerable to rupture. Developments in cellular and molecular biology offer promise of methods to treat vulnerable plaques. If the growth factors and chemotactic agents controlling macrophage infiltration of plaques were fully identified, perhaps the development of vulnerable plaques could be prevented. The overall risk of plaque rupture may be reduced by reducing either plaque vulnerability or rupture triggers. Plaque vulnerability may change with time and treatment. Lipid-lowering therapy could stabilize plaques, reducing their vulnerability, though not necessarily their size. The beneficial effects of lipid-lowering therapy could be due to: (i) stabilization of coronary plaques against rupture by depleting lipid and macrophage foam cells from the dangerous vulnerable plaques; (ii) reduced thrombogenicity of coronary plaques; (iii) regained normal endothelial function including endothelium-dependent vasodilation, and/or reduced systemic thrombotic propensity via beneficial effects on platelets, fibrinogen and plasma viscosity; and (iv) reduced amount of cholesterol entering the plaque. Increasing HDL may also be protective by facilitating cholesterol clearance from the plaque. Modest improvements in lumen diameter may produce significant benefits in reduced event rate. It may also be possible to reduce the effects of macrophage activation. Activated macrophages release lytic enzymes, matrix metalloproteinases that may be responsible for plaque rupture. The transfer of genes encoding natural or exogenous inhibitors of macrophage metalloproteinases into the atherosclerotic plaque constitutes a potential strategy for preventing fibrous cap weakening due to excessive extracellular matrix degradation. Little is known about regulation of matrix metalloproteinases and their inhibitors. Studies suggest that the expression of matrix metalloproteinases may be induced by cytokines, some of which are produced in the atherosclerotic plaque. Inhibition of cytokine secretion within the plaque may therefore have potential clinical relevance for prevention of matrix degradation and subsequent rupture. In an animal model, lipid lowering by diet reduced the number of macrophages and the expression and activity of metalloproteinases, and increased the collagen content of established atheromas, potentially making them less susceptible to rupture.[39]

Recognition of the multifactorial role of the endothelium includes an awareness of its contribution to maintaining a thrombotic balance, ensuring that blood does not normally clot as it passes through the vessels.[40] Antithrombotic products of the endothelium include prostacycline, thrombomodulin, ECTO-ADPase, tissue plasminogen activator, urokinase, and heparin-like molecules. Conversely, prothrombotic products of the endothelium include platelet activating factor, tissue factor, von Willebrand factor, plasminogen activator inhibitor-1 (PAI-1), and other coagulation factors. Oxidized lipoproteins[41] can increase the synthesis of PAI-1 by cultured human endothelial cells. Improvement in endothelial function may be produced by interventions such as the use of antioxidants, administration of appropriate dietary fatty acids, exercise conditioning, avoidance of psychosocial stress, ACE inhibition, blood pressure lowering, and oestrogen replacement therapy. These interventions may be associated with favourable changes such as improved fibrinolytic potential and reduced platelet reactivity. Antiplatelet and antithrombotic agents such as aspirin, clopidogril, Warfarin and other agents have also been shown to be protective against thrombus formation and acute coronary events.

19.4.1 Trigger reduction

If vulnerable rupture-prone plaques are present in the coronary arteries, a rupture might be delayed or prevented by appropriate reduction of the effect of a triggering activity, as might be accomplished through the use of β-blockers, aspirin and possibly ACE inhibitors. Recommendation to avoid certain triggers, such as smoking or cocaine usage, can be readily made. Heavy exertion in sedentary individuals likewise may be associated with increased risk. Emphasis can also be placed on pharmacologic protection during the morning hours for patients already receiving anti-ischaemic treatment. The means of prevention would not be to eliminate potential triggering activities – an undesirable and unattainable goal – but to design regimens that can be evaluated in randomized studies for their ability to sever the linkage between a potential triggering activity and development of a catastrophic coronary event. Therapies should provide protection during the full 24 hours of the day, in particular during the morning period of increased risk.

19.5 Conclusions

The composition, consistency, vulnerability, and thrombogenicity of plaques differ markedly, even for plaques that have been exposed to the same risk factors. More research is needed into the determinants of the initiation and progressive growth of the soft lipid-rich atheromatous core. Although the degree of stenosis is a risk marker for thrombotic occlusion, it is a poor predictor of acute coronary syndromes because the most severe stenoses are often protected by collateral vessels. Vulnerability and rupture risk are not a simple function of stenosis severity. Since stenotic plaques are not necessarily vulnerable and vulnerable plaques are not necessarily stenotic, rupture-prone plaques are easily missed on coronary angiograms. In patients with stable angina, therapies directed against stenotic lesions may relieve pain but do not necessarily improve prognosis, because infarction and mortality depend more on coexisting non-stenotic vulnerable plaques than on angina-producing stenotic lesions. The risk of plaque rupture is a function of both plaque vulnerability (intrinsic disease) and rupture triggers (extrinsic forces). The former predisposes the plaque to rupture, while the latter may precipitate it. The challenge of today is to identify and treat the dangerous vulnerable plaques responsible for infarction and death – to find and treat only angina-producing stenotic lesions is no longer enough. For prevention and treatment, a systemic approach that addresses all coronary plaques will prove to be the most rewarding.

References

1. Davies, M.J. and Thomas, A.C. (1985). Plaque fissuring – the cause of acute myocardial infarction, sudden ischaemic death, and crescendo angina. *Br. Heart J.*, **53**, 363–73.

2. DeWood, M.A., Spores, J., Notske, R., Mouser, L.T., Burroughs, R., Golden, M.S., and Lang, H.T. (1980). Prevalence of total coronary occlusion during the early hours of transmural myocardial infarction. *N. Engl. J. Med.*, **303**, 897–902.
3. Little, W.C., Constantinescu, M., Applegate, R.J., Kutcher, M.A., Burrows, M.T., Kahl, F.R., and Santamore, W.P. (1998). Can coronary angiography predict the site of a subsequent myocardial infarction in patients with mild-to-moderate coronary artery disease? *Circulation*, **78**, 1157–66.
4. Ross, R. (1993). The pathogenesis of atherosclerosis: a perspective for the 1990's. *Nature.*, **362**, 801–9.
5. Zhang, Y., Cliff, W.J., Schoefl, G.I., and Higgins, G. (1993). Plasma protein insudation as an index of early coronary atherogenesis. *Am. J. Pathol.*, **143**, 496–506.
6. Mallat, X., Nakamura, T., Ohan, J., *et al.* (1999). Shed membrane microparticles with procoagulant potential in human atherosclerotic plaques: a role for apoptosis in plaque thrombogenicity. *Circulation*, **99**, 2908.
7. Davies, M.J. (1996). The contribution of thrombosis to the clinical expression of coronary atherosclerosis. *Thromb. Res.*, **82**, 1–32.
8. Falk, E., Shah, P.K., and Fuster, V. (1995). Coronary plaque disruption. *Circulation*, **92**, 657–71.
9. Falk, E. (1992). Why do plaques rupture? *Circulation*, **86**, III30–42.
10. Gertz, S.D. and Roberts, E.C. (1990). Hemodynamic shear force in rupture of coronary arterial atherosclerotic plaques. *Am. J. Cardiol.*, **66**, 1368–72.
11. Small, D.M. (1998). Progression and regression of atherosclerotic lesions. Insights from lipid physical biochemistry. *Arteriosclerosis*, **8**, 103–29.
12. Richardson, P.D., Davies, M.J., and Born, G.V. (1989). Influence of plaque configuration and stress distribution on fissuring of coronary atherosclerotic plaques (see comments). *Lancet*, **2**, 941–4.
13. Cheng, G.C., Loree, H.M., Kamm, R.D., Fishbein, M.C., and Lee, R.T. (1993). Distribution of circumferential stress in ruptured and stable atherosclerotic lesions. A structural analysis with histopathological correlation. *Circulation*, **87**, 1179–87.
14. Burleigh, M.C., Briggs, A.D., Lendon, C.L., Davies, M.J., Born, G.V.R., and Richardson, P.D. (1992). Collagen types I and III, collagen content, GAGs and mechanical strength of human atherosclerotic plaque caps: span-wise variations. *Atherosclerosis*, **96**, 71–81.
15. Loree, H.M., Kamm, R.D., Stringfellow, R.G., and Lee, R.T. (1992). Effects of fibrous cap thickness on peak circumferential stress in model atherosclerotic vessels. *Circ. Res.*, **71**, 850–8.
16. Kannel, W.B., Wolf, P.A., Castelli, W.P., and D'Agostino, R.B. (1987). Fibrinogen and risk for cardiovascular disease. *JAMA*, **258**, 1183–6.
17. Ernst, E., and Resch, K.L. (1993). Fibrinogen as a cardiovascular risk factor: a meta-analysis and review of the literature. *Ann. Int. Med.*, **118**, 956–63.
18. Libby, P. (1995). Molecular bases of the acute coronary syndromes. *Circulation*, **91**, 2844–50.
19. van der Wal, A.C., Becker, A.E., van der Loos, C.M., and Das, P.K. (1994). Site of intimal rupture or erosion of thrombosed coronary atherosclerotic plaques is characterized by an inflammatory process irrespective of the dominant plaque morphology. *Circulation*, **89**, 36–44.
20. Lee, R.T. and Kamm, R.D. (1994). Vascular mechanics for the cardiologist. *J. Am. Coll. Cardiol.*, **23**, 1289–95.

21. Barger, A.C., Beeuwkes, R.I., Lainey, L.L., and Silverman, K.J. (1984). Hypothesis: vaso vasorum and neovascularization of human coronary arteries. A possible role in the pathophysiology of atherosclerosis. *N. Engl. J. Med.*, **310**, 175–7.

22. Muller, J.E., Abela, G.S., Nesto, R.W., and Tofler, G.H. (1994). Triggers, acute risk factors and vulnerable plaques: The lexicon of a new frontier. *J. Am. Coll. Cardiol.*, **23**, 809–13.

23. Millar-Craig, M.W., Bishop, C.N., and Raftery, E.B. (1978). Circadian variation of blood pressure. *Lancet*, **I**, 795–7.

24. Panza, J.A., Epstein, S.E., and Quyyumi, A.A. (1991). Circadian variation in vascular tone and its relation to alpha-sympathetic vasoconstrictor activity. *N. Engl. J. Med.*, **325**, 986–90.

25. Tofler, G.H., Brezinski, D., Schafer, A.I., Czeisler, C.A., Rutherford, J.D., Willich, S.N., Gleason, R.E., Williams, G.H., and Muller, J.E. (1987). Concurrent morning increase in platelet aggregability and the risk of myocardial infarction and sudden cardiac death. *N. Engl. J. Med.*, **316**, 1514–18.

26. Andreotti, F., Davies, G.J., and Hackett, D.R. (1988). Major circadian fluctuations in fibrinolytic factors and possible relevance to time of onset of myocardial infarction, sudden cardiac death and stroke. *Am. J. Cardiol.*, **62**, 635–7.

27. Mittleman, M.A., Maclure, M., Tofler, G.H., Sherwood, J.B., Goldberg, R.J., and Muller, J.E. (1993). Triggering of acute myocardial infarction by heavy physical exertion. Protection against triggering by regular exertion. Determinants of Myocardial Infarction Onset Study Investigators. *N. Engl. J. Med.*, **329**, 1677–83.

28. Mittleman, M.A., Maclure, M., Sherwood, J.B., Mulry, R.P., Tofler, G.H., Jacobs, S.C., Friedman, R., Benson, H., and Muller, J.E. (1995). Triggering of acute myocardial infarction onset by episodes of anger. *Circulation.*, **92**, 1720–5.

29. Muller, J.E., Mittleman, M.A., Maclure, M., Sherwood, J.B., and Tofler, G.H. (1996). Triggering myocardial infarction by sexual activity: low absolute risk and prevention by regular physical exertion. *JAMA*, **275**, 1405–9.

30. Ornato, J.P., Siegel, L., Craren, E.J., and Nelson, N. (1990) Increased incidence of cardiac death attributed to acute myocardial infarction during winter. *Coronary Artery Disease*, **1**, 199–203.

31. Tofler, G.H., Stone, P.H., Maclure, M., Edelman, E., Davis, V.G., Robertson, T., Antman, E.M., and Muller, J.E. (1990). Analysis of possible triggers of acute myocardial infarction (The MILIS Study). *Am. J. Cardiol.*, **66**, 22–7.

32. Muller, J.E., Stone, P.H., Turi, Z.G., Rutherford, J.D., Czeisler, C.A., Parker, C., Poole, W.K., Passamani, E., Roberts, R., Robertson, T., Sobel, B.E., Willerson, J.T., and Braunwald, E. (1985). Circadian variation in the frequency of onset of acute myocardial infarction. *N. Engl. J. Med.* **313**, 1315–22.

33. Hjalmarson, A., Gilpin, E.A., Nicod, P., Dittrich, H., Henning, H., Engler, R., Blacky, R., Smith, S.C., Ricou, F., and Ross, J. (1989). Differing circadian patterns of symptom onset in subgroups of patients with acute myocardial infarction. *Circulation*, **80**, 267–75.

34. Tofler, G.H., Muller, J.E., Stone, P.H., Forman, S., Solomon, R.E., Knatterud, G.L., and Braunwald, E. (1992). Modifiers of timing and possible triggers of acute myocardial infarction in the TIMI II population. *J. Am. Coll. Cardiol.*, **20**, 1049–55.

35. Constantinides, P. and Chakravarti, R.N. (1961). Rabbit arterial thrombosis production by systemic procedures. *Archives of Pathology*, **72**, 197–208.

36. Abela, G.S., Picon, P.D., Friedl, S.E., Gebara, O.C., Miyamoto, A., Federman, M., Tofler, G.H., and Muller, J.E. (1995). Triggering of plaque disruption and arterial thrombosis in an atherosclerotic rabbit model. *Circulation*, **91**, 776–83.
37. Fernandez-Ortiz, A., Badimon, J.J., Falk, E., Fuster, V., Meyer, B., Mailhac, A., Weng, D., Shah, P.K., and Badimon, L. (1994). Characterization of the relative thrombogenicity of atherosclerotic plaque components: implications for consequences of plaque rupture. *J. Am. Coll. Cardiol.*, **23**, 1562–9.
38. Davies, M.J., Bland, J.M., Hangartner, J.R.W., Angelini, A., and Thomas, A.C. (1989). Factors influencing the presence or absence of acute coronary artery thrombi in sudden ischaemic death. *Eur. Heart. J.*, **10**, 203–8.
39. Aikawa, M., Rabkin, E., Okada, Y., *et al.* (1998). Lipid lowering by diet reduces metalloproteinase activity and increases collagen content of rabbit atheroma: a potential mechanism of lesion stabilisation. *Circulation*, **97**, 2433.
40. Rosenberg, R.D. and Rosenberg, J.S. (1984). Natural anticoagulant mechanisms. *J. Clin. Invest.*, **74**, 1–6.
41. Tremoli, E., Camera, M., Maderna, P., Sironi, L., Prati, L., Colli, S., Piovella, F., Bernini, F., Corsini, A., and Mussoni, L. (1993). Increased synthesis of plasminogen activator inhibitor-1 by cultured human endothelial cells exposed to native and modified LDL: an LDL receptor-independent phenomenon. *Arterioscler. Thromb.*, **13**, 338–46.

20 Futures: Diagnosis and therapy of atherosclerosis

Leonard Kritharides, Mark R. Adams, David S. Celermajer,
David T. Kelly and Roger T. Dean
The Heart Research Institute, Camperdown, Sydney, NSW 2050, Australia

20.1 Introduction

In this chapter we will discuss briefly and speculatively the diagnosis and therapy of atherosclerosis. Many of the issues raised here are discussed in more detail elsewhere in the book, in a different context, so references are only provided here when they are not available in other chapters.

Individual cardiovascular disease assessment is currently based on risk factor prevalence. Risk factors do not reveal mechanism, the subject of this book. Risk factors are *population* predictors of disease, and any individual may not develop the clinical manifestation of atherosclerosis expected from their risk factors. Conversely, individuals without any risk factors may have extensive disease.

A diagnostic test for atherosclerosis should ideally determine in an individual their probability of developing clinical disease. Even without improved positive predictive value for major cardiac events, an atherosclerosis diagnostic might confer sociopsychological benefits which follow the reassurance of a negative study and sequential monitoring of disease progression/regression in the individual subject.

Most individuals in western society over the age of 15 have significant fatty lesions characteristic of early atherosclerosis[1] and thus a simple test for atherosclerosis is of limited value. The challenge is to identify quantitative population norms for the whole body burden of atherosclerosis, and to target management in those individuals who exceed these norms. In addition qualitative aspects, such as whether the disease is stable or unstable, and where it is located, should be a diagnostic goal.

The study of any particular mechanistic event in atherogenesis (e.g. oxidation) may help the development of atherosclerosis diagnosis, whether or not it is causal, and whether or not it leads to effective therapeutics. Similarly, diagnosis of certain infectious agents may be of value, but must not be confounded with diagnosis of atherosclerosis. If atherosclerosis is polydisperse in mechanism, so will be diagnostic markers.

20.2 Imaging

Techniques for direct visualization of arteries and their walls have been developed. These include ultrasound, contrast-based angiography, X-ray computed tomography scanning and, magnetic resonance imaging.

Such techniques may be able to assess arterial function and/or structure. Angiographic studies can delineate obstructive stenoses but plaques can exist without any vessel narrowing, owing to remodelling of the unaffected part of the vessel wall opposite an eccentric plaque.[2] Therefore, recent work has focused on imaging the arterial wall directly to measure plaque size, volume, and plaque composition. As atherosclerotic plaques are usually heterogeneous, with lipid pools and areas of inflammation indicating greater mobility to plaque rupture and events, and calcification and connective tissue matrix more associated with stable plaques, the ultimate aim of imaging is to provide this information.

20.2.1 Ultrasound

Ultrasound relies on the reconstruction of images from transmitted and received high frequency sound waves, which reflect from tissue interfaces within the body. In general, resolution and depth penetration are inversely related; that is, high resolution pictures are only obtainable from superficial structures, and deep structures show imperfect resolution.

Therefore, in order to obtain high quality images of the arterial wall, the artery interrogated by ultrasound must be close to the transducer that emits and receives the sound waves, i.e. transducers placed on the skin, can produce high quality images from superficial arteries such as the carotid and brachial vessels, but not from the coronary arteries. Satisfactory ultrasonic recordings from the coronary arteries relies on placement of a transducer close to the vessel in question, either in the oesophagus (transoesophageal echo for the proximal coronary arteries) or near the vessels themselves (intracoronary ultrasound).

External vascular ultrasound is completely non-invasive, free of side effects and relatively inexpensive. It can also be performed serially over a period of time, making external ultrasound measurements potential end-points in clinical trials. By contrast, transoesophageal echocardiography is more invasive and therefore, has less patient acceptability. It can be used, however, to visualize plaque in the ascending and descending thoracic aorta, which some observers have suggested may be a useful surrogate for coronary artery disease, as well as an important risk factor for stroke.[3]

Transvenous ultrasound is a very new modality, where the ultrasound transducer is mounted on a catheter introduced into the venous system of the body, usually via the femoral vein. The catheter has a soft tip, is steerable and can be used to obtain longitudinal images of the iliofemoral system, abdominal aorta, renal arteries, aorta and pulmonary arteries, head and neck vessels as well as detailed intracardiac views. This technology is not routinely used in humans, but may have interesting potential applications for the future, as it might allow screening of the major arteries prone to atherosclerosis, in vasculopathic subjects.

Intravascular (intra-arterial) ultrasound is now also feasible due to advanced catheter technology. Miniaturized catheters, often only 1–1.5 mm in diameter, can be introduced into most arteries of the body, and produce cross-sections of the arterial wall, usually by mechanical rotation of the ultrasound crystal. High rotational speeds of up to 2000 rpm are required to produce adequate frame rates and accurate image reconstruction. High frequency systems give excellent resolution, but only at very shallow depths. Although invasive, intracoronary ultrasound is now regarded by many investigators as the new 'gold standard' for vascular imaging of normal arterial wall and plaque.[4] An intravascular ultrasound from several points along an angiographically visualized coronary artery is shown in Fig. 20.1, demonstrating how intravascular ultrasound images may show significant amounts of atherosclerotic plaque, despite relatively normal angiographic appearances. Areas of inhomogeneity within the plaques will also be appreciated; this relates to plaque composition.

20.2.1.1 Intima–media thickness

Intima–media thickness (IMT) of the carotid arteries is defined as the distance between the lumen–intima interface (often referred to as the 'i' line) and the media adventitia interface (referred to as the 'm' line). This measurement can be made accurately and reproducibly in the common carotid, bifurcation and to a lesser extent the internal carotid arteries.[5]

Carotid IM thickness has been shown to increase with age and to correlate with risk factors.[6, 7] It correlates significantly but relatively poorly with the extent and severity of atherosclerosis in other vascular beds, such as the coronary arteries.[8] Increased carotid IM thickness has been shown to predict the likelihood of subsequent coronary events.[9]

Carotid IM thickness in atherosclerosis has been used as a surrogate end-point in clinical trials. In the Cholesteral Lowering Atherosclerosis Study (CLAS) trial, a randomized placebo controlled trial of lipid lowering, there was a reduction of 0.05 mm in IMT over three years in the treated group compared to a progression of 0.07 mm in the placebo group, a highly significant difference, consistent with an anti-atherosclerosis effect.[10]

Individual centres use different methods to assess carotid IM thickness, e.g. a single integrated measure of the distal far wall of the common carotid artery, or more complex scores of near and far walls in the distal common, carotid bulb and proximal internal carotid arteries. In general, common carotid scores are easiest to measure and are highly reproducible.

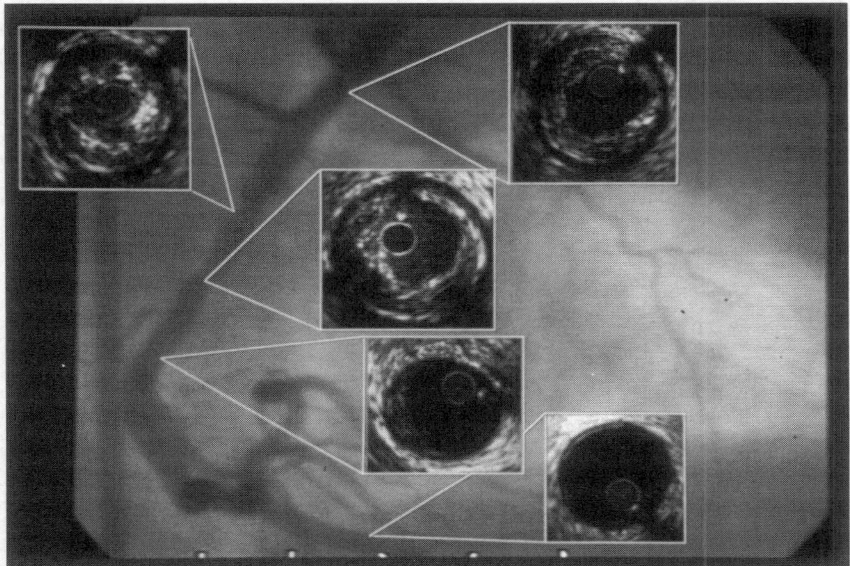

Fig. 20.1 Serial intravascular ultrasound images from the right coronary artery. Although some areas of the artery have a relatively normal angiographic appearance, intravascular ultrasound images demonstrate significant atherosclerotic plaque.

Carotid IM thickness, an ultrasound-based measurement of early arterial wall thickening, appears to be a reasonable surrogate for early atherosclerosis detection.

20.2.1.2 Plaque volume and composition

Plaques have been more difficult to measure with external vascular ultrasound. Those with superficial areas of echolucency may have their size underestimated, and those with prominent calcium are difficult because of shadowing artefacts to the deeper regions of the plaque. In general, plaque volume cannot be measured reliably and reproducibly by external ultrasound techniques.[11]

In contrast, more invasive techniques such as intravascular ultrasound can determine plaque composition. Echolucent lipid rich areas can usually be distinguished from relatively echodense collagen rich parts of plaques, and intraplaque hemorrhage or calcium can also usually be identified confidently.[12] Owing to their invasive nature, these techniques are not widely applicable for the measurement of early atherosclerotic lesions.

20.2.2 Nuclear medicine

The majority of nuclear medicine techniques have proven relatively disappointing in the imaging of pre-clinical atherosclerosis, or even clinically relevant lesions

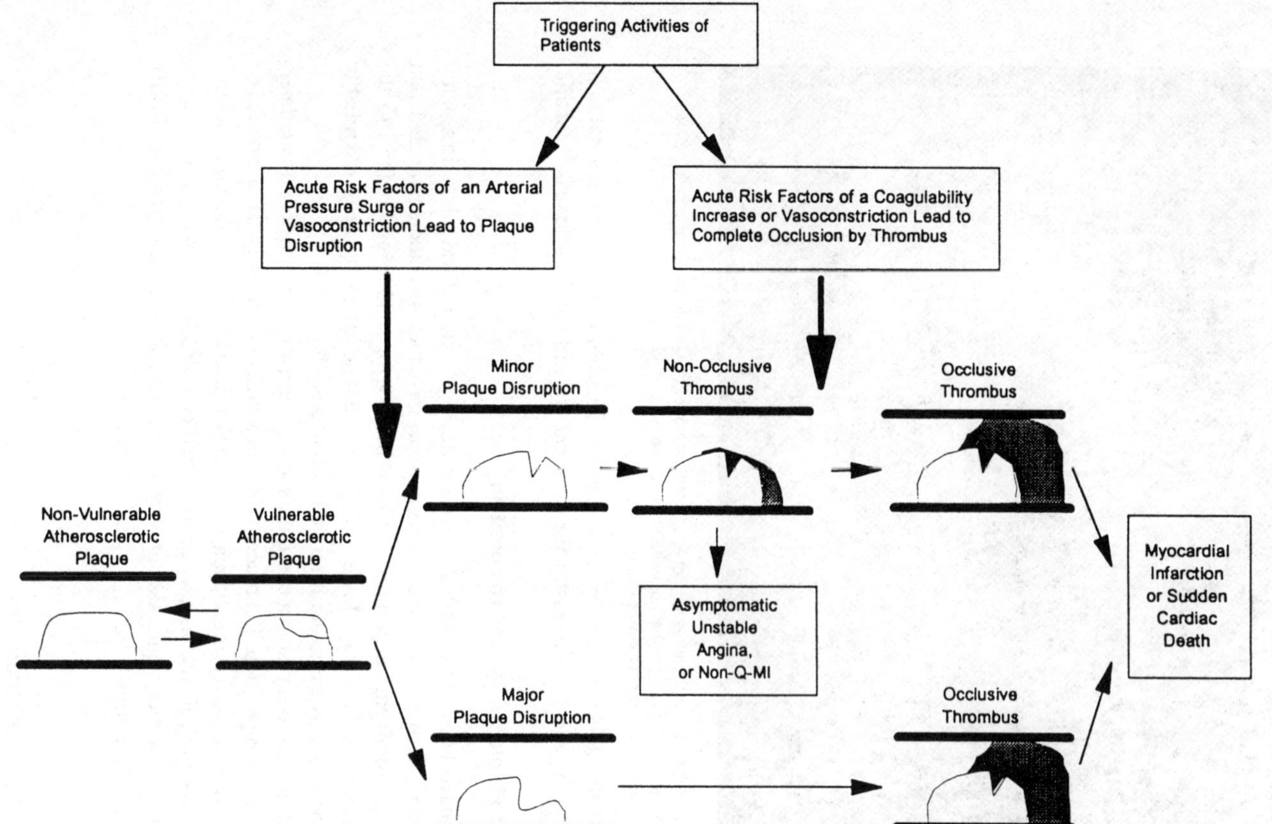

Fig. 20.3 Illustration of a hypothetical method by which daily activities may trigger coronary thrombosis. Three triggering mechanisms are presented: 1) physical or mental stress producing hemodynamic changes leading to plaque rupture, 2) activities causing an increase in coagulability, and 3) stimuli leading to vasoconstriction. The scheme depicting the role of coronary thrombosis in unstable angina, myocardial infarction and sudden cardiac death has been well described by numerous investigators. The novel portion of this figure is the additional of triggers. See text and Muller et al. (1) for detailed discussion. Non-Q-M1=non-Q

themselves. The consequences of obstructive stenoses, such as poor perfusion in the distal organs, can be measured but not plaques. This is related to poor target to background ratio of signal in the coronary and carotid circulations, and the resolutions of nuclear techniques so far available. Certain peptides and monoclonal antibody fragments directed at different components of atherosclerotic plaque and labelled with nuclear agents such as Technetium-99m have shown some promise in early studies; however, these have not reached research or clinical applicability as yet.

20.2.3 Angiography

Selective coronary angiography developed in the 1950s by Sones *et al.*,[13] has been the standard technique for imaging coronary atherosclerosis. Its major disadvantage as a research tool is its invasive nature, with a small risk of morbidity and mortality. Its major limitation from a pathological point of view is that it images the lumen of the artery rather than the plaque. *In vivo* studies and autopsy reports[14, 15] have demonstrated that the degree of coronary atherosclerosis is consistently underestimated by angiography, particularly minimal disease. Where there is uniform narrowing of a segment, coronary angiography is unable to detect the presence of plaque, as angiography relies on the presence of stenoses relative to a normal reference segment, in order to detect disease. Eccentric plaques may be associated with a shift of the normal arterial wall outwards (arterial 'remodelling'); this phenomenon may account for plaques of up to 40% of the lumen diameter not being detected.[2]

Nevertheless, there are some advantages to angiography. It is widely available, quick, relates to anatomical measurements of disease burden[16] and the severity of angiographic disease has been shown to relate to symptoms of coronary artery narrowing.[17] Consequently, angiography has been used frequently in studies of atherosclerosis regression, although the changes noted by this technique have been extremely small indeed.[18]

Angiographic measures of atherosclerosis severity may not predict the risk of subsequent occlusion and infarction, as small plaques which are lipid rich and unstable may be more likely to rupture, resulting in myocardial infarction, whereas large plaques may remain stable for many years.[19]

20.2.4 Ultrafast CT scanning

Ultrafast computed tomography (CT) is a new imaging technique that utilizes either electron beam technology or gated images from a helical CT scanner, both of which allow image acquisition at relatively rapid rates and open up the possibility of using CT scanning for imaging arterial structures. Calcium scoring has been used as an indirect marker of atherosclerosis burden and subsequent cardiovascular risk, and the use of intravenous infusion of dye with rapid slice acquisition for coronary angiography, without cardiac catheterization.

Calcium makes up approximately 20% of plaque volume,[20] and the association of coronary artery calcium with atherosclerotic plaque has been well known for many years. The deposition and turnover of calcium within plaques are active rather than degenerative processes, and cells within plaques elaborate osteocalcin and osteopontin. Despite this, the deposition of calcium within plaques is not uniform; smaller plaques (which may be active) as well as high grade stenoses may completely lack detectable levels of calcium.[21]

With the development of electron beam CT, and possibly even helical CT with appropriate software, accurate quantification of coronary calcium *in vivo* has become feasible. The detection of coronary calcium seems to correlate well with histological changes, and several studies have documented excellent sensitivity for detecting obstructive coronary artery disease.[22] The major problems relate to the specificity of the technique, as many people have coronary calcium without important obstructive stenosis. Further research is required to validate appropriate diagnostic pathways, for asymptomatic or symptomatic subjects who might be screened for coronary calcium with computer tomography and is not recommended, not for routine use.

Serial changes in coronary calcium have been documented in one clinical trial of lipid lowering,[23] but it is not known if these reflect changes in plaque volume. Ultrafast CT may prove to be a useful measure for evaluating progression or regression of coronary atherosclerosis, during therapeutic trials.

Contrast-enhanced CT angiography has also been the subject of a number of recent preliminary reports.[24] Improvements in technology and scanning strategies may allow non-invasive detection or exclusion of obstructive coronary artery stenoses in a majority of cases.

20.2.5 Magnetic resonance imaging

Magnetic resonance imaging (MRI) relies on the detection and restructuring of information gained from relaxation of magnetized nuclei. It has developed into a powerful imaging technique, such as non-invasive angiography, for many structures in the body, particularly those unaffected by motion artefacts for relatively immobile vessels, e.g. those in the brain or the legs. Coronary artery imaging has been much more difficult because of motion artefacts related to cardiac and respiratory cycles. Nevertheless, this is a particularly attractive option, as it might allow simultaneous acquisition of information regarding the anatomy and ultrastructure of coronary arteries, as well as cardiac perfusion, metabolism and function. The use of fast magnetic resonance angiography with breath holding and ECG gating provides relatively poor images of the coronary lumen, although sensitivity and specificity may improve as the technology advances. Problems may remain, however, with the use of this technique, which (like traditional angiography) images the lumen rather than the plaque itself.

In recent years, MRI has been used to measure plaque volume and visualize plaque ultrastructure in arteries *ex vivo*, as well as in experimental animal studies and some human superficial arteries *in vivo*.[25] These studies have yielded important information regarding plaque composition and behaviour under various conditions. Although MRI is costly, the potential of the method to give information about plaque size, composition, and effects on lumen make it an area worthy of attention.

20.2.6 New opportunities

'Older' modalities such as ultrasound, CT, magnetic resonance, and nuclear scanning will undergo refinements to enhance the possibilities for presymptomatic detection of atherosclerosis. With ultrasound, these will include miniaturization of transducers and mainframes, harmonic imaging to improve signal to noise ratios and novel approaches, such as transvenous or intra-arterial techniques. As ultrasound is safe, widely available, and informative, these developments are potentially exciting. With CT and MRI, improved temporal and spatial resolution will undoubtedly result in improved image quality. With MR, direct imaging of plaque is a realistic and exciting goal. Other imaging technologies are also on the horizon, including optical coherence scanning (for arterial anatomy), electron paramagnetic resonance, and temperature maps for plaque characterization (for quantification of plaque 'vulnerability'). Altogether, the aim of establishing an inexpensive, non- or minimally invasive, safe, and repeatable imaging test for atherosclerosis distribution, volume, and characterization remains elusive, and perhaps a combination of modalities will be required.

20.3 Future diagnostic tests

20.3.1 The scope of the problem

There is currently no non-invasive method for diagnosing atherosclerosis. The detection of atherosclerosis by plasma-derived markers could provide information on both the extent (disease burden) and qualitative nature (e.g. stable or unstable) of atherosclerosis. Although as accessible as conventional plasma-derived risk factors (e.g. plasma cholesterol), plasma markers would potentially provide a far more powerful predictor of future coronary events in individuals than risk factors, and may permit the monitoring of changes to the nature and extent of disease. The wide range of potential diagnostic markers are suggested by the diverse processes involved in atherosclerosis.

20.3.2 Distinguishing between risk factors and biochemical markers, and the potential for diagnosing pre-clinical atherosclerosis

As noted earlier in the chapter, risk factors, such as an elevated serum cholesterol concentration, contribute to the overall risk of developing atherosclerotic disease in one population relative to another population without (or with less of) the risk factor. A biochemical marker arises from the disease itself, and, unlike a risk factor, has the potential to assess the extent and nature of disease in individuals.

There are many known risk factors for the development of atherosclerosis in general, and coronary disease in particular. For coronary disease, high LDL cholesterol, low HDL cholesterol, smoking, hypertension, inactivity, abdominal obesity, family history, and diabetes confer increased risk. In a given individual, a good risk factor profile does not preclude the development of coronary disease, an unfavourable profile does not always predict coronary disease.

In populations with clinical coronary disease as few as 10 patients need be treated with cholesterol lowering HMG CoA reductase inhibitors to save one cardiac event in 5 years,[26] whereas in primary prevention (defined as patients with no known coronary disease) 40–50 patients need be treated for 5 years to save one cardiac event.[27] If a biochemical marker of disease re-defined certain individuals within a clinical primary prevention population to a category of pre-clinical atherosclerosis, suitable for a kind of secondary prevention, it is possible the latter would be most likely to benefit from treatment with lipid lowering drugs. This hypothesis remains to be tested.

The best markers will be those which relate to biological and chemical processes specific for atherosclerosis. Their application and identification will be complex because of qualitative differences between different stages of coronary atherosclerosis (fatty streak, fibrous plaque and unstable or thrombotic lesions), as well as differences between coronary and non-coronary atherosclerosis. As well as being pathologically specific, a suitable marker must be able to leave the artery wall. For example, large proteoglycan-lipoprotein aggregates may be pathologically unique to atherosclerotic lesions, but without egress from the intima into interstitial fluid or into the arterial lumen such molecules would not function as markers. Finally, in the absence of another condition which could generate a less specific marker (e.g. C-reactive protein, CRP) even a non-specific marker may be of important prognostic benefit.[28]

20.3.3 Pathology of atherosclerosis and types of markers

Atherosclerosis involves intimal thickening, the subendothelial deposition of lipids, matrix deposition, and cellular proliferation. At later stages, the elastic lamina separating the intima from the media no longer limits the extent of atherosclerotic lesions and the disease (in particular the deposition of a necrotic core) may involve the medial layer. Acute clinical events are usually precipitated by an intramural plaque thrombosis overlying rupture of the shoulder of a lipid rich atherosclerotic plaque or, less commonly a superficial erosion over a fibrotic plaque. Inflammatory cells such as macrophages, lymphocytes, and mast cells co-localize with sites of plaque rupture, regardless of the underlying plaque histology.[29, 30] Thus, as well as defining the extent of disease, which may especially be important for monitoring treatments aimed at reducing the total atheroma burden, markers defining active plaque pathology may selectively predict the short term risk of coronary events.

20.3.4 General approach to marker molecules

Markers could derive from any of the key steps in atherogenesis. These could range from the binding of monocytes to adhesion molecules on the endothelial surface, to the modification of lipoproteins in the arterial wall, to evidence of inflammation preceding plaque erosion and thrombosis. Whether implicated processes such as lipoprotein oxidation or infection of atheroma with *Chlamydia pneumoniae* are atherogenic, or are simply follow atheroma formation is relatively unimportant from the view of a diagnostic marker. More important is whether the presence of these processes in proximity to atheroma components generates novel atheroma-specific molecules.

Specificity and sensitivity will ultimately be the criteria for evaluation of such markers. Markers purported to quantify total atherosclerotic burden in the coronary arteries may, for example, be evaluated against coronary angiography or intra-coronary vascular ultrasound (IVUS). Markers qualitatively evaluating acute plaque inflammation and risk of thrombosis may be evaluated against future coronary events, and will be evaluated against combinations of less specific markers such as CRP. It may be possible to integrate assays of several markers to give an overall evaluation of predominant processes in atherosclerosis which may vary over time.

As any pathological process in atherosclerosis could result in the appearance of diagnostic molecules, it is beyond the scope of this brief review to cover all potential markers. It is, however, worth considering several currently interesting molecules or molecule classes as they highlight the difficulties inherent in the establishment of markers and the limitations of our knowledge to date.

20.3.4.1 Evidence of endothelial activation – soluble adhesion molecules

Plasma concentrations of soluble adhesion molecules, including VCAM-1 and ICAM-1 have been correlated with intima–media thickness in carotid arteries.[31] However, other studies have indicated no clear relationship between adhesion molecule concentration and coronary atherosclerosis.[32] In patients with stable angina pectoris, elevated levels of ICAM and VCAM were associated with future cardiac events,[33] indicating a particular utility in detecting inflammation and anticipating thrombosis. Levels of adhesion molecules in children correlate with some lipoprotein parameters, indicating lipoproteins may be exerting inflammatory effects on the arterial wall during early stages of atherosclerosis.[34] Alternatively, this also highlights that adhesion molecules must also be established as independent of the prognostic effects of atherosclerotic lipoproteins.

Importantly, biological markers of disease may move in opposing directions in response to different interventions. For example, intervention with fish oil lowered concentrations of thrombotic markers tissue plasminogen antigen (TPA Ag) and soluble thrombomodulin, but increased soluble E-selectin and soluble VCAM.[35] As soluble adhesion molecules may inhibit monocyte adhesion,[36] further studies are required to understand their significance in relation to the expression of adhesion molecules on the surface of endothelial cells. At this time, soluble adhesion molecules may at best be considered to indicate some level of endothelial cell activation or inflammation.

20.3.4.2 Serological measures of lipoprotein oxidation – auto-antibodies to OxLDL

A number of lipid-derived and protein-derived oxidation products have been identified in human atherosclerotic plaque. Lipid oxidation products include fatty acid hydroperoxides, hydroxides and keto-derivatives, cholesteryl ester aldehydes and hydroxides, oxysterols, phospholipid oxidation products, F2-isoprostanes, and oxidation products of α tocopherol.[37–39] Protein oxidation products may generate nitrated and chlorinated amino acids, hydroxylated amino acids, cross-linked and fragmented apolipoproteins, and complex lipid–protein adducts.[38, 40] The utility of these or any other oxidation products would depend upon their prevalence in atherosclerotic lesions of different types,[41] their specificity for atherosclerosis as distinct from other conditions of general oxidative stress, and their detectable quantities in plasma. As oxidation of whole lipoproteins has been understood to underlie the genesis of most of these products, plasma autoantibodies to OxLDL (of variable chemical characterization) have been investigated as possible markers of atherosclerosis.

At best, auto-antibodies to OxLDL have inconsistently demonstrated the predictive power of another risk factor (such as cholesterol), and at worst have shown no clear relationship between antibody levels and the extent or presence of disease. Some studies do indicate a correlation between titres to OxLDL and coronary atherosclerosis,[42] but in humans circulating antibodies to OxLDL and malondialdehyde-modified LDL are elevated in a number of inflammatory and autoimmune conditions, and elevated titres can be found equally in patients with and without atherosclerosis.[43–46] Palinski et al.[42] found increased titres of antibodies against OxLDL in LDL receptor-deficient mice; however, there was a very important correlation between antibody titres and the serum cholesterol level, implying that this variable may be highly dependent on serum cholesterol.

One major problem inherent with auto-antibodies to any pathogen, not only to OxLDL, relies on the extent and duration of the secondary auto-immune reaction to a disease component, rather than quantification of the disease component itself. This increases variation and decreases quantitative reproducibility. Future assays measuring oxidation products directly, will probably generate more useful strategies. Such detection may allow demarcation of high risk lesions – e.g. hydroxyeicosatetrenoic acids (HETE) have recently been detected relatively selectively in thrombotic human carotid plaques.[47]

20.3.4.3 Infectious organisms – *C. pneumoniae*

The best investigated infectious pathogen in atherosclerosis is *C. pneumoniae*. There is strong association between atherosclerosis in arteries post mortem and detection of this organism in coronary tissues.[48] There is also an association of *C. pneumoniae* with atherosclerosis as determined by seroepidemiology.[49] Recently, in one healthy population there was no association between prior infection with *C. pneumoniae* and cardiovascular events,[50] and CRP, which had been a positive predictor of events in this cohort, did not correlate with chlamydia serology. Other studies have indicated variably positive relationships, and differences between studies relate to population and disease heterogeneity. Mechanistically discrepancies may relate to dissociation of infection *per se* from the severity of the secondary host immune response.[49, 51] It is also possible that chlamydia or other acute infections may be relevant in defined stages of atherosclerosis, e.g. the precipitation of an acute thrombotic state.[52]

If the effect of chlamydia infection were secondary to the host inflammatory response (e.g. as evidenced by the measure of CRP or IL-1, etc. see below), then chlamydial serology may not add to the qualitative evaluation of atherosclerosis achieved by non-specific inflammatory markers. As with auto-antibodies to OxLDL, indirect determination of chlamydia by the host immune response (and variable approaches to which immune complexes are measured) complicate interpretation as compared with direct isolation of the pathogen or its DNA. For example, one study which compared concentrations of antibodies and specific circulating immune complexes to *C. pneumoniae* in patients with ischaemic stroke versus controls[53] found IgG levels of $>1:32$ were present in 77% of controls and 74% of patients. Specificity was improved by detection of specific IgG antibody-circulating immune complexes, but these were elevated in only 24% of patients and 7.7% of controls, and the relative risk, after correcting for other risk factors was approximately 2.0 (equivalent to many known risk factors). In another study[54] of 51 patients with abdominal aneurysm, 41 patients were seropositive for chlamydia, but only 26 had chlamydia DNA detectable in aortic tissue. This suggests that seropositivity may have been conferred by chlamydia infections other than those in the arterial lesions, a problem which may confound specificity of such assays for atherosclerosis. If a unique infectious organism or a particular strain of the organism unique to atherosclerotic lesions, were identified, then serology to such an organism or its direct detection may indeed be a useful diagnostic marker.

20.3.4.4 Inflammatory markers

Serum markers of inflammation appear to predict atherosclerosis extent, and more particularly, risk of inflammatory thrombotic complications.[55] Pro-inflammatory cytokines such as TNFa and Il-1 are certainly found in human atherosclerotic plaque, and can exert diverse effects on adhesion molecule expression, cell enzyme secretion, and lipoprotein modification. They can also stimulate the expression of secondary inflammatory molecules such as CRP. CRP is a molecule which may have pathogenic roles in stimulating complement activation[28] and has consistently shown prognostic value in anticipating vascular events and mortality, and the potential for medical strategies to modulate disease and outcomes.[56, 57]

Inflammatory markers may have complex interactions with the disease process in the arterial wall. For example, an inflammatory stimulus such as lipoprotein oxidation may induce expression of IL-1 by macrophages, with secondary changes to plasma markers of inflammation by virtue of hepatic secretion of acute phase proteins such as CRP. Less directly, the inflammatory stimulus may induce a less specific stress response, demonstrated by the expression of heat shock proteins (HSP), antibodies to which have recently been found to predict severity of carotid atherosclerosis and 5 year mortality.[58] Thirdly, rather than the primary atherosclerotic lesion being the only source of inflammatory factors, it is possible that inflammation elsewhere may amplify the systemic inflammatory response and aggravate local inflammatory processes, or instigate secondary thrombotic process, as suggested for *in vitro* studies with chlamydia.[52] Such a role for some inflammatory factors would imply that they are important as risk factors for coronary events independent of extent or local characteristics of atherosclerotic lesions.

20.3.4.5 Mineralization proteins – Osteonectin, Osteopontin, Matrix-Gla proteins

Calcification and bone deposition are features of advanced human atherosclerosis[59] and the plasma determination of mineralization proteins may allow quantification of arterial calcification and thus the extent of atherosclerosis. Osteopontin binds calcium (hence is involved in calcification of bone and arteries), is detectable in human atherosclerotic arteries,[60] and is chemotactic. It is upregulated after rat carotid balloon injury, but human vascular smooth muscle cells, unlike rat cells, express osteopontin very poorly, and human macrophages express osteopontin regardless of lipid accumulation[61] as do T lymphocytes.[62] Although atherosclerotic intimal calcification, is related to macrophage content, medial calcification, as occurs in Monckeburgs sclerosis, relates to vascular smooth muscle cell (VSMC) activity.[63] Importantly, proteins such as osteonectin and matrix-Gla protein are constitutively protective against calcification, whereas bone-Gla protein, bone sialoprotein and, collagen II upregulation are characteristic of osteogenesis and chondrogenesis.[64] Thus, differential expression of these proteins may be expected in states of increased or decreased arterial calcification.

Because osteopontin will be induced by macrophage or lymphocyte activity in any part of the body, it may be non-specific for atherosclerosis. For example, osteopontin has also been detected in kidneys (and in renal stones), gastric mucosa, and osteonectin is present in fibrotic liver and brain tissue (astrocytes), cardiac fibroblasts, bone, bone tumours, and lung cancer.[65] Like other markers of macrophage or lymphocyte activity, it may confer useful predictive value in the absence of confounding sources. Finally, fibrinogen and collagen I and collagen IV matrices have differential effects on arterial calcification of vascular cells *in vitro*[66] and the relative expression of these matrices may help measure the extent of arterial calcification.

20.3.4.6 Concluding remarks on non-invasive diagnostics

Delineation of the targets for each potential diagnostic molecule will require tailoring for assessment of disease burden (quantitative), disease type (stable and unstable), and/or disease presence or absence (combination of qualitative and quantitative assays). Assimilation of prognostic information for different molecules may allow diagnostics and monitoring of therapy in ways inaccessible to conventional imaging modalities. As our knowledge of the processes involved in atherosclerosis evolve, so will the potential identification of diagnostic markers for atherosclerosis.

20.4 Future therapeutics

Atherosclerosis therapeutics can be directed towards prevention or reversal of atherogenesis, towards minimization of risk of clinical event from the presence of plaques, and towards the management of the MI survivor. The first three aspects are intertwined and an approach designed to benefit one may impact on one or both of the others. It cannot be assumed safely that inhibition of any single event in atherogenesis will suffice to inhibit the overall process.

Current therapies for atherosclerosis primarily target risk factors, rather than pathogenetic steps (notably hypertension and hyperlipidemia). This is because the relationship between such risk factors and disease is still poorly understood. This book concerns mechanisms in atherogenesis, and so this also remains the emphasis in the present chapter. We can discuss only some selected mechanistic aspects of atherosclerosis and their possible implications for therapy.

20.4.1 Intercellular adhesion and transmigration

The earliest known events in atherogenesis involve the rolling of peripheral blood monocytes on the endothelial surface, their reversible adhesion, and their subsequent active transudation into the intima. The preferential localization of plaques in some but not other areas of vessels indicates that there must be localized factors which influence this process differentially. It has normally been assumed that these factors result from differences in shear-mediated triggering of cellular function amongst the interacting cells, notably the endothelial cells. The search amongst shear-controlled genes could therefore lead to the identification of mediators of transudation which could be controlled pharmaceutically. Clearly, such control would need to be highly selective, since transudation is essential to acute inflammatory responses, and not unique to atherogenesis. Most other approaches to identifying therapeutic targets amongst adhesion molecules and those involved in transudation are grossly impeded by this need for their participation in such acute inflammatory events.

20.4.2 Connective tissue matrix binding of lipoproteins and cells

At least two ingredients are needed for the next stage of atherogenesis, the formation of macrophage foam cells: the retention of the new maturing macrophages and of lipoprotein lipid in the intima. These events are at least uncommon outside atherogenesis. The formation of macrophage foam cells has been observed in other pathological circumstances, but in other organ systems, which could escape alteration by intimally directed therapeutics. The accumulation of extracellular lipoproteins in the intima has been claimed to characterize early atherogenesis, but whether such accumulation is causative for lesions or is itself a consequence of lesion formation is unclear (see Chapter 13). Conversely, there are no other pathologies or sites other than the arterial wall in which massive extracellular, extravascular accumulation of lipoproteins is common. Thus, factors influencing the retention of macrophages and lipoproteins in the intima seem to present very interesting targets for anti-atherogenic drug development. This points us towards molecules on the macrophage surface which are critical to adherence to matrices occurring in the vessel wall, including the possibility that these matrices may be modified by adherence of lipoproteins, or through other mechanisms. These include modifying the nature of matrix molecules expressed in the artery wall that bind and retain lipoproteins less avidly. Whether the accumulation of lipoproteins in the intimal extracellular space is a strong candidate target is less clear, because of quantitative considerations, and because cellular accumulation is probably the critical step for progression, and this might occur regardless of the extracellular local concentration of lipoprotein: bi-directional fluxes of lipoproteins in and out of the incipient foam cells rather than extracellular concentrations may be the critical determinants.

20.4.3 Oxidation

As this book has emphasized, a central thesis in atherogenesis research since around 1981 has been that oxidation of LDL might be a key causal link. One important early suggestion was that oxidation of LDL might create forms in the intima which are taken up into incipient foam cells in an unregulated manner (so that influx might exceed efflux). The evidence that such forms are present in atherogenesis *in vivo* is limited; rather most forms of oxidized LDL which are studied *in vitro* are significantly different from (usually much more oxidized than), those found even in advanced plaques. This thesis has evolved, for example, to include oxidation more broadly, and to view it as a specialized example of the even more classic 'response to injury' theory of atherosclerosis.

It is only recently that direct tests of the theory in animal models of atherosclerosis have been published. These studies have shown that many antioxidants do inhibit atherogenesis in animal models, though data from humans is unimpressive (see elsewhere in this book). Even more importantly, they have indicated that inhibition of lipid oxidation by successful antioxidant intervention in some models and at some tissue locales is accompanied by inhibition of atherogenesis, while in others it is not. From this it seems that atherogenesis may be dissociated in some circumstances from lipid oxidation. It is not yet clear whether inducing oxidation of lipids in the intima is sufficient to induce atherogenesis. It is also unclear whether the antioxidants which have successfully inhibited cholesterol ester fatty acid oxidation *in vivo*, will also inhibit sterol oxidation, and even less obvious whether they will influence protein oxidation (see Chapters 3, 15, 16). Possibilities for therapeutic intervention by antioxidation or more probably co-antioxidation, therefore remain quite strong. These might (i) be specific to some forms and stages of atherogenesis; (ii) require inhibition of oxidation of molecules other than cholesterol ester fatty acids; and (iii) involve action on redox mediators of cell triggering (such as AP1, NFkB), rather than or instead of inhibition of lipid and/or protein oxidation.

20.4.4 Lipid transport

The nett flux of lipid into foam cells is responsible for their formation and maintenance. This may be due to any combination of accentuated uptake or reduced efflux. Thus, oxidation of LDL has largely been considered in the light of creating high-uptake LDL, though it is clear that such forms can be generated *in vitro* at least by non-oxidative means (such as aggregation), which offer other tempting therapeutic approaches. We have put some emphasis on the inhibition of efflux resulting from accumulation of oxysterols within foam cells: 7-ketocholesterol can inhibit such efflux to apoA1. Thus, selective removal of oxysterols from foam cells might constitute an appropriate therapeutic target. This may be achieved perhaps by pharmaceutical vehicles such as tailored molecular imprints, which provide a cavity suitable for selective binding of a relevant molecule. By acting as vehicles to carry oxysterols and/or cholesterol out of the plasma membrane into extracellular depots such as phospholipid, these compounds might enhance efflux and reverse foam cell formation. This approach is probably relevant both to prevention and reversal of atherosclerosis, and might also facilitate the conversion of unstable plaques (characterized by high lipid/foam cell: connective/fibrotic tissue volume ratios) into more durable ones.

20.4.5 Cell proliferation

After accumulation of macrophage foam cells, there is usually migration and proliferation of smooth muscle cells, originating medially, and also some proliferation of macrophages themselves. Localized inhibition of smooth muscle cell proliferation might be a feasible approach to inhibition of atherogenesis, but smooth muscle cells proliferate in other circumstances, and means would be needed to permit organisms to retain this function. On the other hand, proliferation of matured tissue macrophages is not a common physiological event, since macrophage populations are replenished from bone marrow precursors via peripheral blood monocytes. Thus, selective inhibition of macrophage proliferation may be of pharmaceutical interest.

20.4.6 Inflammatory responses

Macrophage proliferation should be seen within the general context that atherosclerosis is an inflammatory disease, commenced by acute inflammatory reactions which are probably defensive, but which in a sense become diverted. Thus, recently increasing attention has been paid to the role of lymphocytes, local immune reactions, and also to other inflammatory cells such as neutrophils. A central issue is whether there is usually a local immunological process perpetuated by responses to modified molecules in the lesions. Local ablation of the response mechanisms, or less desirably, of the effector arm of the system might be feasible. Providing active infectious organism proliferation is not crucial, then ablation of response to local antigens might be acceptable. Otherwise, prevention of certain of the proliferative responses might permit active local immunity, but restrict progression of atherogenesis.

20.4.7 Proteolysis of the extracellular matrix

Once plaques are established, a key issue is what determines their stability. As mentioned above, possibilities exist for control of the quantity of lipid in such plaques, which might ameliorate their physical stability. Arraigned against this is the evidence that proteolysis, perhaps involving both secreted metallo- and lysosomal proteinases may be accelerating the removal of the proteins and proteoglycans of the extracellular matrix. Quantitative information on such events *in vivo* is limited, but the possibility clearly exists of localized inhibition of such extracellular degradation, with the hope of increased plaque stability. Conversely, enhanced synthesis and secretion of matrix components might be envisaged. In both cases it would be critical that the events were controlled only locally, since they are part of cellular or organismal housekeeping (notably, intralysosomal proteolysis and synthesis of extracellular matrix, respectively).

20.4.8 Coagulation

It is generally believed that plaque rupture is the most common precipitating factor for an arterial occlusion, and hence a clinical event. Nevertheless, it seems likely that most instances of plaque rupture do not have such drastic consequences. Many plaque thromboses can occur after the superficial erosion of a fibrous cap with platelet and fibrin deposition, but without underlying hemorrhage. Thus, it remains possible that in areas of blood vessel already severely occluded by plaque, humorally initiated coagulation, involving platelet recruitment and aggregation, may quite often be the mechanism precipitating an infarction. Thromboplastin expression is enhanced on plaque macrophages, but alternative circulating sources may be important on occasion.

From the point of view of atherogenesis and prevention of infarction, the question is whether inhibition of coagulation, especially perhaps local inhibition of thromboplastin generation on plaque cells, might be a worthwhile addition to the armoury. The apparent success of long term low dose aspirin can be viewed in terms of its anti-coagulant activity, but interpretation is not clearcut. In contrast, there is no doubt that control of coagulation is critical post-event, and strategies for this are well developed.

20.5 Conclusions

We have emphasized the heterogeneity of the process of atherogenesis and of its clinical consequences. It is inherent that diagnosis and therapy will be comparably heterogeneous. In humans in clinical practice, unlike animals in experimental studies, intervention will probably involve multiple concurrent therapies. We hope that by exposing the cutting edge of many of the research issues in atherogenesis, we may focus and help the progress towards controlling this major debilitator and killer.

Acknowledgements

We thank Dr Mike Davies for assessing the current literature on EPR imaging.

References

1. Stary, H.C., Chandler, A.B., Glagov, S., Guyton, J.R., Insull, W., Jr, Rosenfeld, M.E., Schaffer, S.A., Schwartz, C.J., Wagner, W.D., and Wissler, R.W. (1994). A definition of initial, fatty streak, and intermediate lesions of atherosclerosis. A report from the Committee on Vascular Lesions of the Council on Arteriosclerosis, American Heart Association. *Arterioscler. Thromb.*, **14**, (5), 840–56.
2. Glagov, S., Weisenberg, E., Zarins, C.K., Stankunavicius, K., and Kolettis, G.J. (1987). Compensatory enlargement of various human atherosclerotic arteries. *N. Engl. J. Med.*, **316**, 1371–5.
3. Fazio, G.P., Redberg, R.F., Winslow, T., and Schiller, N.B. (1993). Transesophageal echocardiographically detected atherosclerotic aortic plaque is a marker for coronary artery disease. *J. Am. Coll. Cardiol.*, **21**, 144–50.

4. Waller, B.F., Pinkerton, C.A., and Slack, J.D. (1992). Intravascular ultrasound: a histological study of vessels during life. The new 'gold standard' for vascular imaging. *Circulation*, **85**, (6), 2305–10.

5. Wendelhag, I., Gustavsson, T., Suurkula, M., Berglund, G., and Wikstrand, J. (1991). Ultrasound measurement of wall thickness in the carotid artery: fundamental principles and description of a computerized analysing system. *Clin. Physiol.*, **11**, (6), 565–77.

6. Howard, G., Sharrett, A.R., Heiss, G., Evans, G.W., Chambless, L.E., Riley, W.A., and Burke, G.L. (1993). Carotid artery intimal-medial thickness distribution in general populations as evaluated by B-mode ultrasound. ARIC Investigators. *Stroke*, **24**, (9), 1297–304.

7. Salonen, J.T., Salonen, R., Korpela, H., Suntioinen, S., and Tuomilehto, J. (1991). Serum copper and the risk of acute myocardial infarction: a prospective population study in men in Eastern Finland. *Am. J. Epidem.*, **134**, 268–76.

8. Adams, M.R., Nakagomi, A., Keech, A., Robinson, J., McCredie, R., Bailey, B.P., Freedman, S.B., and Celermajer, D.S. (1995). Carotid intima-media thickness is only weakly correlated with the extent and severity of coronary artery disease (see comments). *Circulation*, **92**, (8), 2127–34.

9. O'Leary, D.H., Polak, J.F., Kronmal, R.A., Manolio, T.A., Burke, G.L., Wolfson, S.K., Jr (1999). Carotid-artery intima and media thickness as a risk factor for myocardial infarction and stroke in older adults. *N. Engl. J. Med.*, **340**, (1), 14–22.

10. Blankenhorn, D.H., Selzer, R.H., Crawford, D.W., Barth, J.D., Liu, C.R., Liu, C.H., Mack, W.J., and Alaupovic, P. (1993). Beneficial effects of colestipol-niacin therapy on the common carotid artery. Two- and four-year reduction of intima-media thickness measured by ultrasound. *Circulation,* **88**, (1), 20–8.

11. Ratliff, D.A., Hames, T.K., Humphries, K.N., Birch, S., and Chant, A.D. (1985). The reliability of Doppler ultrasound techniques in the assessment of carotid disease. *Angiology*, **36**, (6), 333–40.

12. Nishimura, R.A., Edwards, W.D., Warnes, C.A., Reeder, G.S., Holmes, D.R., Jr, Tajik, A.J., and Yock, P.G. (1990). Intravascular ultrasound imaging: *in vitro* validation and pathologic correlation. *J. Am. Coll. Cardiol.*, **16**, (1), 145–54.

13. Sones, F.M., Shirley, E.K., Proudfit, W.L., and Westcott, R.N. (1959). Cinecoronary arteriography. *Circulation*, **20**, 773.

14. Porter, T.R., Sears, T., Xie, F., Michels, A., Mata, J., Welsh, D., and Shurmur, S. (1993). Intravascular ultrasound study of angiographically mildly diseased coronary arteries. *J. Am. Coll. Cardiol.*, **22**, (7), 1858–65.

15. Hutchins, G.M., Bulkley, B.H., Ridolfi, R.L., Griffith, L.S., Lohr, F.T., and Piasio, M.A. (1977). Correlation of coronary arteriograms and left ventriculograms with postmortem studies. *Circulation*, **56**, (1) 32–7.

16. Gensine, G.G. (1975) Coronary angiography. Mount Kisco, Futura Publishing Company, New York.

17. Proudfit, W.L., Shirey, E.K., Sones, F.M., Jr (1966). Selective cine coronary arteriography. Correlation with clinical findings in 1000 patients. *Circulation*, **33**, (6), 901–10.

18. Crouse, J.Rd and Thompson, C.J. (1993). An evaluation of methods for imaging and quantifying coronary and carotid lumen stenosis and atherosclerosis. *Circulation*, **87** (Suppl. 3), II17–1133.

19. Fuster, V., Badimon, J.J., and Badimon, L. (1992). Clinical-pathological correlations of coronary disease progression and regression. *Circulation*, **86** (Suppl. 6), 1–11.
20. Eggen, D.A., Strong, J.P., and McGill, H.C., Jr (1965). Coronary calcification. Relationship to clinically significant coronary lesions and race, sex, and topographic distribution. *Circulation*, **32**, (6), 948–55.
21. Fiorino, A.S. (1998). Electron-beam computed tomography, coronary artery calcium, and evaluation of patients with coronary artery disease. *Ann. Intern. Med.*, **128**, (10), 839–47.
22. Mautner, S.L., Mautner, G.C., Froehlich, J., Feuerstein, I.M., Proschan, M.A., Roberts, W.C., and Doppman, J.L. (1994). Coronary artery disease: prediction with *in vitro* electron beam CT. *Radiology*, **192**, (3), 625–30.
23. Callister, T.Q., Raggi, P., Cooil, B., Lippolis, N.J., and Russo, D.J. (1998). Effect of HMG-CoA reductase inhibitors on coronary artery disease as assessed by electron-beam computed tomography. *N. Engl. J. Med.*, **339**, (27), 1972–8.
24. Achenbach, S., Moshage, W., Ropers, D., Nossen, J., and Daniel, W.G. (1998). Value of electron-beam computed tomography for the noninvasive detection of high-grade coronary-artery stenoses and occlusions. *N. Engl. J. Med.*, **339**, (27), 1964–71.
25. Skinner, M.P., Yuan, C., Mitsumori, L., Hayes, C.E., Raines, E.W., Nelson, J.A., and Ross, R. (1995). Serial magnetic resonance imaging of experimental atherosclerosis detects lesion fine structure, progression and complications *in vivo*. *Nat. Med.*, **1**, (1), 69–73.
26. Scandinavian, Survival, Study, Group. (1994). Randomised trial of cholesteol lowering in 4444 patients with coronary heart disease: the Scandinavian Simvastatin Survival Study (4S). *Lancet,* **344**, 1383–9.
27. Shepherd, J., Cobb S.M., Ford, I., *et al.* (1995). Primary prevention of coronary events in men with pravastatin – The West of Scotland Study. *N. Engl. J. Med.*, **333**, 1301–7.
28. Lagrand, W.K., Visser, C.A., Hermens, W.T., Niessen, H.W.M., Verheugt, F.W.A., Wolbink, G.-J., and Hack, C.E. (1999). C-reactive protein as a cardiovascular risk factor. *Circulation*, **100**, 96–102.
29. van der Waal, A.C., Becker, A.E., van der Loos, C.M., and Das, P.K. (1994). Site of intimal rupture or erosion of thrombosed coronary atherosclerotic plaques is characterised by an inflammatory process irrespective of dominant plaque morphology. *Circulation*, **89**, 36–44.
30. Davies, M.J. (1997) The composition of coronary artery plaques. *N. Engl. J. Med.*, **336**, 1312–14.
31. Rohde, L.E., Lee, R.T., Rivero, J., Jamacochian, M., Arroyo, L.H., Briggs, W., Rifai, N., Libby, P., Creager, M.A., and Ridker, P.M. (1998). Circulating cell adhesion molecules are correlated with ultrasound-based assessment of carotid atherosclerosis. *Arterioscler. Thromb. Vasc. Biol.*, **18**, (11), 1765–70.
32. Saku, K., Zhang, B., Ohta, T., Shirai, K., Tsuchiya, Y., and Arakawa, K. (1999). Levels of soluble cell adhesion molecules in patients with angiographically defined coronary atherosclerosis. *Jpn. Circ. J.*, **63**, (1), 19–24.
33. Wallen, N.H., Held, C., Rehnqvist, N., and Hjemdahl, P. (1999). Elevated serum intercellular adhesion molecule-1 and vascular adhesion molecule-1 among patients with stable angina pectoris who suffer cardiovascular death or non-fatal myocardial infarction. *Eur. Heart. J.*, **20**, (14), 1039–43.

34. Ohta, T., Saku, K., Takata, K., and Adachi, N. (1999). Soluble vascular cell-adhesion molecule-1 and soluble intercellular adhesion molecule-1 correlate with lipid and apolipoprotein risk factors for coronary artery disease in children. *Eur. J. Pediatr.*, **158**, (7), 592–8.

35. Johansen, O., Seljeflot, I., Hostmark, A.T., and Arnesen, H. (1999). The effect of supplementation with omega-3 fatty acids on soluble markers of endothelial function in patients with coronary heart disease. *Arterioscler. Thromb. Vasc. Biol.*, **19**, (7), 1681–6.

36. Abe, Y., El-Masri, B., Kimball, K.T., Pownall, H., Reilly, C.F., Osmundsen, K., Smith, C.W., and Ballantyne, C.M. (1998). Soluble cell adhesion molecules in hypertriglyceridemia and potential significance on monocyte adhesion. *Arterioscler. Thromb. Vasc. Biol.*, **18**, (5), 723–31.

37. Stocker, R. (1999). Dietary and pharmacological antioxidants in atherosclerosis. *Curr. Opin. Lipidol.*, **10**, in press.

38. Heinecke, J.W. (1998). Oxidants and antioxidants in the pathogenesis of atherosclerosis: implications for the oxidized low density lipoprotein hypothesis. *Atherosclerosis*, **141**, 1–15.

39. Brown, A.J. and Jessup, W. (1999). Oxysterols and atherosclerosis. *Atherosclerosis*, **142**, 1–28.

40. Davies, M.J., and Dean, R.T. (1997). Radical-mediated protein oxidation: from chemistry to medicine. Oxford University Press, New York.

41. Carpenter, K.L., Taylor, S.E., Ballantine, J.A., Fussell, B., Halliwell, B., and Mithinson, M.J. (1993). Lipids and oxidized lipids in human atheroma and normal aorta. *Biochim. Biophys. Acta*, **1167**, 121–30.

42. Palinski, W., Tangirala, R.K., Miller, E., Young, S.G., and Witztum, J.L. (1995). Increased autoantibody titers against epitopes of oxidized LDL in LDL receptor-deficient mice with increased atherosclerosis. *Arterioscler. Thromb. Vasc. Biol.*, **15**, 1569–76.

43. Wu, R. and Lefvert, A.K. (1995). Autoantibodies against oxidized LDL (OxLDL): characterization of antibody isotype, subclass, affinity and effect on the macrophage uptake of OxLDL. *Clin. Exp. Immunol.*, **102**, 174–80.

44. Boullier, A., Hamon, M., Walters-Laporte, E., Martin-Nizart, F., Mackereel, R., and Fruchart, J.C. (1995). Detection of autoantibodies against oxidized low-density lipoproteins and of IgG-bound low density lipoproteins in patients with coronary disease. *Clin. Chim. Acta*, **238**, 1–10.

45. van de Vijver, L.P., Steyger, R., van Poppel, G., Boer, J.M., Kruijssen, D.A., Seidell, J.C., and Princen, H.M. (1996). Autoantibodies against MDA-LDL in subjects with severe and minor atherosclerosis and healthy population controls. *Atherosclerosis*, **122**, 245–53.

46. Uusitupa, M.I., Niskanen, L., Luoma, J., Vilja, P., Mercuri, M., Rauramaa, R., and Yla-Herttuala, S. (1996). Autoantibodies against oxidized LDL do not predict atherosclerotic vascular disease in non-insulin-dependent diabetes mellitus. *Arterioscler. Thromb. Vasc. Biol.*, **16**, 1236–42.

47. Mallat, Z., Nakamura, T., Ohan, J., Leseche, G., Tedgui, A., Maclouf, J., and Murphy, R.C. (1999). The relationship of hydroxyeicosatetraenoic acids and F2-isoprostanes to plaque instability in human carotid atherosclerosis. *J. Clin. Invest.*, **103**, (3), 421–7.

48. Kuo, C.-C., Grayston, J.T., Campbell, L.A., Goo, Y.A., Wissler, R.W., and Benditt, E.P. (1995). *Chlamydia pneumoniae* (TWAR) in coronary arteries of young adults. *Proc. Natl. Acad. Sci. USA*, **92**, 6911–14.

49. Epstein, S.E. and Zhu, J. (1999). Lack of association of infectious agents with risk of future myocardial infarction and stroke: definitive evidence disproving the infection/coronary artery disease hypothesis? (editorial). *Circulation*, **100**, (13), 1366–8.

50. Ridker, P.M., Kundsin, R.B., Stampfer, M.J., Poulin, S., and Hennekens, C.H. (1999). Prospective study of *Chlamydia pneumoniae* IgG seropositivity and risks of future myocardial infarction. *Circulation*, **99**, (9), 1161–4.

51. Epstein, S.E., Zhou, Y.F., and Zhu, J. (1999). Infection and atherosclerosis: emerging mechanistic paradigms. *Circulation*, **100**, (4), e20–8.

52. Dechend, R., Maass, M., Gieffers, J., Dietz, R., Scheidereit, C., Leutz, A., and Gulba, D.C. (1999). *Chlamydia pneumoniae* infection of vascular smooth muscle and endothelial cells activates NF-kappaB and induces tissue factor and PAI-1 expression: a potential link to accelerated arteriosclerosis. *Circulation*, **100**, (13), 1369–73.

53. Wimmer, M.L., Sandmann-Strupp, R., Saikku, P., and Haberl, R. (1996). Association of chlamydial infection with cerebrovascular disease. *Stroke*, **27**, 2207–10.

54. Blasi, F., Denti, F., Cosentini, R., Raccanelli, R., Rinaldi, A., Fagetti, L., Esposito, G., Ruberti, U., and Allegra, L. (1996). Detection of *Chlamydia pneumoniae* but not *helicobacter pylori* in atherosclerotic plaques of aortic aneurysms. *J. Clin. Microbiol.*, **34**, 2766–9.

55. Libby, P. and Ridker, P.M. (1999). Novel inflammatory markers of coronary risk: theory versus practice (editorial; comment). *Circulation*, **100**, (11), 1148–50.

56. Ridker, P.M., Cushman, M., Stampfer, M.J., Tracy, R.P., and Hennekens, C.H. (1997). Inflammation, aspirin, and the risk of cardiovascular disease in apparently healthy men. *N. Engl. J. Med.*, **336**, 973–9.

57. Ridker, P.M., Glynn, R.J., and Henekens, C.H. (1998). C-reactive protein adds to the predictive value of total and HDL cholesterol in determining risk of first myocardial infarction. *Circulation*, **97**, 2007–11.

58. Xu, Q., Kiechl, S., Mayr, M., Metzler, B., Egger, G., Oberhollenzer, F., Willeit, J., and Wick, G. (1999). Association of serum antibodies to Heat Shcok Protein 65 with carotid atherosclerosis. *Circulation*, **100**, 1169–74.

59. Watson, K.E. and Demer, L.L. (1996). The atherosclerosis-calcification link? *Curr. Opin. Lipidol.*, **7**, (2),101–4.

60. Fitzpatrick, L.A., Severson, A., Edwards, W.D., and Ingram, R.T. (1994). Diffuse calcification in human coronary arteries. Association of osteopontin with atherosclerosis. *J. Clin. Invest.*, **94**, 1597–1604.

61. Newman, C.M., Bruun, B.C., Porter, K.E., Mistry, P.K., Shanahan, C.M., and Weissberg, P.L. (1995). Osteopontin is not a marker for proliferating human vascular smooth muscle cells. *Arterioscler. Thromb. Vasc. Biol.*, **15**, 2010–18.

62. Weber, G.F. and Cantor, H. (1996). The immunology of Eta-1/osteopontin. *Cytokine Growth Factor Rev.*, **7**, 241–8.

63. Shanahan, C.M., Cary, N.R., Salisbury, J.R., Proudfoot, D., Weissberg, P.L., and Edmonds, M.E. (1999). Medial localization of mineralization regulating proteins in association with Monckeberg's sclerosis: evidence for smooth muscle cell-mediated vascular calcification. *Circulation*, **100**, (21), 2168–76.

64. Shanahan, C.M., Connolly, D.L., Tyson, K.L., Cary, N.R., Osbourn, J.K., Agre, P., and Weissberg, P.L. (1999). Aquaporin-1 is expressed by vascular smooth muscle cells and mediates rapid water transport across vascular cell membranes. *J. Vasc. Res.*, **36**, (5), 353–62.
65. Chambers, A.F., Wilson, S.M., Kerkvliet, N., O'Malley, F.P., Harris, J.F., and Casson, A.G. (1996). Osteopontin and lung cancer. *Lung Cancer*, **15**, 311–23.
66. Watson, K.E., Parhami, F., Shin, V., and Demer, L.L. (1998). Fibronectin and collagen I matrixes promote calcification of vascular cells *in vitro*, whereas collagen IV matrix is inhibitory. *Arterioscler. Thromb. Vasc. Biol.*, **18**, (12), 1964–71.

Index

ABC-1 (ATP binding cassette
transporter 1) 197
ACE, *see* angiotensin-converting enzyme
acetylcholine, vasomotor responses to
89, 93, 94, 100
acetyl CoA cholesteryl acyl transferase
(ACAT) 177, 180, 188–90
N-acetylcysteine 15, 83
actin, α-smooth muscle 209, 221
actin binding protein 280 (ABP-280)
380–1, 382
acute coronary syndromes 391
endothelial dysfunction 100
platelet-derived NO 167
role of inflammation 394
role of platelets 161–2
triggers 395–7, 398
see also angina pectoris, unstable;
myocardial infarction
acute vascular events, role of platelets
161–2
adeno-associated virus (AAV) 117, 120
adenosine 94
adenosine deaminase deficiency 114–15
adenovirus vectors 117, 119–20, 123
adhesion molecules (cell adhesion
molecules, CAMs) 141, 144–6
in atherogenesis 6, 138, 284–5
circulating soluble 97–8, 415–16
expression in atherosclerosis 145–6
flow-mediated expression 140
inhibition of expression by HDL 145,
283, 286–96
in monocyte recruitment 252
regulation of expression 145, 286
as therapeutic target 419–20
ADP 165, 166
advanced atherosclerotic lesions (types
IV to VI) 4–5
antioxidants in 334, 335, 337–40

macrophage foam cell morphology
177–80
protein oxidation products 314–15
T lymphocytes 235
advanced glycation end-products (AGE)
28, 29–30
interactions with proteins and LDL 32
receptor (RAGE) 30–1, 254–5
receptors 30–1
in uremia and cigarette smoking 37
adventitia 2
gene delivery 121, 122
AGE, *see* advanced glycation
end-products
age
development of atherosclerosis and 3
protein oxidation and 315–16
AIDS 239
albumin
antioxidant activity 311, 336–7
in atherosclerotic lesions 339
endothelial transport 143, 144
alkaline phosphatase 31
alpha 2-macroglobulin receptor/LDL
receptor-related protein (alpha 2
MR/LRP) 271
α_1-proteinase inhibitor 309
Alpha-Tocopherol, Beta-Carotene
(ATBC) Cancer Prevention Study
52–3, 75, 76–7, 81, 82
Alzheimer's disease 312
Amadori products 28
amino acids
oxidation reactions 302–5
oxidized 28, 29
as marker of oxidative stress 317
metabolism 310–11
aminoguanidine 30
amyloid A, serum 394
amyloid β-protein 312

glycated/glycosylated 29, 30, 32, 34
intimal influx 267–8, 269–70, 271
  see also lipoprotein(s), influx
minimally modified (MM-LDL) 146
modified, uptake by macrophages 188, 189
non-oxidative modification 146–7, 272–5
oxidation 9–10, 46, 326–40
  endothelial cells and 147
  extracellular 326–7
  markers/measures 47, 57–9
  platelet-stimulated 161
  products 58
  protective effects of antioxidants 50–1, 54–5, 56, 57–60
  of protein components 312–15
  regulation 329–31
  therapy targeting 420–1
  transition metal-catalysed 13, 35
oxidized, *see* oxidized low density lipoprotein
platelet function and 162
proteoglycan interactions 216–17
in reconstituted HDL 288
retention in arterial wall 267–8, 270–1, 272–5
  endothelial cells and 147
  factors contributing to 273–4
small, dense
  in diabetes 26
  intimal fluxes 270
tocopherol-mediated peroxidation 327, 328–9
uptake by macrophages 186, 188
low density lipoprotein (LDL) receptor
-deficient mice 48, 59
  lymphocytes 238, 240
  OxLDL antibodies 416
-deficient rabbits 54
on macrophages 186, 188
LR11 271
L-selectin 144
lumican 211
lung cancer, antioxidant studies 76, 77
lycopenes 53, 83
lymphocytes 230–44
stability of plaques and 15
total deficiency 242
see also B lymphocytes; T lymphocytes
lysine, oxidation reactions 304–5
lysophosphatidylcholine (lyso-PC) 33, 146, 183

lysosomal acid lipase (LAL) 182, 186–7
inherited deficiency 187
lysosomal enzymes
in macrophage foam cells 177
resistance of oxidized lipoproteins 191
lysosomes
effects of OxLDL 187, 191
free cholesterol accumulation 179–80
lipid accumulation 177–9, 187
lipolysis 186–7

macromolecules, endothelial permeability 142–4
macrophage colony-stimulating factor (M-CSF) 176, 182, 256, 286–7
macrophage foam cells 4, 177–82
formation 6
inflammatory mediator release 182–3
lipid analysis *ex vivo* 180–2
lipoprotein modification and 146
as therapeutic target 420, 421
ultrastructural morphology 177–80
macrophages 6, 176–200, 392
AGE interactions 31
cell death 13–14, 274
cholesterol efflux, *see* cholesterol, efflux from macrophages
in different stages of atherosclerosis 4
human monocyte-derived (HMDM), resistance to cholesterol efflux 199–200
iron accumulation 12–13
lesions 234, 235
lipid accumulation 177–82, 271
  cell proliferation and 183–4
  in cytosolic and lysosomal compartments 177–9
  ex vivo studies 181
  inflammatory mediator release and 182–3
lipoprotein lipid/cholesterol metabolism 186–91
in necrotic core 179–80, 274
plaque stability and 15
PPARα and PPARγ expression 184–5
proliferation 183–4, 421–2
S100A8 and 255
smooth muscle cell phenotype and 218
as therapeutic target 401
see also monocytes
magnetic resonance imaging (MRI) 47, 412–13